Nicolaes Witsen *and* Shipbuilding *in the* Dutch Golden Age

ED RACHAL FOUNDATION NAUTICAL ARCHAEOLOGY SERIES
in Association with the Center for Maritime Archaeology and Conservation and the Institute of Nautical Archaeology

Nicolaes Witsen

and Shipbuilding in the Dutch Golden Age

A. J. HOVING

Translated by
Alan Lemmers

Foreword by
André Wegener Sleeswyk

With an appendix by
Diederick Wildeman

TEXAS A&M UNIVERSITY PRESS
College Station

First edition

This paper meets the requirements of ANSI/NISO Z39.48-1992 (Permanence of Paper).
Binding materials have been chosen for durability.

LIBRARY OF CONGRESS CATALOGING-IN-PUBLICATION DATA

Hoving, A. J., 1947–
Nicolaes Witsen and shipbuilding in the Dutch Golden Age / A. J. Hoving ; translated by Alan Lemmers ; foreword by Andre Wegener Sleeswyk ; with an appendix by Diederick Wildeman. — 1st ed.
p. cm. — (Ed Rachal Foundation nautical archaeology series)
Hoving selected the portions of Witsen's 1671 publication that deal specifically with Dutch shipbuilding and had them translated into English by Lemmers. Hoving then integrated this translation into a new text with commentary. "To reorganize Witsen's text, I have cut his entire work (except the irrelevant chapters . . .) into hundreds of selected fragments and then arranged them in a new sequence that follows Witsen's own description of the construction process."—Introd.
Includes bibliographical references and index.
ISBN 978-1-60344-286-2 (book/hc-hardcover : alk. paper)—ISBN 978-1-60344-404-0 (ebook format/ebook—c) 1. Witsen, Nicolaas, 1641–1717. Aeloude en hedendaegsche scheeps-bouw en bestier. 2. Shipbuilding—Netherlands—History—17th century. 3. Shipbuilding—Netherlands—Early works to 1800. I. Wildeman, Diederick. II. Witsen, Nicolaas, 1641–1717. Aeloude en hedendaegsche scheeps-bouw en bestier. III. Title. IV. Series: Ed Rachal Foundation nautical archaeology series.
VM77.H68 2012
623.82009492'09032—dc22

2011010696

And certainly if by reading books the arts could be mastered,
I would master many more than the few I do now.

—Cornelis van Yk
Master shipbuilder in Delfshaven, 1697

Contents

Foreword

WHOEVER TRIES TO MASTER the text of Witsen's *Aeloude en Hedendaegsche Scheeps-bouw en Bestier* (Ancient and Modern Shipbuilding and Management) soon discovers that the task is akin to finding one's way in a decayed labyrinth, even if he or she reads seventeenth-century Dutch without difficulty. The frequent, rather pointless elaborations in which Witsen indulges resemble shrubbery that has overgrown the paths, and some of the walls seem partly collapsed because Witsen has the disturbing habit of not furnishing the explanations that he has announced. In spite of these shortcomings, since its publication in 1671, Witsen's book has been a most valuable source for our knowledge of shipbuilding in the seventeenth century, even if very few of the many authors quoting the work had really read it from cover to cover.

It is not surprising that it took Ab Hoving fourteen years to arrange that part of Witsen's work which directly bears on the history of shipbuilding in a logically coherent manner, to supplement and elucidate where necessary, and to provide helpful commentary. Of course, these fourteen years were not spent in continuous labor on this task. In this type of research—the work may surely be called such—it is of the utmost importance to pause between the steps of what cannot be anything else but a stepwise approach. These pauses allow one to consider the previous steps critically and to plan the next steps carefully.

Hoving's presentation of this large part of Witsen's work has not only cleared out the labyrinth but also laid out a path with clear signposts that readers can follow to arrive at an understanding of Witsen's construction principles. The most important part of the book is the description of the building of a *pinas,* which Witsen intended to present in 122 consecutive steps.

From Witsen's description, it is apparent that the construction method for this ship was fundamentally different from that described in the other well-known seventeenth-century work on Dutch shipbuilding, *De Nederlandse Scheeps-bouw-konst Open Gestelt* (Dutch Naval Architecture Unveiled), written by the naval architect Cornelis van Yk of Delfshaven (today part of Rotterdam) and published in Amsterdam in 1697. It should be noted that the first phase of construction—which consisted of laying down the keel timber on the stocks, erecting the stem and sternpost, and adding the first strakes on both sides of the keel—was identical in both methods. .

The differences become manifest after this phase. The most salient of them comes to light when we compare the two works. Witsen describes how the shell was first built up to the level of the turn of the bilge, whereas in Van Yk's text we find that the first phase was followed by the erection of the frames. The plank shell described by Witsen was held together by cleats, and frame timbers were subsequently fitted into the shell. According to the method of construction described by Van Yk, which strikes us as much more modern, the planking was fastened to the pre-erected frames.

After Hornell called attention to this fundamental difference in the methods of shipbuilding in 1948,[1] it has become a subject of study the world over.[2] The older system is now known as "shell first," whereas Van Yk describes a system that is characterized as "skeleton first" or "frame first." Since Hornell's days it has become clear that, in practice, the difference is often not absolute—i.e., intermediate forms are commonly encountered. Nevertheless, the distinction is most useful in analyzing modes of shipbuilding.

It is worthwhile to consider briefly the history of this unfamiliar shell-first method of construction. The northern European variation on the method was the clinker system, which results in strakes partly overlapping each other, much as roof tiles do. Between the overlapping parts some sort of luting was often applied, such as tar-drenched moss, after which the strakes were fastened to each other by means of rivets. The natural result of this manner of shipbuilding is a coherent shell into which the floors and futtocks were subsequently fitted. In an older variation on the system the strakes were not riveted to each other but sewn, commonly with rope made from bark. The earliest evidence for this system is furnished by the Hjortspring boat, which dates from 340 BC and was found in 1921 on the Danish island of Als.[3] Early in the twentieth century this method of joinery was still employed in the construction of Skolt Lapp boats, used in Scandinavia north of the Polar Circle. The oldest known riveted clinker planking is that found in the Nydam boat.[4] It dates from the fourth century AD and is now permanently on exhibition in Gottorp castle in Schleswig, Germany.

A much older system of joinery was used in the Mediterranean. It is characterized by tenons set into mortises in the thickness of the strakes, which the tenons join solidly together because dowels traverse both planks and tenons and thus secure them. The excellent quality of these joints rendered caulking superfluous. The earliest shipwreck in which we encounter this pegged mortise-and-tenon system dates from c. 1300 BC.[5] Given this early date, it is not surprising that a passage in the *Odyssey,* the historic context of which must be dated to the eighth century BC, turns out to be a description of this mode of construction.[6] Odysseus builds the ship on which he will depart from the isle of the nymph Calypso. The method was later taken over by the Romans, a landlocked people who took to the sea because there was no other option. They attributed the original invention of the method to the Carthaginians, as may be deduced from Cato's description of the construction of the lid of an olive press, the *orbis olearium,* which made use of these joints; he called them *coagmenta Punicana* or Punic joints.[7] This appellation is correct, for we may regard the Carthaginians as descendants of the Phoenicians.

As it happens, one of the earliest discoveries of Punic joints was made in the Netherlands in 1892.[8] Excavation at the site of the Roman fort Fectio (Vechten) near Utrecht revealed the wreck of a river barge at a depth of six meters. Unfortunately, it was not possible to raise and conserve the vessel at the time, but it could be established unequivocally that the joinery of the planking was made according to this method. This woodworking technique was seemingly forgotten after the retreat of the Roman armies from western Europe.

A factor that contributed to the abandonment of Punic joinery was the metamorphosis that took place in its application in Mediterranean shipbuilding after the second century AD. The distance between the tenons connecting adjoining planks gradually increased until finally the system of carvel planking evolved, in which the strakes

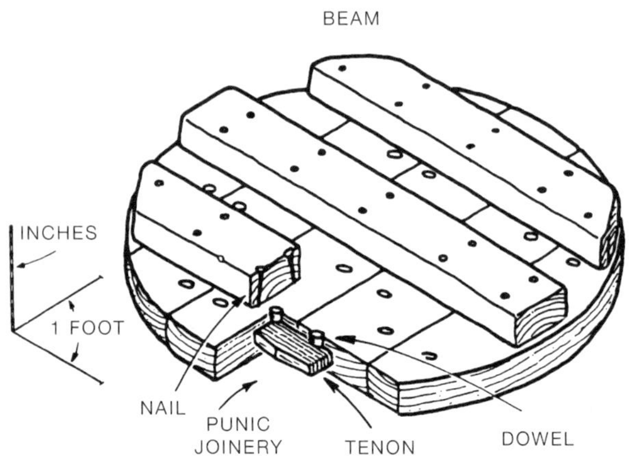

Lid of an olive press, after Cato's description in *De agricultura.* (Drawing by André Wegener Sleeswyk)

Fresco of the building of St. Ursula's ship with the frame-first method, by the "Veneto Master," c. 1350. From Michelangelo Muraro, *Paolo da Venezia* (Philadelphia: Pennsylvania State University Press, 1970). (Reproduced by André Wegener Sleeswyk)

are joined together only by the frames.[9] The Serce Liman wreck, dating from c. AD 1025, represents the earliest known example of the skeleton-first system, which naturally results in carvel planking.

The earliest known depiction of the frame-first method of shipbuilding dates to considerably later, c. 1350, on a fresco by the anonymous "Veneto Master" (1310–58).[10] The expansion of this system of ship construction to northwestern Europe took place about a century later, a step that is sufficiently documented to provide some historical underpinning for the mode of transfer of this technique.

In all probability the transmission is linked to the dynastic marriage of Philip the Good, duke of Burgundy, to Isabella of Portugal in 1430. From that date onward a distinct Portuguese influence in the Burgundian possessions can be traced in a number of fields, including shipbuilding. As part of this cultural transfer, in 1439 Philip ordered the Portuguese naval architect Jehan Perhouse to build in Brussels not only a nao but also a caravel.[11] This type of ship had not been built previously in Flanders. The caravel (*caravela* in Portuguese) gave its name to the method of "carvel-building," in which the frames are erected before the planking is mounted. More than likely Philip issued this order to promote the adoption of this innovation in his territories.

After an "incubation period" of two decades the frame-first method spread to Holland and Germany. According to Jan van Reijgersbergh's *Dye Chronijcke van Zeelandt* (The Chronicle of Zeeland), published in 1551, the first carvel-planked ship was built in the province of Zeeland in 1459 in the small town of Zierikzee, again by a foreign naval architect called Juliaen de Bretoen, or Julian from Brittany. In 1460 a carvel ship was built in the town of Hoorn,[12] according to the town's chronicle, written by Velius in 1604, and in the same year two *kraweelen* are mentioned in a letters of marque issued by Danzig (now Gdansk).[13] These were flush-planked ships built with the skeleton-first method.

Although this new method of construction spread quickly through the north, it did not entirely replace the older shell-first system, as is clear from Witsen's description of the latter as something that was obviously, for him, the standard method in 1671. There are numerous other indications that the older system was only partially replaced. The photograph of the Swedish shipwright who is busy putting futtocks into the completed clinker-built shell, which Casson reproduced in his book on ancient watercraft, is well known.[14] In the Dutch province of Friesland flush-planked boats of a local type called Staverse Jol were built in the same manner; a great number of temporary cleats held the strakes forming the shell together prior to the insertion of frames. The method is still used by some builders of *boeiers* (boyers) and other small watercraft in this province, which illustrates that the skeleton-first method never completely replaced the older shell-first system in northwestern Europe.

No doubt an important reason for preferring the skeleton-first method is that fitting the frames to the hull in a large ship, each of them weighing several hundred kilograms, was a more difficult operation than mounting strakes one after the other on a preexisting set of frames.[15] These frames could be heavier and stronger than those which were inserted in the shell of strakes in the shell-first method. This advantage was an important consideration, even in view of the fact that the new method of construction, when tried for the first time in a town or in a shipyard in the sixteenth century, was often applied to a small ship rather than a large one to lessen the financial risk posed by the new method.

If it is true that the shell-first method is difficult to apply to a large ship, the question naturally arises of how shipbuilders of antiquity made use of the method when building very large ships; in those times, too, the handling of very heavy parts of the many frames of a large ship must have required a relatively large amount of rather difficult labor. It is not only regrettable that only a single description of the building of a large ship in antiquity has survived, but also frustrating that the text contains a lacuna at the very spot where the method of constructing the hull was presumably described. The description concerns the large grain freighter with passenger accommodations, the *Syracusia,* a ship built and launched c. 250 BC for King Hiero II of Syracuse under the general direction of the physicist Archimedes, a citizen of that city-state. The description is from Athenaeus, an author from the third century AD who cites a certain Moschion as his source.[16]

The lacuna in the text is preceded by a description of the materials that were used in building this large ship and is followed by one of its launching, in which a new method devised by Archimedes was employed. In between, so it seems, was the description of the way the hull was built. The remaining fragment of that description, although very short, nevertheless contains an important indication about the sequence of some phases in the mode of construction. The extant fragment describes how the planking of the hull was covered in lead sheathing to protect it against the ravages of the shipworm, *Teredo navalis:* "and each time part [of the planking] was completed it was always covered by tiling made of lead for which three hundred workmen, not counting their helpers, were continually employed."[17]

Before commenting on this passage, I must briefly describe the standard method of mounting the protective metal sheathing on the hull. The latter was first built up at least to the waterline before being subsequently covered by fabric drenched in tar, after which the plating was mounted with short nails of copper or bronze on the wooden hull.[18] The treenails and bronze nails joining the planking to the frames were already in their places, regardless of whether the hull had been built according to the shell-first or skeleton-first method. Once the wooden hull had been built, the difference did not matter.

But the difference was essential while the hull was still under construction. Each time a few strakes were added to the hull, as Athenaeus—or Moschion—appears to say, they had to be fastened to the preexisting frames prior to being covered by lead sheathing. Otherwise, to join the strakes to the frames by treenails, it would have been necessary to pierce the sheathing, which would have negated its protective function. Obviously there would be no need to pierce the protective sheathing after it was mounted if the strakes had been fastened to the frames earlier. One can conclude that this, in fact, must have been the case—i.e., that the frame-first method must have been used.

The reason for proceeding with the planking and the sheathing bit by bit may well have been the poor accessibility of the space under the hull of this large ship once it had been planked up to the waterline. The alternate mounting of a few strakes and a row of "tiles" of protective lead sheathing solved the problem. That the method was mentioned as such in the text may indicate that it was uncommon, as the skeleton-first method (implied by the text) certainly was. We have no other indications of the use of this method of shipbuilding in antiquity. We may surmise now that although uncommon, it was not unknown; perhaps it had been devised by Archimedes. In any case, the *Syracusia* may have stood as a prototype for the later Roman grain freighters, which may have also been built by the skeleton-first method.

We can conclude that Witsen and Van Yk together provide a picture of the dissemination of the methods of skeleton-first and shell-first methods of shipbuilding in late seventeenth-century Holland. The two methods still coexist there and elsewhere; furthermore, it appears that they have done so since the third century BC. In the course of time usage of the shell-first method, which was initially ubiquitous, has become rare, while the skeleton-first method, used in the beginning only for building exceptionally large ships, has become the prevalent one.

—André Wegener Sleeswyk

Acknowledgments

A FEW WORDS OF GRATITUDE are in order. This book is the product of many specialists who contributed to the project with dedication and enthusiasm. In the Dutch version, published in 1994 in Holland, I included a long list of people to whom I offered my sincere thanks, and I will repeat only the names of those who were important for this English edition—which is not to suggest that all the others have become unimportant. Professor André Wegener Sleeswyk wrote the foreword, and Gerald de Weerdt provided some magnificent drawings to illustrate the book as well as a section on sailing with square-rigged ships in chapter 1.

Diederick Wildeman, curator of the Navigation and Library Collections of the Scheepvaartmuseum in Amsterdam, wrote the essay in the appendix on textual variants in the two editions of Witsen's treatise; he has rewritten his original contribution to take into account recent research.

The translation into English was done by my good friend and colleague Dr. Alan Lemmers, a historian and researcher at the Instituut voor Maritieme Historie. We worked together for many years in the attics of the Rijksmuseum in Amsterdam, where we studied and restored the huge collection of ship models and nautical objects known as the Dutch Navy Collection. Witsen's text was difficult to translate for two reasons: seventeenth-century Dutch is not easy to understand, even for the Dutch themselves, and old technical Dutch is an even bigger problem; furthermore, many shipbuilding terms and practices have disappeared since the end of wooden shipbuilding. Alan has done a marvelous job, for which I admire him highly.

An especially thorny problem was finding the right English terminology to fit the text. As in Dutch, many old English shipbuilding terms have been lost in time, and only a few people are capable of providing the correct English equivalents for the terms in Witsen's text. Such a person is Nick Burningham. He was deeply involved in the research for building the replica of the *Duyfken,* the small Dutch yacht that was the first to map parts of the Australian continent in 1616. The ship was built in Fremantle, Australia, a few years ago, and Nick sailed it for a considerable time. In our discussions of terminology we often had conversations about the contents of the text itself, and we were happy to delve into the difficult material with the help of other experts, both inside and outside the field. Olof Pipping of the Vasa Museum in Stockholm contributed to the section on rigging, as did Menno Leenstra, a Dutch sailing instructor with a vast knowledge of seventeenth-century sailing techniques. Frits Scholten assisted with the terminology of baroque architecture, and Jan Piet Puype, one of the world's foremost experts on ordnance, shared his expertise. My original drawings of Witsen's 134-foot pinas were redone in AutoCAD by the ever-productive Cor Emke.

As a result of this collective expertise, the English version of Witsen's material is even better than the Dutch one. Many unclear passages that clogged the original book have been rendered in language that is more comprehensible, even to the lay reader. Sincere thanks also go to three people from the Nautical Archaeology Program at Texas A&M University for making the English edition of this book possible: Donny Hamilton, Director of the Conservation Research Laboratory and George T. & Gladys H. Abell Professor of Nautical Archaeology; Kevin Chrisman, Director of the Center for Maritime Archaeology and Conservation; and especially Filipe Castro, Frederic R. Mayer Fellow II of Nautical Archaeology, without whom this book would never have existed. I also wish to thank Mary Lenn Dixon, editor-in-chief, Texas A&M University Press, for her trust in me and her decision to publish this book, the first serious attempt to make Dutch seventeenth-century shipbuilding available to an English-speaking audience.

Nicolaes Witsen *and* Shipbuilding *in the* Dutch Golden Age

Title page of Witsen's *Aeloude en Hedendaegse Scheeps-bouw en Bestier,* 1671 by Romeyn de Hooghe. In the style of his day, the engraving shows the importance of the shipbuilding industry as an allegory. Shipbuilding, represented by the female figure, is flanked by Mercury and Mars, gods of trade and war, and surrounded by shipbuilders displaying their creations. Heroic figures in the foreground work with shipbuilding tools, and an active shipyard is visible in the background.

ONE Introduction

THIS BOOK IS ABOUT seventeenth-century Dutch shipbuilding, a rather inaccessible field of research up to now. The earliest written sources appear only in the second half of the 1600s, known as Holland's Golden Century.

It is unfortunate that the first Dutch book ever written on the subject is very impenetrable. The bulk of the information that Nicolaes Witsen presents in his *Aeloude en Hedendaegse Scheeps-bouw en Bestier* (Ancient and Modern Shipbuilding and Management, 1671) remains hidden in cloudy formulations and a chaotic structure. Still, it is an extremely valuable book that needs and deserves clarification and attention, which I have attempted to do in the present volume by offering a guide into Witsen's work and the world of his subject—the almost forgotten basics of a craft that contributed as no other to the flourishing Dutch Republic of the seventeenth century.

Nicolaes Witsen

Nicolaes Corneliszoon Witsen was born on May 8, 1641, the son of an influential Amsterdam merchant. He studied law at Leiden University, and in 1664 he visited Russia as a member of Jacob Boreel's legation. Here he collected information on the land and people of Russia and later published a map of "Tartarije" (Russia). After his return to Amsterdam he became a member of the town council. Thirteen times he held the office of burgomaster of Amsterdam. In 1671 he published the first treatise on shipbuilding in Holland, *Aeloude en Hedendaegse Scheeps-bouw en Bestier,* a second edition of which appeared in 1690 (for a comparison of the two editions, see "Variations on Witsen" by Diederick Wildeman in the appendix). In 1672, the "Disaster Year," he became a member of the committee for the defense of the city of Amsterdam, and in 1674 a council delegate.[1]

As a diplomat he was active in England and Scandinavia, and he was the host and mentor of Peter the Great during the czar's visits to Holland. In 1692 he published a second important treatise, *Noord en Oost-Tartarije* (reprinted in 1705), in which he displayed his knowledge of the Russian Empire. He remained an active administrator well into his old age, as is demonstrated by his efforts concerning the placement of beacons on the Zuider Zee. On August 10, 1717, he died and was buried in Egmond aan de Hoef, where he had a country mansion named Huis Tytverdijf (Pastime).[2]

Why a patrician like Witsen chose to engage himself in a subject like shipbuilding—which was probably a trivial pastime in the eyes of his peers, and did so, moreover, at the level of the actual practices of the shipyard—remains an

FIGURE 1.1. (*left*) Portrait of Nicolaes Witsen in the 1671 edition, published when he was thirty-one years old.

FIGURE 1.2. (*right*) Portrait of Witsen in one copy of the second edition (1690), showing the author at the age of sixty; only five copies of this edition are known to exist.

interesting question. It was indeed highly uncommon for a gentleman in his day. But Nicolaes Witsen can be seen as one of the most representative figures of the Dutch Renaissance: an all-round scientist who, in addition to the above-mentioned activities, was an ardent collector, accomplished amateur graphic artist, cartographer, and Maecenas of scientific research. He encouraged his friends who sailed to the East Indies to bring back plants and animals for his collection. His taste for science and technology can be interpreted as an unceasing effort to understand the nature and purpose of creation.[3] His open-minded interest in anything that might lead to a better understanding of God's creation characterizes him as a valuable, albeit unconventional, representative of his time.[4]

The question arises of how Witsen acquired the knowledge that prompted him to publish, for the first time in Holland, such an important work on naval architecture. This can be traced back to three sources. Part of his knowledge came from books, many of which he read in the incomparable Library of the very learned and esteemed gentleman *Isaak Vossius* (53 II 18) in Leiden (now part of the Leiden University Library). A second important source can be found in iconography: the collection of coins and medals of his brother Johan Witsen, secretary of the city of Amsterdam, with primitive depictions of ships, together with the engravings he found in the incomparable treasure of engravings and drawings of the very learned gentleman Mr. *Johan Uitenbogaert,* revenue officer for the city of Amsterdam (24 II 28). Both sources provided material for the historical sections of his book. For his knowledge of contemporary shipbuilding, finally, he interviewed men of the trade. Jan Jacobszoon Vijzelaar, for instance, a former shipbuilder from Harlingen, was the

FIGURE 1.3. Title page of Cornelis Van Yk's *De Nederlandsche Scheeps-Bouw-Konst Open Gestelt* (1697). The allegorical figure Shipbuilding opens a curtain to reveal activities at a shipyard. Below, a ship is being launched in the Dutch way: bow first.

main source for technical information on shipbuilding and later became the curator of Witsen's extensive collection of rare objects and books.

Witsen was a competent draftsman, and the fact the he produced all the illustrations for his book himself shows that he definitely knew his subject, even though inconsistencies in both text and drawings are a characteristic feature of his treatise. Furthermore, he was seldom hindered by obstacles such as copyright, taking much of his text and drawings directly from the works of Oliveira, Furttenbach, Dudley, Crescentio, and Hayward without acknowledgment. Today we shudder at the thought of such plagiarism, but at that time it was common practice.[5]

Witsen and Van Yk

Witsen's treatise has often been compared with that of the Delfshaven shipbuilder Cornelis Van Yk, who published his treatise on shipbuilding in 1697, a quarter century after Witsen. Titled *De Nederlandsche Scheepsbouw-Konst Open Gestelt* (Dutch Shipbuilding Unveiled), the book has been quite justly recommended, not only for being much more structured than Witsen's but also as the work of someone who had a better and closer knowledge of the subject than the aristocrat, diplomat, politician, cartographer, and lawyer Nicolaes Witsen. It is certainly true that Van Yk's book is more consistent and much clearer in its structure; but as for the notion that Van Yk was better informed, I beg to differ.

Anyone who tries to work with both books from the standpoint of practical shipbuilding will quickly find that Witsen goes much farther than Van Yk, however well informed the latter may have been. Van Yk leaves much unexplained, probably because of the tendency of experts to assume too much as common knowledge, whereas Witsen elaborates on every detail with the innocence of an interested layman. His approach is therefore much more useful to us, with our restricted knowledge of the subject, than the approach of his expert fellow writer, Van Yk.

Besides the treatises of Witsen and Van Yk there have been other seventeenth-century works on the subject. Several are anonymous, such as Evenredige Toerusting van Schepen ten Oorlog Bijder See (Proportional Equipage of Men-of-War), a manuscript dated 1660;[6] *Hollandsche Scheepsbouw* (Dutch Shipbuilding) of 1678, which largely reiterates Witsen's treatise; *De volmaakte bootsman . . .* (The Perfect Boatswain) of 1680; *Taackelasie boeckje . . .* (Rigging Booklet) of 1690; and *Nieuwe Hollandsche Scheepsbouw* (New Dutch Shipbuilding) of 1695, followed in 1697 by the qualitatively much superior work of Van Yk.

Witsen and Other European Authors

Although Witsen's book was the first treatise on naval architecture to be published in the Netherlands, it was not among the earliest such works in Europe. There have been manuscripts and books on the subject since the fifteenth century.[7]

The first known English manuscript was that of Matthew Baker, Fragments of Ancient English Shipwrightry, which dates from the sixteenth century and is now in the Pepys Library of Magdalene College, Cambridge. The sixteenth century also saw the first Spanish work on shipbuilding, Diego García de Palacio's *Instrucción Náutica* (1587) as well as the famous Portuguese manuscripts of Fernando Oliveira, Ars nautica (c. 1570) and Livro da fábrica das Naos (c. 1580).

In Italy the leading works included Bartolomeo Crescentio's *Nautica Mediterranea* (1607), Pantero Pantera's *L'Armata navale* (1614), and the Englishman Robert Dudley's *Dell' Arcano del Mare* (1646). Joseph Furttenbach published his *Architectura navalis* in Germany in 1629, and in France Ithier Hobier (*De la Construction d'une gallaire et de son équipage,* 1622), Jean Boisseau (*Description d'un Navire Royal*, 1637), and Georges Fournier would be the first three to dedicate a book to the subject of shipbuilding; Fournier's *Hydrographie* (1643) was reprinted many times. Sweden's contribution to shipbuilding literature is relatively late, with Åke Rålamb's *Skeps byggerij* from 1691.

But England most certainly tops the list of discourses on shipbuilding with the works of John Smith (1626), Thomas Heywood (1637), Simon Smith (1641), Henry Bond (1642), Henry Mainwaring (1644), Thomas Miller (1660), Edward Bushnell (1664), and Edward Hayward (1665).[8] Bushnell's work was already into its third printing when Anthony Deane wrote his *Doctrine of Naval Architecture* in 1670, and it is this work that bears the closest resemblance to Witsen's book.[9] Like Witsen, Deane describes the building of a ship (in his case, a third-rate man-of-war) by elucidating its design step by step. There are additional similarities between the two: Deane, like Witsen, became mayor of his native town (Harwich), and, like his counterpart in Holland, he was Peter the Great's mentor when the czar came to England to study shipbuilding.

There is no doubt that Witsen took his bearings in several libraries before starting on his own book and, as noted earlier, drew generously from the material he found there. But it would be nonsense to claim that his book consists mainly of plagiarized information. He wanted to collect whatever was known on the subject and publish his findings in a comprehensive historical work. That we are more interested today in the material relating to specific construction techniques of Witsen's own period is partly due to our different perspective toward the information he presents. In chapter 2 I have tried to select and make accessible the information in Witsen's treatise that is relevant to Dutch shipbuilding in his time.

Aeloude en Hedendaegse Scheeps-bouw en Bestier

Nicolaes Witsen's *Aeloude en Hedendaegse Scheepsbouw en Bestier* is a much-quoted source of information. It is mentioned in every study of seventeenth-century Dutch shipbuilding, and quite rightly so. The book was not only the first Dutch work to be published on the subject of shipbuilding, but in addition to countless building specifications and curiosities, it contains the earliest known description of the building method used in those days. Even so, it has actually been read by only a few, as can be deduced from the fact that only a small number of passages (always the same) are quoted, and they are mostly misinterpreted and misused. However, this is not so surprising: aside from its many intriguing qualities, the book has one major handicap—it is almost unreadable, due to its totally chaotic structure, which consists of numerous layers of discourse moving in endless concentric circles around a main theme, shipbuilding.

As indicated by the title, Witsen's book contains two main parts, one about shipbuilding, the other about the management and operation of ships. The latter part deals with such general matters as the organization and functions of the crew, orders and how they are issued, food and rations, harbors, Dutch maritime victories, and so forth. We shall ignore this part and concentrate on the part about shipbuilding.

Of the almost three hundred pages that Witsen devotes to the subject of shipbuilding, we will also ignore sections that are irrelevant to our understanding of seventeenth-century Dutch methods. For instance, the first seven chapters, comprising some fifty pages, deal with shipbuilding in antiquity, mainly the Roman era. The last of these chapters—concerning after which measurements ships were built one and a half centuries ago (p. 42)—is actually Witsen's translation of Oliveira's imperfectly understood Latin work, *Ars nautica*. While these chapters might interest scholars in other fields, they fall outside the scope of this book. Likewise, chapters 14, 15, and 16 have been excluded, as they deal with Mediterranean galleys (translated from Furttenbach and Crescentio), European shipbuilding (mainly from Dudley and Fournier), and, finally, Indian sailing craft.

The remaining two hundred pages on Dutch shipbuilding are presented in English in chapter 2 (which takes its title from Witsen's chapter 8, How Ships are built in Holland today), but they have been rearranged in such a way as to provide the modern reader with a much more transparent and accessible version of the original edition. I have provided commentary on Witsen's text in order to share the knowledge I have gained by working with Witsen's book and the cognate literature for some fourteen years after my purchase of the 1979 facsimile edition of Witsen.

What is the use of rewriting an old book? It is not that Witsen's original text is without logic—he methodically begins with an illustrated glossary of all technical terms and follows this with the formulas used in the shipyards to determine the shape of the ship's components. He then applies these proportions to the building of a 134-foot pinas ship, which serves as an example of all the ships of his time (for more about this type of ship, see "The Pinas as Example" later in this chapter). One could, according to Witsen, build any ship after this example, provided the proper rules of construction are observed. After listing a large number of examples of building specifications, Witsen continues with a sequentially numbered description of all the phases of the building process, followed by practical directions regarding all the stages up to the launching. This section is followed by descriptions of others vessels, Dutch as well as foreign and seagoing as well as inland craft, then a treatise on ship measurement, and finally a scientific discourse on several phenomena that arise in shipbuilding and their "explanation." Witsen then adds an inventory of necessary items for a medium-distance voyage—such as to *Curaçao, Aleppo, Guinea* or elsewhere (280 I 5)—in a pinas ship.

Witsen's approach seems to offer a relatively logical, if somewhat arbitrary description of the process for building a pinas. And it would have been satisfactory if only Witsen had stuck to his devised plan. But straightaway, with the introductory descriptions of the ship's parts, he loses himself in the measurements, arrangement of the cabin, and the construction of the complicated taffrail of the pinas. Moreover, crucial data for the construction of the lower hull are hidden between the proportions of the components. Among the measurements of the pinas we find extensive contemplations on the way the quarter galleries were decorated before Witsen's time, while armament details must be deduced from the inventory at the end of the treatise. If these were the only sorts of inconsistencies, the readability of the work would not be seriously affected, but Witsen sometimes changes subject two to three times in one paragraph, often even in one sentence.

The structure of the book thus becomes chaotic, and whoever tries to read the text, even if well versed in the subject, almost immediately loses the thread of Witsen's reasoning. At some places it is clear that his sources, such as Harlingen shipbuilder Jan Jacobszoon Vijzelaar, the Amsterdam master shipbuilders Jan Ysbrantsz Hoogzaet and Jan Dirrikze Grebber, or the "famous master Shipwright" Dirck Raven, have dictated large sections of the text. We can almost see Witsen diligently taking down their statements and raising his head at certain points to interrupt the shipbuilder with questions on matters unclear to him—these are then inserted in the text. Later even more changes are inserted and joined with actual observations and measurements of the pinas, which Witsen must have seen firsthand. What remains is an impenetrable lump of half-digested information, digressions, and ruminations, all presented to the reader as a book.[10]

As a result, it becomes altogether unclear when the text is dealing with general information or with specific data on the pinas, and we can no longer find our way.

This explains why almost always the same passages are quoted in modern literature: writers quote one another rather than risking the maelstrom of Witsen's confusing conglomeration of data. Understandably, no one ever made an attempt to translate the book until now.

Others, discouraged by the book's impenetrability, simply discard it as nonsense, written by a man who did not understand what he was writing about. I hope to demonstrate that this is not true. Witsen's book is a preeminent historical source, containing much more information than one might suppose at first sight.

How This Book Originated

In the 1980s I was an amateur model builder with a strong interest in creating a model of a seventeenth-century pinas based on clear, consistent data rather than a contemporary ship model or a well-preserved archaeological find. This urge led me to Witsen's *Aeloude en Hedendaegse Scheeps-bouw en Bestier* and its forty-page description of the 134-foot pinas, which seemed to meet my requirements. This model of Witsen's pinas (see fig. 1.10), which took three years to build, is exhibited at the Noordelijk Scheepvaart Museum (Northern Shipping Museum) in Groningen.

The Drawings

To build my model, it was necessary to make drawings, which entailed sorting out and isolating Witsen's text concerning the pinas, its dimensions, and the relationships of its parts. Due to the impenetrability of Witsen's discourse, this task proved to be so toilsome that it took years to achieve any results. Making the drawings of the pinas alone cost three years of spare-time labor, almost as much as constructing the model later on. Yet I am confident that the drawings are accurate representations of the ship Witsen must have envisaged when he took the trouble of putting its dimensions on paper. My various drawings of the ship's structural components incorporated throughout chapter 2. Large, detailed plans of the ship's lines, compartments, rigging, and other equipment are shown in drawings 1 through 5 (see appendix).

The Building Method

With the data concerning the pinas distilled into a coherent whole, the remaining text proved to yield even more information. When the building process described by Witsen is compared with that described by Van Yk, what had hitherto gone unnoticed becomes evident: Dutch shipbuilding methods must have differed significantly according to region. I will return to this topic shortly.

Contract Specifications

The use of contract specifications (*bestekken*) also appears in a completely new light. These contracts, written up by a lawyer and stipulating the specifications of the ships to be built, were actually legal agreements between shipbuilder and commissioner. A contract contains data on the length, breadth, and depth of the ship, its keel, stem, and sternpost, and any more or less important part of the construction; the contract also states when the ship was to be finished, how much money it would cost, and when it was to be delivered. Unlike drawings, with which we are accustomed to working nowadays, these building specifications remain a closed book to us—until we combine their data with the building method described by Witsen. Then it suddenly appears that much more can be known about the ship in the contract than previously assumed.

Apart from contracts, so-called *certers* have also survived from those days. They contain the shipwright's notes on a specific ship or type of ships—for instance, the number of planks needed for a 100-foot-long fluyt or the costs for a 120-foot-long ship. In addition, we can glean information from the *charters,* specific rules that the large entities like the Dutch East India Company and the Dutch Admiralties laid down for their ships, divided into classes, called "charters" (rates). Together these archival documents form an inexhaustible source for researchers, once we understand what we are dealing with. Contracts as historical sources will be discussed in chapter 3, which also includes a section on model building as historical research and a selection of contracts that Witsen presents as examples.

Selection Principles

To reorganize Witsen's text, I have cut his entire work (except the irrelevant chapters, as explained earlier) into hundreds of selected fragments and then arranged them in a new sequence that follows Witsen's own description of the construction process: on pages 144 to 146 of his book he breaks down the building of the pinas into 122 consecutive stages, and I have used this structure as the basis for reshaping the whole text, raking together all the fragments relevant to each stage under numbered headings in chapter 2 that correspond to those in Witsen's original enumerated list.

Not all the text was used; among the more theoretical passages only those relevant to our understanding of the subject have been included verbatim. Thus, pages 141 to 144 (***How to set the wind to the Sails***) are relegated to a special section with commentary by Gerald de Weerdt (see "Sailing Square-Rigged Ships" later in this chapter); pages 234 to 249 (***Mathematical treatise, on how much water rests against the Ship's side***) and pages 250 to 279 (***Explanation of the shape of many parts of the Ship***) have likewise been incorporated only partly or in summary. In these instances Witsen tends to lose himself in theoretical speculations and arguments, which might be of interest to Witsen researchers but do not add anything to our knowledge of seventeenth-century Dutch shipbuilding. It also happens that Witsen states the same thing two or three times in different places but with different phrasing; for the sake of completeness I have provided all these passages in full.

How to Use this Book

Chapter 2 includes two kinds of text: (1) Witsen's translated text about every constructional part of the ship or building stage; and (2) brief remarks, germane references to other sources (mainly Van Yk), and accounts of the author's experiences in building the pinas model. Witsen's text is printed in a historical seventeenth-century font, equal to the original book and the modern text in corresponding sections has a modern font. Moreover, Witsen's text is divided into two categories: (1) general information about shipbuilding techniques and considerations relevant to each construction part of the ship; and (2) quotes that explicitly refer to the pinas ship that is used as an example for the shipbuilding method. The latter is placed in boxes, to differentiate from the general text and help model builders easily find the necessary data.

For example, the section 'First: One makes the keel' contains not only the specifics about the construction of a keel in general, the formulas used to calculate its size, and also the way in which it was assembled, but also presents the data about the keel of the pinas, placed in a box. In a different font the author's notes and explications follow under the same section name.

Each subsequent section traces the building process, stage by stage, as defined in Witsen's book and with all related details grouped together for easy reference. Following the sections devoted to Witsen's 122 steps or stages for building the pinas, the author has added 16 additional sections at the end, containing miscellaneous information such as ladders, bolts and nails, anchors; details Witsen must have overlooked in his 122 building stages.

Original illustrations are placed along the relevant sections, with Witsen's labels and keys translated in the caption, while modern illustrations show a reference to their sources.

The author's reconstruction drafts of the pinas (drawings 1 through 5) can be found in the appendix; keys to these drawings can be found in table 4, preceding the drawings. Portable document format (pdf) files of these drawings and keys can also be accessed from the Centre for Maritime Archaeology and Conservation-J. Richard Steffy Ship Reconstruction Laboratory website located at Texas A&M University—http://nautarch.tamu.edu/shiplab/AbHoving.htm. A glossary of technical terms is provided at the end of the book.

Although this book cannot be read with the ease of a novel, I do not feel particularly apologetic since the original work did not meet this standard, either.

In translating the Dutch text into English, we lost some interesting phenomena. In the seventeenth century Dutch orthography had not yet become fixed, unlike modern Dutch. For example, in Witsen's original text I counted as many as seven different spellings for the same word, with some words spelled differently within the same sentence. All this colorful orthography is lost in the English translation, which uses standard modern spellings throughout. On the other hand, Alan Lemmer's translation solved many problems in the difficult phrasing and vocabulary and in the misty formulations, which have become clearer in translation, despite the loss resulting from the need to choose a single interpretation over other possibilities. This means the translation of Witsen's text requires much less explanation than the Dutch edition of this book did. In some cases, however, there is no suitable English equivalent for a term used by Witsen, and the original Dutch word is retained. Whereas Witsen frequently states dimensions with units of measure in singular form—for example, "115 *voet*" (foot) instead of "115 *voeten*" (feet)—the translation adopts standard usage with numerical agreement. To give readers the flavor of the original, much of Witsen's unpredictable and sometimes startling use of punctuation and capitalization, another feature of his text, has been retained wherever clarity was not compromised.

Shipbuilding Methods in the Seventeenth Century

A comparison of the shipbuilding process as discussed in the treatises of Witsen and Van Yk reveals that method in the region known as the "Zaankant" (on the Zaan River and in Amsterdam) was quite different from that in "de Maze" (the area around the Meuse River, i.e., Rotterdam and the South). It was only natural that Witsen, writing in Amsterdam, would describe local practices, just as Van Yk in Delfshaven was focused on the shipbuilding industry centered in Rotterdam.

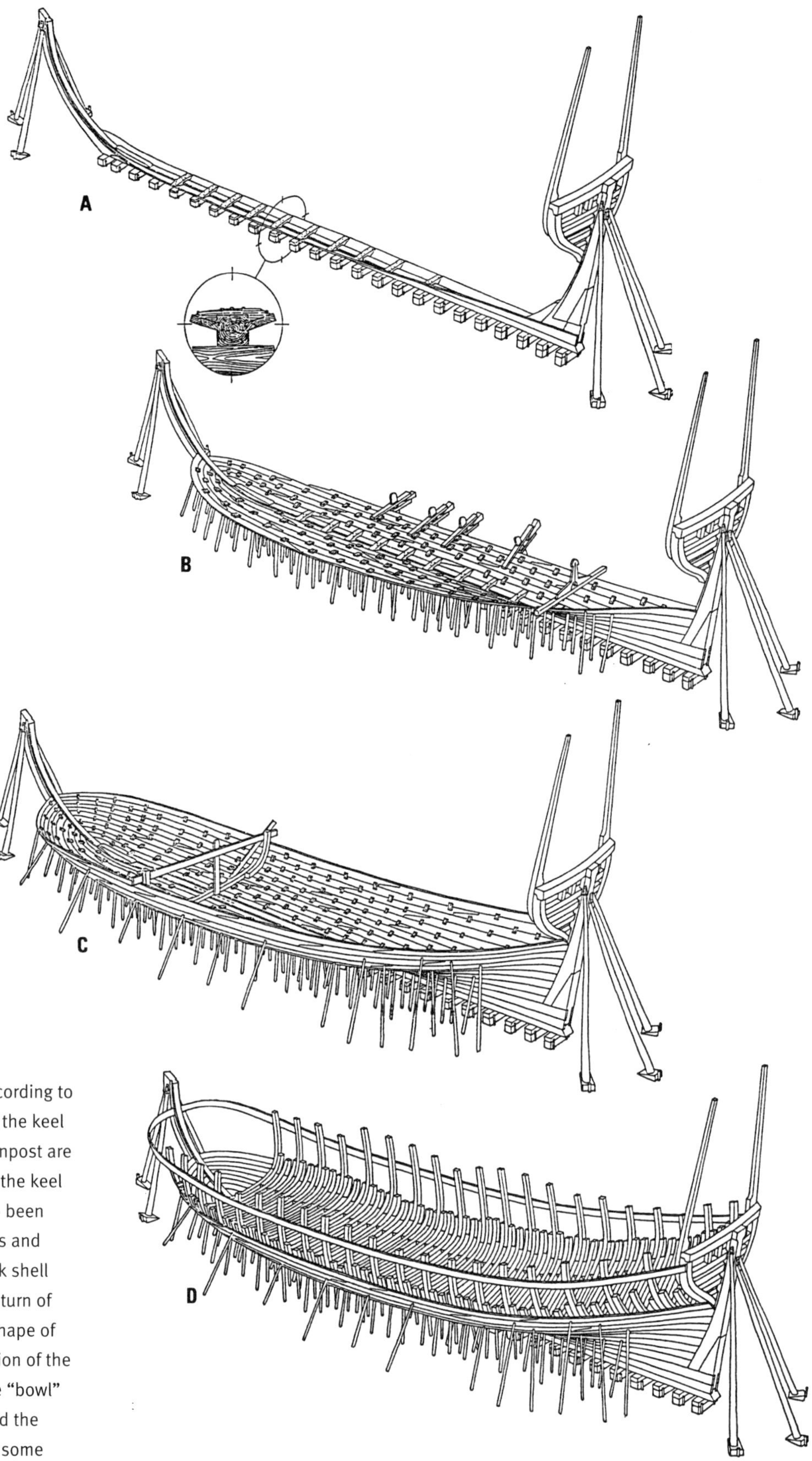

FIGURE 1.4. Shipbuilding according to Witsen: (*A*) The keel is laid on the keel blocks, and the stem and sternpost are erected; the first strake along the keel (the garboard strake) has also been fixed. (*B*) With the aid of cleats and planking tongs the lower plank shell of the ship is shaped. (*C*) The turn of the bilge is made, using the shape of two bilge futtocks at the location of the main frame as a mold. (*D*) The "bowl" is filled with frame timbers and the master ribband is attached to some futtocks to indicate the maximum width and the sheer of the ship. (Drawing by Anton v.d. Heuvel)

Witsen's Shipbuilding Method

The building method described by Witsen is called the shell-first method, as distinguished from its counterpart, the frame-first method. With the former method construction begins with the planking or skin of the hull, after which the frames are fitted into the shell formed by the outer planking; with the latter method the frames are raised first, after which the planking is applied to them.

The shell-first method is very old. In the reclaimed IJsselmeer Polders in the central Netherlands, wrecks of cog ships have been found that were built, without any doubt, using the shell-first method, while Scandinavian excavations in Nydam, Kvalsund, Oseberg, and Gokstad, among others, have shown that this method was practiced in northern Europe during and even before the Middle Ages.[11]

Originally the shell-first method was used for clinker-built ships only; both the northern European ships, like the viking ships and medieval cogs, were built this way, either entirely or in part. What is interesting is that Witsen describes the use of this method for carvel-built vessels.

The shell-first method represents one of the simplest ways to construct a hull; yet this does not imply that it is a primitive method—quite the contrary. Seventeenth-century shipbuilders had a precise understanding of the effects produced by any variation in the traditional construction formulas, even though they were not capable of making scientific calculations of their adaptations. It was precisely this freedom to vary the standard formulas that makes Dutch shell-first shipbuilding of that time unique.

Another fascinating aspect of this method is that there was no distinction between design and execution. The ship was not designed on the drawing board but was shaped during the building process, not on the basis of an engineer's calculations but through the master shipbuilder's active engagement in the building process on the yard. As Witsen notes:

(265 II 50) The outward shape of the ships is made with the eye and approval of the master.

Archaeologists use the following criteria to determine whether a northern European ship was built shell-first (these criteria do not necessarily apply to other shell-first traditions):

1. The framing timbers lie in a relatively arbitrary order, and the available timber determines the way the parts are joined.
2. The frame timbers are not connected to one another to make complete frames but are fastened to the planking and the ceiling.
3. Small dowels (*spijkerpennen*) are present in holes made by nails of the cleats, which held the lower hull planks in place in the first stage of the building process.[12]

In my view, however, the most important feature of this method is that building took place without preparatory drawings: the design came into being in the yard through the actual construction of the ship itself. During building the traditional formulas were not used in an absolute way. On the contrary, the shipbuilder had a sense of the effects produced by small adaptations in the application of the traditional rules: a slightly wider ship was more stable, which was useful for a man-of-war; a flatter bottom and a less raking stem gave more loading capacity for a freighter, while a bottom with greater rising and a more raking stem and stern caused the ship to be faster, which was desirable for frigates. It was not possible for the shipwright to calculate the effects of small changes in the formulas, but his experience, it was hoped, led him to the best results. It was this game of playing with the traditional rules that established the quality of the shipwright. Witsen offers this advice:

(262 II 38) It is not my intention that one should observe these proportions exactly to absurd precision: think of it as a guideline, from which one is not to diverge too far, and an assurance against awful blunders as long as one follows the rules.

Shipbuilding as Described by Van Yk

A completely different method of shipbuilding is described by the Delfshaven master shipbuilder Cornelis Van Yk. Two preliminary remarks are in order.

The first remark concerns terminology. Although Amsterdam and Rotterdam are separated by a mere 100 kilometers on the map, the terminology for parts and timbers diverges quite significantly in the two treatises. A floor timber is a *buikstuk* for Witsen, but in Van Yk's vocabulary it is a *legger;* Witsen's *kielgang* (garboard strake) is a *zandstrook* for Van Yk; for Witsen a carling is *karvielhout* but *klamaai* for Van Yk; a keelson is called *kolsem* by one, *zaathout* by the other. The terminology gets really confusing when Van Yk uses the same words as Witsen but in reference to different parts—such as *buikstuk,* meaning not the floor timber, as in Witsen, but the curved futtock in the bilge—which in Witsen's vocabulary is a *zitter.* I cannot explain this difference.

The second remark concerns the stocks. Practical as he is, Van Yk starts his shipbuilding story with the setup of the shipyard. It is striking that the stocks or keel blocks

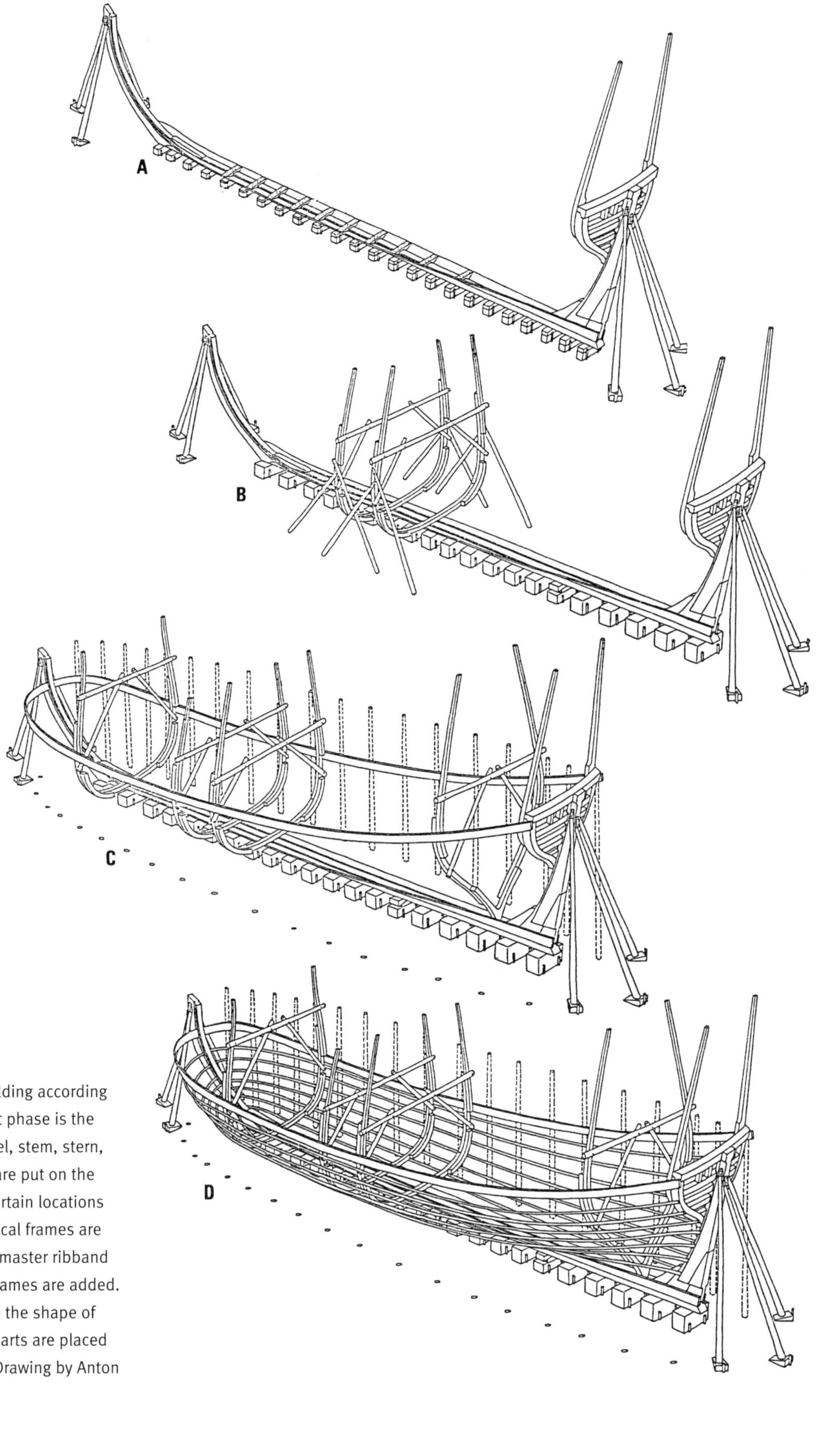

FIGURE 1.5. Shipbuilding according to Van Yk: (*A*) The first phase is the same as Witsen's: keel, stem, stern, and garboard strake are put on the keel blocks. (*B*) On certain locations on the keel two identical frames are erected. (*C*) After the master ribband is placed, two more frames are added. (*D*) Ribbands indicate the shape of the hull; then frame parts are placed in the construction. (Drawing by Anton v.d. Heuvel)

he describes are notably higher than they appear to be in Witsen's text, being "three and a half feet, forward a little less, aft some more, to facilitate working underneath the hull."[13] Witsen does not describe the stocks, but from illustrations such as II.45 and II.46 it is apparent that he pictured them lower. The reason for this difference can be found in Van Yk's statement about working under the hull: with Witsen's method working underneath the floor is not necessary, and for caulking the ship is laid on its side, whereas with Van Yk's method the planking is fastened to the frames at a much later stage, as we will see.

Both building methods are identical up to the raising of the stem and sternpost and the application of the garboard strake, the first plank next to the keel. Witsen then continues with the rest of the planking of the lower hull, but Van Yk first raises two complete frames on precisely determined places on the keel. Between these two frames the shape of the hull is uniform. We will discuss how these frames are shaped later.

A master ribband (*scheerstrook*) is then fixed to poles, which the master shipwright has raised around the outer shape of the hull. This temporary ribband determines the height of the maximum width of the ship. Witsen also mentions this master ribband, but for Van Yk it is much more important, as it already shows something of the shape of the hull.

A third frame is then raised on the scarf of the stem and keel, being more or less identical to both main frames but a little smaller, as the hull narrows forward and aft and the frame is raised a little higher than the others. The shape of the aft frame is deduced from the shape of the forward frame, a process for which there is a simple rule (see fig. 1.6). The aft frame is placed as far from the back of the sternpost as the fore frame is placed from the front of the stem. Thus, according to Van Yk, the shipbuilder is able to determine the exact trim by the stern.

Then ribbands are applied from stem to sternpost over the four frames. The master shipbuilder models his ship by moving these ribbands until he is satisfied with the shape of the ship. Then the rest of the frames are positioned in the basket of ribbands, braced and planked internally and supported by shores. Next the deck beams are fitted, and finally the outside is planked.[14]

All this gives the impression of being a somewhat more advanced method than the one Witsen describes. Yet nothing indicates that one could have evolved from the other. One of the building contracts presented by Van Yk as an example, clearly based on his building method, dates from 1629, evidence of a respectable age.

The origin of Van Yk's building method remains unclear. Some vocabulary, such as the word *cent* for "ribband," might point to contacts with Iberian shipbuilders (cf. *cintas,* meaning "master ribband"), possibly before and during the Eighty Years' War against Spain. But I cannot offer any definitive answer to this question.[15]

English Methods

In his edition of Deane's *Doctrine on Naval Architecture,* Brian Lavery makes some interesting points. As in Holland, contract specifications in England were customary, and not every shipwright worked from drafts. A system called "whole molding" was practiced, whereby master molds were used to construct the hull with the same curves over the whole length. Even today wooden sloops and boats are built this way in England and North America. The results are easy for experts to recognize, as the shape in the bow and stern is not ideal. Deane, however, reveals that members of a small group of top designers, all using drafts, applied curves with different radii, which

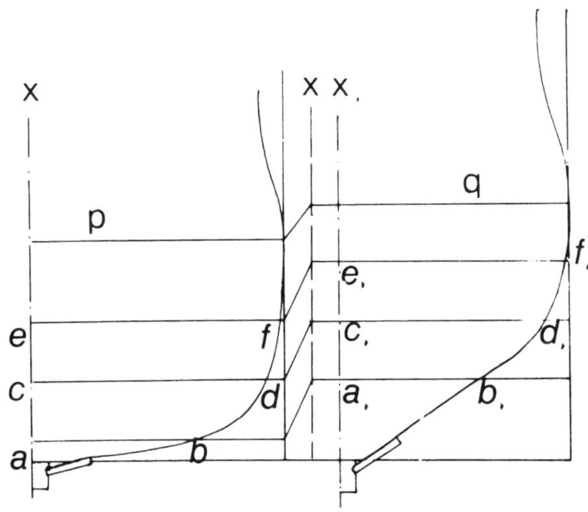

FIGURE 1.6. Construction of Van Yk's aft frame. *Left:* The fore frame, its centerline indicated by *X*. *Right:* The aft frame. The waterlines are drawn as a_1–b_1, c_1–d_1, and e_1–f_1, increasing in height by as much as the ship sinks in the stern. Because the ship's beam narrows aft, the centerline shifts to X_1. A line connecting b_1, d_1, and f_1 indicates the shape of the aft frame. The lines *p* and *q* are drawn at the height of the master ribband. The garboard strake twists to a steeper angle of rise. (Drawing by A. J. Hoving)

FIGURE 1.7. Model of the East Indiaman *Prins Willem,* built in 1649. In the First Anglo-Dutch War (1652–54) it served as a man-of-war; it sank in 1662 with all hands on board during one of its return trips from Asia. (Courtesy Rijksmuseum, Amsterdam)

greatly complicated designing but significantly improved the shape of the ship.

The English shipyards took more time to build a ship, but their ships lasted longer. Deane's successful *Resolution* was lost thirty-eight years after its launching without giving the impression of being an old ship, and *Sovereign of the Sea* lasted almost one hundred years before a knocked-over candle finished its career. In Holland ships were usually worn out in twenty years. Abel Tasman's *Heemskerck,* built in 1639, was broken up eleven years after its launching, and *De Zeven Provinciën,* De Ruyter's famous battleship, lasted twenty-nine years. Of the Dutch ships mentioned in Vreugdenhil's list for the period 1660 to 1670, which were eventually broken up, the average age is twenty years.[16]

The Pinas as Example

It is almost impossible to make any hard and fast statements about seventeenth-century ship types and their names—not only because these types evolved slowly over the course of the century, but also because the meaning of the names appear to have changed, although no reason for this can be found in the ship's outward shape or construction. The pinas is an example of this.

The return ship (*retourschip*) of the Dutch East India Company was one of the largest vessels in the seventeenth century, although it existed in several sizes. Common characteristics were the square tuck and other features that made it well suited for repeated voyages to the East Indies, such as having a large-cargo capacity and being heavily armed. An example of such a ship is the *Prins Willem,* a model of which is displayed in the Rijksmuseum in Amsterdam. This size of return ship was so heavily armed that in times of war the Dutch East India Company hired them out to the Admiralties as men-of-war. Thus the *Prins Willem* briefly served as the admiral ship of Witte de With, one of the Dutch admirals, in the Battle of Kijkduin of 1652. Six gunports were added on the main deck, and its armaments were increased from twenty-four to forty guns.

Apart from large, armed merchantmen, the Dutch East India Company had smaller ships, which were not built for a return voyage. These were mainly small merchantmen, such as *fluiten* (fluyts, sometimes called "flutes" or "flyboats"), and later mainly *hoekers* (or *quellen*) and galliots, which maintained the regional commerce between the East Indian, Philippine, and Japanese isles and with the Asian continent. Furthermore, the East India Company employed warships for military display as well as real action against the Portuguese, British, or local pirates—and more often than not to persuade the indigenous peoples to conduct their trade through the Dutch. Besides their military accoutrements, these ships often had some additional cargo capacity.

The warships or war yachts (*oorlogsjachten*) were called *pinassen,* after a fast Basque rowing and sailing ship. The name thus applied to an armed ship, built for speed rather than for trade. *Heemskerck,* the ship with which Abel Tasman discovered New Zealand, was of this type. It had a loading capacity of 60 last (about 120 tons).[17]

At the end of the seventeenth century a lawyer by the name of Van Dam was commissioned to write the history of the Dutch East India Company. Published in 1701, his book was called *Beschryvinge van de Oostindische Compagnie* (Description of the East Indian Company). In this book the term *pinas* emerges for the first time in text covering the year 1652, when Van Dam mentions that pinases are replacements for fluyts. (The round-sterned fluyts soon developed major leakage problems, as their sterns were more vulnerable to drying out in the tropical sun. This is why they were ultimately replaced by galliots and *hoekers.*) In the section for 1662 he gives the dimensions of "jachten of pinassen" (yachts or pinases), an indication that the terms were more or less synonymous at that time.[18]

According to Van Dam the large pinases before 1650 measured 128 feet, the smaller ones 116 feet. But by the end of the century the East India Company called all return ships "pinases." Like the return ships, they were made

FIGURE 1.8. Model of a pinas, or war yacht, from the mid-1700s. The model was hung in a church for decoration, which is why the underwater part of the hull is not to scale. Yet it gives a fair image of the pinas of that period. (Courtesy Het Scheepvaartmuseum, Amsterdam)

Figure 1.9. Model of a pinas, built by the author from information in Witsen's treatise. Originally a small war jacht, the pinas developed into an armed merchantman of medium size in the middle of the seventeenth century, later into a heavy East Indiaman. (Courtesy Cees de Jonge, The Visual Art Box)

in three rates (*charters*), measuring 130, 145, and 160 Amsterdam feet (1 Amsterdam foot = 28.3 centimeters). Whether the return ship was simply renamed "pinas" or the advantages of the above-mentioned warship were fused with the older, slower merchantman remains unclear. It is noteworthy, however, that in the 1690s, when this shift occurs, we find the frigate mentioned by the Dutch East India Company for the first time, probably a replacement for the old term *pinas,* which then acquired another meaning.

A comparison of Witsen's 134-foot pinas with the smallest rate of 130 feet, established in the resolution of the East India Company of April 4, 1697, reveals that the first, being four feet longer, is also four feet narrower than the second. We can then ask ourselves whether the use of the name *pinas* in the earlier case did not in fact refer to a faster ship. Unfortunately, we cannot be so quick to draw this conclusion. In 1660, for instance, the East India Company commissioned a ship, the *Diemermeer,* with exactly the same dimensions as Witsen's pinas (134 × 29 × 13 feet); yet this ship was a fluyt. Measurements alone, then, are not conclusive in defining ship types.

The evolution from return vessel to pinas can be compared with the shift to frigates in the war fleet. When the Dunkirk pirates became such a nuisance in the 1630s that Holland decided to deal with them in an organized way, it soon became apparent that the small, fast pirate ships had many advantages over the large "ships of violence" of the Dutch Admiralties. This is the main reason why the frigate was introduced as a smaller, faster man-of-war next to the large warships. The frigates proved such a success that soon the larger rates were "frigated" (*gefregateerd*), meaning that they were reduced in size and were built more for speed. This shift elicited Witte de With's lament that the entire Dutch fleet operating in the First Anglo-Dutch War actually consisted of frigates.[19]

It is possible that this change, induced by the Admiralties, occurred with some delay in the merchant fleet as well, but it is difficult to find proof to substantiate such a hypothesis. One indication is that in the written sources on the Admiralties, the terms *frigate* and *pinas* were almost interchangeable. It is clear, however, that in Witsen's day the pinas was a relatively fast and well-armed merchantman, capable of taking on enough cargo to make it a profitable business undertaking but also armed in a way that it could effectively defend against enemies. This must have been why Witsen chose the pinas for his example: it was a design that could illustrate aspects of both the man-of-war and the merchantman.[20] As Witsen explains:

(263 I 28) The ship which is built here in our mind is neither the widest, nor the narrowest; which measure is taken with premeditation, to show a man-of-war as well as a merchantman. Those sailing for freight only, and cannot defend themselves, are narrower above and broader below: those going to war only, are broader at the top and narrower at the bottom.

(53 I 30) Of which the measurement and shape are well understood, that from these one can easily derive and fabricate ships of any length and usage (mutatis mutandis, change what has to be changed), both large and small; because all rules, symmetries and proportions remain the same, on all the distances on the keel; when the ship has a keel of one hundred and eighty, or of only sixty feet.

So if we master the information on the pinas, according to Witsen, we have the key to all other types of ships. But Witsen is careful:

(53 I 45) It is true, on the other hand, that to apply the mathematical measurement to its full exactness, is not always possible, as with the many and varying curved and crooked shapes, that one has to apply to the ship's timbers.

(53 II 36) Material, place and opportunity often cause deviations. The use, which is of endless variation, forces the building master to bend rules and measures.

(53 II 26) The more precisely the proportions and symmetry in shipbuilding are observed, the more perfect the ship will be, graceful, strong and well-sailing.

(54 I 12) No one expects that all parts of the ship will be shaped to mathematical detail: this will only be done for those parts, which are the most important in the ship, from which the others will follow without exactness.

Shipbuilding Theory in the Seventeenth Century

A seventeenth-century ship was designed and built by applying a number of proportional formulas (see table 1 in the appendix), which we will go into extensively in chapter 2. A table of such formulas, which Witsen says he obtained from the "famous master shipwright Jan Dirrikze Grebber," demonstrates that, using these formulas, it is possible to generate a kind of recipe for ships of all sizes (fig. 1.10). Within the boundaries of these formulas the shipbuilder had some leeway, as we shall see. Of course, he started with a general mental image of the vessel, but until the erection of the top timbers he was always open to the possibility of changes, depending on the availability of timber, alterations in plans of the builder or the commissioner, or other considerations. He also knew pretty well how far he had to deviate from the standard rules to obtain a bulky carrier or a sharp frigate. This freedom to deviate from the rules was gained solely through experience, either personal or that of others, in constructing successful ships (or failures). There is no evidence whatsoever that shipbuilders were capable of calculating a ship's performance beforehand. Even the draft of the vessel could be only estimated,[21] which Witsen's following remark so plainly demonstrates:

(180 II 49) Today ships are smeared with a mixture of resin and tallow as far as the estimated draft of the vessel.

In Holland at the time, there was a dawning notion of shipbuilding theory, but only in the more intellectual circles, as exemplified by Witsen and earlier by Simon Stevin, and not among actual shipwrights.[22]

Although concepts such as metacenter and center of balance did not yet exist, Witsen formulates quite complicated arguments for issues like the mechanical effect of the mast on the ship—the mast is a lever . . . (p. 250)—and calculations of the displacement of the pinas ship. I will return to this in the discussion of tonnage calculations (see "Ship Measurement" later in this chapter). But for most theoretical questions, such as the equilibrium of floating bodies, Witsen does not get beyond theorizing and does not arrive at calculated predictions, as we can see in these examples:

Het Schip is langh over steven. / The ship's length between stem and stern.	Wijt binnen de huit op zijn uitwaetering / Width inside planking on the height of the watershed		Hol op zijn uitwaeteringh. / Depth at the watershed		't Vlack is wijdt. / The floor is wide		Het rijst. / It rises		Wijt op het boeyzel, of op de kimmen. / Width at the top of the bilges		Hol op zijn boeyzel, of op de kimmen. / Depth at the top of the bilges		De voorsteven hoogh. / The stem is high		Hy valt. / It rakes		De achtersteven hoogh. / The stern is high		Hy valt. / It rakes		Het hek langh. / The transom is long		Het heck dick, breet, en heeft bocht, maer altijt maeckt men 't 2 duim breder als de maet. / The transom thick, wide and curved, but always 2 inches more than according to the formula		De barrighouten zetten voor op. / The wales rise afore		De barrighouten zetten achter op. / The wales rise abaft		Het verdeck hoogh aen boort op zijn hals. / The upper deck high in the side on the main frame		De oplangen hangen over 't boeizel. / The second futtocks are wider than the bilge		De rantsoen-houten staen van 't heck na beneden toe, / The fashion pieces are from the transom downwards		De heck-stutten zijn boven 't hek langh, / The stern timbers are long above the transom		De heck-stutten zijn boven wijt, / The stern timbers are wide at the top		De stutten komen op de hals in, / The tumblehome of the top timbers at the main frame	
Voeten / Foot	voet / Foot	duim / inch	voet / Foot	duim / inch	voet / Foot	duim / inch	voet / Foot	duim / inch	voet / Foot	duim / inch	voet / Foot	duim / inch	voet / Foot	duim / inch	voet / Foot	duim / inch	voet / Foot	duim / inch	voet / Foot	duim / inch	voet / Foot	duim / inch	voet / Foot	duim / inch	voet / Foot	duim / inch	voet / Foot	duim / inch	voet / Foot	duim / inch	voet / Foot	duim / inch	voet / Foot	duim / inch	voet / Foot	duim / inch	voet / Foot	duim / inch	voet / Foot	duim / inch
60	15	0	6	0	10	0	0	6	13	10	2	5½	11	0	10	5½	10	5½	1	8½	10	0	0	6	1	5½	4	5½	0	0	0	6	5	5	10	5	5	0	0	0
65	15	5½	6	5½	10	3¾	0	6½	13	10¼	2	6½	11	5½	11	0	10	5¾	1	9¼	10	3¾	0	6½	1	6½	4	9½	0	0	0	6¾	5	5½	10	5½	5	3	0	0
70	17	5½	7	0	11	7½	0	7	16	2½	2	7	12	0	11	5½	12	2½	2	0	11	7½	0	7	1	6½	5	2¼	0	0	0	7	6	2	12	2	6	0	0	0
75	18	8½	7	5½	12	2½	0	7½	17	4	2	5½	12	5½	12	0	13	7½	2	9	12	2½	0	7½	1	6¼	5	4	0	0	0	7½	7	0	13	7	6	3	0	0
80	20	0	8	0	13	3¾	0	8	18	6	2	8	13	7½	13	0	14	0	2	3¾	13	3¾	0	8	1	7½	6	0	0	0	0	8	7	4	14	6	6	8	0	0
85	21	2½	8	5½	14	1½	0	8½	20	3½	2	9	14	9	14	0	14	6	2	4½	14	1½	0	8½	1	7½	6	5½	4	6	0	8½	7	8	15	0	7	2	1	2
90	22	5½	9	0	15	0	0	9	21	1½	3	0	15	10	15	0	15	8	2	7½	15	0	0	9	2	2½	6	8¼	4	8	0	9	8	0	15	5	8	0	1	8
95	23	8½	9	5½	15	9	0	9½	22	¼	3	1¼	16	8	16	0	16	4	2	8½	15	9	0	9½	2	3¼	7	3½	4	10	0	9½	8	4	16	5	8	3	1	10
100	25	0	10	0	16	7½	0	10	23	3	3	5	18	1	17	8	17	5½	2	9	16	7½	0	10	2	6	7	5½	5	6	0	10	9	0	17	6	8	8	1	10½
105	26	2½	10	5½	17	3½	0	10½	24	8¼	3	7	19	1	18	5½	18	1½	3	2½	17	3½	0	10½	2	5½	8	2	6	0	0	10½	9	5	18	1½	9	0	2	0
110	27	5½	11	0	18	3¼	1	0	25	5½	3	9	19	6	19	0	19	3	3	2	18	3¼	1	0	2	6¼	8	3	6	1½	0	0	10	6	19	3	9	3½	2	¼
115	28	8½	11	5½	18	9½	1	½	26	8½	4	0	20	3	20	0	19	9½	3	3½	18	9½	1	½	2	7½	8	5¼	6	2	1	½	10	6	19	9	9	5	2	½
120	30	0	12	0	20	0	1	¼	27	9½	4	0	21	3	21	0	21	0	3	5½	20	0	1	1	3	0	9	0	6	3	1	¼	10	8	21	0	10	0	2	1
125	31	2½	12	5½	20	7½	1	1	28	9	4	4¼	21	5	21	3	21	4	3	7½	20	7¾	1	1½	3	1	9	4	6	4	1	1	11	0	21	4	11	0	2	1¼
130	32	5½	13	0	21	7	1	½	30	1½	4	5¼	22	8½	22	8	22	8½	3	9½	21	7	1	2	3	3	9	8	6	5½	1	2	11	8	22	8½	11	5½	2	2
135	33	8¼	13	5½	22	5½	1	1	31	3¼	4	5½	23	4¼	23	4	23	3	3	10½	22	5½	1	2½	3	4	9	8½	6	5¾	1	2½	11	10	23	3	12	0	2	2¼
140	35	0	14	0	23	2¼	1	1½	32	7	4	7	23	10½	23	10	24	5½	4	1	23	2¾	1	3	3	5	10	5½	6	6	1	3	12	5	24	5½	12	5½	2	3
145	36	2½	14	5½	24	2	1	2	33	8½	4	7½	24	3	24	0	25	½	4	2	24	2	1	3½	3	5½	10	6	6	6½	1	3½	13	0	25	0	12	10	2	3½
150	37	5½	15	0	24	10½	1	2½	34	8½	4	9	25	4	25	4	26	3	4	4	24	10½	1	4	3	8	11	3	6	7	1	4	13	6	26	3	13	6	2	4
155	38	8¼	15	5½	26	1½	1	3	35	10½	4	10	26	0	26	0	26	9	4	5	26	1½	1	4½	3	9	11	3½	6	7½	1	4½	13	10	26	9	13	10	2	4½
160	40	0	16	0	27	4	1	4	37	3	5	0	26	6½	26	6	28	0	4	8	27	4	1	5	3	9½	12	0	6	8	1	5	14	5	28	0	14	9	2	5
165	41	2½	16	5½	27	5½	1	4½	38	2¼	5	2	27	2	27	2	28	5½	4	8¾	27	5½	1	5½	3	10	12	¼	6	8½	1	5½	14	9	28	5½	14	10	2	5½
170	42	5½	17	0	28	3¾	1	5	39	3½	5	5	27	8½	27	8	29	8	4	9	28	3¾	1	6	3	10½	12	8	6	9	1	6	15	0	29	8	15	0	2	6
175	43	8¼	17	5½	29	3¾	1	5½	40	6½	5	5½	28	3	28	0	30	6	5	0	29	3½	1	6½	4	0	12	9½	6	10	1	6½	15	6	30	0	16	10	2	6½
180	45	0	18	0	30	0	1	6	42	1	5	7	29	5½	29	5½	31	2½	5	2	30	0	1	7	4	5½	13	5½	7	0	1	7	16	0	31	0	17	0	2	7½
185	46	2½	18	5½	30	10	1	6½	43	1¼	5	9	30	1½	30	0	32	0	5	4	30	10	1	7½	4	6	13	6	7	1	1	7½	17	0	32	0	17	5	2	8
190	47	5½	19	0	31	7½	1	7	44	½	6	0	30	8½	30	8	33	4	5	6½	31	7½	1	8	4	6½	14	3	7	2	1	8	17	5	33	4	17	10	2	8½
195	48	8¼	19	5½	32	5½	1	7½	45	2½	6	2	31	6½	31	6	34	0	5	8	32	5½	1	8½	4	7	14	5½	7	3	1	8½	18	0	34	0	18	0	2	9
200	50	0	20	0	33	4¼	1	8	46	4	6	5	32	8	32	8	35	0	5	10	33	4¾	1	9	5	2½	15	0	7	5	1	9	18	0	35	0	18	0	2	10

FIGURE 1.10. Grebber's table of proportional formulas (p. 114). Specifications in feet and inches are shown for the major parts of ships ranging from 60 to 200 feet in length (although we have no indications that there were any ships longer than 180 feet in Holland in the seventeenth century).

(273 II 51) It is hard to determine the vertical balance of the ship, if it has been corrupted by a supervening weight: the center of gravity, as maintained by many, should be a little aft of the center of the ship.

(257 I 44) The center of gravity of a floating Ship [. . .] is a point in the middle, around which all parts are in balance, unchangeable. But having thoroughly consulted *Philosophy,* although this does not make any difference to practice, this will apply only as long as the body is parallel to the horizon; but should it tilt, the center of gravity will also change; the lowest part after the shift, will be found the heaviest and the center of gravity will incline toward this; of which the reason will be clear to whoever understands the constitution of the whole; but because such would not fit here, I will refrain from further explanations.

Even so, in Witsen's time model tank-testing experiments were already being carried out to determine the ideal hull-shape, but they were directed more toward speed than stability. And there were no means to translate the outcomes of these experiments to full scale.[23] The emphasis on testing for speed is seen in this remark:

(274 I 3) All shapes of ships have been tested by towing slices of wood through the water, to determine which would sail the best: the ones going fastest through the water by means of the pulleys, were thought to be the best sailors.

In chapter 18 Witsen presents explanations for a number of issues, such as the shape and size of the sails and the proportions of the ship's body, but they are based on experience and have no theoretical foundations. His remarks always come down to warnings about the unpleasant consequences of extreme deviations from the standard. An example:

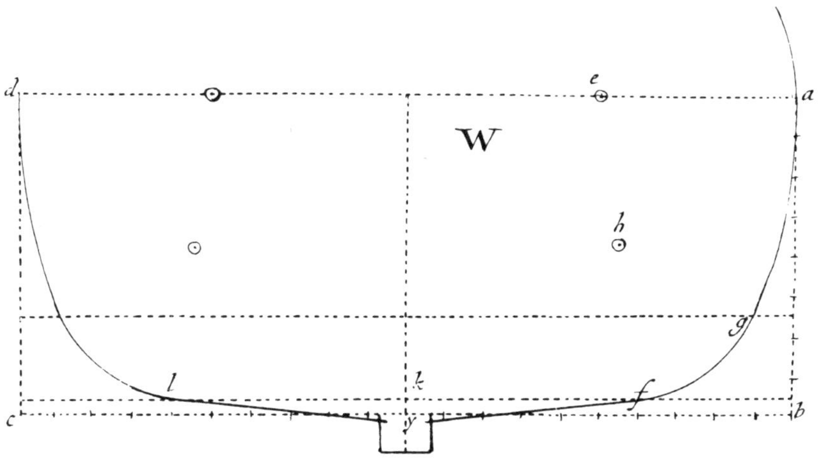

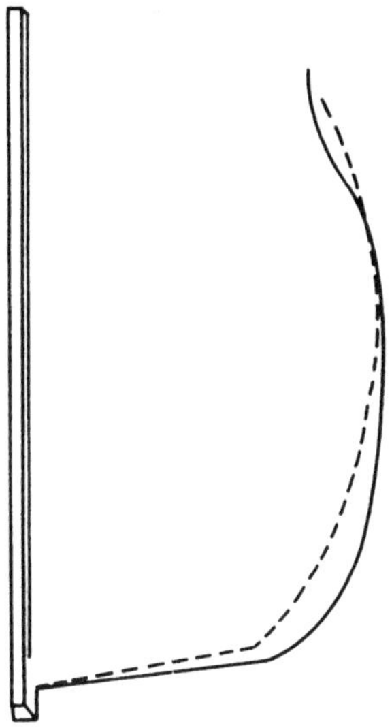

FIGURE 1.11. PLATE LII W

(151 I 46) The figure at the letter *W,* above, shows how to make the construction on paper, before starting on the Ship. *a, d,* is the main deck, from *e* and *h* arcs are drawn, *g* the intersection in the turn of the bilge, *f* and *ı* the highest points of the rise of the floors, *y* the keel, and *k* the place on which the keelson and the floor timbers will come, as indicated in the print.

But to make sure of the rake of the futtocks and the curve of the bilges, one makes a model or drawing on paper: one draws a line on the depth, and divides the width into four parts, puts the point of the compass on the third of the four parts, and draws a circle a little lower than two thirds of the depth, as shown at *g* in the hollow of the bilges; and if the floor is wide, and rises three fifths, as at *f,* then one puts the foot of the compass on *f,* makes a scratch at *h,* and another at *g;* then again a scratch at *h,* and puts the foot of the compass on that center point, and draws the curve of the bilges, as one can see done from *f* to *g.* From *f* to *ı* and from *y* to *k* one has the rise of the floor.

FIGURE 1.12. If Witsen's theory of frame construction is applied to the main frame of the pinas, his impressive method does not seem to fit: the solid line represents the pinas, the dotted line is the product of the theory. Witsen has apparently invented a scientific method of frame construction (Drawing by A. J. Hoving)

(254 I 43) One can fairly ask why is a mast not made any longer than the width and the depth of the Ship twice? [The formula for the length of the mast was twice the width of the ship plus its depth.] The answer will be, that masts which are higher, come down easily, for which the reason is that the wind can apply more force to the top, which is far from the fixed end, than to the lower part of the mast.

(254 II 32) If the mast were to be made smaller than mentioned above, one will experience that too much wind is lost, and that the shortage of the longest arm or the mast will unnecessarily prevent progress.

So the rules were always determined by practical experience and were never a product of theory or calculations. Even with the dawn of theoretical notions about the various forces at work on a sailing ship, scientific formulas regarding stability, loading capacity, and other aspects of physics were completely unknown.

This does not mean that every ship launched from the yard was a complete surprise. As mentioned above, the shipbuilder was very much aware of the ways in which deviation from the rules influenced the performance of the vessel, but it was knowledge gained by practical experience. Upon the death of a shipbuilder, his knowledge was buried with him.

As a result, shipbuilding was an extremely static trade in which changes took much longer than in fields using a method based on mathematical formulas and calculations. The high costs involved also discouraged risky experiments. Most changes were in fact the result of accidents. A successful ship was taken as an example and imitated as closely as possible.

This static quality was very probably one of the main reasons why Dutch shipbuilding in the seventeenth-century gradually lost its lead to the other western European nations. In England, and later also in France, shipwrights had begun designing ships on the drawing

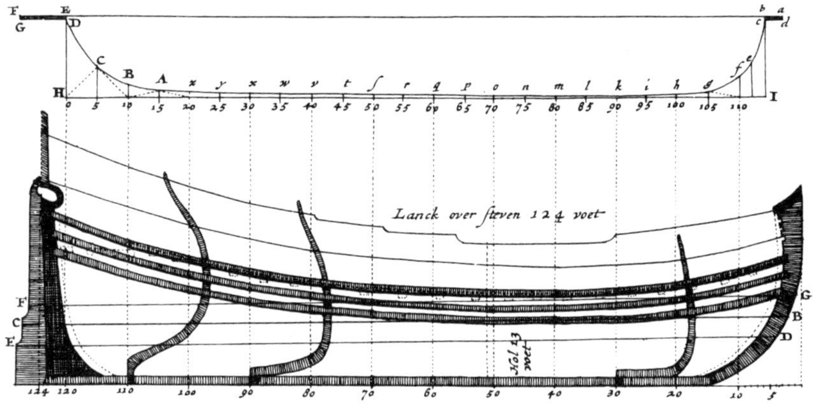

FIGURE 1.13. PLATE XCIII
Witsen's illustration of a method for measuring capacity. First the shape of the hull is recorded at the height of the line *E–D*. After the ship has been loaded to *F–G*, the same procedure is followed. *C–B* is the average of the first two measurements; it is then multiplied by the height *E–F* to yield the weight of the cargo if multiplied by the weight of water.

board and on a theoretical basis, which produced a manifold increase in the rate of change. At the beginning of the eighteenth century the Amsterdam Admiralty even found itself forced to import foreign expertise to modernize its shipbuilding techniques, and English shipwrights were appointed to the Amsterdam Admiralty.[24]

Although nothing points to the use of mathematical practices on the yards, Witsen himself insists on demonstrating such a method to obtain the shape of the main frame (see fig. 1.11).

We find no trace of the use of such geometrical methods in contemporary sources. In contracts and descriptions of ships of that era there is never any mention of the lengths of sweeps or the locations of the centers from whence the sweeps had to be struck. And if applied to Witsen's pinas, the method produces a completely different shape (see fig. 1.12). In all probability it was Witsen himself, aware of the geometrical methods applied in other European countries, who devised this method, either to raise the status of the building method he advertised, or from a need to find explanations for the phenomena he encountered.

Even the sober Van Yk, when dealing with the shape of the (main) frames, writes: "What figure these frames are to take, depends largely on the width of the ship, and the depth, and further on the eye and the judgment of the master shipbuilder. But the shallowness of the Dutch estuaries is often the reason that he cannot take into account that, which otherwise he would think best" (p. 69). Concerning the fore and after body, he states: "I have never found that our shipbuilders have any secure, or fixed rules regarding the shaping of these frames" (p. 76).

If Witsen's geometrical method was used by shipbuilders after all, it can only have been to design a frame on paper. In the yard it would have been useless, for with the shell-first method, no entire frame was ever assembled to be raised on the keel; instead, section after section was fitted in the hull, following the building stages.

Ship Measurement

In chapter 17 of his treatise Witsen calculates the **amount of water resting against the sides of a Ship.** He means the upward pressure of the water, or, according to Archimedes, the weight of ship plus cargo. Again the 134-foot pinas is Witsen's example, and the method he uses is to divide the ship into calculable geometrical surfaces and bodies, leaning heavily on Simon Stevin's "never fully praised Water-weight."[25]

With a supposed draft of 10 feet, Witsen arrives at $19.718\frac{1}{3}$ cubic feet of capacity for the submerged part of the hull. The weight of one cubic foot of seawater (from the island of Texel) had been accurately determined to weigh 46 pounds 18 *lood,* or 22.9 kilograms (an Amsterdam pound was 494 grams, a *lood* 15.4 grams). Witsen's total displacement then comes to 435.5 tonnes. One foot of extra draft would increase the burden with 81.5 tonnes; one foot less would decrease the weight by 70.5 tonnes. These figures reasonably agree with those of the Scheepsbouwtechnisch Ingenieurs Bureau (Bureau of Naval Engineers) in Bloemendaal, calculated on the basis of my reconstruction drawings of the pinas: 480 tonnes at 10 feet of draft.

In a comparable fashion Witsen then calculates the weight of the pinas, and here he makes a curious mistake. Without giving any reason, he again supposes the draft to be 10 feet, which is much too much for an empty ship, and then arrives at a total of 600 tonnes. By calculating the weight of all the parts, I have been able to establish the actual weight at only slightly more than half of that. It is curious, and perhaps indicative of Witsen's approach with these completely theoretical calculations, that he was apparently unable to integrate his calculations and discover the impossibility of his final result. Plainly, with the weight of the ship so much greater than the upward pressure, it would sink on the spot!

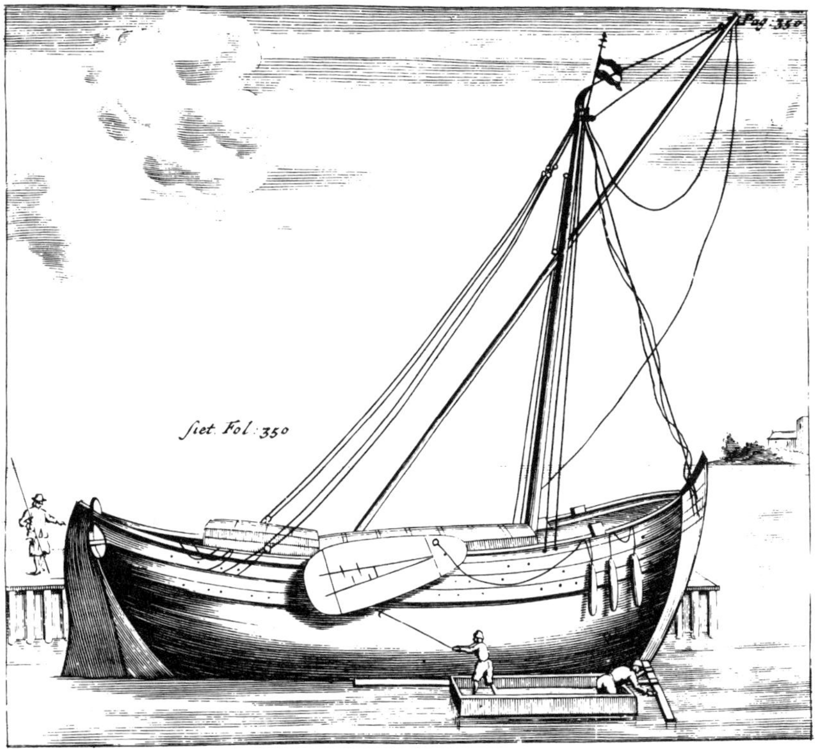

FIGURE 1.14. Van Yk's illustration of a surveyor determining the shape of a vessel being measured.

The purely speculative nature of Witsen's calculations becomes apparent when we compare his method with the actual methods used to estimate a ship's burden in the seventeenth century. The main point of interest was the cargo capacity or, in the case of warships, the carrying capacity in terms of ordnance, supplies, and crew. Nobody was in the least interested in superfluous figures such as displacement or actual weight of a ship.

How then was the carrying capacity determined? We know of three methods. The first and simplest one was purely empirical: an empty ship was loaded with cannon balls until it reached the load line. As the weight of the cannon balls was known (in pounds), one could easily deduce the weight in lasts (*lasten*) by dividing by 4,000, as the seventeenth-century last was equivalent to about two tons.[26] Both Witsen (p. 242) and Van Yk (p. 320) mention this simple but labor-intensive method. As a result, its chief drawback was that, rather than measuring each ship separately, one supposed the same capacity for all ships of the same size and type, which could lead to considerable errors due to difference in build.

The second method is mentioned by Witsen when quoting his friend Johannes Hudde (1628–1704), fellow burgomaster of Amsterdam and renowned mathematician (pp. 241–47). Van Yk also describes this method (pp. 249–51). It consisted of measuring the difference in the draft of a ship when empty and when loaded, and then calculating the capacity of that difference. This method required that the inspector of weights and measures performed the measurements from a boat alongside the ship. Although less labor intensive than the cannon-ball method, it was still too cumbersome to be ideal.

The third method was an early application of the block coefficient, the ratio between the actual volume of the submerged part of the ship and a block of the same length, breadth, and depth. Length, breadth, and depth were multiplied, which gave a total volume in cubic feet that was much greater than the actual volume of the ship. This was therefore divided by a factor that had been predetermined by one of the two methods mentioned above and that varied for each type of ship. The factor adjusted for notional block coefficient and converted the volume from cubic feet into lasts. For a variety of inshore vessels the factor ranged from 170 to 240, but factors were also determined for large seagoing ships. Yet, as with inshore vessels, it was impossible to apply the same factor to all seagoing ships. In the anonymous manuscript Evenredige Toerusting, which deals with men-of-war, the factor is 250. Hendrick Decquer, who performed an inquiry into the measurement of ships for the Dutch East India Company, arrived at the same number in his report (c. 1690), but when we compare ships for which both measurements and the number of lasts are known, we find that there were actually wide differences in the ratios (see table 2 in the appendix). As can be expected, fluyts, with their small crews and light armament, show the most advantageous factor: about 200. Most East Indiamen and warships have a factor of 250, whereas war yachts, which had little cargo capacity and a burden devoted largely to armament, sometimes reach 350.

On pages 34 to 36 of his appendix Witsen describes an experiment from 1647 in which the measurements of three ships of different ages—new, middle-aged, and old—were compared. The factors for these ships were determined to be 231⁶⁄₇, 239⅕, and 256 respectively, an average of 242, after which the Skippers were handed their Measurement certificates, and the Tolls [. . .] had to be paid (appendix, 36 II 16). The measurements were used in calculating the tolls to be paid and were branded on the stem, sternpost, or main beam.

This leaves us with the question of why Witsen presents us with his impressive calculations of the upward pressure and weight of the pinas. As we have seen, the erudite Witsen had considerable knowledge of foreign authors writing on shipbuilding, and Fournier's *Hydrographie* seems to have impressed him most of all. It is possible that, to cast a scientific aura over Dutch shipbuilding in general and his own book in particular, he deemed it necessary to demonstrate his theoretical knowledge—only half-digested—with these sample atypical calculations.

At the beginning of the seventeenth century both the Dutch East India Company and the Dutch Admiralties registered the size of a ship in lasts rather than its length. Many have tried to establish a method for deducing length from a given measurement in lasts, but such endeavors can prove treacherous. In his article on the dimensions and units of Dutch ships, Van Bruggen lists five ships whose length he gives as 90 feet, but the number of lasts ranges widely: 68, 75, 80, 150, and 180! Likewise, three East Indiamen mentioned by Van Dam, 130, 138, and 150 feet in length, all measured 300 lasts.[27] And what about Abel Tasman's ships? The war yacht *Heemskerck* had a capacity of only 60 lasts and was 106 feet long, whereas his second ship, the fluyt *Zeehaen,* with a capacity of 100 lasts, was only 100 feet in length.[28]

It seems, then, that not only were the units of measurement inconsistent in those times but so too were the motives in establishing the cargo capacity. To try to determine the length of a ship for which only the number of lasts is known remains a hazardous undertaking.

From the displacement diagram made by the Scheepsbouwtechnisch Ingenieurs Bureau in Bloemendaal and the calculated weight of the pinas as a whole, computer calculations put its displacement at 635 tonnes at a draft of 12 feet (which must have been about the maximum).[29] If trimmed by the stern with a maximum draft of 12 feet, the tonnage comes to 586.

The ratio between cargo capacity and displacement can then be established as 0.55, which means that the pinas could carry almost as much cargo as its own weight:

(317 II 37) A ship, says Fournier, a French Author, carries as much as she weighs herself.

Sailing Square-Rigged Ships

GERALD A. DE WEERDT

Witsen himself was neither a shipwright nor a sailor. Yet his remarks on matters such as rigging and sail handling are far from theoretical fantasies. Apparently he was well informed by real experts.

Sails

The sails of seventeenth-century square-rigged ships were not cut with optimal aerodynamic curvature like they are today. The square sails were simply composed of straight widths of textile (the so-called cloths), resulting in a completely plane canvas. For the greatest efficiency the wind was allowed to push some curvature into the sails, for which the head rope was given some slack when it was tied to the yard:

(139 I 50) The Sails should not be too tight nor too flat against the Yards, but in such a way that, the wind blowing into them, it will give it some roundness; and thus the wind is better caught in the Sail, and can transmit more of its movement to the canvas, and the Ship's progress shall improve.

It was the common opinion of sailors that the foremost sails, those of the foremast and the bowsprit, yielded the best effect—which was probably true, as the ships were built and rigged to sail on a reach of no more than two compass points (22.5 degrees) closer to the wind than a beam reach. Many paintings and engravings show that the foresails were the last to be reduced.

(139 II 5) The Sailors agree, that the foresails are the ones giving the Ship most of its momentum; giving for a reason, that the wind will apparently draw forth the Ship; and will only push aft and in the middle: also that the wind in the foresails will cut through the Sea in a better way, the waves being nearer.

(The last argument is of little value and should not be taken too seriously.)

It was also generally believed that the topsails were more effective than the lower ones. This was not untrue, as wind speeds increase with height and the wind is less hampered by turbulence from the ship and the waves; however, the topsails were frailer and easily damaged despite their smaller size and could not be used in heavy weather, as Witsen explains:

(139 II 17) The topsails are believed to give more progress to the Ship than the lower ones, and such, because they are removed from the weight. Yet the topsails have this disadvantage, that they more easily come down, and cannot be used in hard weather.

Sail Handling

The sails were used as follows. With a half wind it was advisable to set all sails. When running, the aft sails would shield the foresails, which would receive no wind. In that case, the mainsail was usually furled to allow the wind to the foresails. This was why the ship would sail faster with a quarter wind or half wind than when running with a free wind.

(139 II 38) With a half wind all the sails will work, which cannot be with a free wind; with a free wind the mainsail is clewed up, to allow wind to the foresails.

This is also why a Ship will have more speed, with a wind from aside, than with a free wind, though not completely favorable.

(140 I 23) Ships make the most progress, when there is a grab-line wind, or quarter breeze, and such with two or three compass points, from a head wind: because then all the sails work.

When sailing close-hauled, the spritsail topsail was not used. First of all, this gave the ship a lee helm; furthermore, this sail had no tacks or bowlines that could be set forward, which made it impossible to set the sail in the required fore-and-aft position.

(140 I 28) When sailing a close haul, the Spritsail topsail cannot be used, because in such a case it flutters too much, and cannot be set straight enough.

In fact, the course of the ship was determined by the setting of the sails. The narrow rudder was only good for tiny course corrections once the ship had attained some speed. Although Witsen does not mention it, experiments with replicas have shown that, with no sails set, these ships have a tendency to lie abroad the wind. By hoisting the foresails, the ship would start running, and for heading up, more sails had to be set aft.

(140 I 50) With the Foresails the Ships are turned from the wind; and with the aft sails into the wind.

The reason for this is, that while the Ship turns around its middle point, when the stern is moved, the foreship is pushed the other way: and likewise with the foresails. From which it follows that the Mizzen will take ships into the wind, and the Spritsail topsail will bear them away.

The mizzen was mainly used for steering. Except when running straight before the wind, the mizzen was usually close-hauled to the wind, as a kind of fantail keeping the ship in a steady position relative to the wind direction and thus enhancing the stability of the course.

(139 II 47) The aft triangular sail will help the rudder, in times of danger, because this sail can best be set to the course, which is wanted, and the Ship necessarily needs to be steered from behind.

The effect of the rudder was limited, and it was even less when the ship came into shallow waters like the Dutch harbors and the Zuider Zee. The effect of the rudder was then diminished so much that steering was done mainly with the sails.

(140 II 28) In the shallows, Ships will answer their helm badly: and then the Ship is to be steered with the sails, which is done with hoisting, dropping, setting thus or that way, in the art of the experienced Sailor.

When the wind hardened, the topsails were the first removed. The next step in reducing the sail area was to lower the foresail and mainsail yards. Besides reducing the effective sail area, this also reduced the risk of damage to the yards, as their movement decreased, along with the forces exerted on the masts and shrouds.

The mizzen was one of the easiest sails to handle. It could safely be carried in almost all weather conditions, which was general practice, except when scudding (i.e., running before the wind in a storm).

(140 I 38) In a storm the Mizzen is struck, and the Foresail tied to the bow. The lower the Sails, the better they can be handled, and the less danger of them carrying away. However tempestuous the winds, the Mizzen can always be carried: and seldom is fouled.

Materials

Wood

Before we delve into Witsen's discussion of ship construction, it seems appropriate to say something about the

FIGURE 1.15. Materials for shipbuilding were imported to Holland from all over Europe (Drawing by A. J. Hoving)

building materials he mentions. Wood is the most important of these materials, and Witsen deemed it necessary to warn the shipbuilder against false economy:

(263 II 31) In my building I will use wood without meanness: give each part its due, according to the dimensions it requires, sparing no wood, knowing well that in building, the masters often make the parts more delicate and weak, trying to save material: but am conscious of the fact, that avarice costs many a man his neck.

Timber for shipbuilding was obtained from several regions in Europe: straight oak trunks were floated down the Rhine, forming gigantic rafts on which hundreds of people lived, to be sold in Dordrecht. Crooked oak was imported via Hamburg. Timber was always named after the harbor where it was purchased, seldom after its region of origin.

Oak and pinewood were imported from "Koninkxbergen" (Königsberg in Prussia, now called Kaliningrad). Pinewood for masts and spars was obtained from Riga as well. Norway supplied pinewood, too, particularly deal (deal is pinewood sawn into planks). Pegs for treenails came from Ireland. Domestic beech, elm, and ash were used for pumps, blocks, carvings, handspikes, and tops.

(178 II 41) That the material from which one builds Ships is wood, everyone knows. *Nec in navi cadentem lignum attingere,* falling in a Ship without touching wood, as the ancients said, when discussing a case which seemed impossible to them.

In this country it is mostly *Oak* from which Ships are built, which comes to us from the Rhine and Westphalia. This wood for its density admits little water. Westphalian crooked timber and Rhineland straight timber are much recommended.

Ships are also built of *Pinewood,* but it is light and not as strong as oak, for which reason it is seldom used for Ships of war, or ships exposed to much violence; yet for the upper and inner works, where not much force is exerted, it is very apt, as for Ships destined for loads; because pinewood is light, and Ships made of pinewood float high on the water, and therefore can take more cargo. This wood is brought to us from Norway and the East, as is deal, which is like pinewood, but lighter and more fragile. Both kinds of wood are difficult to bend, and have much sapwood. Masts from Norway and Muscovy are judged the best in our country, and are the most used. Koninkxbergen supplies the best planking in pinewood as well as oak, Norway the best deal.

(180 I 3) *Elm* and *Lignum Vitae* are good for making Ship blocks and sheaves.
Domestic *Beech* is recommended above domestic oak.

(180 I 41) *Alder* and *Lime* grow fast, but yield a light and fragile material: but Alder will grow tough and strong if left in water, or in the ground for many years.

(182 I 36) *Elm* does not take treenails well. *Lime* is recommended, because it does not suffer tearing and cracking so easily.

(182 II 23) *Fir* was chosen for yards, and to make sprits; it is pliant, but rots in saltwater, which is why it is better used on sweet rivers than in sea. It is used on Ships above the water, where it is adequate for its lightness; the trees are distinguished in male and female specimen, of which the male is thought to be stronger and sounder.

(183 II 33) *Lime* and *White Poplar,* are good for pumps and other tubes, because they are soft inside and hard on the outside.

(183 II 25) Much more could be said here on the quality of wood and trees, but seeing as much of it is of no importance to Shipbuilding, we shall not pursue it beyond stating that *Lignum Vitae* and *Walnut* are good for sheaves, because they are hard and strong.

Witsen mentions other kinds of wood, such as Brazilian, Biscayan, Irish, and English wood, and especially many sorts about which the ancients gave their opinion, such as fig wood, the pine tree, the juniper tree, and so on; yet he makes the following clear:

(181 I 12) For Shipbuilding oak tops everything, above all other trees, because it is tough, bends well, is strong and not too heavy.

Witsen also discusses the quality of wood:

(180 I 53) The shipbuilder should mind the quality of wood used for building, and for keel and frames he chooses the best wood; for carvings and decorations bad wood is adequate, and all the inner paneling may be light and badly constructed, as long as it does not interfere with the firmness of the ship: but sapwood and rotted wood he must avoid everywhere.

(179 II 28) It is important to see to it, that the wood used for constructing the Ship is dry and not wet; moist wood, when it dries when fastened to the Ship, produces crevices and cracks and often causes great discomforts when at sea.

(181 I 20) The fewer knots there are in the wood, and the longer the grain, the more useful it is for building.

Wood which contains the most resin, gum or turpentine, is the most water-resistant. Here fir and deal surpass oak.

Wood of a brown color indicates that it is wet and moist, which is why one should choose the yellow color. The heartwood and center of the tree, gives the best timber, which is why one should choose trees, which have the widest heartwood.

(183 II 47) Wood which has lain between wind and water, is liable to much rot. The quality of masts lies in their thickness, roundness and length, and also in being without knots and long-grained.

Never should a tree be cut as long as it bears fruit: it is with trees as with women, weak when they carry. Take notice of the Moon, and cut the trees when she is absent, for it is believed she increases the moisture in the trees.

And also do not let wood dry too much, so that it does not crack and disintegrate.

(184 I 46) It is also best to fell a tree after a long spell of quiet and dry weather, because rain wets the wood, and wind closes it, such that the water inside cannot seep out.

(179 I 34) Wood cut in winter, when the leaves have fallen and the tree is closed, is the strongest. Before felling trees it is helpful to pierce them across the base four or five days beforehand, so that the living juices can ooze out and the wood can dry.

(183 II 8) This can be taken for a general rule, that one does not choose wood of the largest size for shipbuilding, because large trees are old trees, and old trees are like old people, weak and brittle.

Apart from wood and its qualities, Witsen also gives full attention to the way shipbuilders should handle the wood, offering these extremely practical suggestions:

(179 I 27) In building and cleaving wood one should pay attention to the grain, the more one works with the grain, and the less one saws, the stronger the work will be: as dry wood is to be preferred to wet wood, which is solid and liable to split.

(179 II 48) It is a sound rule amongst carpenters, that when they have both good and bad wood to build a Ship from, they will sort the good from the bad, and use the bad for the inside.

(180 II 9) One has to watch carefully the wood from which the ship's treenails are made: because brittleness has been the cause of many a lost ship. Dry and young wood is the best suited, and much of it comes to us from East Ireland and elsewhere; pins turned from Knots, which are very hard, in my opinion are strong and good enough for this.

It is also important to watch, that the holes, in which the pins are struck, are made with a sharp drill, and are evenly round, smooth and not shriveled, when the holes are rough and uneven water will leak in through there, and bugs, jumpers (a species of small animals) will eat through the wood much more easily. This is why, on a ship, one sees these holes rot sooner in the sides where the drill cut against the grain, than above or below where it cut along the grain.

(184 I 50) When one burns the wood to bend it, and it gives fine coal, this is a sign that it will bend well and is pliant; coarse coal gives stiff and unbending wood. It is well to pay heed when using birch or other woods in Ships, to always take it 4 to 6 inches longer than the measured circumference would suggest, because otherwise one will be short.

I will conclude this subject with the following list, which shows Witsen's fascination with comprehensive details:

(184 II 8) *Special conditions of Norse wood.*

Balks [large, roughly squared lengths of timber] from Vlecker [Flekkerøy] are ordinarily about 23 to 24 feet long.

Those from Groenwijk in the Vlecker are about 24 to 25 feet in length.

Balks from Maer usually are 22 feet long.

Balks from the East Rijssen [present-day Ryen?]have a length of 22 to 23 feet. Those from West Rijssen reach a length of 25 feet, yet are weak and thin.

At Koperwijk, where it is usual to count 1,100 parts for one thousand, they are broad about 10 inches.

In Anslo 63 parts are reckoned to a shock [a bundle of twelve sheaves], yet two shocks make one hundred.

In Drontem [Trondheim] and Norme one counts 60 to a shock.

Spruce-fir spars are counted 11 for ten.

Balks from West Kiel are in general 26 to 27 feet long.

Balks from Langezont [Langesund] are 27, 28 to 29 feet long. Beams from Koperwyk are 21 to 22 feet long, or 29 to 30 feet.

In Frederikstad planks are counted 63 to a shock, and more often than not are hard and cracked.

At the Sande there is all sorts of rubbish, bad planks, spars, balks and all kinds of buckets [translation questionable].

At Holemstrant and Witsteen all kinds of good wood are found, but the planks are of inferior quality. Helves are sold by the tult, twelve go into a tult.

Staves for buckets come by the shock, a shock consisting of 60.

Iron and Coal

Iron was imported from Spain, Sweden, and the Harz Mountains in Germany. It was often bought as partially processed material, such as "Nuremberg steel," "Gothenburg bars," and "Hans Musiker steel." Through the alloying of Spanish and Swedish iron, the properties of the material could be changed to make it more suitable for special uses. The quality of the intermediate product was indicated on the material with an assayer's marks. Coal was imported from Scotland, England, and Liege in Belgium.

(119 I 26) While not only anchors, but many other parts of the Ship are made of iron, it will not be without importance to mention here, which is the best, so that the shipwright, when building the Ship, will make the best choice; and so make his Ship stronger.

The flat and square bars, which are marked with the letter F, are taken to be the best: the crowned H is the next best; the uncrowned H is third.

The Orgel-gront-yzer [high-quality iron from Öregrund on Sweden's northeastern coast] is taken for the same prime quality as the iron marked with the F.

Stockholm iron, in general is not as good, as the Spanish iron.

Danzig iron, which comes in long bars, is better than Stockholm iron.

As the square and long Gothenburg bars are better than the Stockholm ones.

Nuremberg steel, marked with a fir tree, is the best; that carrying an hourglass follows, then the halberd, then the trefoil leaf, straight or bent.

Swedish steel is valued ten to a hundredfold less, than Nuremberg steel.

Hans Musiker steel, which is brought to us from Bergslandt and stamped with two cogwheels, is estimated in between Swedish and Nuremberg steel.

Ship anchors are made with rude iron, mixed with Spanish iron: which blend very well together. Spanish iron alone, is too soft, and has no stiffness.

(119 II 24) The coal with which this iron and steel is welded and heated, is brought to us from Newcastle, which is thought to be the best: Scottish coal in general is more meager, yet some is quite as good. Liege coal is thought of the same quality as Newcastle. It

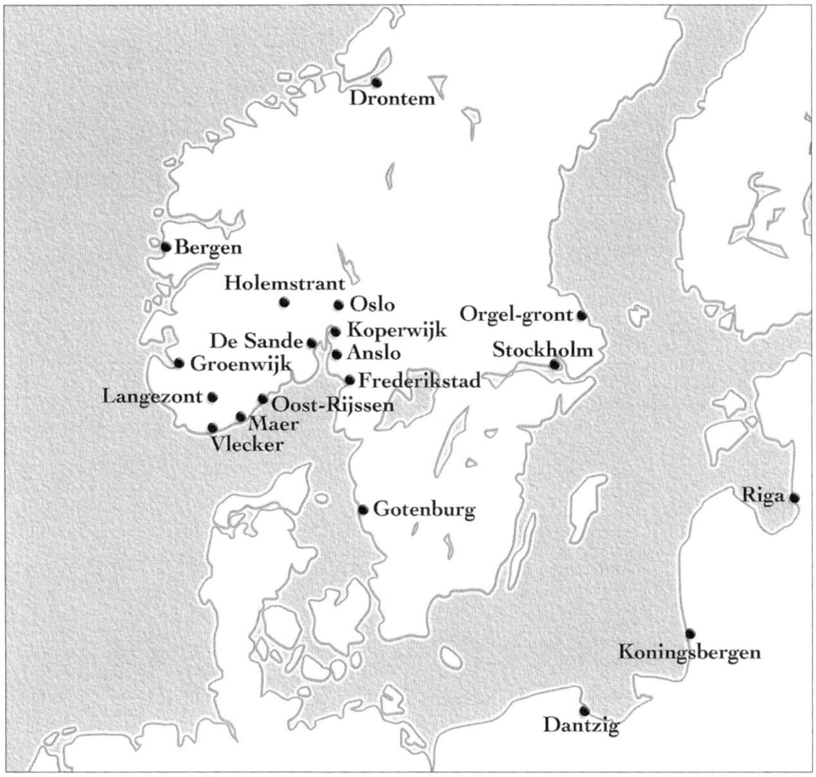

FIGURE 1.16. Scandinavian and northern European cities mentioned by Witsen as centers of shipbuilding materials. The Dutch gave their own names to many towns: Drontem (Trondheim), Groenwijk (Grønvik), Holemstrant (Holmestrand), Koperwijk (Kopervik), Langezont (Langesund), Orgel-gront (Öregrund), and Vlecker (Flekkerøy), among others (Drawing by A. J. Hoving)

is measured by the hoet, a hoet being 28 bushels. [A bushel is 8 gallons or the volume of 8 gallons.]

To test coal one takes a handful, squeezes it strongly, and opens the hand, and if it falls apart on the hand, it is good, but if it sticks together in a lump, it is no good. Or if thrown in water and sinking, it is no good, because it has to be like fat, floating on the water at first, to be good.

Hemp

Hemp was imported from Russia and Italy. These species of hemp had long fibers, unlike the modern cultivated species. For the building of the replica of the *Batavia* in Lelystad, long-fibered hemp had to be imported from Italy.[30]

(134 | 23) *On rope and Hemp.*

The Stays and main shrouds, as well as all standing rigging, are best made from Local hemp, as this does not stretch, yet it is somewhat coarser. Hemp from Italy is thought to be the best from which to lay rope, after which comes hemp from Riga, and Russian hemp is third in quality.

Hemp must be cleansed, and must not have been heated, when used for laying rope.

Hemp of fine fiber yields the best rope; long, well-combed and retted; ripe, to be judged by the stubbles; of a white-green color; not too thick nor too fine, the yarn is to be spun.

Rope is twisted from three strands, be they thick or thin.

Rope, which is laid too stiffly, will easily crack, that which is laid too loose, is too brittle. [Witsen seems confused: loosely laid rope is not brittle; it is weak and snaps easily.]

After the cable is laid as specified, it is always unwound about four twists, so that it will lie better in the remaining twists, and will not snap from being wound too tight.

When one tars rope, it is important to watch that the tar is not too hot, and to let the rope suck in sufficient tar before removing it from the cauldron.

When rope is stewed or heated, which is done to dry and enhance the penetration of the tar, it is important to watch, that the kiln is neither too hot nor too cold.

When stretching the yarns of which the rope is laid, pay attention that all yarns are strung at an even length and that when uneven they do not snap while laying.

Rope has to be dry when laid and spun.

When laying rope, men with sticks help the

FIGURE 1.17. A seventeenth-century signboard for a shipyard in the Amsterdam area. Many of the tools depicted here can be recognized in Witsen's illustrations. (Courtesy Rijksmuseum, Amsterdam)

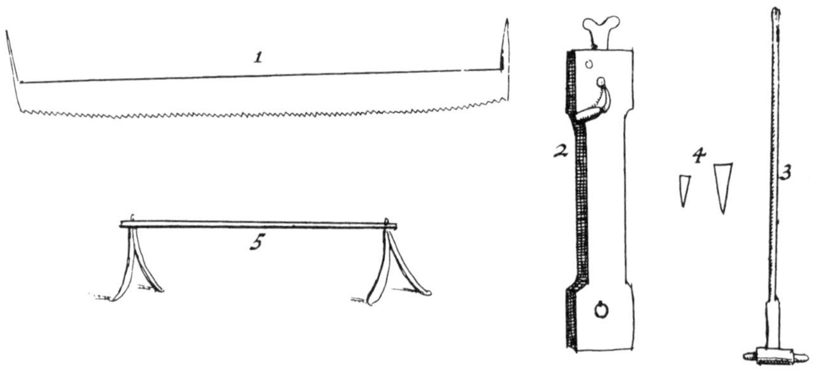

FIGURE 1.18. PLATE LXXIII (185 | 9) No. 1. A Saw or Whipsaw.
2. Jackscrew.
3. Maul
4. Iron wedges.
5. Burning iron.

windings about, because the wheels cannot handle very long and heavy ropes. While laying, the ends of the cable are held with heavy weights, to prevent the cable from crinkling and interrupting the work.

Of sailcloth, the last material discussed, Witsen does not say much more other than to note that it is woven from finely beaten hemp.

Tools

Now we shall have a look at the building tools listed by Witsen. In addition, he provides a series of plates showing the various tools, each labeled with a number that identifies it in the key to the illustration (see figs. 1.19 to 1.30). Readers can refer to these drawings wherever Witsen mentions a tool in his text concerning the pinas.

The list is divided into two groups: tools present in the shipyard and tools that the shipwright or carpenter was supposed to carry in his own toolbox. Of course, the large cross-cut saw (*1*) does not fit into this category; although almost all timber delivered to the yard was already sawn to thickness, some large pieces still had to be faced or ripped (sawn lengthwise, along the grain, as when a heavy crook was sliced to obtain knees of even thickness). This was done on a pair of trestles (*36*), with one man standing above the timber and another one below.

The jackscrew (*2*) was used to lift heavy loads or to clench heavy planks. The wood was cleft with mauls (*3*) and iron wedges (*4*). Planks were heated for bending over an iron rod (*5*). Straw fires were burned under the plank.

Timber and crooks were moved in the water with a wood hook (*6*). With the chips pick or hook (*7*) shavings

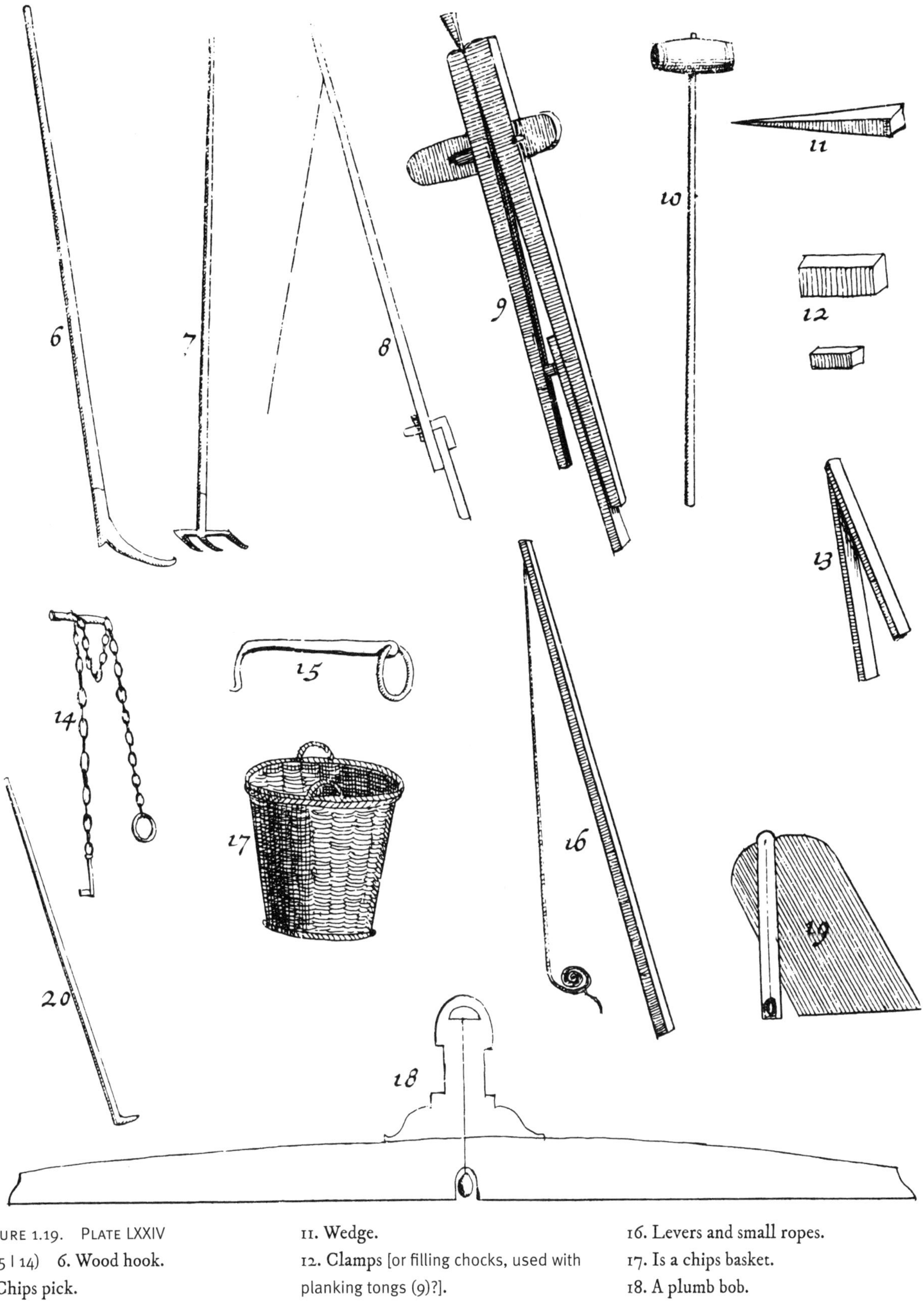

FIGURE 1.19. PLATE LXXIV (185 | 14) 6. Wood hook.
7. Chips pick.
8. A Klaes Jacobszen.
9. Planking tongs.
10. Treenail mallet.
11. Wedge.
12. Clamps [or filling chocks, used with planking tongs (9)?].
13. Shows the shape of the bevel.
14. Hook and chain.
15. Cant hooks.
16. Levers and small ropes.
17. Is a chips basket.
18. A plumb bob.
19. Small level.
20. Crowbar.

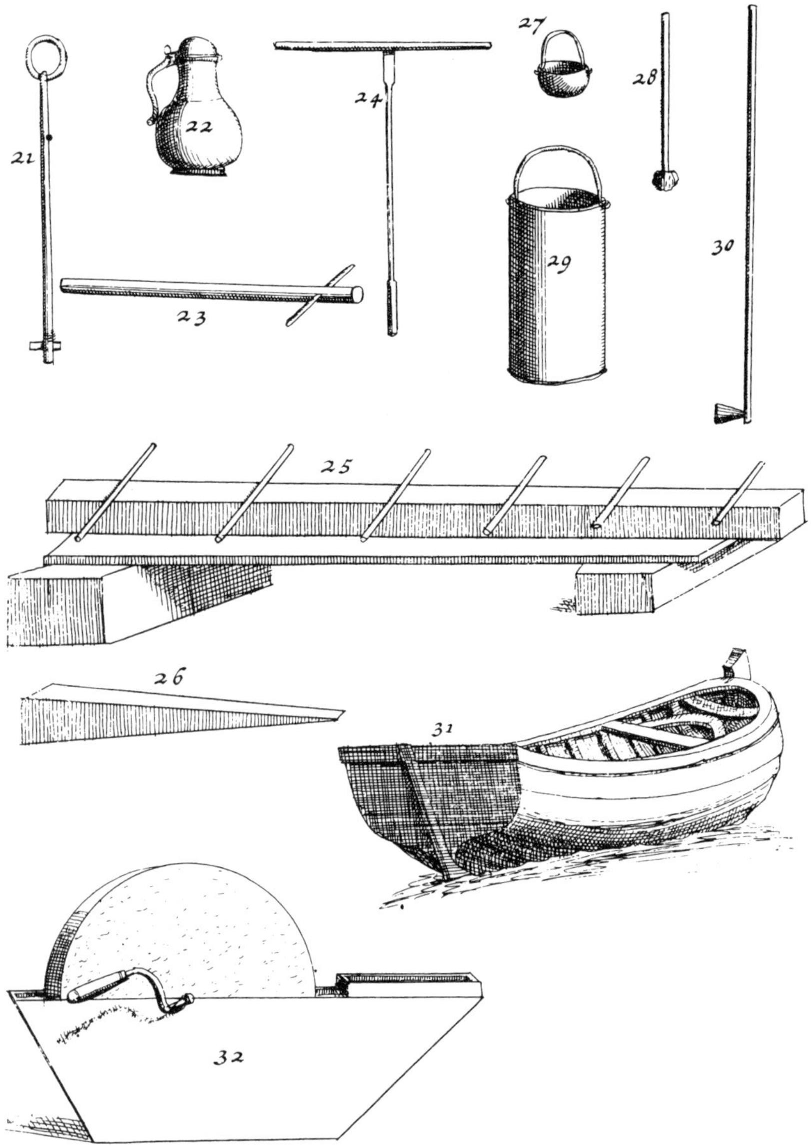

Figure 1.20. Plates LXXIV, LXXV, LXXIV, LXXV
(185 | 29) 1. Shows the shape of ringbolts with ring and wedge.
22. Is a tankard with a lid.
23. An iron ram.
24. An auger.
25. A wooden ram.
26. A large wedge for the ram.
27. A grease pot.
28. Grease brush.
29. A tarring cauldron.
30. Tar cloth and tar brush.
31. A boat of the yard.
32. A grindstone or a whetstone.

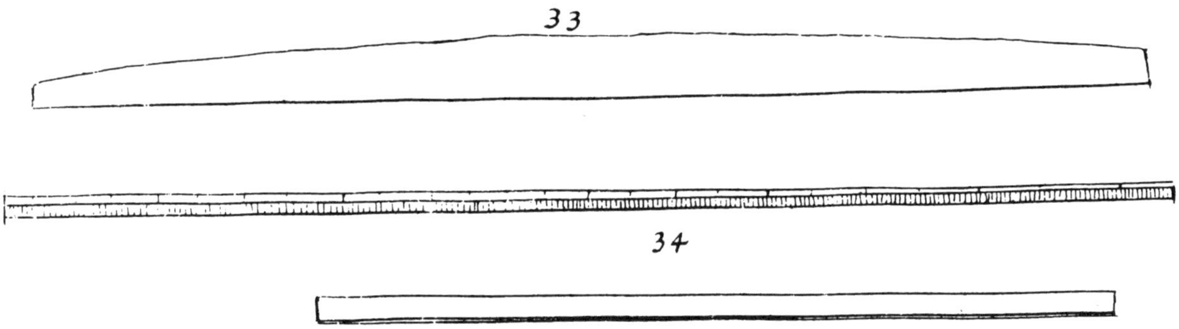

Figure 1.21. Plate LXXVI
(185 | 42) 33. A mold.
34. A spline [or flexible batten (shown below)].

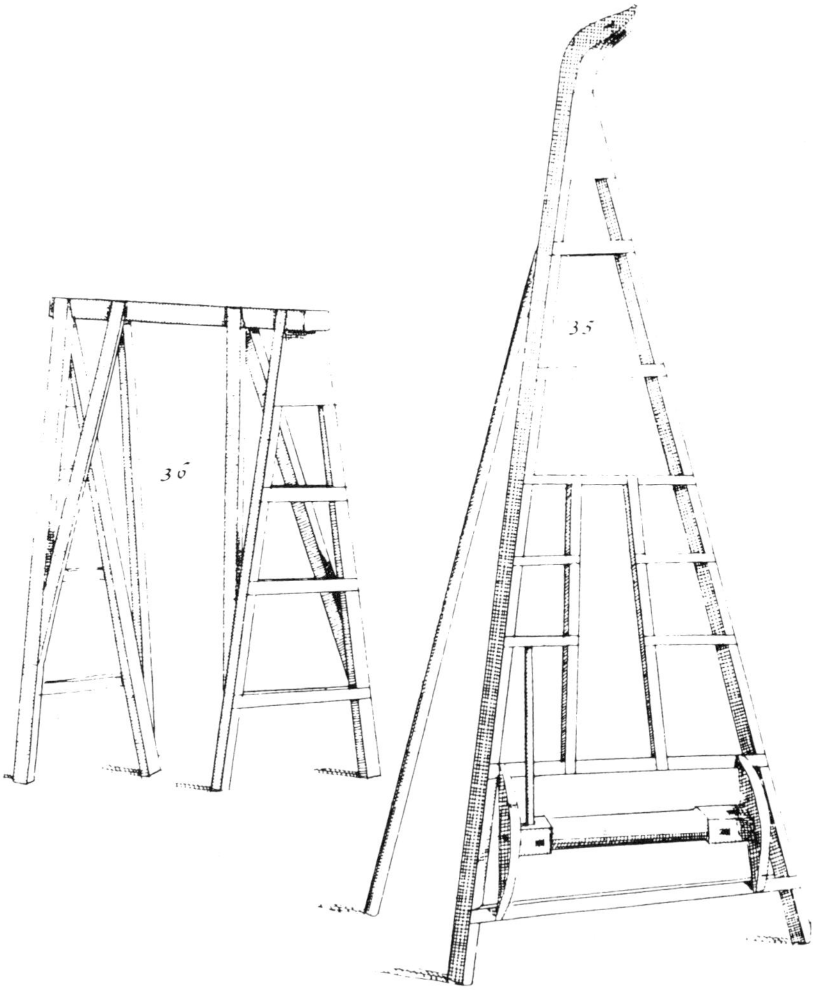

FIGURE 1.22. PLATE LXXV (left)
(185 I 44) 35. A gin.
36. A trestle. There should also be a measuring rod of 20 feet [measuring rod shown as 34].

FIGURE 1.23. (above) The gin depicted by Witsen can be oriented in two directions. This model from the collection of navy models at the Rijksmuseum shows the alternate way. (Courtesy Rijksmuseum, Amsterdam)

and chips were gathered by the chip pickers in the yard.[31] The chips were collected in a chip basket (*17*) and used as fuel.

A Klaes Jacobszen (*8*) was an instrument with which the garboard strake, if slightly awry, was forced into place. This tool was also called a Houten Klaas (Numb Hand), which in common speech meant someone who never does anything spontaneous or unexpected.

A type of clamp known as tongs or *boeitang* (*9*) was an important tool in shaping the bottom planking. It was used with the hook and chain (*14*) to squeeze the planks tight and keep them from jutting upward. A wedge (*11*) and chocks (*12*) were used together with the tongs (see fig. 2.45). A lever (*16*) was also used for this.

Treenails were hammered into the planks with a wooden mallet (*10*), as an iron hammer would damage the nails.

The bevel (*13*) was used to copy angles. Plumb bobs and levels (*18, 19*) were used to check the symmetry of the ship at various building stages. The familiar crowbar (*20*) was used to pry things apart, as it is today.

Ringbolts (*21*) were not tools, strictly speaking, but were drilled through the hull and anchored with a wedge or forelock. Thus the bolt could always be undone. The ring could be used, for instance, to hitch the breeching of the guns. The holes for the bolts were drilled with an auger (*24*), and the bolts were hammered through with an iron ram.

The wooden ram (*25*) was used to drive a number of wedges (*26*) underneath the keel when the ship was launched.

Other tools included a mold (*33*), such as for a futtock, part of a frame, which did not dramatically change in shape over the length of the ship. It was easy to plane it down to another shape, if necessary.

The ribband was a temporary but important device with

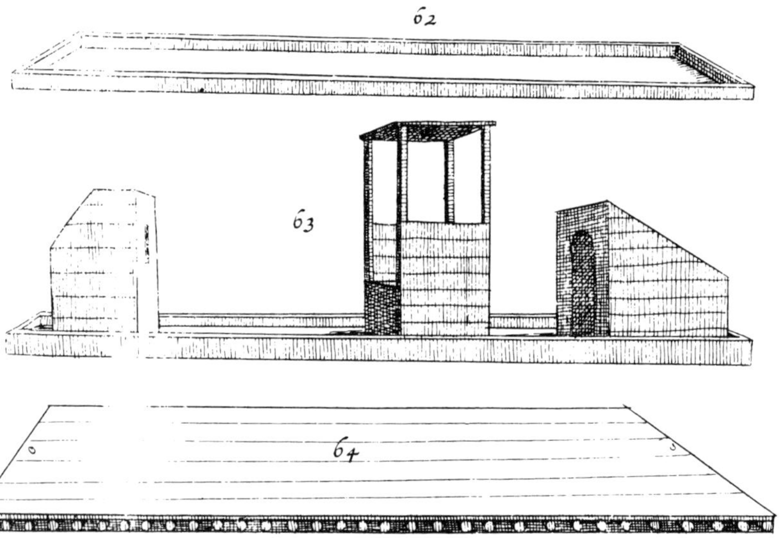

FIGURE 1.24. PLATE LXXVI
(185 I 47) *62*. An open Punt to sail with; used for pitch and for resinous rope. As shown with its upper structures at No. *63*.
64. Rafts.

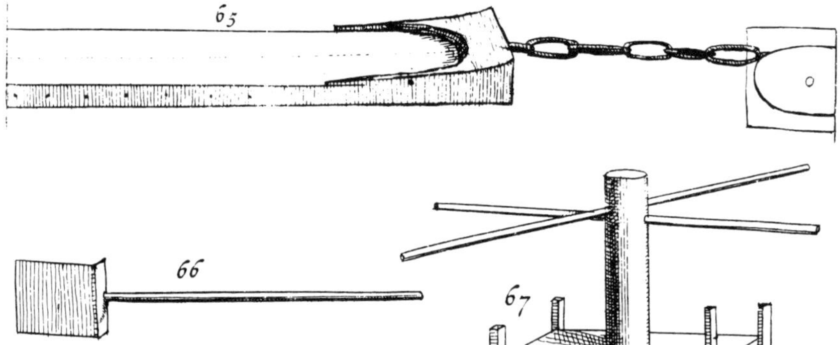

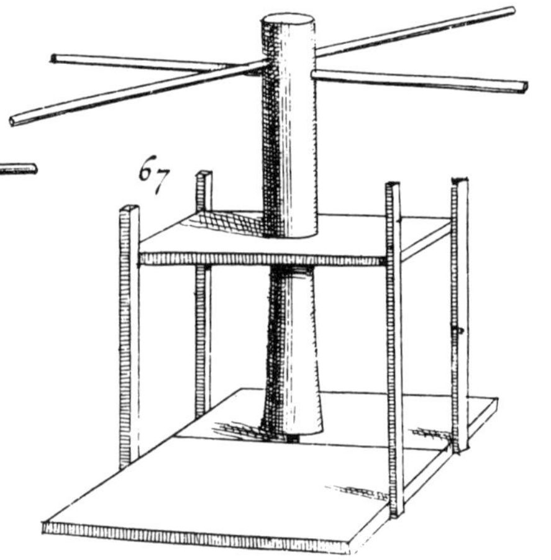

FIGURE 1.25. PLATE LXXVI
(185 I 53) *65*. A Cradle, used to haul ships of any size on land.
66. A lateral cradle.
67. A Capstan.

Following are the tools which the shipwright's mates are obliged to bring for themselves, when coming to work and earning a day's wages.

which the shipbuilder faired his strakes. The spline (*34*, below) was a flexible batten used for several purposes, like drawing curved lines and taking measurements for planking.

The twenty-foot measuring rod, unnumbered in Witsen's text, is mistakenly mentioned with the trestle (*36*) but depicted with the spline (*34*, above).

The gin (*35*) is a small hoist for lifting weights. Curiously, Witsen has the crooked head pointing the wrong way.

The punts and rafts (*62*, *63*, *64*) speak for themselves, as does the capstan (*67*).

Witsen describes the cradle (*65*) and lateral cradle (*66*) extensively:

(185 II 13) When preparing to bring a Ship back onto the yard, one shoves this beam or plank under the keel of the Ship from the fore end to the rear, steering it with hooks at the sides and at the rear, in such a way that it lies straight underneath and against the keel. Then the Cradle is tied to both sides against the Ship with loose ropes: for which it is perforated at numerous places as said, that it does not slide away. To the rear end of the Cradle a handspike or beam is shoved in the hollow groove touching the sternpost, and nailed down through the hole across which is above; and so the sternpost is fixed; and thus wound up over numerous sheaves, the entire Ship can be hoisted on land, the bottom of the Cradle as the slipway being

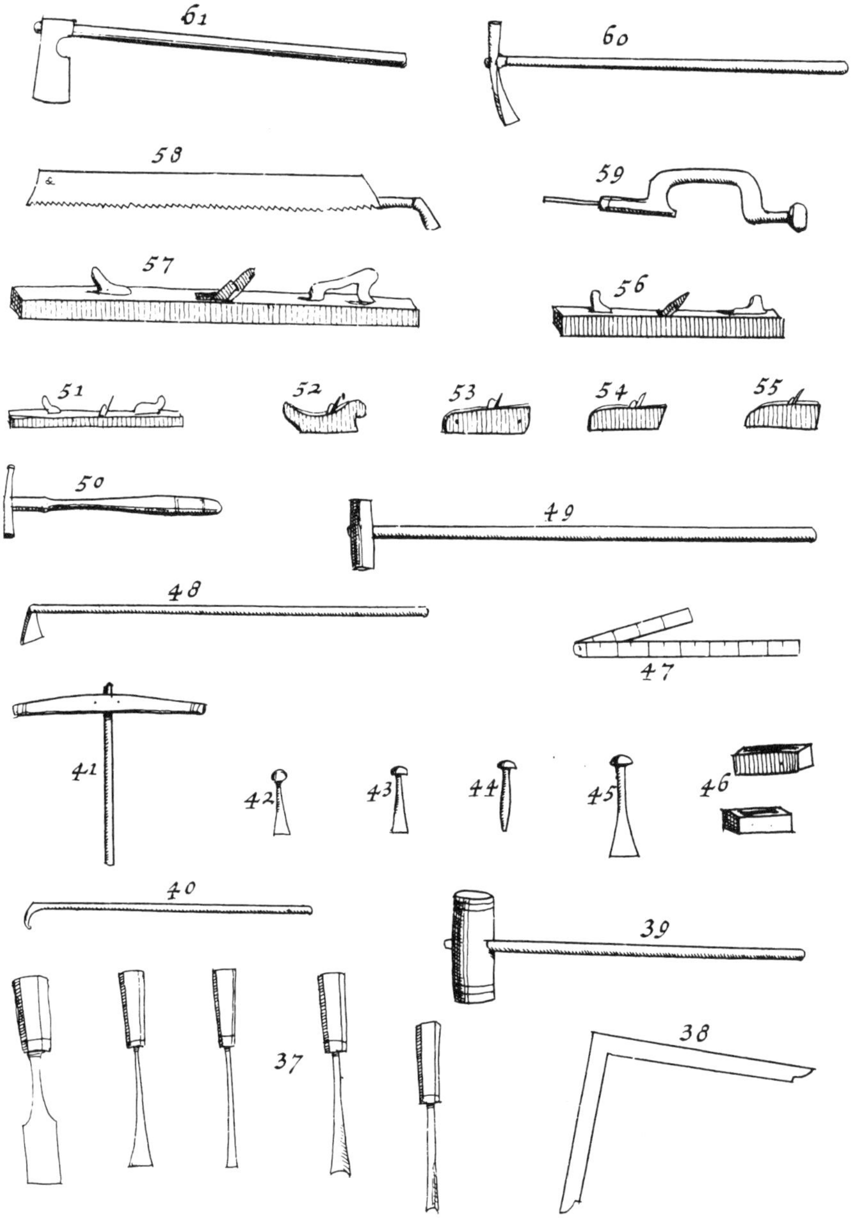

Figure 1.26. Plate LXXIII

(186 | 8) 37. Chisels of various shapes are shown on the plate, such as one-inch chisel, mortise chisel, ripping chisel, narrow, broad, etc. Also several Gouges, narrow, broad, curved and flat.
38. A Try Square.
39. A Mallet.
40. A seam hook.
41. Caulking mallet.
42. Making iron.
43. Caulking iron.
44. Nail punch.
45. Ordinary chisel.
46. Are small Water trays.
47. A Foot rule.
48. A Scraper.
49. A Maul.
50. Small hammer.
51. A Jack Plane.
52. Plane [for making a hollow or round profile].
53. Rabbet plane.
54. Odief [plane with a half-hollow, half-round knife (Gothic)].
55. Plogen [plane for making grooves in the sides of planks].
56. A great carpenter's plane [used for smoothing a rough plank].
57. A wood fore plane.
58. A Handsaw.
59. A Drill for the nails of the chocks, ten or eleven inches bitt, six, seven, eight-inch bitt, for tough middle-size nails, single and double, for nails called 10-pounders and a drill for nails with small heads.
60. An Adze.
61. An Axe.

well greased, the cradle being a little hollow, the slipway a little stooping, so that it touches even better. The topside of the Cradle needs to be zoor [Old Dutch: lit., dry, withered] and not smooth.

66 [in fig. 1. 26] Shows the lateral cradle; this appliance serves to hoist the Ship onto or against the land sideways; for this beams are shoved sideways underneath the ship, fixed into the ground, along which chains and ropes are led over sheaves: the lateral cradle, well greased, is then laid against the keel on the crossbeams, with which the Ship is controlled. One must carefully watch, when a ship is wound up the yard sideways, that the chain does not snap, nor the water level sink too low, and the Ship tumbles over, to which round ships are the most prone.

The shipwright's manual tools (38 to 61) are ordinary tools that can still be found in every carpenter's workshop. Some are special tools (39 to 43) for caulking, the filling of the seams between the planks with unlaid rope and tar; others are planes of various sizes and profiles (51 to 57).

Etching of a shipyard by the seventeenth-century maritime painter Reinier Nooms.

TWO How Ships Are Built in Holland Today

THE SEVENTEENTH-CENTURY Dutch shipbuilding method was a proportional system in which shipbuilders applied traditional formulas inherited from their predecessors. The treatises of Witsen (1671) and Van Yk (1697) provide two sets of such formulas, and their accuracy can generally be confirmed from the many ship specifications that are still in existence today.

The formulas determined the ship's main dimensions as well as the sizes of almost every structural component. Although terminology in the formulas may have differed from shipyard to shipyard, the outcomes, if compared with each other, were always practically identical. Thus, there was a direct connection between the main dimensions of the ship and those of its parts.

We can distinguish two kinds of formulas. In the first category are those formulas that reflect fixed principles of proportion and govern the sizes of each major part relative to the whole; these formulas were subject to very few adaptations until the end of the wooden shipbuilding era. On the other hand, we find formulas for parts that changed over the years due to fashion trends (like the width between the counter timbers or the length of the beakhead), ship function (a man-of-war versus a merchantman or a fishing vessel), a new "theoretical approach" (such as the introduction of three-deckers), or the characteristics of particular bodies of water (e.g., some fluyts were especially designed for the shallow waters around Taiwan and of the Ganges in India). The formula for the length of the beakhead is a good example of one that changed over time. Early in the seventeenth century its size was one fifth of the ship's length; by the end of the century it had shrunk to one eighth. A comparison of Witsen's and Van Yk's shipbuilding formulas are presented in table 1 in the appendix.

Witsen distinguishes 122 steps in the building process, which we will follow one by one in this chapter after we acquaint ourselves with his remarks about ship dimensions. There are also a number of phases in the method he describes; these phases—which required a choice in certain proportions or dimensions that would affect the ship's performance in ways that were evident to the shipbuilder—will also be highlighted.

Ship Dimensions

(262 II 46) Circumstance, and differing practices, change the build and shape of ships: with which I do not mean to say that, though these proportions may be good, that other proportions should not be equally good, although I know of no other, let alone better.

(95 II 24) The measurements in feet as in inches which are handled here, all are from Amsterdam, which in size are one foot in forty smaller than the measurements of Enkhuizen, Zardam, Hoorn and Edam, and one inch smaller to each foot compared to the Maas foot.

. . .

(263 II 22) Ships of medium size are the best by far to sail the sea: ships which are too large all lie ill at sea and easily run aground; and in case of accident are badly damaged. Ships too small carry small loads only; are easily flushed by the sea and are ill suited for bad weather.

. . .

(273 I 51) Long and narrow ships are faster than short and wide ships, because they meet less resistance. They are also faster because their sails can be trimmed far better and stand farther apart, than on shorter ships. And particularly because they cut two to three waves simultaneously, and do not get stuck in between the waves, as with the short ships, which have to rise against the waves in front, before they will be propelled by them. They turn less easily however than short ships, needing a larger radius, for which their increased speed is to blame.

. . .

(65 I 30) To define the proportions of the smallest trifles for the shipwright, would be useless and much too laborious: because they follow automatically when the large parts meet their proportional requirements.

For a start only the length is given, from which the width and the depth are derived.

. . .

(71 II 31) The length of ships, is measured across the stems, which is on the outside, while the depth is measured at one third from the front, that is, on the main frame, where the ship is at its lowest: the width is measured at one third from the bow as well, at the bottom, at the top of the bilge planking and at the ribband set up at the height of maximum breadth, where the main deck is to be.

. . .

(265 I 26) Ships are at their widest at one third of the length measured from the front: it then narrows toward the stern, and thus makes good speed: if one were to place the widest section farther forward, the ship would not cut through the water well enough: if one were to place it farther aft, the water aft will be oversailed, and steering will become difficult: this is why one third from the front is the right place for the ship to be widest. From observing beams in the water one can deduce that bodies which are thicker forward, make the best speed.

. . .

(265 I 17) *The ship is wider forward than aft;* and such because if wider aft, it would not heed the rudder, produce too much wake, preventing progress, and also pushing the water downward in such a way that almost none touch the rudder; which is why such vessels are prone to yaw.

. . .

(262 I 26) Most of the other parts of the ship I deduce from the stem: this represents the basis and the measure of the same: for which however one could have chosen any other part. Its thickness is derived from the length of the ship, taking into account a lesson from experience: that the proportion laid down here should meet the requirement that the thickness should be neither too heavy, nor too light.

. . .

(262 II 28) It is also true, that experienced Shipwrights do not derive such measurements directly from the stem: through long experience they are able to determine the right measurements of any major part of the ship without deducing them from the stem: yet in doing so they apply, although unwittingly, the same rules.

. . .

(65 II 25) *About the width derived from the length.*

2 The length of the ship divided by four, the width is taken as a fourth part. For example, 100 feet length gives 25 feet for the width at the main frame. When a ship is less than 100 feet long, the width should be less than a quarter length: as the master feels right. Yet others make the ship 3 or 4 feet wider, than the quarter length: in order not to be crank. And a merchantman may well be 2, 3, 4 or 5 feet slimmer, depending on the length. To ride easily on the anchor cable: and to be steered lightly, with a small crew, ships are often made less wide than would otherwise be the case. Ships of 165 to 170 feet can be 6 feet less wide than a

quarter of the length. For fluyts a ratio of one fifth of the length for the beam.

. . .

(263 I 50) Besides completely round or square, and so short that the sails cannot be arranged, or so long that it is weak in the loins, one sees ships in every shape today; which is allowed, as long as everything is carried out according to the rules, the infinitely diverse use giving infinitely diverse shapes for the ships: wide ships get the broadest and heaviest timber: ships which are too wide are weak in the sides, which is caused by their wide spread: long and narrow ships, although of one frame shape, float better than wide and short ships; which is why they are better to build for sea: the reason for which being, that long ships have the more flat timber on the water. A wide bow above, well cut below is held to be more graceful and suited for the sea. This increases the ship's support; too sharp a cut makes ships not go about easily: it is also elegant for ships to be round above the turn of the bilge.

. . .

(263 I 14) Ships which are too wide, lie laborsome on the water and lurch without measure; which is because wide ships, which are stiff, resist the waves which want to lift them and circumvent them, which is why they make these ships shake and shudder. Whereas slim ships lie demurely on the water and follow the waves up and down softly, without shuddering and shaking too much. One will see many a mast come down on wide ships, which would keep standing on slimmer ships.

. . .

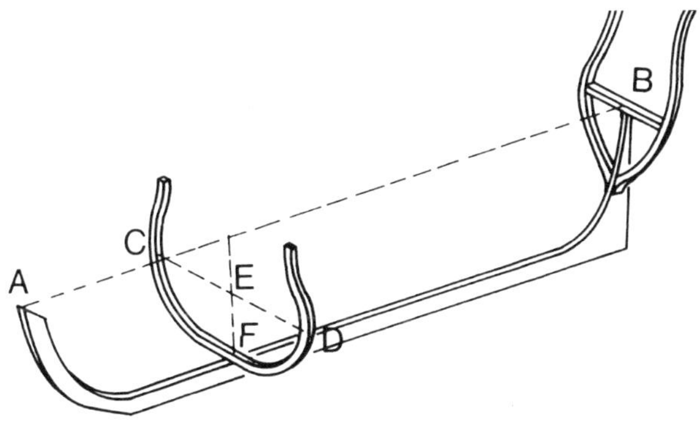

FIGURE 2.1. Measuring points. Length is measured over *A–B*, beam over *C–D*, and depth over *E–F*. (Drawing by A. J. Hoving)

(267 I 26) Too narrow ships take little cargo, and are unsuited to bad weather. Too long ships are weak at the loins and are easily hurt when they fall dry. Short ships are slow sailors and lurch without measure.

(65 II 48) *About the depth derived from the length.*

3 To get the depth from the length, take one foot of depth for each 10 feet of length. For instance, 100 feet length comes to 10 feet depth at the main frame, but for this depth one can find ships 4 feet longer, or 2 feet shorter, yet seldom.

(53 II 24) Let us take a pinas ship, one hundred and thirty-four Amsterdam feet long, for an example. Which (built in our thoughts only) will be discussed in all its parts.

(74 I 38) 9. *About the depth, and width, at the height of the watershed.* [Literal translation; what is actually meant is the height of the "master ribband," indicating the underside of the lower deck where the scupper's upper ends are.]

1. At the height of the watershed the depth is 13 feet.
2. Wide 29 feet.

Ship Dimensions

The first phase in the decision-making process of building of a ship was determining its dimensions.

According to Witsen, an Amsterdam foot of eleven inches is measured and calculated as being 0.283133 meter and 39/40 of the unit used in Enkhuizen and the River Zaan area. The inch is 2.574 centimeters. Other important measures used at the time were the Hoorn and Edam foot of only ten inches (0.275804 meter) and the Rijnlandse (Rhineland) or Maas (Meuse) foot of twelve inches (0.313947 meter).

The main dimensions of the pinas ship are 134 × 29 × 13 Amsterdam feet (37.92 × 8.21 × 3.68 meters).

The Length

The length of a ship was measured over stem and stern (i.e., from the front of the stem to the back of the sternpost. Other methods existed—like taking the length of the lower gun deck, the length of the waterline, or the length overall—yet we are left with the impression that these measurement methods were used only for ships already built.[1]

It was difficult, of course, to measure over the stems if the ship was completely paneled and fitted, and other points of measurement had to be found. But for a ship being constructed, shipbuilders generally used the measurement over the stems, as few other points of measurement were available: technically, it was easiest to tie a rope between stem and stern and measure that. Thus, the beakhead and the overhanging counter and stern gallery were not counted with this method.

The length of the ship was taken as the main measurement for building a ship in Witsen's time. In the archives ships are often indicated by length. But this was not always so. At the beginning of the seventeenth century ships were referred to by their number of lasts, a measure of the ship's cargo capacity and equal to approximately two tons (see "Ship Measurement" in chapter 1). As the client was generally interested only in the loading capacity of the ship he had ordered, the length was determined by the shipbuilder. A similar method based primarily on the vessel's carrying capacity was used in the sixteenth century by the Basques, who measured ships according to *tonnelados* (barrels of a standard measure): a ship of four hundred *tonnelados*, for instance, was constructed so that four hundred barrels could be fitted into it.[2]

It is unlikely that such an exact correlation also prevailed in the Netherlands, but in the 1630s the practice of referring to ships according to number of lasts was abandoned in favor of giving their length in feet. Admiralty ships were usually referred to by their number of guns or their rate—we read about an eighteenth-century man-of-war of not 160 feet but 74 guns.[3] Merchantmen, however, were measured in feet between the posts until well into the nineteenth century.

With the length established, the performance of the ship in the Dutch shoal waters was already largely determined: a ship too large was difficult to handle and was liable to run aground due to its ensuing greater depth. Once grounded, it was often impossible to get afloat again. Small ships were popular but offered relatively small cargo capacity, took in a lot of water over the sides, and could accommodate only small crews when many men were sometimes needed to defend the vessel against evildoers.

It is not clear what the average size of ships was in Witsen's time. Witsen calls his 134-foot pinas a ship of average size. As mentioned in chapter 1, ships on Grebber's list (fig. 1.11) range between 60 and 200 feet, but we have no indication that ships of 200 feet actually existed. In Van Yk's work we find a contract from 1629 for a ship of 172 feet, and Witsen mentions a ship of 180 feet, but elsewhere he states: **The largest Rates for men-of-war ships used today, are long 175 feet between stem and stern, and wide 43 ft.** (92 II 17). There is some allowance for commercial vessels, but the existence of ships larger than 180 feet seems unlikely for that period.

The Main Frame

The width (breadth) and depth (in the hold) was measured on the main frame, which was on a third of the length taken from the front of the stem. At this point the ship was at its widest. The width was measured over the frames, excluding the outside planking. Only where the real width was important, as with inland craft, which had to pass through locks and bridges and thus were held to a maximum width, was the width measured overall.

The depth (also measured at the main frame) was the distance between the upper side of the keel and the upper side of the lower deck beams, without including their camber.[4] The main point of reference in determining the size of many of the ship's parts was the thickness of the inboard face of the stem, and this key dimension is the basis of almost half of Witsen's shipbuilding formulas.[5]

Of course, not every shipbuilder used this particular reference point. Van Yk, for instance, used it to a much lesser extent, and Witsen himself admits that there are other possible points of reference yielding approximately the same result. One can easily verify this with an example: the lower wales according to Witsen's method should be of the same width as the inboard thickness of the stem and half as thick. Van Yk, on the other hand, calculates the width of the wales by taking 12 inches for 100 feet of the ship's length, for every 10 feet more length half an inch extra, and the thickness half of that. In the example of the pinas this comes to 13½ inches, which is only slightly more than we would get with Witsen's method—13¼ inches. For more comparisons between Witsen's and Van Yk's formulas, see table 1 in the appendix.

The Width

The width of the ship is derived from the length. Usually a quarter length was taken. But to adjust the design to its purpose, this rule was often adapted: a man-of-war, which had to provide a stable platform for its guns, would be made wider, whereas fast ships would be slimmed down. The pinas, had it followed the rule, should have been 33½ feet wide; yet it was only 29 feet wide.

Too wild a deviation from the rule would of course result in difficulties: ships that were too wide were difficult to maneuver and would lurch wildly, sometimes resulting in broken masts, while narrow ships could hold little in the way of cargo and were crank and more vulnerable to breaking up when running aground.

The Depth

The depth was also calculated from the length: one tenth. As with the width, the shipbuilder could play with these proportions to influence the performances of the ship. Thus it made a big difference whether the ship was built for corn or for wood; in the case of corn (which is relatively heavy), the depth would be reduced by one or two feet to avoid overloading the ship.

First: *One makes the keel.*

(66 I 4) *About the Keel.*

4. The thickness of the keel, is ¼ more than the thickness of the inner side of the stem, and the breadth 1½ times the width of the stem, which is at the main frame at one third from the bow, where the ship is at its widest; forward and aft it will meet the dimensions of the stem and sternpost.

. . .

(149 I 3) When making a keel, one takes one fourth of the thickness of the stem for its thickness, and broad it shall be one half broader than the stem: at the ends it shall become somewhat thinner to meet the thickness of the stems.

. . .

(72 I 30) The breadth of the keels forward and aft often varies, as the timber is peaked, and to adjust to the stem, always becoming thinner and narrower.

. . .

(72 I 5) A Ship of 180 feet length over the stems, needs a keel broad 2½ feet in the middle, as many Shipbuilders will teach. Wherefrom one can proportionally derive, that the keel for 175 feet shall be 2 feet 4½ inches broad in the middle.

For	170	2 feet 4½ inches.
	165	2 feet 2½ inches.
	160	24 inches.
	155	23½ inches.
	150	22½ inches.
	145	2 feet.
	140	21 inches.
	135	21 inches.
	130	19½ inches.
	125	19 inches.
	120	18 inches.
	115	17½ inches.
	110	16½ inches.
	105	16 inches.
	100	15 inches.
	95	14½ inches.
	90	13½ inches.
	85	12½ inches.
	80	12 inches.

. . .

(72 I 1) Short keels are of one piece; long keels of 2 pieces; the longest keels of 3 pieces.

. . .

(72 I 37) With a keel of 3 pieces, one chooses the best and soundest timber for the middle section, and joins the other pieces to its ends, these scarfs are to overlap generously. The lower scarfs are always to point to the rear of the ship, and bolted well together with flat-headed bolts, especially in large ships, the bolts being clinched on the inner side of the keel, and the ends of the scarfs nailed too with large nails, according to the thickness of the scarfs. One holds such keels, made of 3 pieces, to be stronger than the ones made of two: because with such a keel of 3 pieces the scarfs are forward and aft, while in the middle, where the keel has to endure the most pressure from the main mast, it is free of scarfs; which is very good, and makes the ship strong; although the main step is usually set at the middle of the keel, the broadest part of the keel is still at one third of the length from forward.

. . .

(148 II 54) As with QQ [fig. 2.3] at *c* and *d* showing how the Scarfs of the Keel are to be.

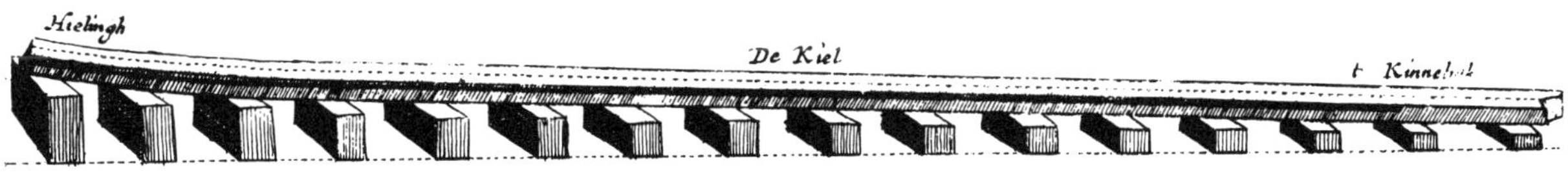

Figure 2.2. Plate XXVIII. The keel. *Left,* heel (i.e., aft end of keel); *right,* boxing scarf (lit., "jawbone")

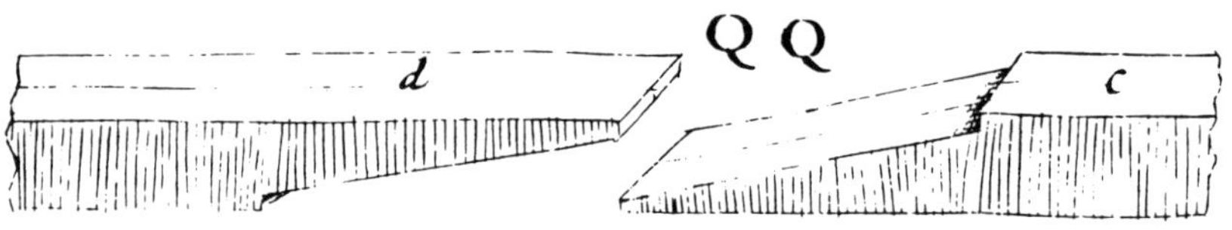

FIGURE 2.3. Plate LI (drawing QQ)

(72 II 11) A keel should be well planed below, so as not to hinder or hamper the launching of the ship.

. . .

(264 I 8) The heel, protruding behind the keel, supports the rudder, and prevents anything from getting stuck between the sternpost and the rudder, which would then foul the rudder.

> (71 II 31) 1. The keel broad 2 feet, long at the bottom 104 feet, thick 16 inches.
>
> (71 II 47) 2. The Scarfs of the keel, long 8 feet, the lips of the scarf, thick 3½ inches, the scarf fastened with eight bolts, nails, treenails.

First: One Makes the Keel

The keel, together with the posts, is the foundation of the ship. According to Van Yk, the scarfs in the keel had to be five inches long for every inch the keel was wide, and for each foot the scarf was long, at least two clenched bolts were required. The keel of the pinas was two feet broad; the scarf then should have been ten feet (2.83 meters) long and bolted with twenty bolts. Witsen, on the other hand, states only eight feet (2.26 meters) and eight bolts.

The length of the keel, in fact, represents the second phase in which the character of the ship was determined. This was closely related to the inclination of the posts forward and aft, or the rake of the posts. It should be noted that the rake of the stem would diminish gradually throughout the seventeenth century, becoming more vertical and requiring a proportionally longer keel. Van Yk notes that this adaptation, which apparently stemmed from necessity, also greatly increased the performances of the ship.[6]

Forward, the keel would end with a typical scarf, called the boxing of the stem, making the joint between the stem and keel as strong as possible. At the rear, the keel was sawn off obliquely (the so-called heel) so that objects like seaweed or cables would not get entrapped between rudder and sternpost. As a result of the trimming of the stern (whereby ships lay deeper in the water aft than forward), it was by no means unthinkable that the vulnerable rudder would scrape the sea bottom at shallows and pick up substances that would obstruct it.

Witsen does not mention a false keel, a plank fastened below the keel that, in case of damage, could easily be removed and replaced without having to repair the costly keel itself. Van Yk, however, does mention it, which could indicate that the false keel was an invention of the last quarter of the century; or possibly Witsen did not know about it or did not consider it worth mentioning.

2. The Stem.

(73 I 7) The stem is the guideline from which all proportions in the ship are derived: it in turn is derived from the length of the ship: they are named such because they give the ship its strength. [The word *steven,* "stem," is perhaps related to *stijven,* which means *verstevigen,* "to strengthen."] The stem is usually made of two parts, the sternpost in one piece.

(66 I 14) *About the Stem.*

5. The stem, thick 10 inches on the inside [an example for the 100-foot ship that Witsen has mentioned previously], this thickness is found from the length of the ship: example, 10 feet length, 1 inch thick.

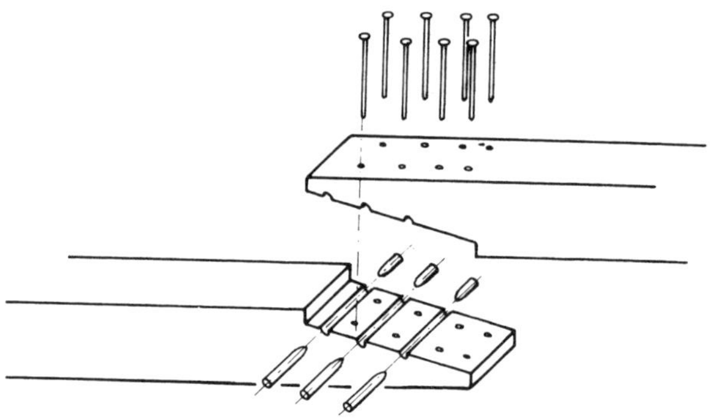

FIGURE 2.4. Joint in the keel. The joint was bolted through with eight bolts, and through its surface stopwaters were drilled; these were filled with moss and plugged with wooden dowels. (Drawing by A. J. Hoving)

Or another method: (66 II 9) ¼ Parts of the height of the stem, in the perpendicular, also gives the thickness of the inside of the stem, if one multiplies with the numerator, and divides by the denominator, and takes the result to stand in inches. On ships with forecastles this thickness is made more than on ships without forecastles.

. . .

(66 I 19) 2. The thickness of the foreside is ⅗ of the thickness of the inside.

. . .

Or: (66 II 29) and ¾ of the thickness of the inside of the stem is the thickness of the outside.

. . .

(66 II 18) *Breadth of the Stem.*

1. In the middle the stem is 3 times as broad as it is thick: above and below it is broader.

. . .

(66 II 26) According to some, the breadth of the stem above is twice the thickness inside.

. . .

(150 II 14) The Stem should be quite deep above, because one bolts into it from the outside.

. . .

(68 I 33) The stem should be a little more, than a third broader below than above.

3. For the height of this stem one adds the depth, the rise of the deck forward, and what is to be above, like the cable tier, the forecastle, &c. Example, 10 feet depth, 2 feet rise of the sheer, one could also take 3 feet for this, 6 feet upper deck, comes to 18 feet for the height of the stem, and the same for the rake.

. . .

(66 I 30) When taking the height of the stem, one needs to know beforehand if the ship shall have a forecastle, or not:

should there be no forecastle, then the stem should be much longer than the height of the cable tier. Or one takes two eleventh parts of the length of the ship over the stems for the height of the stem in the perpendicular. Others also take eleven sixtieth parts of the length for this.

(66 II 37) *The Rise of the deck.*

6. The Rise forward is 2¼ inches for every 10 feet length, and aft 6½ inches.

. . .

(72 II 20) The rake of the stem is measured from the boxing of the stem to the hanging line or perpendicular from the top of the stem.

. . .

(66 I 40) For the rake of the stem one takes 28 twenty-ninth parts, of the height of the stem, in the try square.

. . .

(149 II 54) When a Stem is made to bear Weight, one may make it two feet higher than its rake, or even 3 or 4 feet, and also a little more curved than usual. The less the Stem rake, the longer the keel should be, the longer the keel, the larger the bottom, the larger the bottom, the more the ship will carry. The stem of frigates have a large rake, and one tends to make them crooked above and of an easy curve.

. . .

(66 I 28) A fluyt has less rake, a Frigate more.

. . .

(149 II 4) When making the *Stem,* one takes both parts and lies one on top of the other in the try square [a 90-degree angle], in accordance with the required height and the rake of the stem, then one makes the Scarf, which is to be marked above and below; the ends of the scarf one makes thick, to one fourth of the total thickness: then one marks the forward end of the boxing and the lower [boxing] Scarf, which comes onto the keel, and when the Scarf is made and joined, one is to line it with Moss before fastening, then the upper end is nailed, and then fastened with four bolts, clenched at one end with plates, and nailed: Then make the front, two fifths thinner than the inside, after which the inside face and the outside are made. After this shape the inside and next the outside and to make this, fit the two pieces together to know whether it is right, put a Ruler [*rey*] on its side on each end and in the middle and check the Rulers on all three ends on both sides and if anything is not right one way or the other, mend it by eye and scribe the scarfs.

(149 I 48) *How the Stem is made.*

a On the figure at *P* [fig. 2.5] is the Scarf.
b The Scarf coming onto the keel.
c The boxing scarf [*kinnebak*, lit., "jawbone"].
d The small Bolt, clenched with the plate.

The other points are bolts, treenails, and nails with which it is fastened to the Scarf.

(149 II 35) To continue the making of the *Stem*, the pieces are laid onto each other as said, and one checks if the curve is as required, if so, then one lays a batten from *a* to *b*, which is the height of the stem, as can be seen on the plate at *Q* [fig. 2.6], and a batten from *c* to *b* which indicates the rake of the stem: then one takes the try square and holds it at the inside of the battens at *b*, where it is in the right angle, and draws the joint of the boxing in the square at *c* and *e*, as from *d* to *c*, which is the overlap on the keel, and when it is drawn from *d* to *c* and from *c* to *e*, this will give the rake of the stem; the bolts in the scarfs of the stem are then hammered in from above and below.

. . .

(56 II 39) *The boxing scarf joins the keel.*

(72 II 18) 2. *Of the stem.*

1. The stem, high 25 feet, has a rake of 26 feet.

2. Thick on the inside 13¼ inches, at the front 9 inches thick, broad 3 feet below, above 2 feet, inner curve 5 feet.

3. The boxing scarf, long 8 inches.

4. The Scarf, long 6 feet, and thick 3½ inches, the scarf clenched with 4 bolts.

5. The overlap on the keel, long 5 feet and 5 inches, and thick 4 inches.

2. The Stem

The molded dimension of the stem, viewed from the side, was, according to Witsen, three times its thickness; according to Van Yk, it was only twice its thickness. If we check this dimension with the measurements of the pinas, Van Yk's formula is closer to the result. In other contracts the breadth appears to be around two and a half times the thickness, a bit more above and below, and it is clear that some liberty was taken with this proportion.

The height of the stem was obtained by simple addition: first the depth of the ship was taken, to which the rising of the deck forward was then added, and finally the height between lower and upper deck. The total gave the height of the stem, but only in the case of an open forecastle. With the so-called closed forecastle, the forecastle

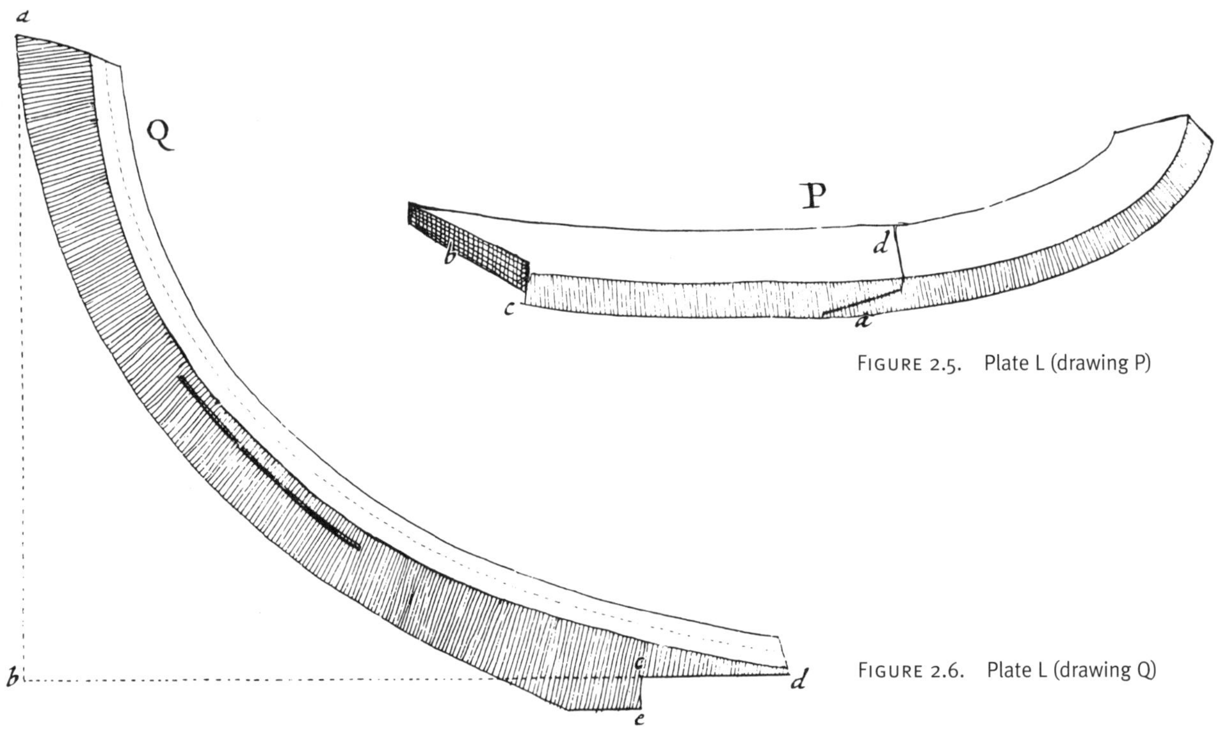

Figure 2.5. Plate L (drawing P)

Figure 2.6. Plate L (drawing Q)

deck would run all the way to the stem and the height had to be extended to the height of the forecastle.

Witsen supplies a formula for the rake of the stem: $^{28}/_{29}$ of the height, or 96 percent. This is extremely remarkable, as this formula is not to be found in any contract in Witsen's own work or in that of Van Yk. It is not even very logical to provide a formula for this measurement because at this point the shipbuilder could influence the shape and character of the ship as nowhere else. With a "neutral" ship one could perhaps state that the rake of the stem corresponded approximately to its height. But, as mentioned before, the rake of the stem decreased over the century. According to Van Yk, this was partly due to the increasing length of the ships, which would require two expensive curved timbers and force up the price, but also because sailing performance with a steeper stem (and, as a result, a longer keel) proved to be superior.

In general, we can say that, if a shipwright wanted the ship to be fast, like a yacht or frigate, he would choose a strongly raked stem; a slower vessel, built for carrying capacity more than speed, would warrant a steeper stem. Witsen's formula might well be taken as the neutral middle option. Of the thirty-four contracts found in the treatises of both Witsen and Van Yk, the average rake in relation to the height was 90 percent, with extremes ranging from 48 percent to 158 percent.

While trying to draw a number of stems found in contracts, I soon discovered that the points of measurement were not always consistent. Van Yk, giving sketches of the construction of a stem with much rake and one without, takes the rabbet of the stem above for the apex of the triangle formed by the height and the rake. The front and backside of the stem were then simply "drawn around" the rabbet. For the construction of the stem of the pinas, it is clear that Witsen took the top of the stem for the apex of the triangle, after which he "drew in" the rabbet of the stem.

The use of large compasses in the shipyards is very hypothetical. As far as I can see, any curved line needed in the construction of the ship was produced by bending a flexible batten, the *rey* (or *rij*), and never by using compasses.

3. The Sternpost.

(73 | 15) Of the Sternpost a prominent Shipwright says the following:

The sternpost is the most important part of the

Figure 2.7. (*below*) Construction of the stem of the pinas. The measurements were taken from Witsen's pinas data. (Drawing by A. J. Hoving)

Figure 2.8. (*right*) *Top:* Open forecastle; the upper deck lead through a small foredeck to the beakhead. *Bottom:* Closed forecastle; the beakhead can be accessed only through the forecastle deck. (Drawing by A. J. Hoving)

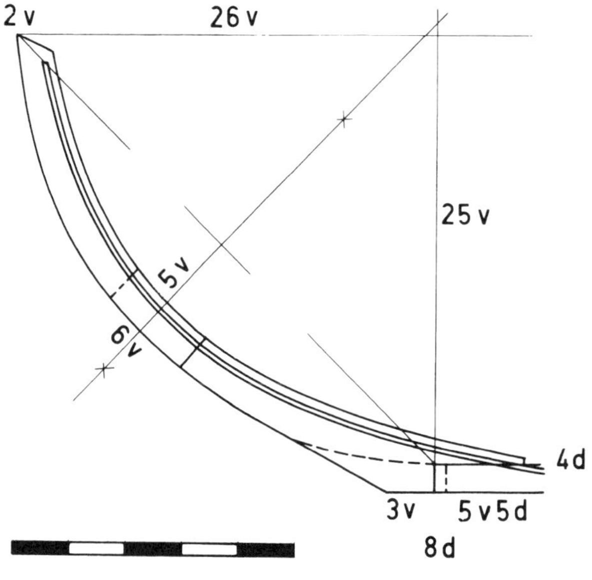

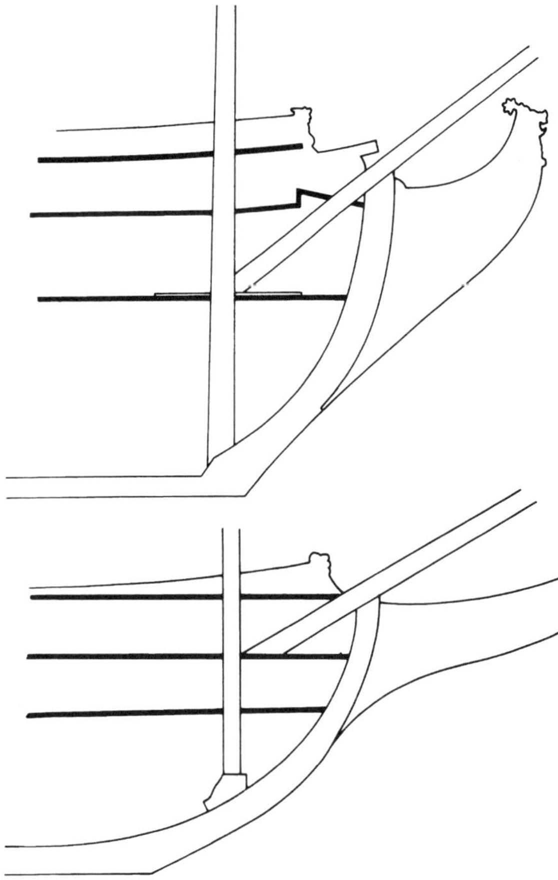

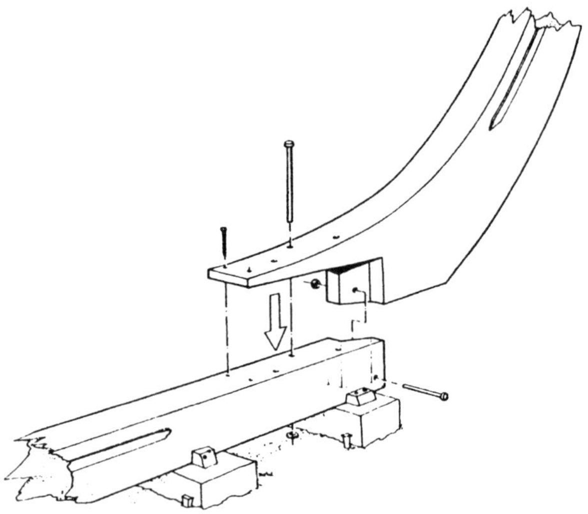

Figure 2.9. Joint of stem and keel. (Courtesy G. A. de Weerdt)

ship aft, and should be well made, of one piece of timber, which is good, sound and strong, as it will have much to suffer and the rudder will be hung to it, which in stormy winds is subjected to much impact from the water.

. . .

(72 II 32) The Sternpost is measured as the stem, on the end of the heel.

. . .

(66 I 45) *About the Sternpost.*

1. For the length of the sternpost, one adds up the depth, the rise of the sheer aft, and what is above that. For example, count 3 feet for the gun room, depth in hold 10 feet, rise of the sheer aft 5 feet 10 inches, this sternpost will be 18 feet 10 inches high. Others take three quarters for every 10 feet length for the rise of the sheer, which here would give 10 feet rise, or for 10 feet length 7 inches rise of the sheer.

. . .

(73 I 25) This sternpost is generally made as high as the stem is on the inside, or according to the depth of the ship, or after the first full deck is laid. One has to mind though, that the gunport in the tuck, below the taffrail, should stand a few feet ample above the lower deck, because of which the sternpost will have to have enough length.

The sternpost of a ship of 180 feet, should at the inside in the ship be 5 quarter thick, and at the back side, one foot at the top.

(66 II 8) 3 The thickness is as of the stem, inside and outside.

(66 II 22) 2 The sternpost, being 1/5 broader than thick above, and below 5 times broader than thick.

. . .

(73 I 43) One makes the lower part of this sternpost such that it will stand with as much as 6 or 7 feet (according to the breadth of the timber) on the keel, and at the rear, one or one and a half feet from it, one leaves a pin of its own wood, which is let into the keel in a hole as thick as the pin, such that the sternpost comes to stand so much more steady aft: forward it is fastened with nails.

. . .

(73 I 40) The sternpost is to remain square as far as the transom is wide.

. . .

Figure 2.10. Plate XLVII. Sternpost. *Bottom:* The pin to fit the keel.

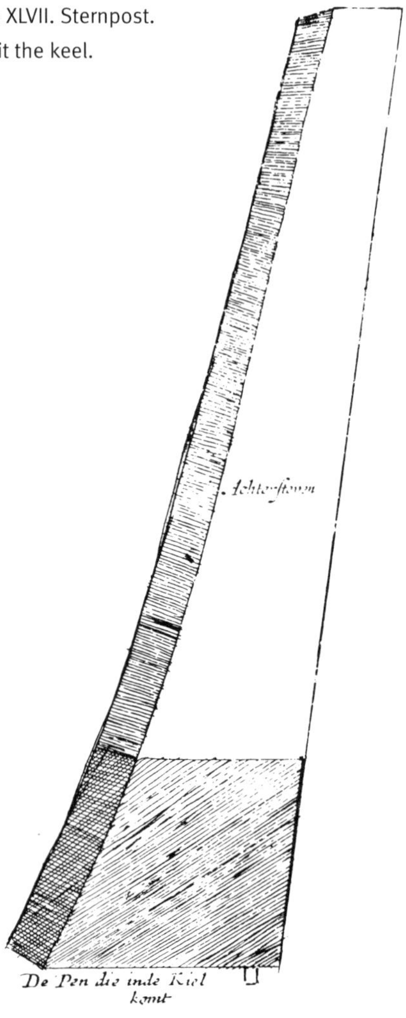

(66 II 5) 2. The Rake of the sternpost is thus, every 6 feet of height must rake 1 foot, so the rake is 3 feet.

. . .

(72 II 49) To be sure about the rake of the sternpost, one foot of rake is taken for each 6 feet of the length of the sternpost, and should someone wish it to have more rake, this would be bad for the ship. All exaggeration of the rake will have to be made up for by a loss in the middle, because the wake must preserve its properties. On the sternpost numbers are applied, from which the draft of the ships can be read.

. . .

(146 II 17) *In the figure A* [fig. 2.11] *one can observe the shape of the Sternpost, standing on the Keel.*

Ab The rake of the Sternpost.
Ac The height of the sternpost.
be The breadth of the Sternpost below.
cf The breadth of the sternpost above.
gb The length of the Skeg.
h The Rabbet.
bd The heel of the keel, and thickness.
e The Knee of the sternpost on the Keel.
y The stopwaters.
o The Marks for the Stopwaters.
k The pin of the Sternpost standing in the keel.

(146 I 22) When making the ***Sternpost,*** take a piece of timber of the required length, breadth and thickness: extend a Measuring Rod [*Rey*] from the upper outside end, at *a,* and another Rod at the bottom end at *f* and *e:* cross the Rods as at *b,* perpendicularly, then make a notch at *d,* and then take a line from *d* and lower it to the Rod at *c* which will make the rake of the Sternpost; from *f* to *e* is the breadth of the lower end of the sternpost, from *a* to *g* the breadth of the upper end of the Sternpost, as can all be seen in the Figure *C* [fig. 2.12].

To determine the rake of the Sternpost, for every six feet of length of the Sternpost one foot is reckoned for the rake: if the distance from *a* to *b* is taken for the length and from *b* to *c* for the rake, then make the back from *a* to *c* straight with a Rod from *a* to *b,* and to adjust the Sternpost in the right angle hold the try square at *b,* and make marks according to the Measuring Rod at *f* and *e* [probably meaning *c*]. For cutting and shaping the separate parts one has molds or Patterns, on which the timber is laid down to shape it accordingly, just as

FIGURE 2.11. Plate XLVIII (drawing A)

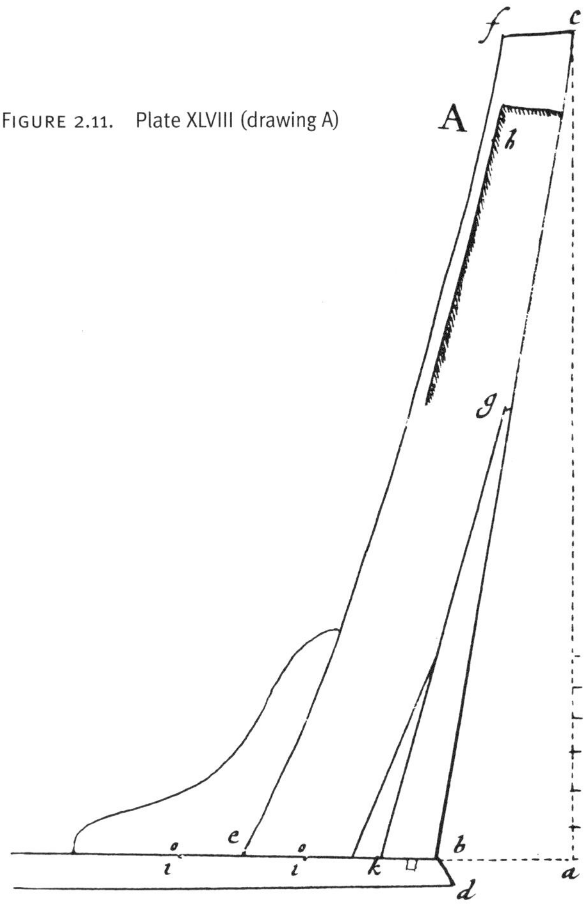

the master ribbands are determined to shape the sheer and sides of the Ship as wanted.

After the Sternpost has been put in the right angle, the inside and outside cut according to the line and in perpendicular, as can be seen from *a* to *b* and *d* to *c,* then the sternpost is laid with the inside facing upward and a center line is drawn as from *g* to *h,* and to each side the required thickness of the inner side of the Sternpost is measured and hewn down to the Rabbet, which is to come in the sternpost, in the perpendicular, from *d* to *c* and also from *e* to *f.* When this is done, the sternpost is turned around with the outside face facing upward, and the skegs are fastened to it, as can be seen from *a* to *i* in the plate at *B* [fig. 2.13].

. . .

(146 II 33) When then the ***Sternpost*** stands with its back facing up, then the Skeg *k* is first made on the Sternpost, at *l,* then other pieces are fitted underneath, as *m* and *n,* after which the back is made straight and a center line is drawn onto it, as from *o* to *p,* and then it is cut to thickness above and below as with *g* to *k* and from *i* to *h,* from *f* to *d* and from *c* to *e.* The Pin is

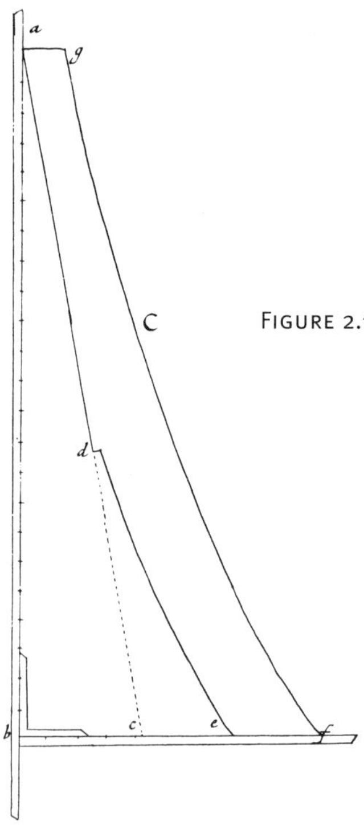

Figure 2.12. Plate XLVIII (drawing C)

made about as thick as the back side of the Sternpost, long one third of the thickness of the keel, and broad one third of the Sternpost, as can be seen at *q*, from *c* to *k* it is so much thinner as the planks of the Garboard strake which are to come there are thick: And this fades to naught at the inside at *r*. Then the Skeg is fastened with nails and bolts, as can be seen at the dotted line. . . , at *a* and *b* is the inner face, which should be curved a little, as shown from *s* to *t*; the lower the longer the bolts have to be. [. . .]

The inner piece is necessarily made of one timber, as also the outer Skeg: but the inner Skeg may be made of two parts. All this can be seen at the letter *D* [fig. 2.14], but many shipwrights do make the Sternpost a little thicker than shown here.

(72 II 35) 1. The sternpost, high $24\frac{1}{4}$ feet; rake 4 feet.

2. The back thick 9 inches, or outside beneath the square on the inside thick $13\frac{1}{4}$ inches. [The square is the top of the stern over the total thickness of the wing transom, where it is kept square and not decreased at the back side like the rest of it.]

3. Above broad 19 inches; below broad 6 feet.

(147 I 5) The wide part aft is long 16 and $\frac{1}{2}$ inches, is called the square, is thick 13 inches, below the square it is thick 9 inches.

3. The Sternpost

I have not previously discussed the rabbet in the keel, stem, and sternpost, in which the planks are fitted, because these rabbets will be dealt with in connection with section 15 (**Then the garboard stern rabbet is made, with the keel rabbet**). Yet the garboard stern rabbet must be mentioned briefly here because of its effect on the thickness of the sternpost.

On each side, the ends of the lowest two or three strakes slowly turn to the vertical and run right through to the aft face of the sternpost instead of terminating with hood ends in rabbet, as was done according to English practice. To keep the stern from becoming unreasonably thick as a result, the sternpost was slimmed down accordingly. In the case of the pinas, this was $1\frac{1}{2}$ inches on each side, up to a height of $5\frac{1}{2}$ feet (1.56 meters).

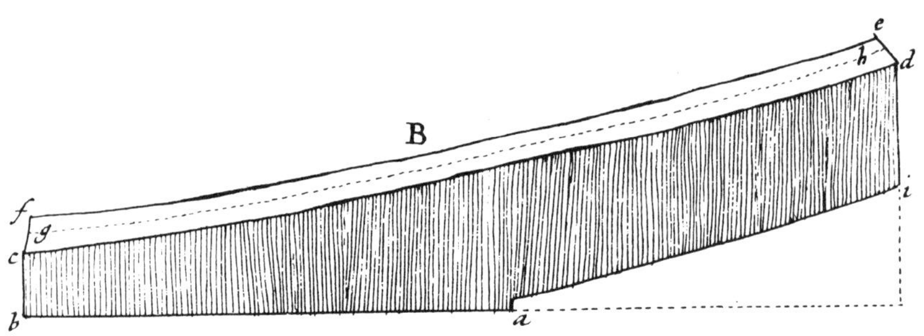

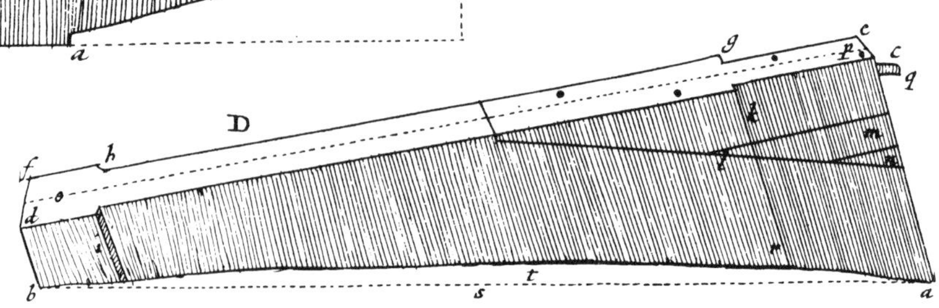

Figure 2.13. (*top*) Plate XLVIII (drawing B)

Figure 2.14. (*bottom*) Plate XLVIII (drawing D)

4. *The wing transom.*

(73 II 23) In the figure No. 20 [fig. 2.16], a is the wing transom, b are the transoms, c the broekstuk [only the top of the *broekstuk* is visible in this sketch; cf. fig. 2.33], which joins the lower ends of the fashion pieces: d is the fashion piece, e the chocks of the gunports in the tuck.

(66 II 40) *Of the Wing transom.*

7. In order to obtain the length of the wing transom, one takes ⅔ of the width of the ship.

(68 II 13) The length of the wing transom, equals the breadth of the bottom.

. . .

(66 II 44) 2. For the thickness, breadth and the curve, take as many inches as the wing transom is broad in feet.

. . .

Or: (68 I 36) The wing transom is also arranged in thickness, breadth and curve, as the inside of the sternpost.

. . .

(55 II 12) The Wing transom is radial, outward.

. . .

(269 I 25) The *Wing Transom* stiffens the fashion pieces, and thus holds together the entire ship.

. . .

(147 I 17) *Next follows the Model of the Wing Transom as seen on the Plate at letter E* [fig. 2.18].

a Is the upper side, *b* the back, *e* the inside face, *f* the underside, at the back it is made with a hollow, underneath which a rabbet, in which the planks of the Tuck are made fast; at *d* is the rabbet, *c* is the hollow, and when made thus, the back is laid at the inside of the Sternpost, close to the upper end, and fastened with a dovetail. It has to be level, on the inside as on the upper side, on which the Fashion pieces are set, which ends are held on the sternpost. The lower side should come level with the rabbet, and the sternpost does not protrude above the wing transom. When this wing transom is made, the camber is made after the length it has, and also the breadth and the thickness, as is to every ten feet of length eight inches to the camber, and to the breadth every foot of length one inch of breadth, and the thickness slightly less.

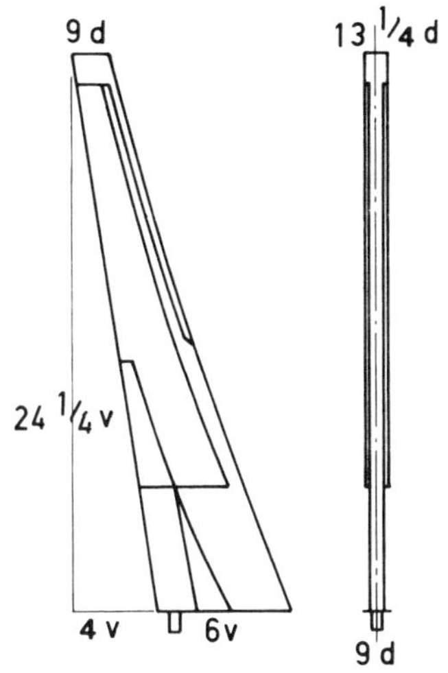

FIGURE 2.15. (*top*) Construction of the sternpost of the pinas. The measurements were taken from Witsen's pinas data. (Drawing by A. J. Hoving)

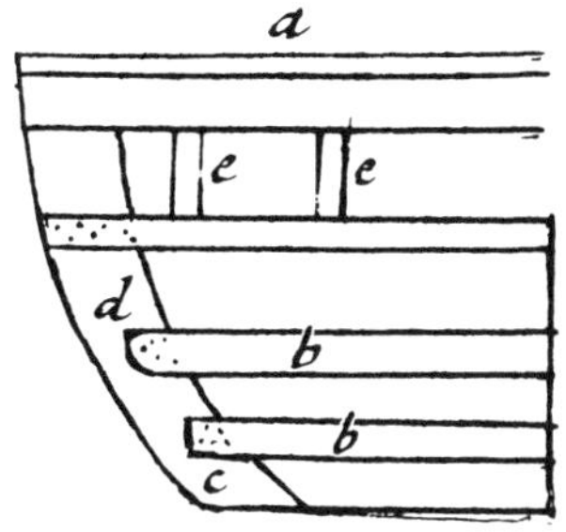

FIGURE 2.16. (*bottom*) Plate XLI

(73 II 6) 4. *Of the Wing transom.*

1. The wing transom be 19 feet 6 inches long, thick 1 foot 2½ inches, broad 1 foot 3½ inches.

2. The wing transom has a curve of 1 foot 2 inches, the rabbet long 3½ inches, deep 3 inches.

3. The hollow above the rabbet long 3 inches, deep ½ inch. All that it is more, remains square, it's joined half into the sternpost, and secured with a dovetail, two bolts are struck through, which also go through the sternpost but for one inch.

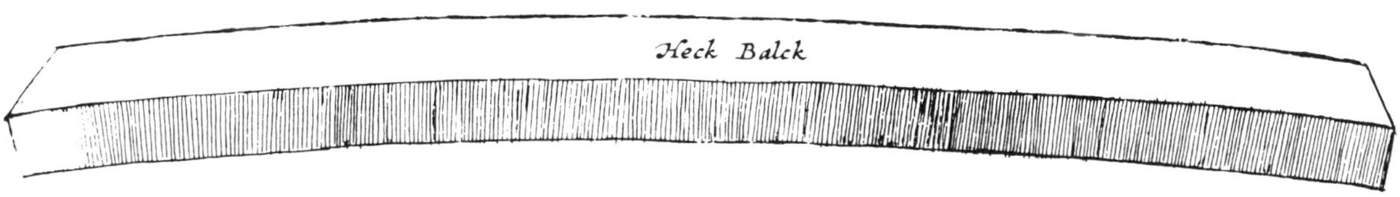

FIGURE 2.17. Plate XLVIII. Wing transom.

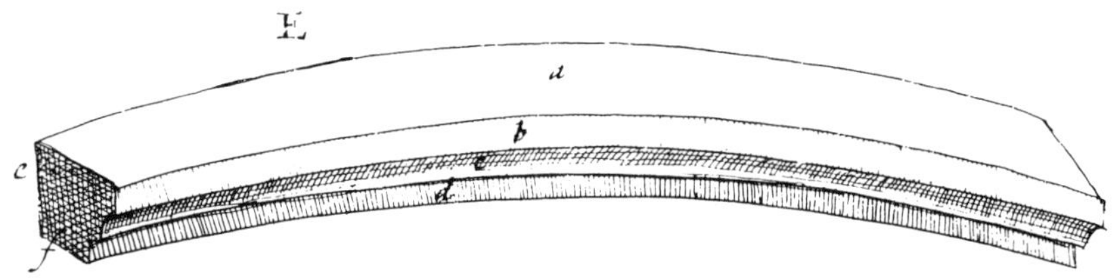

Figure 2.18. *(left)* Plate L (drawing E)

Figure 2.19. *(below)* Plate XLI

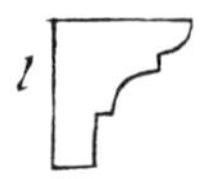

4. The Wing Transom

At the upper extremity of the sternpost, athwart the longitudinal direction of the ship, a heavy, somewhat curved beam was set: the wing transom. This beam was the upper edge of the square tuck (on the fluyts, with their round stern, this feature was absent, as are the parts described in sections 5 to 9 below). On the aft face the wing transom was beveled in a profile and also had a rabbet into which the planks of the tuck were nailed.

The length of the wing transom bears on the final shape of the vessel and the amount of room in the stern. Here we run into another decisive phase for the shipwright. According to Witsen, the length of the wing transom was two thirds of the width of the ship, but opinions concerning the length of this part varied during the century. Van Yk, writing twenty-five years after Witsen, reckoned it should be three fourths of the width. The wing transom of our pinas was 67.4 percent of the width, slightly more than two thirds of the width. Early in the century the rule was 50 percent.[7]

5. The Fashion pieces.

(56 II 37) The Fashion pieces make the width of the ship, and the tuck, aft.

. . .

(67 I 7) 2. The thickness is 5/6 parts of the thickness of the wing transom.

. . .

(67 I 20) 10. The fashion pieces, must have a roundish curve.

. . .

Or: (68 I 41) The fashion pieces are thick half the sternpost.

. . .

(67 I 7) 3. The breadth is 2 times the breadth of the wing transom.

(68 I 29) To be sure about the position of the fashion pieces, they are put with their back or outside one foot above half the sternpost.

. . .

(66 II 51) 8. 1. The fashion pieces are put halfway the length of the sternpost at the back. With Freighters these timbers come down lower than with Frigates.

. . .

(147 I 36) The lower end of the Fashion piece should follow the curve of the rabbet.

. . .

(73 II 35) The fashion pieces are fastened to the wing transom with a kind of tapered halving joint and are lapped onto the sternpost.

. . .

(147 II 11) *Of the Fashion piece at F* [figs. 2.20 and 2.21].

When making a Fashion piece, first make a try square, as from a and e to *c* and *d,* put the compass with one foot at *e,* and measure half the length of the Wing transom, as from *e* to *b,* draw the circle to *c,* and let a little wood or room inside the circle between *f* and *g,* and at *b* outside the arc of a circle the same yet a little more, the one to obtain the roundness, the other the hollow curve, from *c* to *d* at the lowwer end the breadth follows, as from *a* to *b:* and this is the proportion of the inner side of the Fashion piece. (But in general one has molds after which the fashion pieces are cut.) Then one lays it with one end on the sternpost, with *c* and *d,* and with the end *a* and *b* at the end of the Wing Transom, and fastens the ends with a lap joint on the sternpost as also on the Wing Transom, as seen at *G* [fig. 2.22]. *a* Is the scarf of the Fashion piece on the wing transom.

(73 II 32) The fashion pieces wide 1 foot 9 inches, thick 8 inches: the ends reach down to 1 foot below the middle of the sternpost.

5. The Fashion Pieces

The fashion pieces were two very large, expensive, curved timbers that made the curve of the tuck. For ships built for speed, the point where the fashion pieces were joined with the stern was placed higher than for carriers. This was another moment when the shipwright could influence the properties of the vessel.

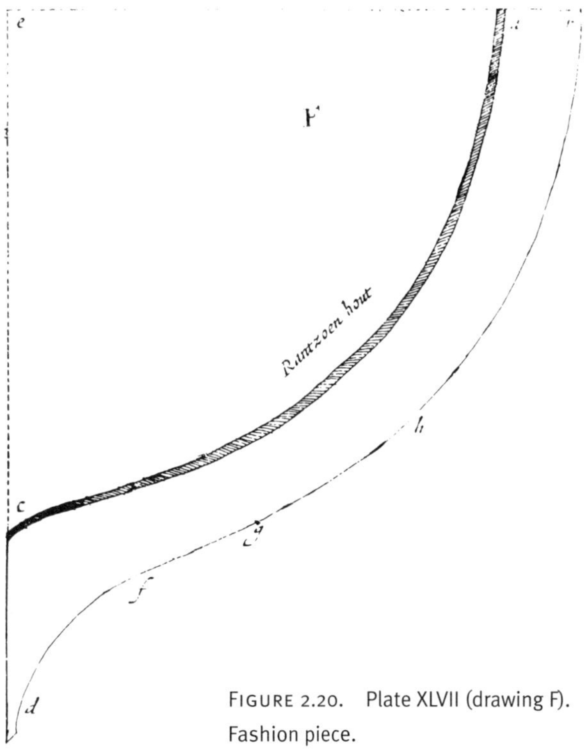

Figure 2.20. Plate XLVII (drawing F). Fashion piece.

6. The Chock underneath it.

(73 II 49) but the chock, which comes in the corner of the fashion pieces, is sharp to one side.

6. The Chock underneath It

This chock seems to fill the remaining space between the fashion pieces. I have never seen it in shipwrecks or in models, and it is not mentioned by Van Yk.

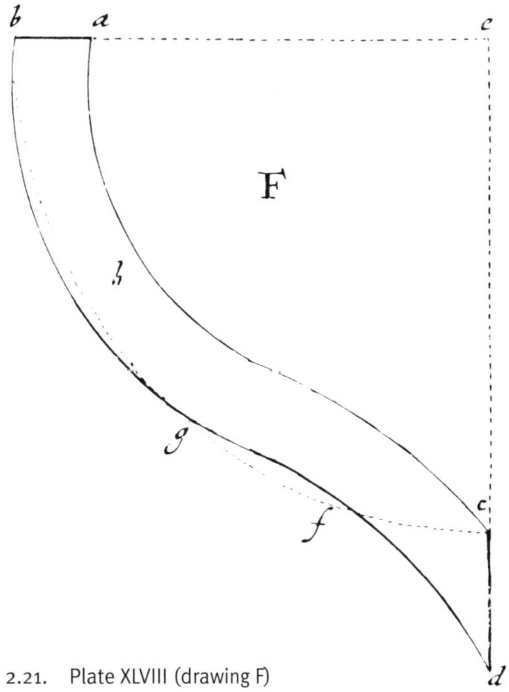

Figure 2.21. Plate XLVIII (drawing F)

7. The Broekstuk.

(54 II 33) The broekstuk [. . .] come in the tuck, for the same of firmness.

(67 I 11) 5. The broekstuk 1 inch thinner than the sternpost.

(73 II 46) The broekstuk is thick 7 inches, and covers the lower ends of the fashion pieces.

7. The *Broekstuk*

The translation of this word is difficult, because there seems to be no equivalent in English. It is a plate of wood, one inch thinner than the sternpost, that firmly joined the lower ends of the fashion pieces. Later, in section 40 (Make the hanging Knees, fit the ceiling in the Bilge, make the crutches and the Keelson, and the Mast Step), we shall see that the *broekstuk* was also the basis for a number of knees, which strengthened the stern.

8. The transoms.

(147 I 46) In the Figure *S* [fig. 2.25], *a* Is the *Wing Transom,* and *b* one side of the Fashion Piece, *c* and *e*

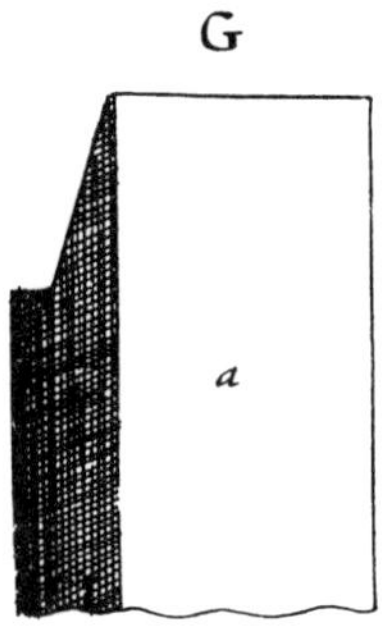

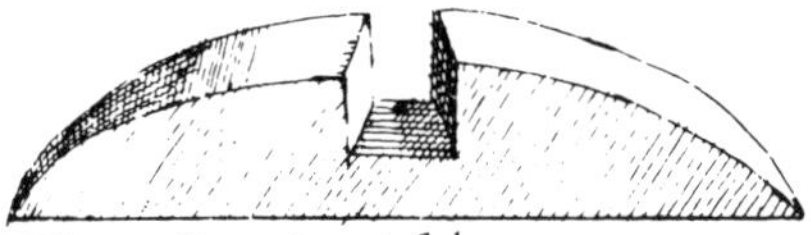

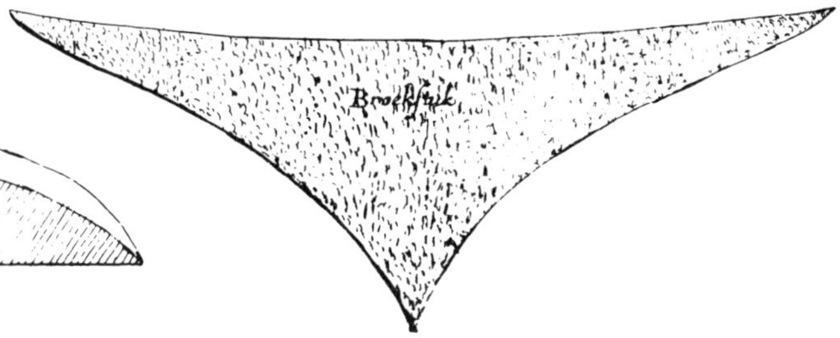

Figure 2.22. (*left*) Plate XLVIII (drawing G)

Figure 2.23. (*center*) Plate XXVII. Chock below the *broekstuk*.

Figure 2.24. (*right*) Plate XXVII. *Broekstuk*.

in T [fig. 2.27] are the Transoms in the Tuck, *d* is the hole, or dovetail, in which the ends of the transoms are lain, these transoms come across the Sternpost with a shoulder; and the Wing Transom is fastened to the sternpost with a dovetail and the same, as also the transoms, is secured with two bolts to the stern-post, at the ends or dovetails that are bolted with nails until the stern timbers are fastened over it. When the Stern is raised or put upright, the bolts and some nails are hammered straight through the stern timbers, and nails at the lower end.

. . .

(68 | 39) The transoms are 1/3 thinner than the wing transom.

. . .

(67 | 9) The transoms are 1/5 thinner than the stem.

. . .

(55 | 39) Between the transoms in the tuck are to be the gunports.

. . .

(147 || 36) *Model of a Transom, turned on its front side, seen at letter H* [fig. 2.28], *on the plate alongside.*

The dovetail at *a, b,* and *c* is the shoulder on the sternpost. When the joins of a transom are made, one takes a plumb line and measures to what extent the timber is off plumb (vertical), then take half the thickness of the Fashion piece at the inside, letting it taper to naught at the outside. Take off half of the transom, and at the end as much as one cut from the transom at the inside of the fashion piece, and this will make the dovetail; but at the inside the hole of the fashion piece should be half an inch narrower as on the outside. The transoms are to be equal, and they are to be hollowed out as deep as the hole of the dovetail, which makes the joint of the transom, as can be seen at B. The transoms are not fastened before the sternpost is raised; 4 holes are drilled through the wing transom for the bolts, 2 in the transoms, 13 or 14 in the Broekstuk, 8 in the Fashion pieces 5 or 6 in the sternpost, to keep together the lower ends of the Fashion pieces, 10 or 12 nails are hammered into the Fashion pieces, the Rabbet in the sternpost tapers to naught below, the bolts of the Transoms come but for two inches through the sternpost.

As can be seen at I [fig. 2.29], the Transom is put plumb, and marked with chalk, the wood from *f* to *g* is removed, or taken away, as from *e* to *d,* the part from *h* to *g* cut, and the joining of the dovetail in the Fashion frame, which is shown in figure K [fig. 2.30] at *k*.

(73 || 38) 2. The upper transom lies 2 feet from the wing transom, according to the height of the gunports. The transoms are thick 13 inches. The transoms come as low as the lower end of the rabbet and the fashion pieces, and are let in at the ends with a dovetail: they are of even thickness and breadth as the wing transom, the upper lying 2 feet from the wing transom.

8. The Transoms

The transoms joined the fashion pieces and the stern-post. These timbers strengthened the tuck laterally and

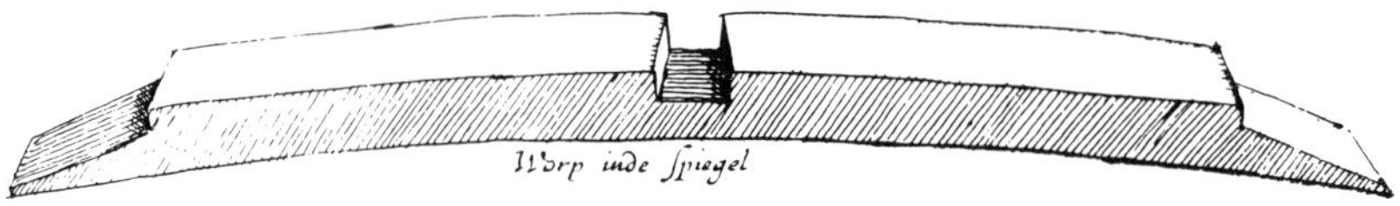

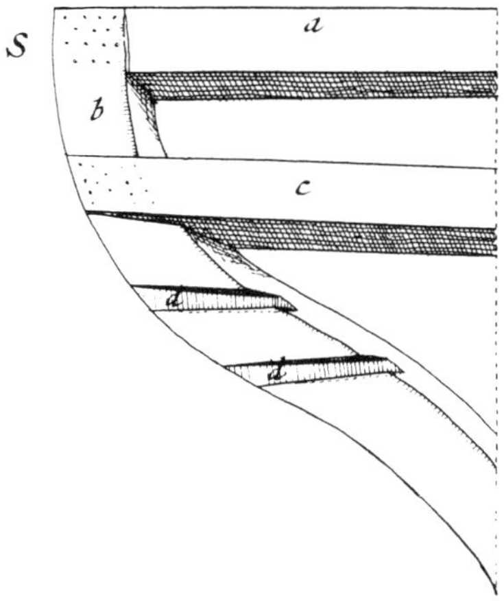

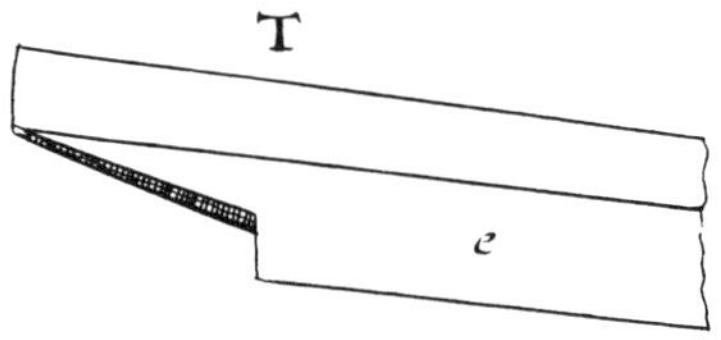

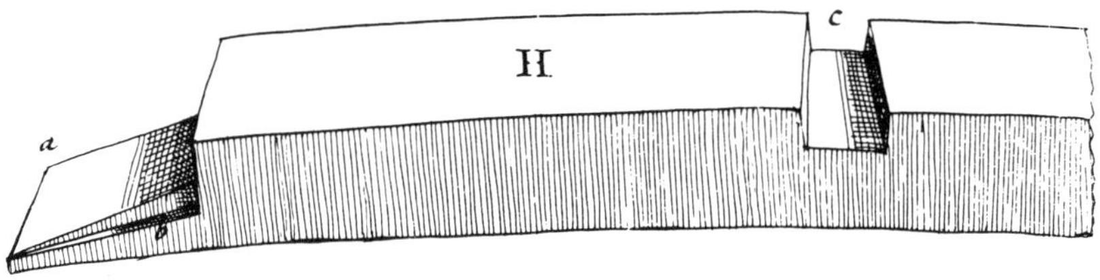

FIGURE 2.25. (*top*) Plate XXVII. Transom in the tuck.

FIGURE 2.26. (*center left*) Plate L (drawing S)

FIGURE 2.27. (*center right*) Plate L (drawing T)

FIGURE 2.28. (*bottom*) Plate XLIX (drawing H)

also provided a structure to which the planking was fastened.

9. The Stern timbers.

(55 II 9) Stern timbers: are timbers that come above the tuck, and typical for Pinases.

. . .

(74 I 3) The lower parts of the stern timbers are as broad as the fashion pieces, come up as much above the sternpost as the sternpost is long, and are 2 feet broader than half the wing transom: they lie straight across the transoms, and flat on the fashion pieces, fastened with nails and bolts.

. . .

(67 I 13) 6. The stern timbers thick $2/3$ the sternpost.

7. Agreeing with the curve of the fashion pieces below.

8. The upper ends as long as the wing transom is far from the keel.

. . .

Or: (68 II 7) The stern timbers as far above the transom as the sternpost is long.

(66 II 47) The stern timbers are mostly made as far above the wing transom as the sternpost is long.

(67 I 18) 9. Above wide $3/5$ the length of the wing transom.

> (73 II 51) 3. The stern timbers long 23 feet above the wing transom, above wide 10 feet 6 inches, thick 8 inches, below broad 20 inches, above thick 6 inches, broad 7 inches.

9. The Stern Timbers

The last stage of stern construction was making the stern timbers, which determined the amount of room above the main deck. Their length, including the part overlapping the fashion pieces, was equal to the height of the wing transom—that is, according to the formula. It is clear that as the height of the stern was reduced in the course of the century, the length of the stern timbers diminished accordingly.

The top ends of the stern timbers were angled toward each other. The amount of inclination changed with the

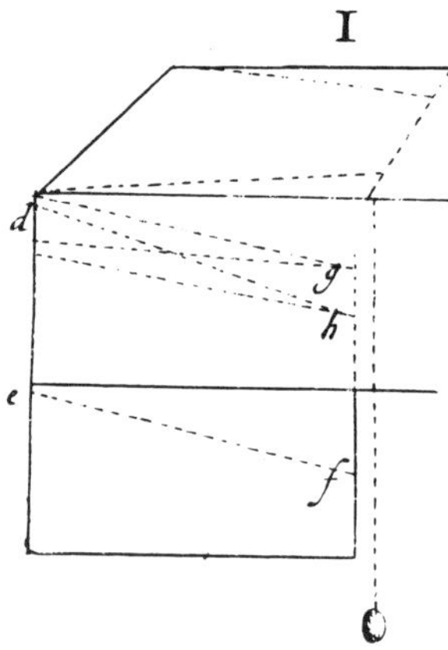

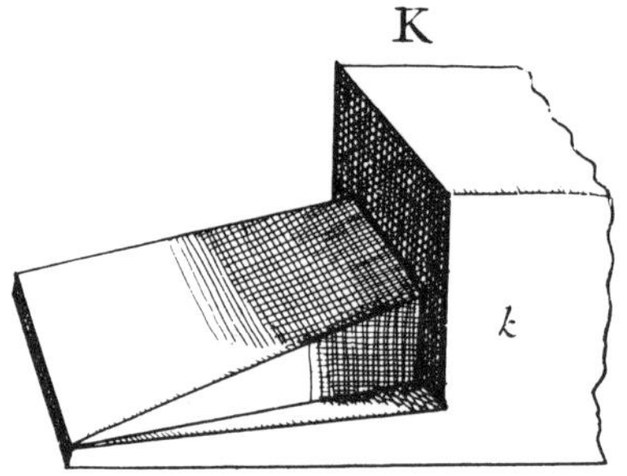

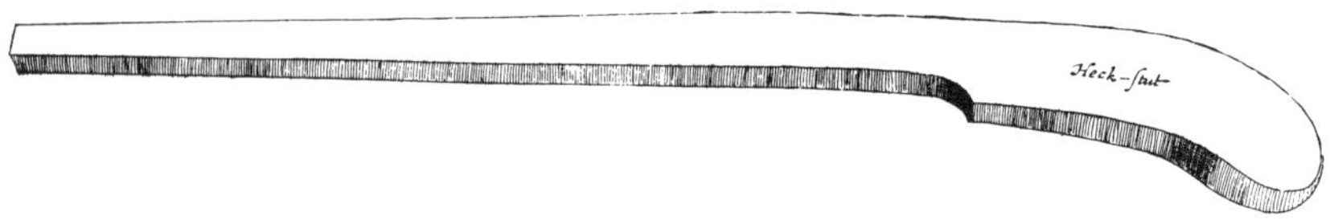

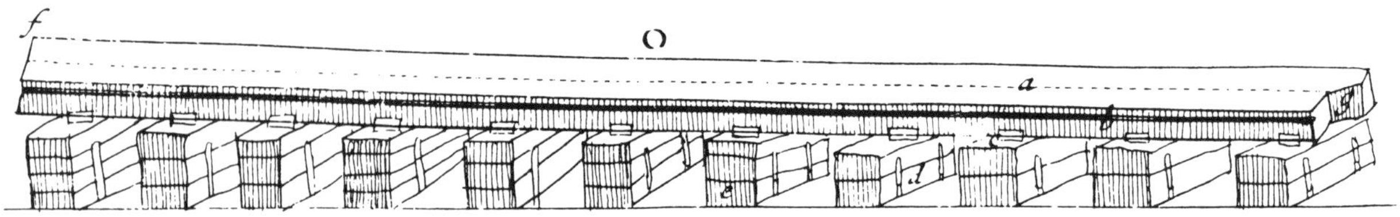

Figure 2.29. (top left) Plate XLIX (drawing I)

Figure 2.30. (top right) Plate XLIX (drawing K)

Figure 2.31. (center) Plate XXVIII. Stern timber.

Figure 2.32. (bottom) Plate LI (drawing O)

times: at the beginning of the seventeenth century the upper works of the stern were long and narrow; later they became broader and lower. Witsen gives two formulas for calculating the distance between the tops of the top timbers: two feet wider than half the wing transom, or three fifths of the length of the wing transom. For the pinas this calculation would yield either 10 feet ½ inch or 11 feet 8 inches. However, the actual measure was 10 feet 6 inches wide.

10. Then the keel is laid.

(55 II 6) [. . .] the keel is laid on blocks ready for further building.

. . .

(72 II 11) A keel is laid higher at the end than at the forward end, for the ship to be launched better, for the yard needs to be firm.

. . .

(149 I 18) [. . .] and once the keel scarf is made, the keel is laid, forward and aft, half the thickness of the stem higher, and kept as much lower in the middle, and then on each pile of blocks a cleat is nailed against the keel, to prevent the keel from moving when planking the ship, then the Boxing Scarf is made at the fore end of the keel, and a middle mark is made along the keel, from which the middle can always be measured. 4 or 5 bolts are nailed through the knee standing on the keel and against the stem: all the seams below are filled with moss, against seepage. One should heed, that the keel is laid a little longer than its required length, for the boxing scarf and the heel to be made at the ends.

- *a* Under letter *O* [in fig. 2.32] is the Middle mark.
- *b* The Rabbet.
- *c* The cleat against the keel.
- *d* A stake with which the keel blocks are fastened.
- *e* The Keel block.
- *f* The Heel of the keel.
- *g* The Boxing Scarf.

(72 I 4) 3. The keel sagging 6 inches.

10. Then the Keel Is Laid

Up to this stage all the parts had been made but not yet assembled. The assembly started with laying the keel on the keel blocks, with the keel sagging a little in the middle (half the thickness of the stem). The intention of this sagging, or rocker, was that, once the ship was in the water, the upward pressure of the water in the middle of the ship would straighten the keel again.[8]

Witsen does not elaborate on the keel blocks, contrary to Van Yk, who gives such details as height, number and distance of the blocks, and the position of the so-called *dompblokken*. We will come back to this in section 58 (Finish it all around, to prepare it for launching, and, when ready, launch it), where we find only slightly more information on the method of launching.

---

11. Remove the Stern timbers or [and] *transoms.*

---

11. Remove the Stern Timbers and Transoms

The stern, whose components are described above, was partly disassembled (the fashion pieces, wing transom, and the *broekstuk* were already fastened). Of course, this was done to make the entire construction as light and convenient to handle as possible. The stern timbers and transoms were replaced as soon as the stern was raised and fastened to the keel.

---

12. Raise the Stem.

(66 II 31) The stem is put on the keel, with its scarf as long as possible and as long as the scarf will permit.

. . .

(148 II 30) As can all clearly be seen on the adjoining plate at letter *M* [fig. 2.33], where

a is the Katte-block.
b A Gin block.
d Single-sheave block.
e The Shear Mast.
f The plate.
g One of the posts, to which the stays are attached.
h Stays.
i The Hook and chain.

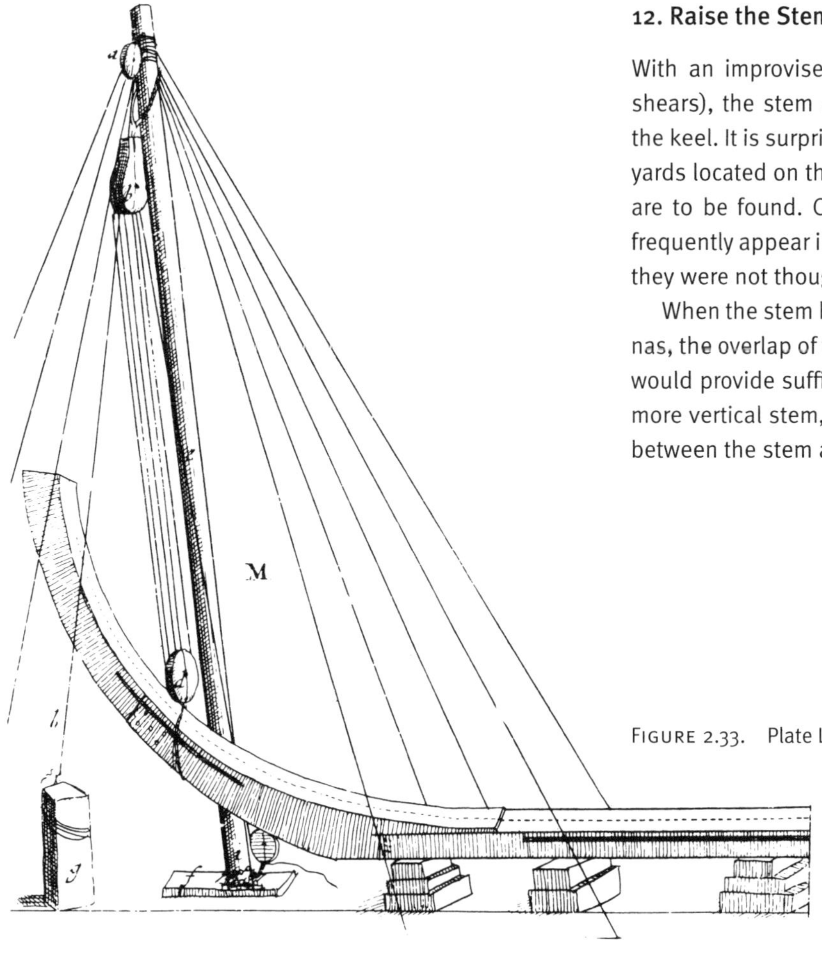

12. Raise the Stem

With an improvised crane (the so-called shear mast or shears), the stem and sternpost were raised and set on the keel. It is surprising that in pictures of shipyards, even yards located on the same spot for a long time, no cranes are to be found. Cranes did exist at the time, and they frequently appear in depictions of harbors. But apparently they were not thought necessary in a shipyard.

When the stem had a considerable rake, as with the pinas, the overlap of the scarf could be made quite long and would provide sufficient connection onto the keel. With a more vertical stem, however, a separate knee was placed between the stem and the keel to reinforce the scarf.

Figure 2.33. Plate LI (drawing M)

13. Raise the Stern, insert the transoms, and attach the Stern timbers.

(148 II 42) *At N* [fig. 2.35] *one can see how the stern with the broekstuk etc. is raised.*

Where the shear tackle block is *a,* which is made fast to the sternpost with a strop, above the strop a bolt goes through the sternpost, so that the strop will not slip while erecting, *b* is the strop, *c* is the tackle of the shears.

(148 I 50) *When raising the Stern;*

Take a mast about twice the length of the sternpost, lash a block to the end above, and another onto the sternpost, which should be fastened at the point of balance of the stern with the stern timbers and transoms removed. Also attach six or seven stays to the upper end, with which the mast is stayed upright and secure; the shear mast stands on a plank below, which is surrounded by pegs, and around which an iron chain is fastened, to which the lead block is attached, which is wound by the capstan. When the stern stands, the transoms are inserted, then the stern timbers above them, and fastened, when hoisting these up then take a cat block and slip it onto the upper end of the shear mast. Then hammer bolts into the transoms and also into the stern timbers, and so fasten the stern, below on the inside with large nails, and behind with stern shores, which stand on planks which are secured in the earth and above to the stern. The planks should be laid athwart the tuck, for the sake of strength.

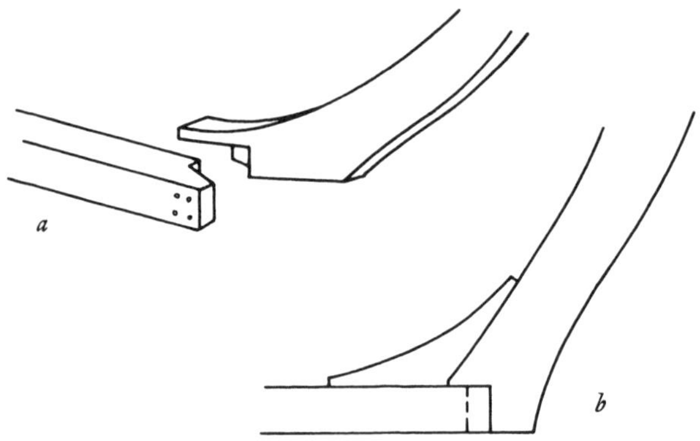

FIGURE 2.34. Two methods of connecting the stem to the keel. (*a*) Stem with a considerable rake. (*b*) Stem with less rake; the knee on the joint was called *slemphout* (rising wood). (Drawing by A. J. Hoving)

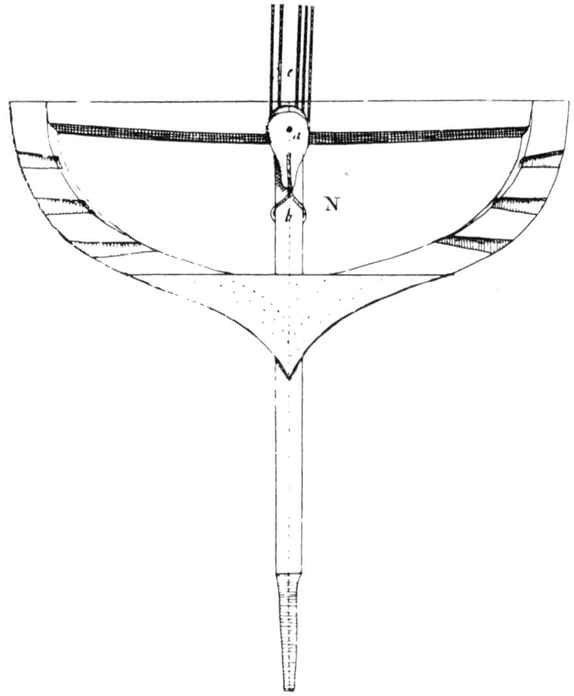

FIGURE 2.35. Plate LI (drawing N)

(148 I 29) *Model of the Tuck on the inside, to be seen at L* [fig. 2.37].

Of which the names are as follows: *a* is the Wing Transom, *b* the Transoms, *c* the Sternpost, *d* the Broekstuk, *f* the Fashion Pieces, *g* the notch of the stern timbers, *h* the width between the upper ends of the stern timbers and also the stern timbers themselves. The Broekstuk keeps the Fashion pieces together. Against the Fashion pieces below a Chock is also applied; the broekstuk is attached with 4 bolts at either side of the sternpost, and apart from these in the middle a few more are hammered in, straight into the Sternpost. All members of the Stern, such as stern timbers, transoms, broekstuk etc., are made ready below, then hoisted up, and inserted.

13. Raise the Stern, Insert the Transoms, and Attach the Stern Timbers

14. Make a knee on the keel, and against the Sternpost.

(73 I 54) Forward of the sternpost a knee is set, which is fastened to the keel as well as to the inside of the sternpost, so that the stern will be well secured.

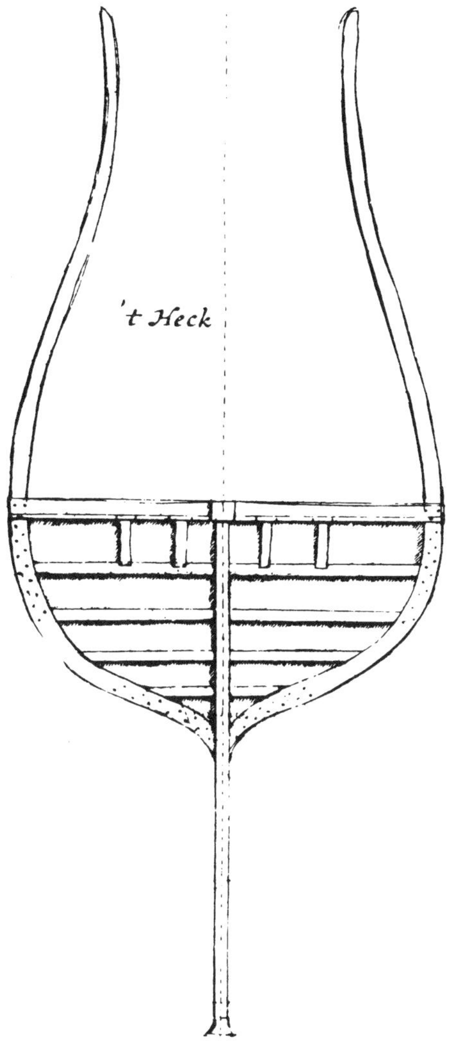

FIGURE 2.36. Plate XXXIX. Transom

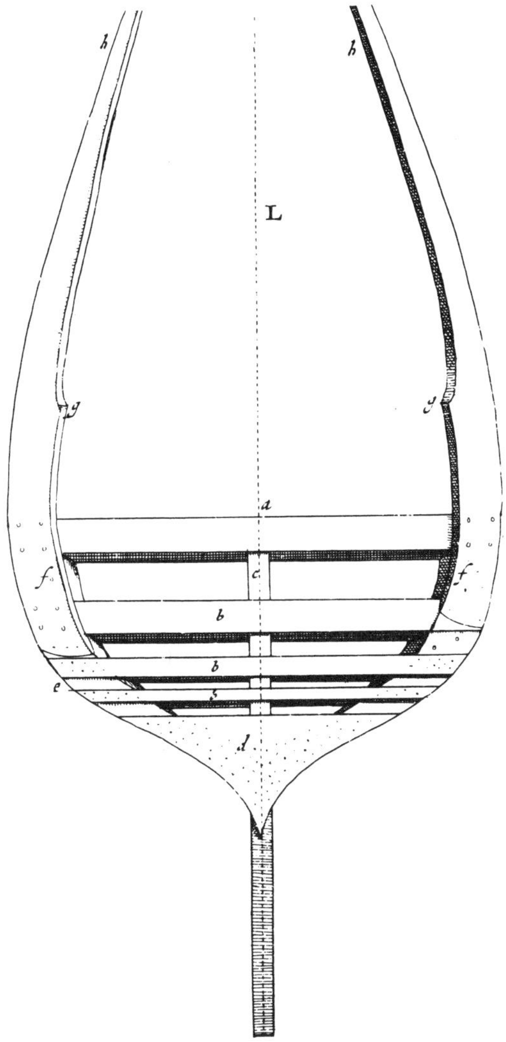

FIGURE 2.37 Plate XLIX

(74 I 10) *6. Of the Knee on the keel, at the sternpost.*

1. The knee on the keel, at the sternpost, thick 1 foot 3 inches, broad 50 inches: the ends long 6 feet.

2. Through each end 4 bolts are hammered: in some cases also knees are applied to the sternpost above. Through the scarf of the keel 8 bolts, and through the scarf of the stem 4 bolts are clenched with small rivet plates: through keel and sternpost 6 bolts, first going through the knee which stands on it, and then through the scarf of the keel.

14. Make a Knee on the Keel, and against the Sternpost

Compared with British shipbuilding practice of the time, the knee was only a minor timber. In England a considerable deadwood knee was fastened to the keel to form the base for the frames in a far sharper hull. With the Dutch flat bottoms such a large knee was not needed.

15. Then the garboard stern rabbet is made, with the keel rabbet.

(148 II 50) On the Plate at PP [fig. 2.40] one can see the rabbet of the keel in *a* and equally in *b,* and the way the planks are set in there.

. . .

(66 I 10) The rabbet, thick, broad and deep, as the planks are thick, the planks ¼ the thickness of the stem.

. . .

(66 II 34) At ¼ from the inside of the stem, the planks are entered in the stem.

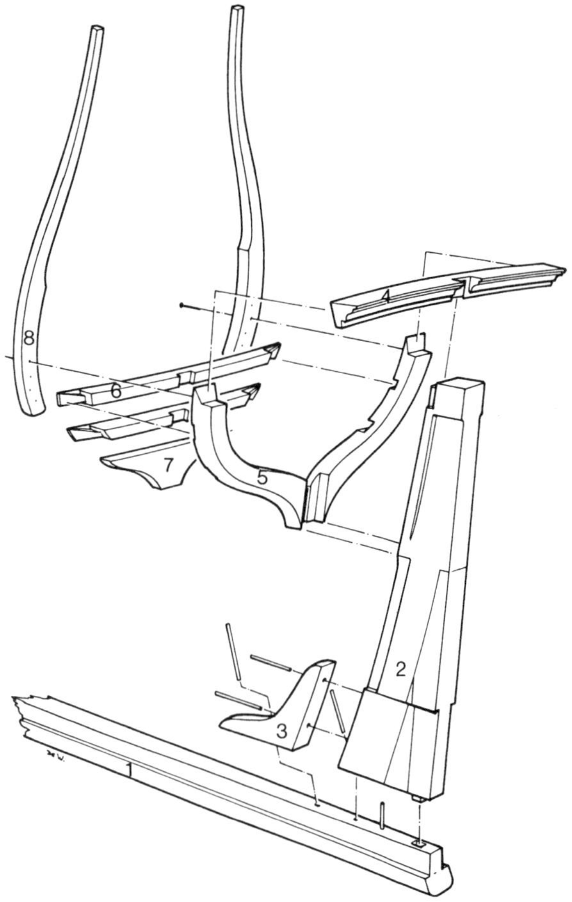

FIGURE 2.38. Exploded view of the stern construction. The rabbet on the mid-section of the front side of the post differs slightly from Witsen's description and is based on the wreck of a seventeenth-century trader that was recovered in the former Zuider Zee. *1*, keel; *2*, sternpost; *3*, knee; *4*, wing transom; *5*, fashion pieces; *6*, transoms; *7; broekstuk; 8*, stern timbers. (Courtesy G. A. de Weerdt)

FIGURE 2.39. (below) Plate XXVII. Knee on the keel against the sternpost.

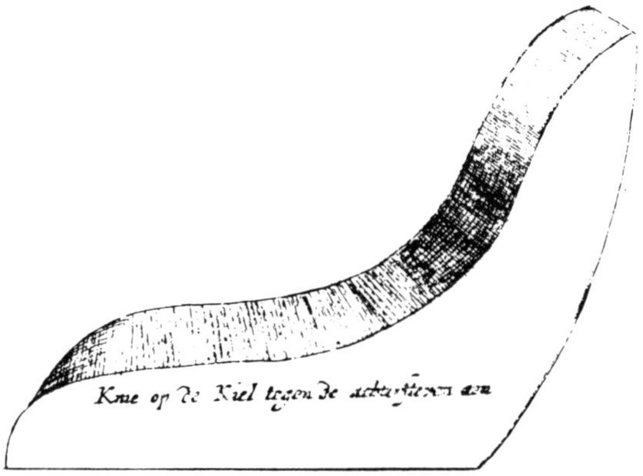

(149 I 9) Make the rabbet one fourth the thickness of the stem, from the upper side of the keel, also make it about one fourth deep, and as broad, at the rear end it should be kept straight down, [. . .].

. . .

(66 II 25) 3. The rabbet here is little less than in the keel.

. . .

(150 I 12) Then a part of the ***Rabbet in the Stem*** is made, as is shown at f on the same plate [fig. 2.41; f is missing, but Witsen's heavy black line indicates the rabbet], and when the stem is made fast on the keel, then a spline [*rey* or *rij*] is taken and bent from the end of the rabbet in a curve as wished, the upper side is copied from the spline, and with a compass with a piece of chalk also the breadth of the Rabbet is drawn from the spline, then cut the Rabbet and make the overlap of the scarf and the boxing scarf well fastened.

(72 II 42) 4. The rabbet from the side 4½ inches; the rabbet broad 3½ inches, the rabbet deep 3 inches.

5. The garboard stern rabbet long or high 5½ feet, thick outward 6 inches, the square thick 13 inches.

(71 II 42) The rabbet, which turns at the rear in as much as the ship is planked flat or sharp, is broad 3½ inches, deep 3½ inches, the edge above the rabbet broad 4 inches.

15. Then the Garboard Stern Rabbet Is Made, with the Keel Rabbet

The garboard stern rabbet, mentioned previously in section 3 (The Sternpost) and made in the sternpost as well as in the keel, was a rabbet in which the planks were set. The dimensions of the rabbet depended on the thickness of the planks as a matter of course. In general, we can say that the upper edge of the rabbet was as much removed from the upper side of the keel as the breadth of the rabbet itself, leaving about half the keel outside the planking.

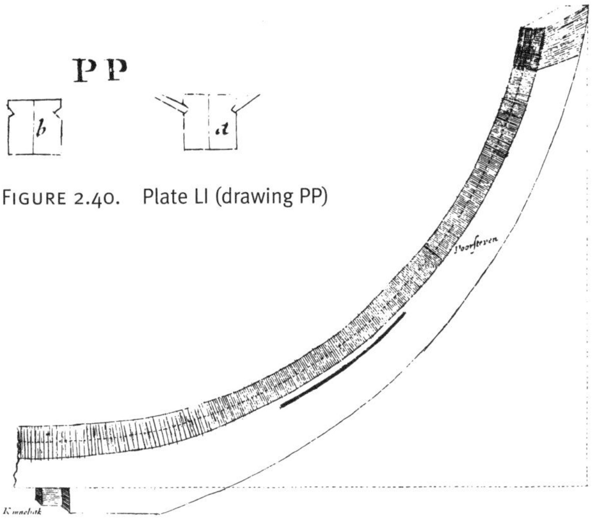

Figure 2.40. Plate LI (drawing PP)

Figure 2.41. Boxing scarf, stem

Of paramount importance for the eventual hull form was the place where the rabbet started to turn for the planks to come to vertical at the stern. Therefore, only the master apprentice or the master shipwright himself cut the rabbet. If the rabbet was cut in such a way that the garboard strake, the first plank, ended up standing too vertical, this was easily repaired using the Klaes Jacobszen (no. 8 in fig. 1.20). If the plank ended up not vertical enough, one could only cure this by bending it upward by means of the same tool, which of course would result in a gaping open seam below, between the plank and the keel, which could hardly be caulked.

Witsen describes a trick the shipbuilder applied at the stem: one part of the rabbet, halfway up the stem, was already cut when the stem was still lying on the ground. After joining stem and keel he would span a natural curve between the rabbet of the keel and the already existing segment in the stem with a straight flexible batten, giving it a good natural roundness. By following the ribband with a chalk, the curve was copied onto the stem and could then be cut. With very vertical stems this line would go over the stem knee, but only just.

16. Drill the Stopwaters in the Stem and in the Keel.

(149 I 14) and when the scarfs are joined, then in the rabbet a stopwater is drilled from below, that no water may come through there.

(74 I 24) The seam between the sternpost and the keel, is sealed with stopwaters.

16. Drill the Stopwaters in the Stem and in the Keel

Because the surfaces of the scarfs between the different construction elements were large and it was therefore impossible to get a watertight fit, holes were drilled straight across the seams. According to English practice, the scarfs were filled with tarred flannel or sometimes tar and hair, but the Dutch practice for make the joints watertight consisted of drilling holes through them, filling them with moss, and plugging up the outsides with wooden dowels. The water seeping in would swell the moss, making the scarfs watertight (see fig. 2.4).

At this point the first phase of the building process was complete, and the yard personnel would be treated to *stevensbier* ("stem-beer" or wine).

17. Make the Garboard strake.

(150 II 33) Here in Amsterdam, as soon as the Stems and Keel are made, the planks next to them, which make the body of the ship, are put in place, but at the Meuse they leave an opening at both sides of the keel, which is shut only in the last instant, when the Ship is to be launched. Which is done to easily get rid of all that was left like shavings, chips etc. while building through the bottom. When the keel and stems are in position a line is strung fore and aft, to know the middle of the ship and apply an equal amount of timber to each side.

. . .

(150 I 28) When the Spant [the stem-keel-stern construction] is standing, then a start is made with the garboard strake, with planking, and with the turn of the bilge. When making the garboard strake, here and there a hole is chiseled in the plank and with a compass it is checked if it is of the same breadth inside and outside. Then with the compass along the keel, whatever is to be removed of the plank is drawn onto it, then the plank is removed, and with a line first what is to be removed from the inside is indicated, as the chiseled holes indicate, and so the plank turns where it is removed, and then the line is drawn where the

compass could not reach. Then the first side is hewn off to nothing, and every 7 inches a hole is drilled, one inch from the side, outside and at the inside to nothing. On the hewn side then moss is laid, which is laid on with pitch, then the plank is put in position, and fastened to the keel. The outside of the plank is then lined as desired, or hewn off and then another is put against it. After this with a scraper the uneven places are marked even, below and also above, and after these have been removed, hewn off, and put back on again, and it fits well, then the cleats are taken with which the planks are held together, put inside and out about three or four feet from one another, and so done, here and there on the keel and garboard strake also a cleat is hammered, coming above the keel, and closing on the keel as well as on the garboard strake, and so fastened it on the keel.

17. Make the Garboard Strake

The only reference in Witsen's book to the difference in the building methods of the Meuse and Amsterdam regions is the remark in passage 150 II 33. It is curious that he does not mention the difference in building method (frame-first versus shell-first), but only the fact that in Rotterdam the garboard strake was left out until the last minute just so the rubbish could be removed through the hole in the bottom!

Van Yk, on the other hand, is more explicit in his comment on the northern building method: ". . . where one is used to shape the bottom of the ship not with ribbands, as is done at the side of Meuse, but with the planks themselves which go around the ship" (Van Yk, p. 70). Whether the difference was unknown to Witsen or he simply did not consider it worth mentioning is not clear.

A line was stretched between stem and stern. This was the most important point of measurement during the rest of the building process. Every shipwright (and model builder) knows how difficult it is to take measurements in a ship, where no straight line is to be found. This also led to many problems in the interpretation of the measurements of the pinas. Understanding or establishing the points of measurement was often a problem, which was solved mainly by means of the model.

The shipbuilder would first lay down the garboard strake, the first plank at each side of the keel. Anyone who reads the description of the way in which the shipwright checked how well the garboard strake fitted into the rabbet will readily put aside all notions of Witsen as someone who did not understand his business. Though he is not always entirely comprehensible, Witsen must have had at least a more than average interest in the subject to come up with a description like that found above in passage 150 I 28, which shows an understanding of the difficulties to be solved. The method is, in principle, similar to the way flooring crews still work, laying the flooring along a wall and scribing the shape, using the wall as a reference. Witsen not only displays interest and understanding; he also deemed the practice important enough to pass on to posterity.

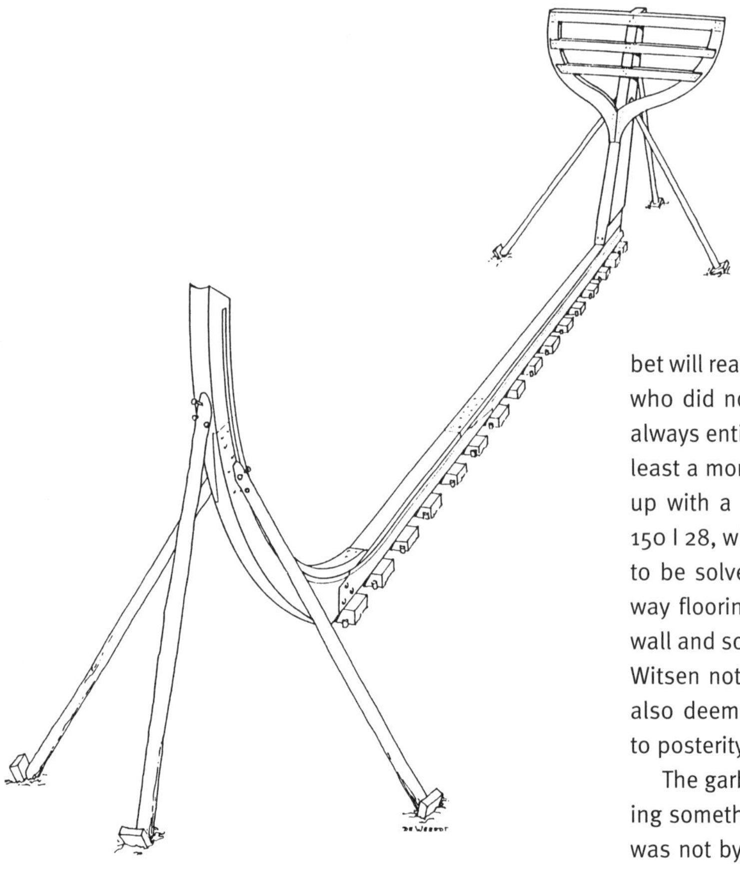

FIGURE 2.42. Stage 1 in the building of the pinas: stem and sternpost are erected on the keel, the rabbet is hewn, and the joints are furnished with stopwaters. (Courtesy G. A. de Weerdt)

The garboard strake was angled slightly upward, forming something like a gutter at either side of the keel. This was not by chance: this way, the bilge water could accumulate and be pumped out more easily. The outer side of the garboard strake had to come up to the height of the upper face of the keel.

Forward, the strake ended in the rabbet of the stem; aft, it turned, as described above, to the vertical and ended in the garboard stern rabbet, the part of the stern-post that was made thinner at both sides for this purpose.

All scarfs in the strakes were tapered, stopped scarfs, which spanned at least four frames between the stopped ends.

18. Make the Bottom, and make it level.

(67 I 29) *About the Bottom.*

9. 1. For the width of the bottom take ⅔ of the entire width, which is deck height. Example, 25 feet wide comes to 16 feet 7½ inches, for the width of the bottom.

2. The bottom should rise ½ inch to every foot on the main frame. The rest of the width of the ship is to be added at each side of the bottom equally: up to main deck height. When the ship is wider than the proportions given here, then the last bottom strake is given a little more rise than would otherwise be the case, going well into the turn of the bilge, and not be like a trough. [. . .]

The width of the bottom mentioned above; should be understood as on the place of the main frame, where the ship is its widest, which is at one third from the front.

The bottom furthermore, rises and descends, is made sharp or flat, to one's pleasure, and according to the use of the vessel.

. . .

(68 II 15) The bottom planks one fourth part of the thickness of the stem.

. . .

(68 I 43) The planks thick ¼ of the stem.

. . .

(68 II 10) More often than not the planks are taken a little less than one fourth of the stem.

. . .

(151 I 7) The planking of the turn of the bilge, and farther up in the sides of the Ship, that have to be bent, are tightened with Jackscrews, after they have been heated.

(55 II 14) Boejen [or *boeien*]: is putting on the planks underneath, outward up to the top of the bilges, where the side of the ship is called the boeizel [or *boeisel*].

. . .

(67 I 22) The planking should be fashioned in the way the futtocks bend outward.

. . .

(72 II 47) When attaching the planks to the stems, often they are tarred well around.

. . .

(150 II 17) *How the first planks are applied, is shown in the Plate at letter R* [fig. 2.45].

a Is the Keel block.
b The Keel.
c The Garboard strake.
d The plank next to the Garboard strake.
e The Planking tongs with which the Planking is held, forced and bent.
f The Chock.
g Wedges with which one makes the planking tongs tight.
h The Hook and chain.
i The Lever.
k The Lever rope.
l The planking cleat.
m The cleat on the keel and the Garboard strake.

. . .

(151 I 12) When the bottom is planked, it is hewn off to the well-lined proportion, it is made equal height on each side using a level with a plumb bob [no. 18 in fig. 1.20] and then small shores are placed underneath, for it to remain unalterable. Then with the chisel all important places are marked, to be found again when needed. As can be seen for the strut *d* on the plate at *V* [fig. 2.46].

. . .

(151 I 1)
k The Garboard strake.
c The Keel.
i The chocks between which the Keel lies.
h The Keel block, underneath the keel.
a The floor timbers.

(74 I 29) The planks thick 4 inches

Figure 2.43. Siewert van der Meulen's late seventeenth-century engraving of the bottom of a ship on the stocks. The planking tongs are clearly visible.

(265 II 53) It is certain that ships, of which the bottom rises the least forward and aft, are the best carriers, and those of which the floor rises most, go the fastest and sharpest through the water, where in general also the best sailors are the ships which have the most ballast in the middle [. . .].

It is the eye of a good shipwright which gives the ship a proper shape. The bottom of the ship proposed above [i.e, the 134-foot pinas] rises less in fast-sailing frigates, and more than plump carriers: it is true though, that a more rising and rounder bottom is more pleasing to the eye, but the use prevails before beauty here, and an average rise was chosen, not the highest nor the lowest, because this ship is meant for two purposes, as stated above.

. . .

(67 I 50) The bottoms of merchantmen are made wider than for frigates.

. . .

(265 II 39) The bottom [of the pinas] rises over half a foot, which I think is enough for ships which are not built for speed, as long as the entry and stern are of a decent shape, going in and out as wanted, which shape can in my view not be predetermined to perfection, nor expressed in words.

(267 I 38) Ships which are too sharp in the bow, will sink too deep afore. Too snub a bow does not cut through the water, and slows the progress.

(74 I 26) 7. *About the Bottom.*

1. The bottom wide 21 feet.
2. Rising 6 inches.

(67 II 6) In this proposed ship the bottom is wide 21 feet, at the place of the main frame, where the first floor timber is laid: 12 feet forward it will be one inch narrower, or 1½ inches: at the main frame it rises 5 inches, and forward somewhat more: 11 feet forward it rises 4 inches, where it is 2 inches narrower. 12 feet and 5 inches aft of the place of the main frame it is two inches narrower, yet rises more: another 11 feet and 8 inches farther aft, 2 to 2½ inches narrower: again 11 feet 10 inches farther aft it narrows with 3 to 4 inches, yet rises more, after which it narrows 8 to 10 inches again and rises, and thus ever narrower and deeper.

18. Make the Bottom, and Make It Level

The shipbuilder continued with the bottom. Plank after plank was fitted against the preceding one and then squeezed tightly against one another with the hook and chain (no. 14 in fig. 1.20) and nailed together temporarily

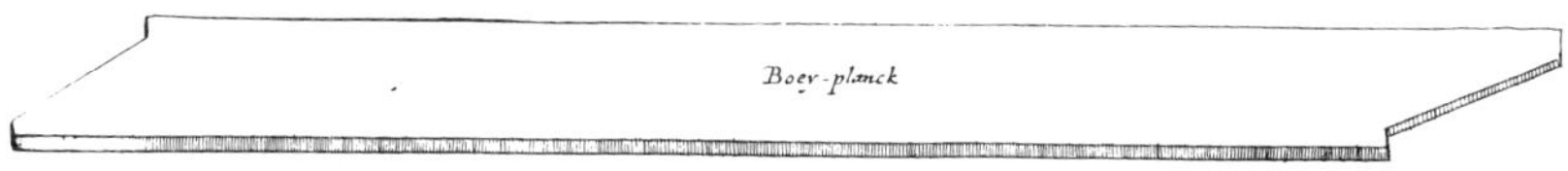

Figure 2.44. (*top*) Plate XXVIII. Outside plank.

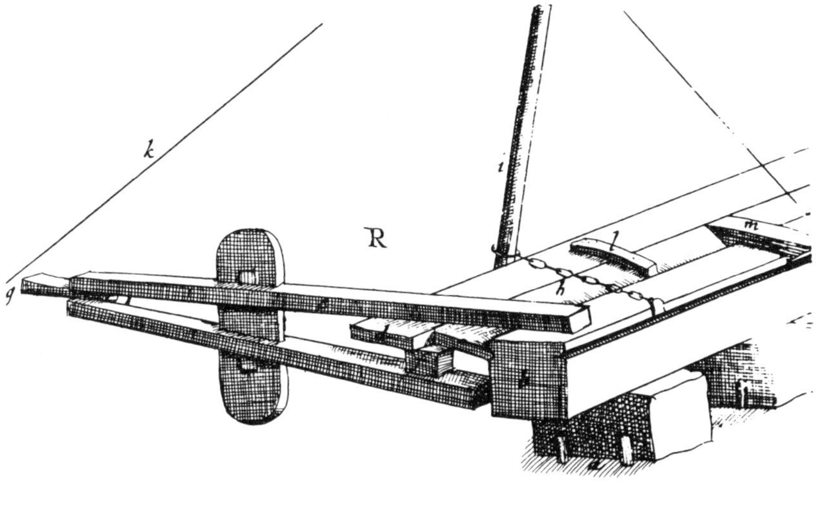

Figure 2.45. (*center*) Plate L (drawing R)

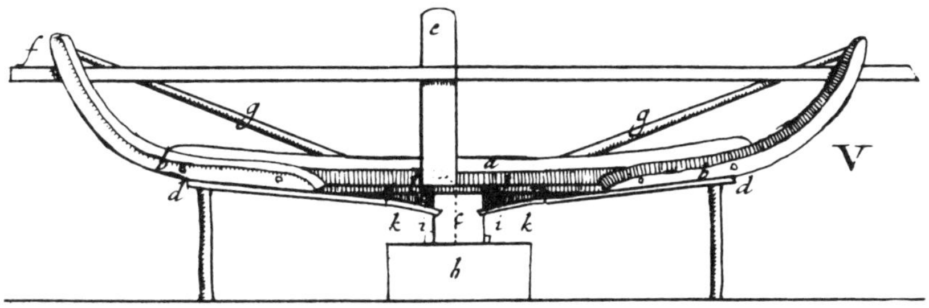

Figure 2.46. (*bottom*) Plate LII (drawing V)

with cleats. The cleats were placed approximately a meter apart, but they must have been set closer together in the bow and stern, where they had to check a lot more strain due to the curvature.

Anyone who tries to squeeze planks together with their thin sides against each other will notice that the material tries to escape the pressure sideways—the planks tip up. This was countered with the planking tongs (no. 9 in fig. 1.20), which, much like large pincers, held the planks rigid and kept them from buckling at the edges. As a result, the planks, in section, lay in a straight line; this is confirmed by all the sections drawn by Witsen. Yet this feature cannot be observed on any of the remaining contemporary models. It is even more difficult to determine from wrecks, and on paintings and engravings this part of the ship is submerged.

Toward the bow the bottom planks were led to the stem in an increasingly wider curve and fitted in the stem rabbet; this is the case at least for the pinas and probably also for other relatively sharp-cut vessels. For fluyts and other carriers it was unusual to have all the strakes run right through to the stem; the first or second bilge strake would then close off the bow by going round the ends of the bottom planking and reaching the stem.

Toward the stern the bottom strakes followed the garboard strake, filling the garboard stern rabbet and ending higher in the sternpost rabbet.

Planks that had to bend were heated over fires: they were laid over a trestle, weighted down at the right places, and heated from below for long hours while keeping the upper side wet. The process demanded a lot of experience from the shipbuilder to give the planks their required bend without charring the inside too much. He had to mind the knots of the wood, too, which could influence the bend unexpectedly.

The ship's bottom was supported from underneath by shores (little poles), which kept the shape from sagging. Meanwhile the symmetry had to be observed meticulously, since mistakes at this stage tended to multiply in the course of the building process.

The width of the bottom, measured at the main frame, was two thirds of the width of the ship—a little more for carriers, a little less for yachts. The bottom could rise toward the sides (deadrise), from relatively a lot for fast ships down to almost horizontal for carriers. Together with the width of the bottom, setting the deadrise was another of the typical design moments in which the properties of the vessel were determined. The form of the bottom planking, the curvature toward the bow and the stern, had important consequences for sailing performance, but in an ordinary contract this could not be specified and was left to the eye of the master. It is unique that these features for the pinas are given in Witsen's book, which provides us with the exceptional opportunity to study the shape of a seventeenth-century ship apart from distorted wrecks or contemporary models, which are unreliable in this respect.

In addition to the measurements of the bottom, as given in the pinas text, the reader can find data important for the shape of the ship in sections 21 (And make it quite even in height, when it is shored), 22 (Then make the Frame, the Futtocks), 24 (Then make the Master Ribband around, and make it level), 38 (With the Top Timbers), and 39 (Make the Ribbands around, Shores and spalls). Together they supply the almost complete measurements of eight frames. The first frame is the main frame, located at one third of the total length from the front, with frame 2 at 12 feet forward and frame 3 another 11 feet 4 inches farther forward. Frame 4 is 12 feet 5 inches aft of the main frame, frame 5 another 11 feet 8 inches farther back, frame 6 at 11 feet 10 inches from frame 5, frame 7 at 18 feet 5½ inches, and the last, frame 8, at 14 feet 6½ inches from their respective preceding ones. Why the distances between these frames were so irregular is not clear.

Witsen supplies us with four to seven different kinds of measurements for each frame at various measuring points on the hull; with these data the frames can be reconstructed without difficulty (see table 3 in the appendix and fig. 2.47). I used these measurements in a set of drawings to check whether they are actually fair. In figure 2.48 the length is shortened by a factor of 500, while the height is reduced by 100. With the contracted hull shape, I could instantly verify whether the lines were fair. My first attempts to join the measurements as an acceptable drawing failed miserably, as there appeared to be no way of constructing a fair line between the given coordinates. Only when the sagging of the keel was incorporated in the calculations for the drawing (sagging plus or minus 15 centimeters in the middle) did the coordinates allow for the lines to be connected in a flowing curve. This is additional proof that Witsen did not give just any measurements but derived them from a ship that was actually built.

It is interesting to notice at the same time that the Dutch shipbuilder did not use any circles in the construction of the longitudinal shape but curves that were nongeometrically obtained (i.e., without geometric tools) and that were pleasing to the eye. His English colleagues, on the other hand, started with arcs of circles, as can be verified in many cases.[9] As noted above, a compass was used in constructing the stem and fashion pieces; but even then it was not used as the ultimate shaping instrument—the shipbuilder would always deviate from the arc of circle.

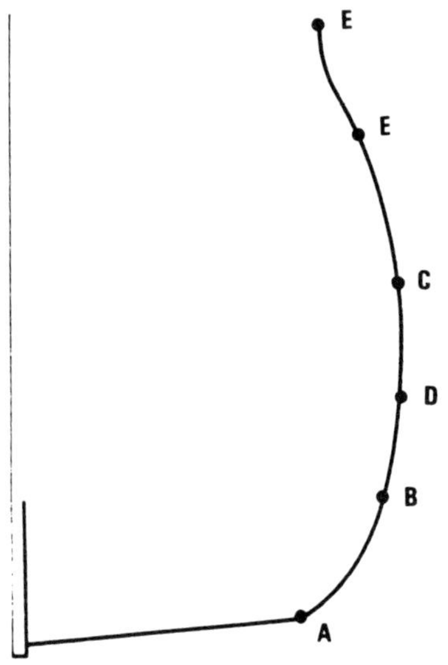

FIGURE 2.47. Measuring points on the hull: *A,* the start of the turn of the bilge or chine; *B,* the top of the bilge planking; *C,* the watershed, or height of the master ribband or height of breadth; *D,* the curve of the futtock (a midpoint between *B* and *C*); *E,* the tumble home of the top timbers. (Drawing by A. J. Hoving)

Inevitably there were some minor mistakes and imperfections in Witsen's measurements for the frame shapes. For frame 5 the height of the top of the bilge planking was missing; for frame 6 the height of the master or breadth ribband was missing. The height of the breadth ribband of frame 4 was wrong (9 feet 6 inches was given by Witsen instead of the correct 8 feet 6 inches), and some measurements of the top of the bilge planking and top timbers were missing, but none of these mistakes and omissions stood in the way of an acceptable reconstruction of the hull.

With the use of the figures obtained, the frames or sections could be drawn. This resulted in a rather confusing fact: the aft section lines crossed the midship section lines in the topsides! In this the pinas deviates from all known drafts taken from models, reconstructions, or wrecks from the seventeenth-century. Further research uncovered, however, that Chapman drew the same crossing of lines in a drawing of an eighteenth-century fly-boat (fig. 2.50), and the simple draft of a "fluyt or carrier" (*Fleit eller Lastdragare*) supplied by the Swede Åke Rålamb (fig. 2.51) also shows the same phenomenon. We could speculate on the reason, but it is probable that this crossing was a result of the building method, in which, as we will see later, top timbers of an almost identical shape and curve were placed on the shell made of the bottom and turn of the bilge, which already rose forward and aft. The widest point of the frames would rise accordingly forward and aft, which in the body plan causes the lines to cross.

With these data the shape of the hull can be projected, albeit this projection will still have the lines of the bottom, turn of the bilge, height of breadth, and the sheer rail plotted as if they were straight waterlines or bow-buttock lines, although they curve in both planes. From this drawing to a modern lines plan with straight sections in all three planes is only a small step. The reader can compare the body plan to the lines plan (see drawing 1 in appendix) for the pinas.

19. Also a floor timber at the main frame [hals], *with a Futtock on either side.*

(151 | 22) When the Bottom is made fair, then at the main frame the bilge frame is made from a floor timber and a Futtock, through each end of the floor two bolts are passed, which go through the Futtock, to hold it fast and immobile, as can be seen at *b* on the same plate [i.e., fig. 2.46].

Then a plank as shown at e is attached to the Floor timber, standing with one side on the middle of the keel and of which one side is vertical, from which to measure the even width of the futtocks on the inside of the planking.

Then a thin batten is attached to the ends of the Futtocks, straight across, as shown by letter *f.*

After that supports are placed as shown in *g,* so that the futtocks will not move when the planking is applied.

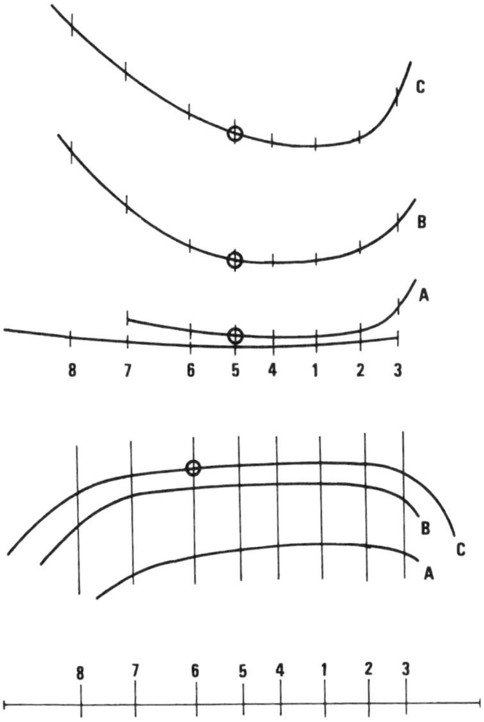

Figure 2.48. (*top*) Frame shapes based on Witsen's measurements. *Above:* All measurements of ship length have been divided by 500, the others by 100. The keel is slightly curved downward in the middle. Numbers 1–8 are the initial frame locations. Mistakes in Witsen's text are marked with a small circle. *A,* the start of the turn of the bilge, seen from the side; *B,* the top of the bilge planking; *C,* the master ribband or height of breadth. *Below:* Same lines, seen from above. (Drawings by A. J. Hoving)

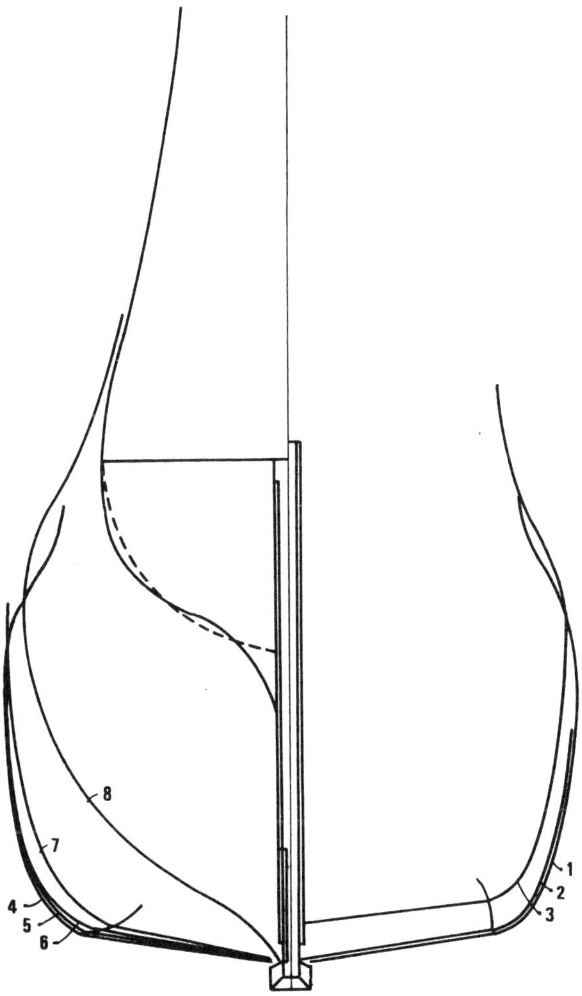

Figure 2.49. (*bottom*) Reconstructed frames of the pinas. The intersection of frame lines due to the steeply rising height of breadth in the stern is clearly visible. (Drawing by A. J. Hoving)

(55 I 54) Floors [*buikstukken*] lie straight across the bottom.

. . .

(55 I 43) Bilge futtocks [*zitters*] lie in the turn of the bilge next to the floors.

. . .

(60 II 52) The bilge futtock is a crook, which is placed in the turn of the bilge. The turn of the bilge is the lowest or last strake of the bottom, which is, on the outside of the ship.

. . .

(54 II 52) Bilge futtocks are beams, which lie on the floor of the ship.

. . .

(54 II 56) Two bilge futtocks, a floor, and two futtocks, are called a frame.

. . .

(67 II 32) 11. *Of the thickness of the frames.*

1. These are on the keel ¾, at the height of breadth half, as the top timbers on the master ribband ⅖ of the thickness of the inside of the stem.

2. Forward and aft more light, above and below heavier.

. . .

(68 II 3) The floors 2 inches thinner than the stem is thick inside, or on the keel ¾ of the stem, at the master ribband half, at the top ribbands ⅖.

. . .

(68 I 44) The floors three-fourth parts of the stem. The futtocks against the planking half the stem.

(74 I 44) 1. The floor timbers, on the keel, thick 9½ inches, as the transoms. In the bilge, thick 9 inches, in the turn of the bilge thick 10 inches.

2. The bilge futtocks at the top of the planking of the bilge, thick 7¼ inches.

(74 II 8) The floor timbers are placed 9½ inches apart, and are as broad, as are the top timbers.

19. Also a Floor Timber at the Location of the Main Frame, with a Futtock on Either Side

A floor was then placed in the bilge at the main frame. This was the horizontal part of the frame. A crooked timber, a bilge futtock, was attached at either end, which had the shape of the required turn or bend. The shape of the bilge futtocks was thus another part of the design of the ship. These bilge futtocks were attached to the floor timber with bolts, unlike all other frame timbers, which were attached only to the planking and the ceiling and not to one another. The whole assemblage of floor and futtocks was nailed to the hull planking to prevent shifting.

20. Then plank the turn of the bilge.

(60 II 5) because the ship is wide 27 feet on the top of the bilges, and 29 feet wide in all, overhanging one foot on each side. Or 27 feet is taken for the width, and every side hangs out 1 foot, comes to 29 feet.

20. Then Plank the Turn of the Bilge

As the building of the shell continued, the use of the planking tongs became somewhat more complicated at this point, for the roundness of the bilge had to be created with the same straight tongs as were used for the bottom, so little chocks of wood, called *oorhouten* (clamps), were utilized (see no. 12 in fig. 1.20).

Because the first bilge strake was at an angle with the bottom planking, the transition from bottom to turn was clearly visible. This was most obvious on very flat-bottomed ships; on sharply rising floors it was hardly noticeable.

When a ship had a very wide and flat bottom, the outermost bottom strake was often canted a little so the ship would not be like a trough (67 I 29). Angular constructions underneath the waterline were not favored because they were vulnerable and could result in leakage. Van Yk comments:

> But in the Noorderkwartier [the northern part] of Holland . . . I have seen that some Masters still build Ships which, in the rise or the edge of the Bilge have a Plank, which goes stoutly over the other Planks, and thus gives a bilge Seam. Such that the Bottom is clearly distinguished from the Turn of the Bilge.
>
> But this Method is not practiced at the Meuse [Admiralty of Rotterdam], where the Bilges and Floor curve together, which is thought the best way. (Van Yk, p. 70)

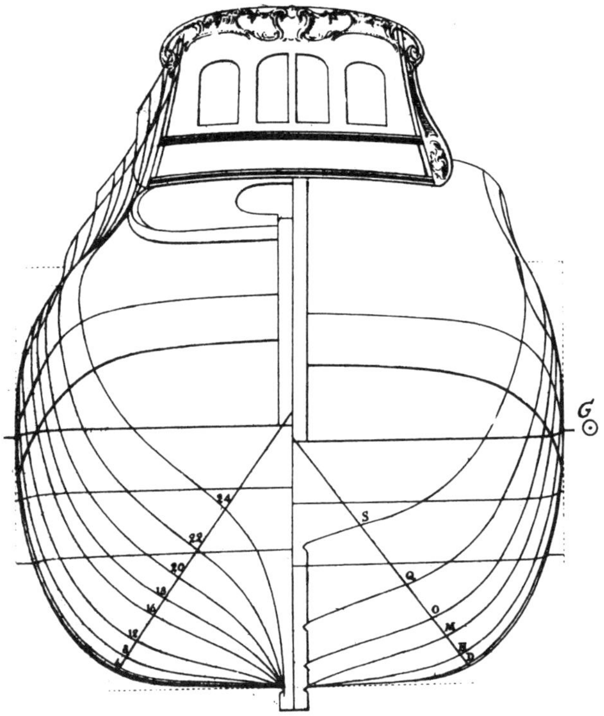

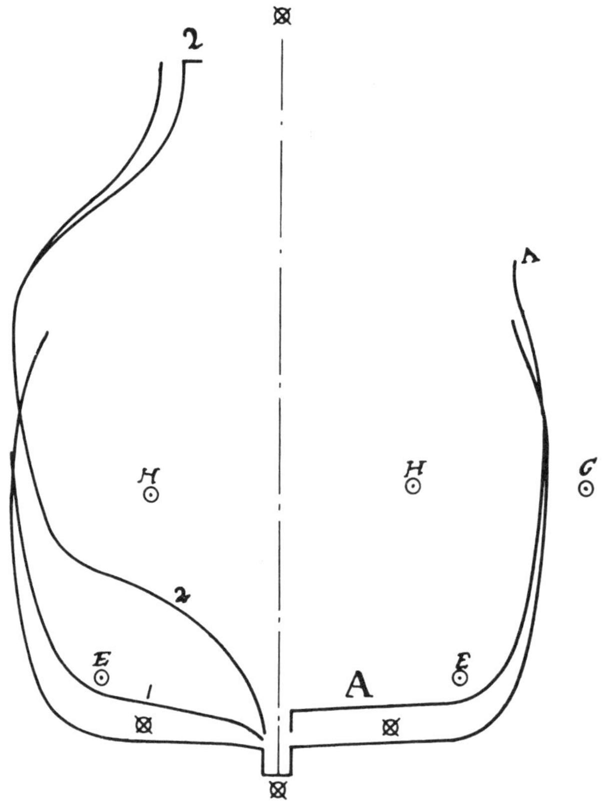

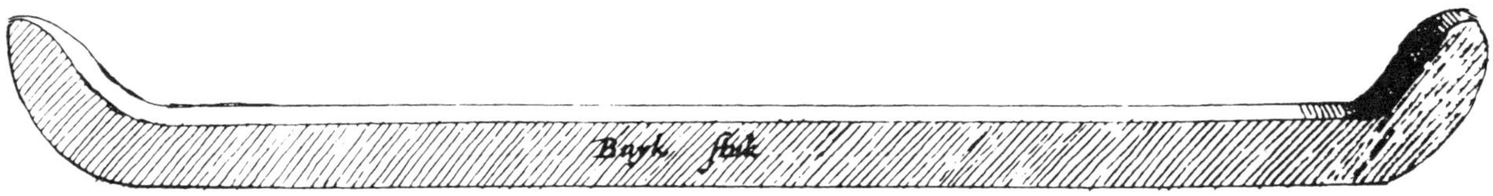

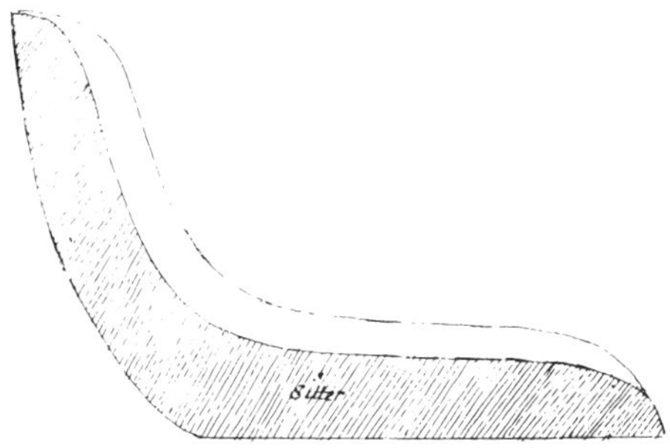

FIGURE 2.50. *(top left)* The same phenomenon of crossed lines can be seen in Chapman's illustration of an eighteenth-century Dutch "fly-boat." From Chapman, *Architectura Navalis Mercatoria (1777),* plate LII. (Drawing by A. J. Hoving)

FIGURE 2.51. *(top right)* Drawing of frames of a fluyt. From Rålamb, *Skep byggerij,* plate G.

FIGURE 2.53. *(right)* Plate XXVII. Bilge futtock.

FIGURE 2.52. *(center)* Plate XXVII. Floor timber.

This marks the end of the shell-first phase of construction. As we shall see, the rest of the ship was built according to the frame-first method.

21. *And make it quite even in height, when it is shored.*

(67 I 25) On the plate No. 20 [fig. 2.56] a is at the watershed level, and at b are the bilges or the top of the planking of the bilges, through which the futtock projects outward, or hangs over the Nail, c is the bilge strake, and the outside of the bottom, with the beginning of the turn of the bilge, d is the keel.

. . .

(60 II 53) The chine is the lowest or last strake of the bottom, meaning at the outside of the ship.

. . .

(67 II 22) *About the Kimmen*
[planking of the bilges].

10. About the depth and width of the planking of the bilges.

1. The depth is one third of the total depth at the place of the main frame. Example depth 10 feet, comes to 3 feet 3⅓ inches.

2. About the width of the turn of the bilge, for 10 feet of the ship's length make the bilges 1 inch narrower than the total width, such that the futtocks hang out 10 inches over the top of the bilge planking.

. . .

(61 I 1) The bilge strake is a plank made at the outer edge of the bottom, three or four bilge strakes are usually applied to the ship upon one another; depending on the depth of the ship.

. . .

(266 I 29) The *Bilge Strakes* are planks, which stand between the bottom and the rising sides of the ship, are roundish and stand transversely in relation to the waters surface, and such, because when the ship heels over considerably, or touches ground, it will sit flat and will be less harmed: the roundness is also useful because a chined bilge is more liable to leak and is more vulnerable, when it hits the ground. The bottom ceiling, which comes against it, is much heavier, as the ship carries more load here, and better be prepared for bumping.

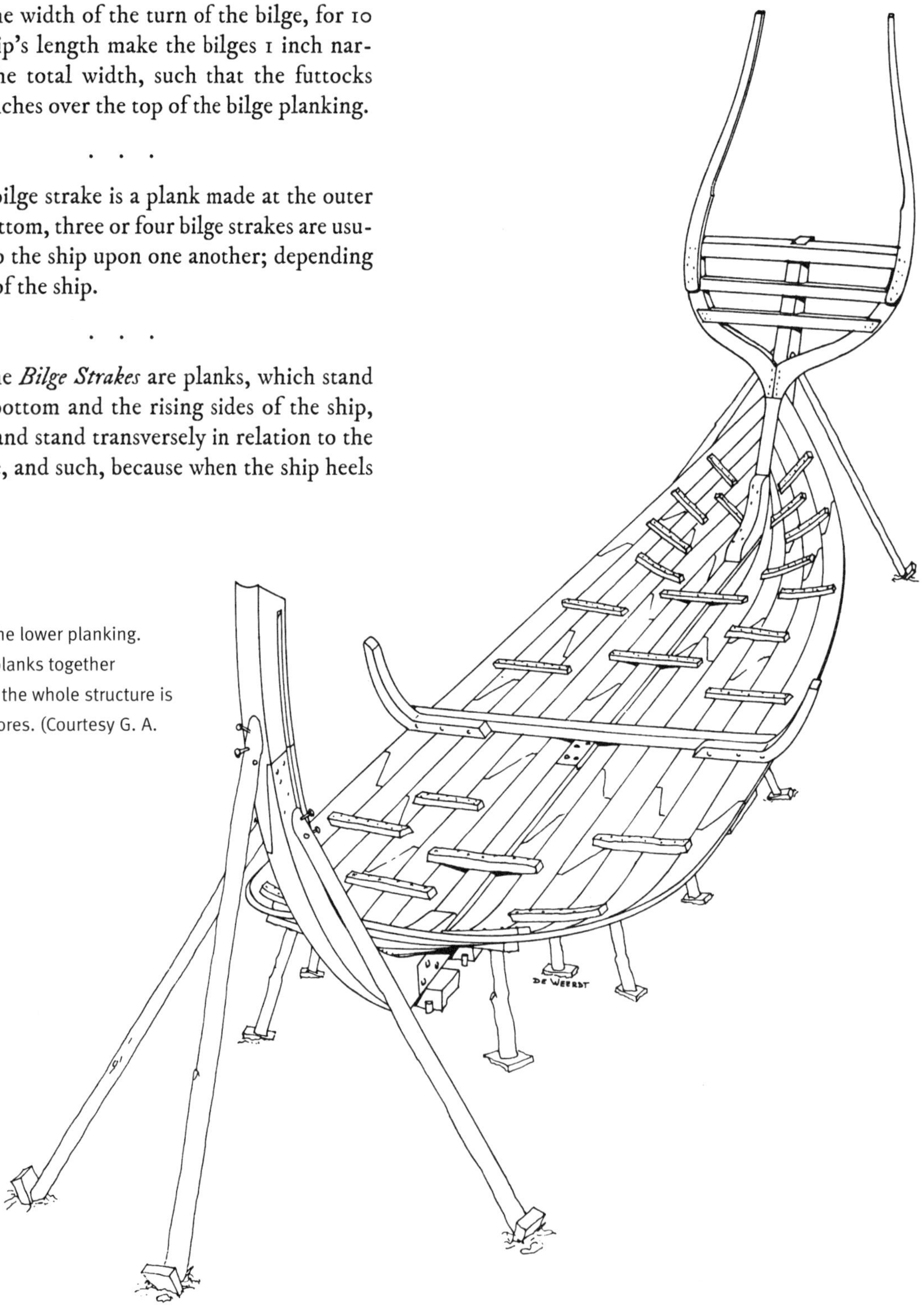

FIGURE 2.54. The lower planking. Cleats hold the planks together temporarily, and the whole structure is supported by shores. (Courtesy G. A. de Weerdt)

(267 I 41) Square and angular ships are weak, and when they pound, they easily crack.

(150 II 47) When the turn of the bilge is planked, lay the *Tingel against the Keel;* usually it is laid somewhat farther from the keel, than the keel rises above the garboard strake, and this is called the Limber hole, or Watercourse, as can be seen at *l* on the plate at *V* [fig. 2.46].

. . .

(55 II 1) *Tingel* [. . .] is a length of wood, between the keel and the first plank at the keel, to level the bottom; here a hole is left open, through which the water runs.

. . .

(269 I 21) The *Tingels,* near the keel, are strips of wood, with which the watercourse on the floor is determined, serving also to obtain a level bottom.

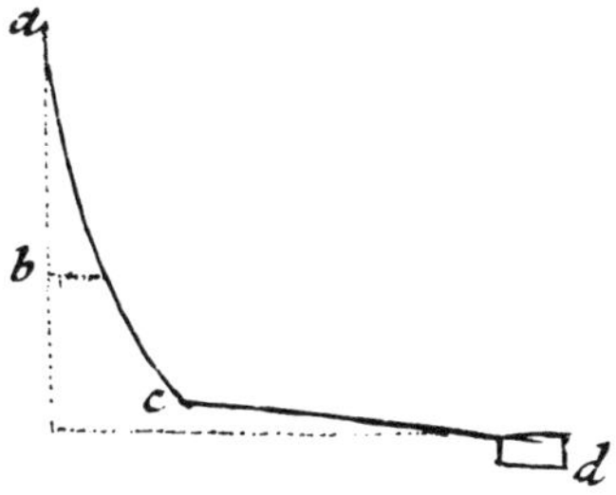

FIGURE 2.56. Plate XLI

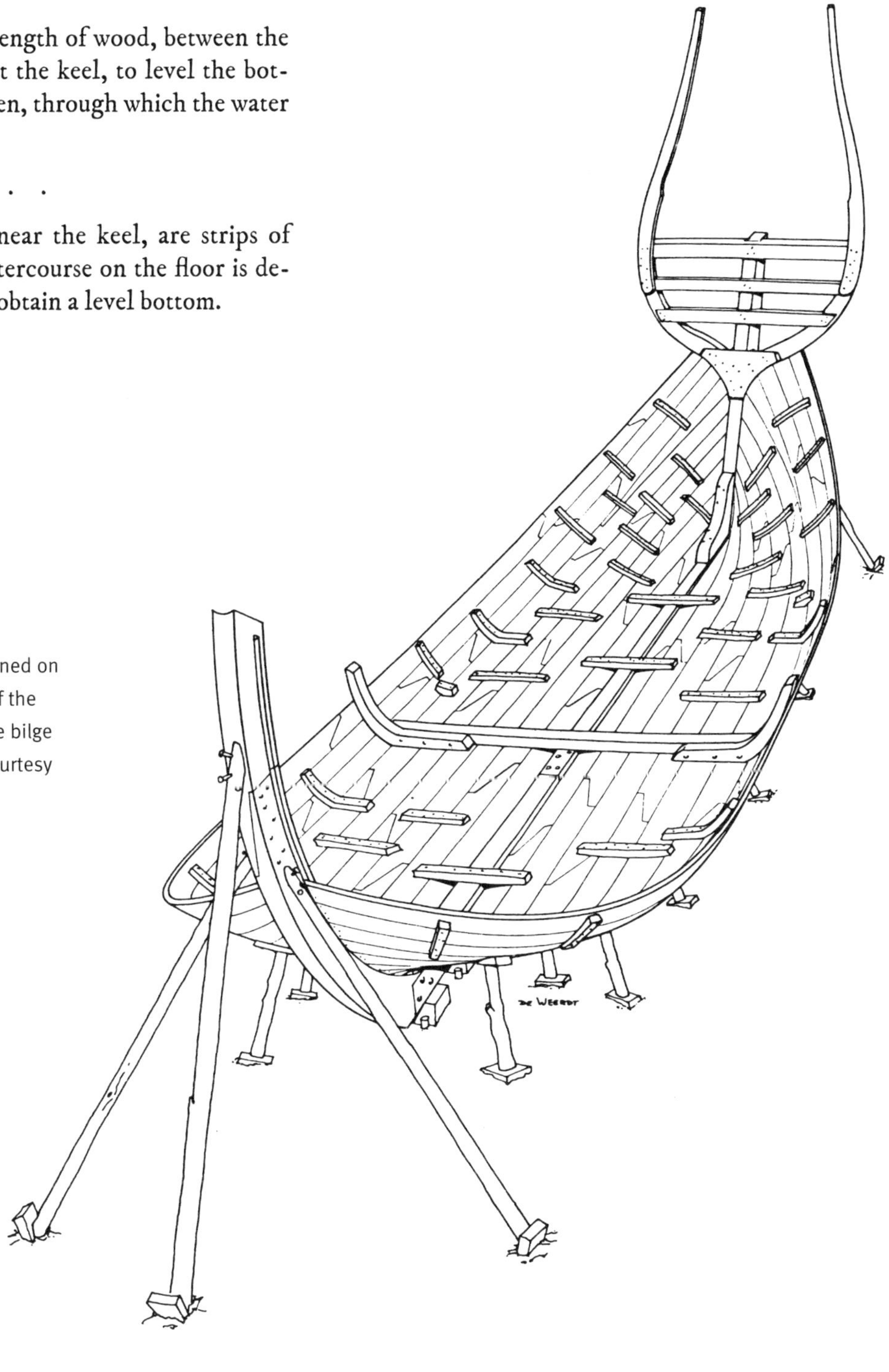

FIGURE 2.55. The hull is widened on both sides with the planking of the turn of the bilges, for which the bilge futtocks dictate the shape. (Courtesy G. A. de Weerdt)

FIGURE 2.57. Plate XXVIII. The *tingel* shapes the limber hole. (For some unknown reason, Witsen draws his *tingel* with a square instead of a triangular cross section.)

(74 I 31) 1. The turn of the bilge planked to a depth of 4 feet 5 inches.

(93 II 29) The hull planks are thick 4 inches.

(81 II 10) *About the planking of the Turn of the Bilge.*

At the place of the main frame these are planked, to a depth of 4 feet 5½ inches.

12 Feet forward planked to a depth of 4 feet 8 inches.

11 Feet forward planked to a depth of 4 feet 4 inches.

12 Feet and 5 inches aft, it is planked to a depth of 4 feet 5½ inches.

11 Feet 10 inches aft it is planked to 5 feet 1 inch.

18 Feet 5½ inches aft, there the depth is planked to 7 feet ½ inch.

14 Feet 6½ inches aft, the depth is planked to 9 feet 9 inches.

(81 II 42) *About the width of the planking in special places, measured from the middle line. How wide the turn of the bilge is planked.*

At the main frame it is wide 13 feet and 6 inches.

12 Feet forward there it is wide, 13 feet 5 inches.

11 Feet 4 inches forward, 12 feet 7 inches.

12 Feet and 5 inches aft, behind the frame, 13 feet 6 inches.

11 Feet and 8 inches aft, 13 feet 5½ inches.

11 Feet and 10 inches aft, 13 feet 5½ inches.

18 Feet 5 inches aft, 13 feet 1½ inches.

14 Feet 6½ inches aft, 11 feet 6½ inches.

This width should be maintained at the height of the turn of the bilge mentioned before.

(74 I 34) 3. The tingel at the keel, thick 2 inches: the other side is sharp at the side of the keel, broad 6 inches, lying from the keel 4½ inches.

21. And Make It Quite Even in Height, When It Is Shored

As with the bottom planking, the bilge strakes were supported by shores to prevent the whole from sagging. And here, too, symmetry was of paramount importance. To achieve symmetry, the line that was stretched between stem and stern was used as well as the large level (no. 18 in fig. 1.20).

When fashioning a ship's shell this way, it becomes apparent that obtaining symmetry, notably in the bow and stern, is very difficult. This was a general problem in shipbuilding, so it is not surprising to read that many a ship, once launched, had to be leveled with ballast while afloat. This was even true for the *Victory,* Nelson's proud ship that now lies at the Portsmouth dock, despite having been built according to methods much more advanced (using drawings) than the Dutch method described here.

Symmetry was not the only difficulty. It was very important that the shape of the bow or stern should not angle in too much or flare out, for this would result in grave difficulties later when the planks of the "skin" between the turn of the bilge and the wale (planking called *huiddicht,* lit., "skin-closed") were to be joined with flowing lines. This required a lot of experience and "the eye of the master" for a successful result.

22. *Then make the Frame, the Futtocks.*

(60 I 52) The futtock: this makes the width, and depth of the ship, for instance, when setting up a mold of the futtock, then put a nail at the depth of the ship and suspend the plumb from there, and measure the overhang at the turn of the bilge.

. . .

(54 II 38) Futtocks, which are placed here and there in the ships sides, at some intervals, to fair the framing, or obtain the ships body: are also called Frame timbers, because they frame the ship, to which the other timbers are added.

. . .

(82 II 28) Forward and aft against the posts, at the sides, futtocks are always placed: which are to flank the posts and keep them secure in place.

. . .

(55 I 45) And futtocks on the floor timbers. Futtocks tight on the floor timbers make strong ships.

Figure 2.58. Plate XXVII. Futtock.

(68 | 48) The futtocks are thick 2 fifths parts of the stem at the height of breadth.

. . .

(71 || 20) No fixed measures can be given of the bends and curvatures in the ribs of the Ship, such as the timber, futtocks, &c., because they change according to the use of the ship: they are to be made by eye and should decrease slowly.

(74 | 50) The futtocks, at the height of breadth ribband, thick 6 inches, stand wide 29 feet, on its curve from the other across, on the sheer they are 5½ inches thick, are broad 9 inches and stand apart as much: those standing aft or forward are lighter, thinner and narrower.

(82 | 44) *About the curve that the Futtocks have: between the Height of Breadth and the turn of the bilge Planking:*

The Futtock, which stands at the place of the main frame, has a curve of 3 inches.

The next one forward, 3½ inches.

The foremost 4½ inches.

The first behind the main frame, 3 in.

The next one after, 4 inches.

The third one aft, 4 inches.

The one that follows has a curve of 6 inches.

The one following that one, has a curve of 7 inches.

The ribband that comes above the height of breadth, is made according to how much one wants the top timbers to tumble home.

(82 || 10) *On the overhang of the Futtocks, over the bilge strakes at the last height, and above the Height of Breadth.*

The futtock at the place of the main frame hangs 1 foot over the boeizel [upper bilge strake].

And 2 feet [should be "12 feet"] forward, 1 foot 1 inch.

11 Feet 4 inches forward, 1 foot 8 inches.

12 feet 5 inches aft of the main frame, it hangs 1 foot ½ inch over the bilge strakes.

11 Feet 8 inches, 1 foot ½ inch.

11 Feet 10 inches aft, it hangs over 1 foot½ inch.

18 Feet 5½ inches, 1 foot 1¼ inches.

14 Feet 6½ inches aft, 1 foot 9 inches.

22. Then Make the Frame, the Futtocks

Futtocks were then raised at several places, interspersed over the length of the ship. These were the lower rising parts of the frames. For the largest part of the ship they all had the same curvature. At most, the builder would slightly plane his mold, the plank he used to mark off the frame parts, working from the main frame forward and then aft.

Witsen's detailed description for two illustrations of the hull may serve by way of explanation (see figs. 2.59 and 2.60).

23. And fasten Putlogs thereto, on which the Scaffolding comes.

(151 || 20) When the bottom and the turn of the bilge have been planked, the bilge futtocks are placed on the level marks, and a plank is laid on top of the planking of the bilge, and outside planks across on which one can stand; as is shown at *r* on the plate at *X* [fig. 2.61].

Then a chock is nailed to the outside of the futtocks, with which it stands on the bilge-planking, as in *e*. Then a line is drawn from the stem to the sternpost, on the center line, and pulled down with another line, attached, on the center line, as in *b* and *g*, and from there the transverse symmetry of the futtocks is checked as in *f*: then a line with a plumb is lowered from a nail for the whole depth of the hold, as in *f* to *h*, letting it hang outside the bilge planking, in order to make the facing futtocks equal in overhang, as shown from *d* to *e*. Then to each Frame timber a Putlog is attached, as in *p*, besides 2 Supports underneath it, as at *n* and *q*, to support the scaffolding that will have to be raised when building up the ship on the outside, besides a Strut as at *o*.

m Is a Bilge shore on the plate at *X*.
l A Bottom shore.
k The Keel [a keel block, in fact].
t and *v* Chocks in which the keel lies [t is missing in the drawing].
a The Keel.

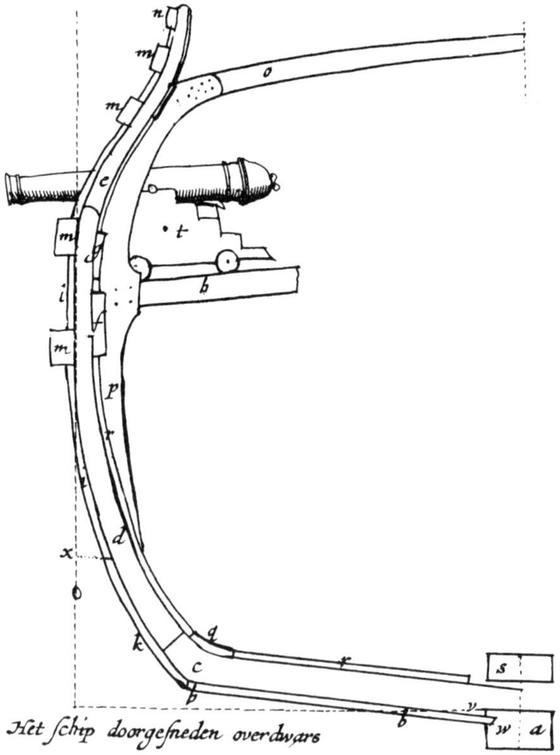

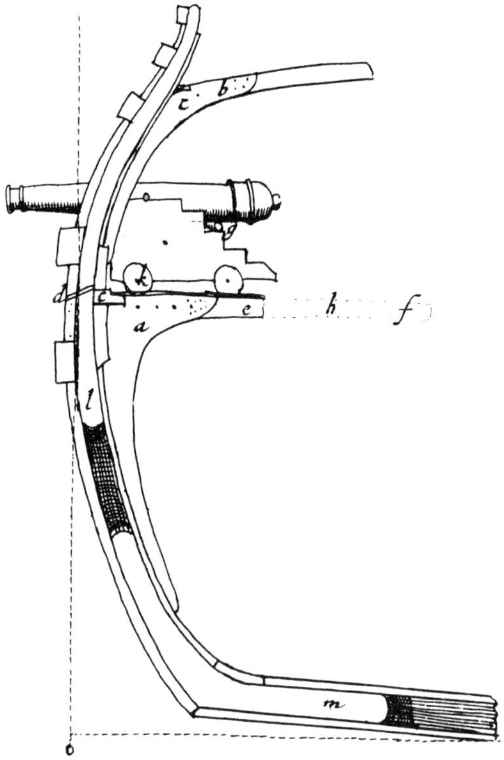

FIGURE 2.59. (*left*) Plate XLIII. The ship transected.
(60 I 48) In ***the Ship, transected,*** the first plate shows the keel at A. B the bottom of the ship, beginning at B and ending at B. C is a floor timber, lying in the bilge, across the keel, from C to C. D is the futtock [. . .] E is a top timber or top futtock. F a deck clamp: herein the deck beams lie as in a foundation, driven with dovetails. G is the lower deck spirketing. H the lower deck beam. I is the planking; between the turn of the bilges and the lower wale. K is the turn of the bilges. L strakes between wales [mistakenly marked *i*]. M the wales outside against the Ship. N is the sheer rail, the upper edge of the ship, not unlike the wales. O is the upper deck beam. P the futtock rider. Q is the bilge stringer. R the ceiling of the bilge, and inside the plank shell. S is the keelson. T a gun carriage. V is the tingel, lying across the watercourse or limber hole. W the rabbet in the keel, or the score, in which the garboard strake comes. The Point X projects from the planking of the turn of the bilges one foot; which is at the top of the turn of the bilge, or the end of the turn of the bilge; and such, where these two lines cross each other; and this makes the width of the entire ship, at the height of the scuppers.

FIGURE 2.60. (*right*) Plate XLIII
(60 II 33) On ***the second plate of the transected ship*** A is a knee in the hold. B a knee to the upper deck beam. C is the waterway. D is a scupper. E is the binding strake. F planks on the main deck. G the stool block underneath the gun. H the lower deck beam, or beam in the hold. K a truck of the gun carriage. L is the top timber. M a bilge futtock. [. . .] The letters E, H, F, in this second plate of ***the Ship in cross section,*** stand outside the Figure, to which I was forced by the size of the plate; it is enough to know that they belong to the main deck.

---

i	The Garboard strake.
f	The Limber hole.
b	The Floor timber.
c	The Futtock.

When the width of the Futtocks at both sides has been measured, then a line is attached to the sheer line at *g*, and pulled down to the center line on the keel and attached there, and then measure the width at the futtocks and the Sheer line, at *g*.

23. And Fasten Putlogs Thereto, on Which the Scaffolding Comes

The fact that Witsen goes to greats lengths to explain how the scaffolding for the workers was built is further evidence of his depth of knowledge.

24. *Then make the Master Ribband around, and make it level.*

(152 | 14) When the Frame timbers have been set, then the Master Ribband is fixed around, at the height of the scuppers, and furthermore the other floor timbers, bilge futtocks and futtocks are made, place the futtocks, after which the bend strakes are faired, and then the stringers and ceiling are fitted in the turn of the bilge. As can be seen at *b* in figure *Z* [fig. 2.62]. *m* is the Bilge ceiling. When the bilges have the ceiling fitted, then take a plank and push it, with its side resting on the bilge strakes, between the futtocks as in *g,* with two trusses underneath as in *h,* lest the planks should break, and lay planks over the entire Ship at 6 or 7 feet apart as in *g, h,* to form a scaffolding, from which the Deck clamp is raised, and then the deck beams are laid.

. . .

(53 || 43) Northern timber carriers load the most, when they incline toward the square. Corn ships, and those aiming at piece goods, are roundish; and rise a lot. East and North traders, who load coarse goods, are generally larger than ships for piece goods, wines and such: similar also to the Salt ships. Masts are best stowed in holds, that are wide below, and angular.

. . .

(263 | 3) A ship built solely for cargo carrying, is best off with a wide hold, and narrow above, to be able to stow many goods, and be sailed with a small crew.

(81 || 25) *How to place the master ribband: and how high it is to be measured from the bilge strakes.*

At the place of the main frame the master ribband stands 8 feet and 3½ inches above the bilge strakes.

12 Feet forward above the bilge strakes, 8 feet.

11 Feet 4 inches forward, 7 feet 8 inches.

12 Feet and 5 inches aft of the main frame, 9 feet 6 inches.

11 Feet 8 inches aft, the master ribband stands 9 feet above the bilge strakes.

11 Feet 10 inches aft, 9 feet 4 inches.

18 Feet 5½ inches, 9 feet 5½ inches.

14 Feet 6½ inches, 8 feet 9 inches.

(82 | 13) *Width of the Futtocks measured on the Master Ribband.*

The first is measured wide from the center line, at the outside of the frame timber, as follows: The first main frame, 14 feet and 7 inches.

The second forward, 14 feet 6 inches.

The third forward, wide from the center line, 14 feet 4 inches.

The fourth, but to aft, 14 feet 6½ inches.

The fifth aft 14 feet 6½ inches.

The seventh aft 14 feet 2½ inches.

The eighth aft, from the center line, 13 feet 4¼ inches.

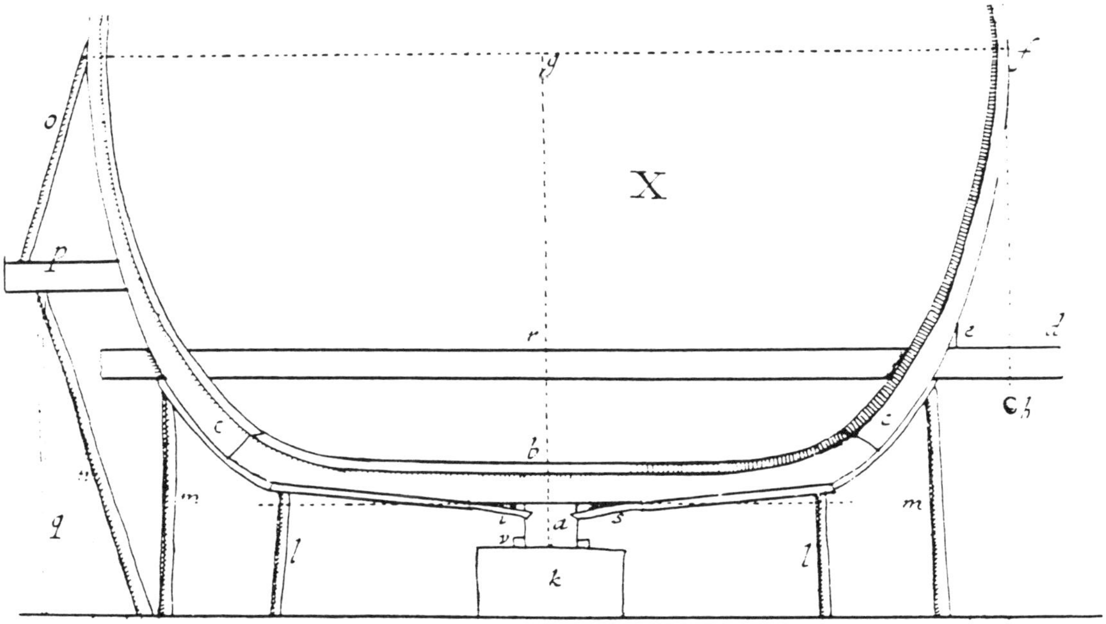

Figure 2.61. Plate LII (drawing X)

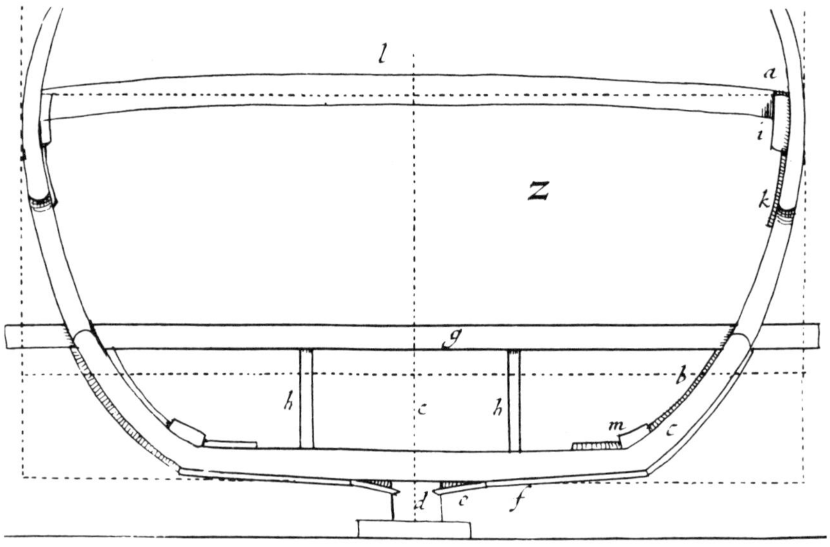

FIGURE 2.62. Plate LII (drawing Z)

24. Then Make the Master Ribband (or Height of Breadth Ribband) Around, and Make It Level

The word *scheerstrook* is difficult to translate, but "master ribband" perhaps comes closest. The master ribband was a temporarily applied batten that, when fixed to a few standing futtocks, indicated the maximum width of the ship all around. It started at the stem and ended at the extremities of the wing transom.

The height of breadth of the ship rose forward and aft, as mentioned before, so the master ribband also indicated the sheer. On this ribband, which was removed later, the locations of the deck beams, hatchways, masts, gunports, and the like could also be marked. To some shipwrights the master ribband was of such importance that they would save it in case they wanted to build other similar ships.

At this stage the master ribband—and with it, the shape of the hull—could still be changed. This moment, then, was extremely important, as it determined the width, the rising of the sheer, and the shape in the bow and stern. As stated earlier, a ship built too wide would become "laborsome" (it would ride with a heavy motion, and excessive strain would be put on the rigging), while a ship that was too narrow would be "tender"; a square ship had more loading capacity while a nicely flowing shape made a "good sailor."

25. And also the Shores and Spalls are set.

25. And Also the Shores and Spalls Are Set

The shores and spalls were components of the scaffolding (see *n, o,* and *q* in fig. 2.61).

26. The Floor timbers, bilge Futtocks, Steekers, Crutches, and Transoms.

(55 I 8) Steekers are timbers, in the stern, lying against the sides.

. . .

(54 II 54) The Crutch is low in the stern in a ship.

. . .

(55 I 50) *Crutches* [*zogstukken*]: are pitchforks that come against the posts forward and aft, instead of separated timbers.

(74 II 11) The steekers forward on the bilge strakes are thick 6½ inches, at the stem 9 inches, at the height of breadth 5½ inches, on the ribband above the height of breadth 5 inches. They lie apart at equal distances, below on the floor 9 inches.

26. The Floor Timbers, Bilge Futtocks, *Steekers*, Crutches, and Transoms

Now the shell could be filled up with frame timbers, and all the floors, bilge futtocks, and other futtocks were fitted. The shape of the floors and bilge futtocks could be taken from the "plank shell," but the shape of the futtocks must have been taken from molds.

In the stern, where the planking was bent to fasten it to the sternpost, the shape of the floors changed toward a big, naturally formed V, and even further to a Y shape.

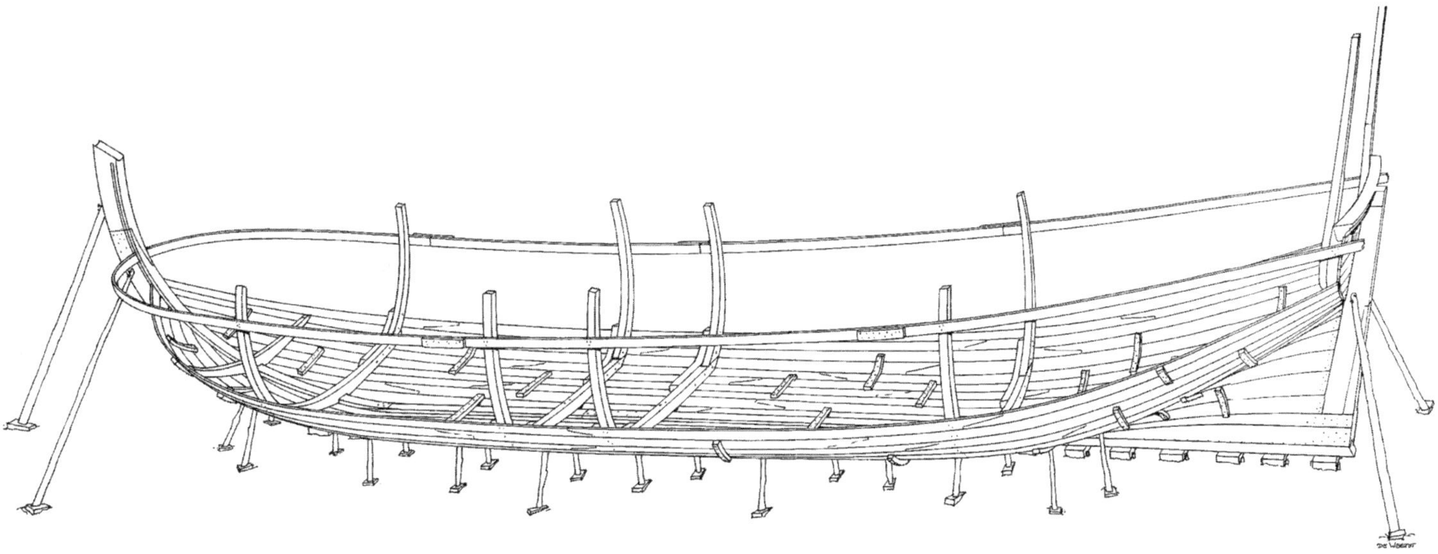

Figure 2.63. A master ribband attached to some erected futtocks. The master ribband indicates both the widest point on the hull and the sheer of the ship. (Courtesy G. A. de Weerdt)

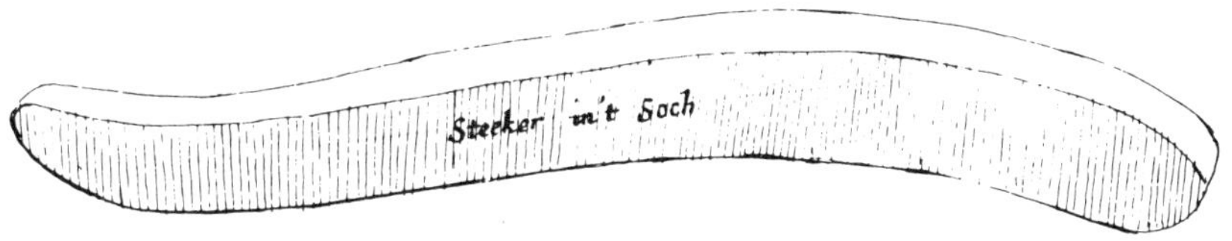

Figure 2.64. Plate XXVII. *Steeker* (futtock).

Such a crutch was called a *pykstuk,* while the futtocks that were of a reversed shape were called *steekers*.

Witsen mentions the transoms here by mistake; he has already discussed them in section 8 in connection with the tuck.

27. The Calves, and Toonen.

27. The Calves and *Toonen*

With "calves," Witsen means the vertical timbers in the tuck between the upper transom and the wing transom, making the vertical sills of the portholes in the tuck. I was not able to determine what he means by *toonen*.

28. The Beams ready.

(67 II 47) *About the Beams.*

13. About the beams in the hold.

1. 10 Feet of the ship's length give $1\frac{1}{8}$ inches of the thickness and breadth of the beam.

2. For the curve, every 10 feet of the ship's length 1 inch. Example, 100 feet length, 1 foot and $\frac{1}{2}$ inch comes to the thickness and breadth.

10 inches for the curve.

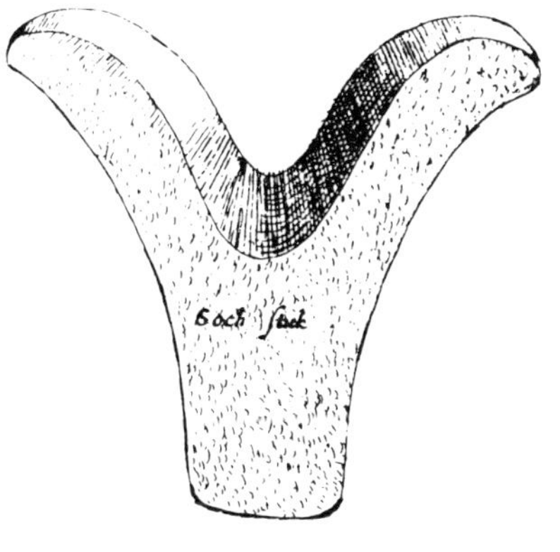

Figure 2.65. Plate XXVII. Crutch.

The lower beams are often made as thick as the stem is on the inside, and the beams of the cabin one third of the stem.

(76 II 37) The beams of the main deck 9, 10, 11, or 12 inches, the curve 8 inches. Their number is according to the strength the ship is supposed to have and the number of guns it will carry: the upper beams are lighter than the lower beams, at each end of the beam a hanging knee is set, 20 beams would lie in this ship: a breasthook is laid in the bow of the ship.

. . .

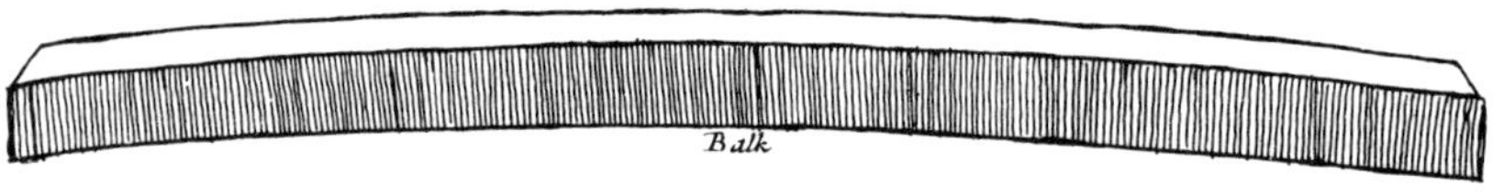

FIGURE 2.66. (*below*) The shell with floors and bilge futtocks in place. (Drawing by G. A. Weerdt)

FIGURE 2.67. (*left*) Plate XXI. Beam.

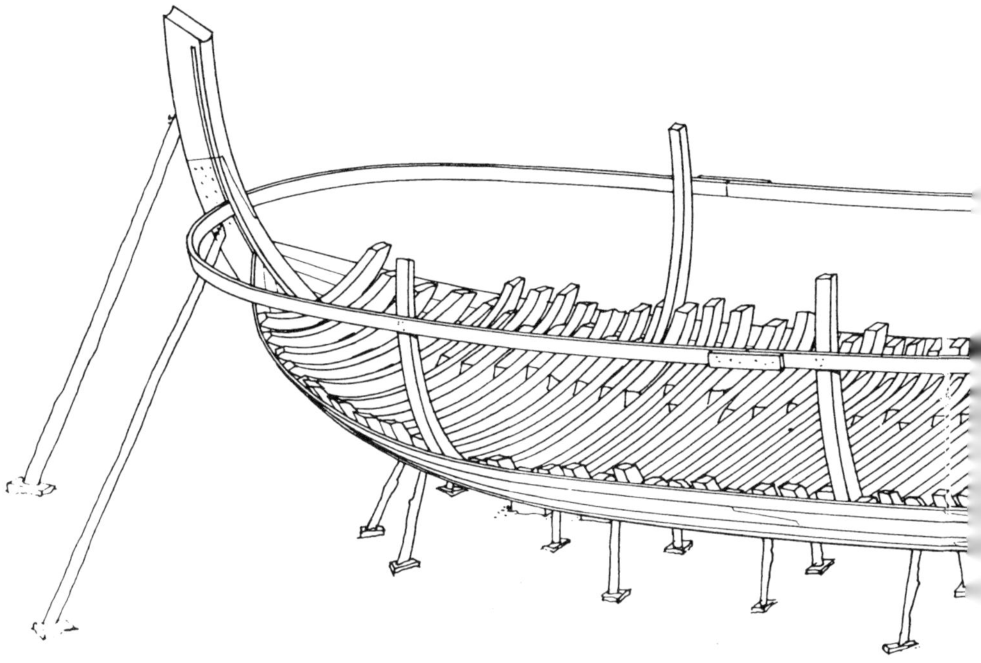

(68 | 10) The load beams are often 3½ feet lower than the other beams, on a depth of 13½ feet: in many ships an Orlop Deck is made here. Where the cable tier is behind the mast, there the Soldiers sleep. And the Orlop Deck is highest.

> (75 | 19) 13. *About the Beams.*
> 1. The beams broad 13 inches, thick 12 inches, more or less.
> 2. Have a curve of 8 inches.
> 3. Lie 3 feet 3 inches apart, also more or less.
> 4. The main hatch wide 7 feet.
>
> The forward beam lies behind the main frame one and ½ feet, with its back; the other lies 7 feet farther aft.
>
> The ends of the beams lie approximately 5½ inches above the deck clamps, dovetailed.

28. The Beams Ready

The beams of the main deck contributed substantially to the strength of the hull. They joined the sides of the ship, which was so essential that, in large and deep ships, an extra row of beams was often laid about one meter underneath the main deck beams, on which a so-called orlop deck was made. On eighteenth-century models we can see that the planks of the orlop deck were laid down in grooves between the beams and were not fastened with nails, as was done on the main deck, which is discussed here. The orlop deck was where cargo was stowed that had to be kept dry. Soldiers traveling to the Indies to serve the East India Company also were quartered there.

The pinas was not built with an orlop deck.

The dimensions of the beams were derived from the length of the ship: 1⅛ inches for every 10 feet of length. The main deck beams of the pinas should be, according to this formula, 15 inches thick; yet they are only 13 inches. The camber of the beams, making the deck higher in the middle than in the sides (for drainage toward the scuppers), was calculated at 1 inch for every 10 feet of length. For the pinas, then, this should be 13½ inches; yet Witsen allows only 8 inches.

Before continuing with the fitting of the beams, Witsen inserts a couple of paragraphs on other jobs, probably to emphasize that several aspects of the building process were carried out simultaneously. He does this at other points in the text as well, disturbing its coherence.

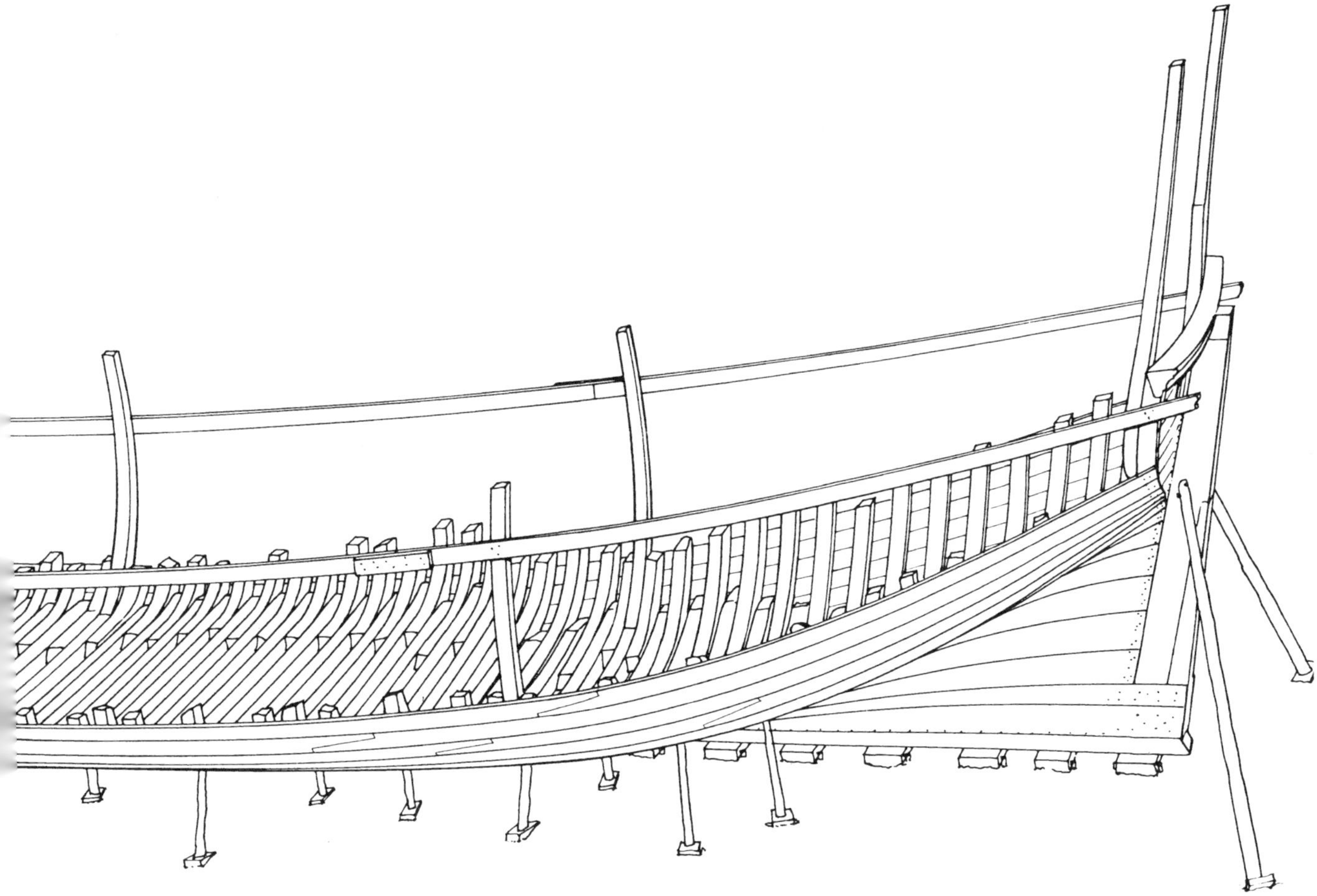

29. Make and level the Bilges.

29. Make and Level the Bilges

"Leveling the bilges" means that the inner surface of the floor timbers, bilge futtocks, and futtocks were smoothed with an adze to prepare them for the ceiling (inner planking). Because the thickness of the timbers always varied considerably, this was necessary for tying the ceiling and stringers together effectively.

30. Fit the ceiling in the Bilges.

(67 II 40) 12. *About the Ceiling, in the bilges and the bottom.*

1. The bilge stringer is a little thicker than outside planking of the bilge.

2. The ceilings in the bilges, and on bottom, are somewhat thinner than the bilge strakes.

(256 II 27) The Ceiling planks in the bilges are thicker than anywhere else, because there the pounding of the water is the most.

(74 II 26) *About the Ceiling Planks.*

Before applying the ceiling, one should have dubbed off.

The bilge stringer is laid over the bilge strake, which often goes beyond the boxing scarf, but not all the way forward to the stem: on which the ceiling is put, which goes to the stem, and first 4 ceiling strakes are made above the bilge stringer and 3 below it, before the bottom is also given ceiling, etc. as can all be seen at the illustrations.

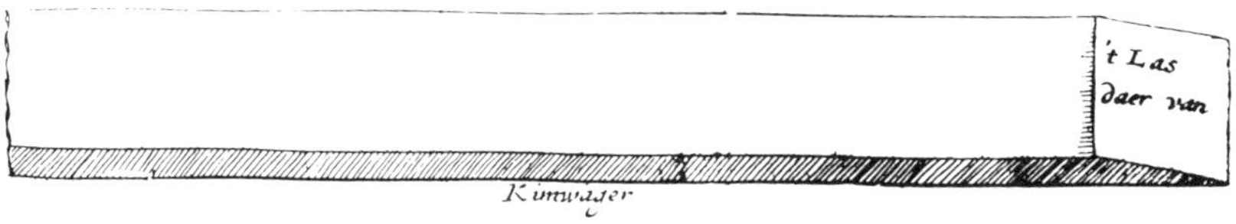

Figure 2.68. Plate XXVII. A bilge plank and its joint.

FIGURE 2.69. Ceiling planks in the wreck of a seventeenth-century trader. Apparently the builder was not very careful in the choice and the neatness of his material. (Photograph by A. J. Hoving)

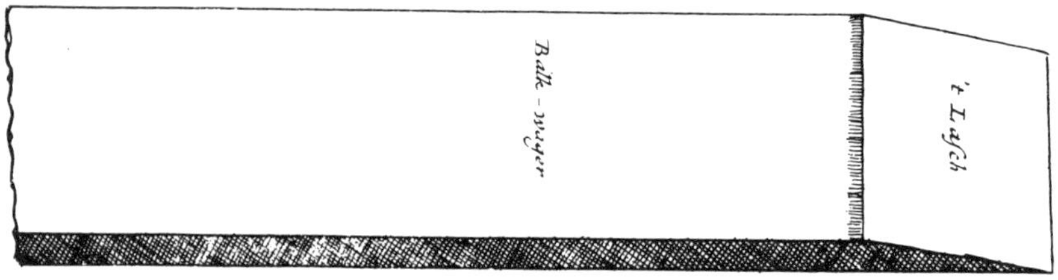

FIGURE 2.70. Plate XXI. Deck clamp and its joint.

(74 II 20) 11. [ceiling planks].

1. The bilge stringer, broad 1 foot 8 inches
2. Thick 4½ inches.
3. The other ceiling planks thick 4 inches.
4. Broad 1 foot 10 inches.

30. Fit the Ceiling in the Bilges

The bilge stringers were planks fitted internally in the turn of the bilge. Often they were somewhat thicker than the outer planking because the turn of the bilge was considered a vulnerable spot. They did not continue all the way up to the stem. The other, slightly thinner ceiling planking above it did continue forward.

31. Make the scaffolding to raise the beams.

31. Make the Scaffolding to Raise the Beams

This scaffolding was constructed inside the ship to allow workers to reach the level where the deck clamp was to be fixed (see fig. 2.62).

32. On this the Futtocks are leveled, where the deck clamp is fastened.

32. On This the Futtocks Are Leveled, Where the Deck Clamp Is Fastened

Men stood on this scaffolding to dub off or fair the futtocks where the deck clamp was fastened. It was important for the deck clamp to join all the futtocks well in the same way as the bilge ceiling.

33. Fasten the Deck Clamp.

(55 II 31) Deck clamps are planks, in which the deck beam lie.

. . .

(68 I 17) *About the Deck clamp.*

14. About the deck clamp in the hold.

1. For the thickness of the deck clamp, take ½ the thickness of the inside of the stem, like 10 inches, 5 inches.

. . .

(152 I 37) When applying the Deck clamp, then the height where the Deck clamp is planned should first be faired, then the Deck clamp is fastened with a bevel scarf [Witsen says *plat lasch* (*plat las,* flat scarf)] with

nails, and also two thin ceiling strakes are applied underneath, before laying the beams, as at *k; i* is the Deck clamp [see fig. 2.62].

. . .

(264 | 20) The main deck, on men-of-war, must, to be fit for fighting, be flat for the better part: and sometimes it is made to rise a step forward, to make the ship look better forward.

. . .

(267 | 45) The gun room should not be too low; because then the water, which comes over the ship, will come into it and bring much misfortune.

(75 | 4) *About the Deck clamp, and others.*

1. The Deck clamp, broad in the middle 2 feet 6 inches, aft 1 foot 10 inches, forward 1 foot 8 inches.

2. Thick in the middle 5½ inches, forward wide 1 foot 8 inches, aft 4 inches [should be 1 foot 4 inches].

3. The Aft end lies 2 feet beneath the first transoms: there it is broad 1 foot 10 inches, at the main frame 5½ inches beneath the foot of the quarter galleries, as also 26 feet before the tuck, where it starts to drop, and at the tuck it lies 3 feet lower than the wing transom, 2 feet beneath the first transom. [Witsen writes the two last measurements the other way around but apparently makes a mistake: the transom is beneath the wing transom.]

4. The ceiling underneath 3 inches thick.

(74 || 39) At the main frame the deck clamp in my Ship lies 5½ inches below the height of breadth ribband.

12 Feet forward, 6½ inches beneath the height of breadth ribband.

11 Feet forward, 8 inches beneath the height of breadth ribband.

12 Feet 5 inches aft, meaning, behind the main frame, 5½ inches below the height of breadth ribband.

11 Feet 8 inches aft, 5½ inches.

11 Feet 10 inches farther aft, 5½ inches beneath the height of breadth ribband.

18 Feet 5 inches, below the height of breadth ribband aft 6½ inches.

14 Feet and 6½ inches aft, and below the master ribband, which is, beneath the upper side, 1 foot and 5½ inches.

33. Fasten the Deck Clamp

The deck clamp was an exceptionally broad plank, half the stem's thickness, to which the deck beams were secured with dovetails. The height of this component was measured from the height of breadth ribband.

In the stern of Witsen's pinas, where the rising of the deck would make the walking surface much too inclined if it continued to follow the sheer, the deck was made flat from the point where the bulkhead of the gun room would be. In some ships, such as *Prins Willem*, the gun room was made lower than the main deck and only accessible by way of some steps.[10] Later, as with the Dutch two-decker, the main deck was made almost horizontal, and the place of the gun room no longer had to be taken into account when applying the deck clamp.[11] (For a comparison of these lower-deck designs, see fig. 2.71.)

Before the deck beams were laid, a couple of ceiling planks were placed underneath the deck clamp to prevent it from sagging.

34. *Lay the Beams, with the Deck clamp underneath.*

(152 || 47) When the Deck clamp is put in place, then the beams are laid, with dovetails in the Deck clamp as at *a* [in fig. 2.62], *l* is the beam.

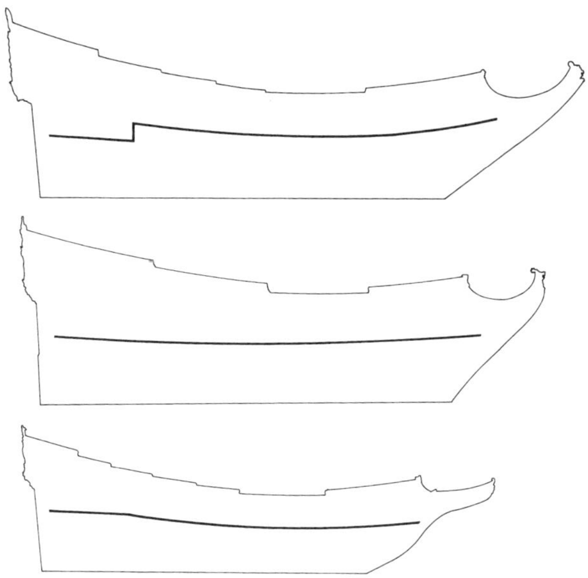

FIGURE 2.71. Lower-deck runs. *Above:* The run of the lower deck of the Prins Willem (1649); the gun room is placed deeper that the rest of the lower deck. *Center:* The run of the lower deck of a Dutch two-decker; the deck is almost entirely flat. *Below:* The run of the lower deck of Witsen's pinas; there is a break at the location of the bulwark for the gun room. (Drawing by A. J. Hoving)

When the beams are in place, then the lower scaffolding is made on the beams, and the frame trusses are placed as shown at *n* in figure *X* [fig. 2.61].

34. Lay the Beams, with the Deck Clamp Underneath

As Witsen states, the deck beams were let into the deck clamp with dovetails, resulting in a very firm construction for both inwardly and outwardly directed forces.

Usually the beams would lie about three and a half feet (one meter) apart in the ship, except where the main hatch was, which always had to be seven feet to allow for the largest barrels.

---

35. Raise the Scaffolding.

---

35. Raise the Scaffolding

To prepare for the raising of the beams of the upper deck and to give support for the installation of the first top timbers, the scaffolding was raised to the beams of the main deck. It is noteworthy that large scaffoldings with poles driven into the earth, as described by Van Yk, are not mentioned by Witsen. The ship itself was almost always used to support the scaffolding.

---

36. Place the Molds.

(262 II 54) It fits a man-of-war to be wide above, for the managing of the gun, and to enhance defense.

. . .

(266 II 38) That our men-of-war then are curved above, is done for prettiness and because a large crowd of men can better fight from such a ship, than from narrow ships, and because their better ability to fight is preferred over their speed in sailing.

. . .

(266 I 51) Most of the ships, which in this country are built for war, are wider above than the ship, described before: or, to speak more clearly, the curve given to the frames where they rise above the lower body of the ship is pulled more inward, than is usually done with men-of-war; because although on today's men-of-war it may go somewhat inward, yet a little farther up it goes outward again and thus makes that the ship is fairly wider, than the ship I describe [the pinas.]. So it is with my ship, because we trust that ships, having their topsides slightly curving inward, are more easily managed and defended with less crew; which is useful on a ship that has other purposes besides fighting. Narrowness above gives little windage: it benefits the sails, which is also why narrow and low cabins are made on ships with little tumble home.

. . .

(265 II 20) Between the upper deck and the main deck, especially on men-of-war, there should be as much room for a man to walk easily, unless, to obtain more space in the hold, it is made slightly less deep.

(266 II 44) Ships are best narrow above, for then they are difficult to board; because her bulging bellies make the side difficult to jump over: they are less praiseworthy because, when lying on one side, they heel over more and more easily take in water in the sides; of which the curved sides, along which the water rolls, are the cause; they heel over more, because the shrinking sides do not fall flat on the water, as the bulging bellies below, and so do not help the ship to float. Ships that are too wide above have this discomfort, that they become crank and easily capsize.

36. Place the Molds

These molds would not have consisted of much more than planks sawn in curves, but their purpose was clear: this was the moment when the ship above the waterline was shaped. Witsen does not provide any details, but in practice, while I was working on the model, it became clear that with top timbers or molds taken over from the eight initial frames, the shape can be determined perfectly by eye. The stern timbers mounted on the tuck, which at an early stage predicted the shape of the hull to some extent, no doubt served as a point of departure in this. After placing a top timber at the bow and one in the middle, the upper works were well shaped by using ribbands and by adding more top timbers or molds as one went along.

---

37. Make the Ceiling between the bilges and the beams.

(152 II 7) When the beams are laid, then one goes ahead with the ceiling between the Bilge ceiling, and underneath the Deck clamp, and by dubbing off, as is shown with the man standing on the scaffolding at *a* in the plate *AA* [fig. 2.72]; then also the knees for the hold beams are made, as can be seen at *b*, and then the Keelson is laid down, as at *e*, and the top timbers are raised, as at *g*, and the waterways are laid, as at *c*, and the Binding strakes, as seen at *d*.

37. Make the Ceiling between the Bilges and the Beams

In the meantime the ship was further planked on the inside until all framing had disappeared behind the ceiling.

38. With the Top Timbers.

(60 II 50) The top timbers are the timbers with which the full height of the ship is obtained.
(68 I 50) The foot of the top timbers equals a quarter of the stem.

. . .

(68 I 22) About the tumble home of the top timbers.

15. The tumble home of the top timbers at the main frame.

1. With the height of the upper deck in the side one and ⅔ for the tumble home. For example, the upper deck is high 6 feet in the side, then it should tumble 2 feet and ⅓ for the rake.

. . .

(68 II 9) The top timbers lean inward, mostly one third of the height of the upper deck.

. . .

(152 II 49) The top timbers in place, the tops are connected athwartship and also put spalls at the inside, with one end attached to the beam, and the other to the top timbers, and then these spalls are made around where the upper wale and sheer rail are to be made, and also the railing above, and then all the top timbers are placed, the Waterway is laid against the top timbers, and half (which is half the waterway or half the thickness of the waterway) is let into the beam, then the Binding strakes are laid, and the outside is further finished.

(76 I 19) Where the top timber tumbles at the location of the main frame 2 feet, that is with a height of 6½ feet, and with the height of 8½ feet, it tumbles 3 feet: the foremost, with a height of 6 feet, tumbles 1 foot 10 inches: with a height of 9 feet, it tumbles 3 feet, and with a height of 12 feet, it tumbles 3 feet 6 inches: that is aft of the main frame and there is tumble home too. The seventh top timber leans inward, on the ribband 2 feet 8½ inches, and these stand on the futtocks.

The fourth top timber from the front leans inward 2 feet 4 inches, at a height of 6½ feet; at a height of 8½ feet it tumbles 3 feet.

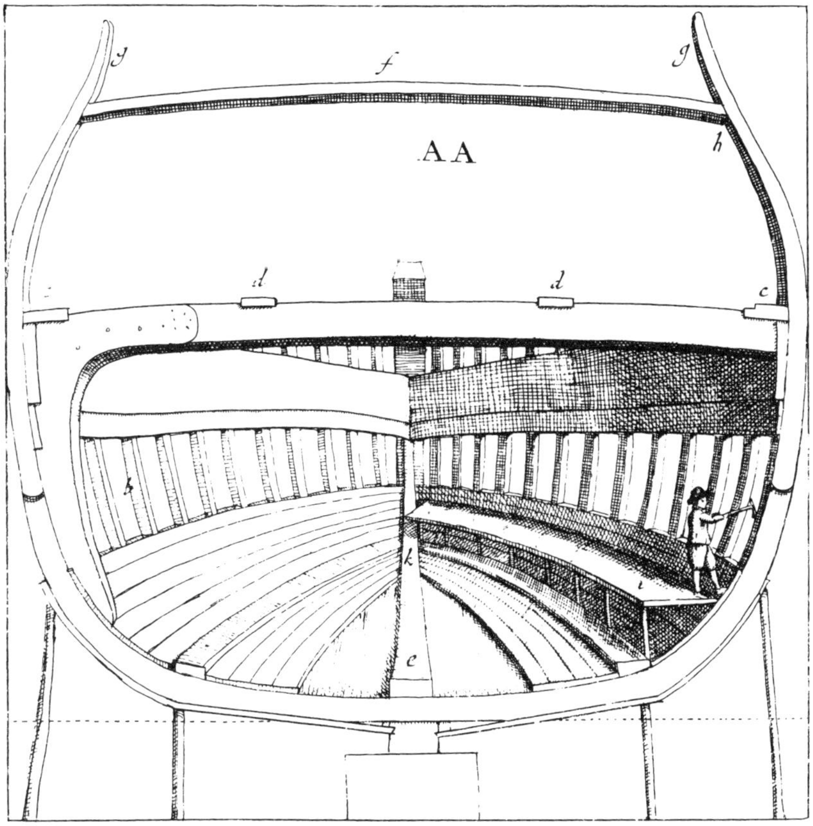

Figure 2.72. Plate LIII (drawing AA)

FIGURE 2.73. Plate XXIII. Top timber at the location of the main frame; the top portion is called a *hanger*, the bottom a *steeker*.

The fifth top timber at a height of 6½ feet tumbles 2 feet 6 inches. At 8½ feet, let it tumble 3 feet.

The eighth top timber tumbles 5 feet at a height of 14 feet measured from the deck beam, 4 feet at a height of 10 feet, 2½ feet at a height of seven feet.

And thus one can curve the top timbers, and the ship will not be badly shaped, although everyone can curve them according to the shape he wants the ship to have.

(76 I 11) 21. *About the top timbers and the upper deck.*

The top timbers tumble 2 feet at the top, are 4½ inches thick, on deck level as well as at the location of the main frame.

38. With the Top Timbers

The top timbers, as noted previously, are the uppermost parts of the frames. Their lower ends were fitted between futtocks, and they ran up to the railing. Their thickness was about a quarter of the stem.

The top timbers were curved inward (the tumble home), making the upper deck narrower than the main deck. Witsen supplies a formula for the tumble home: one third of the height of the upper deck. But clearly this was the last phase in which the shipwright could intervene in the shaping of the hull, so he was free to vary from the rule. For carriers a lot of tumble home was preferred, making boarding a difficult undertaking; on men-of-war the upper deck was wider to supply ample space for fighting.

39. Make the Ribbands around, Shores and spalls.

(74 II 6) The ribband comes in one inch on the bilge strakes: forward and aft it is half an inch thinner, than in the middle.

(82 I 30) *About the Ribband of our Ship.*

The Ribband at the top of the futtocks is 1 foot 10 inches above the height of breadth ribband at the stem.

At the first frame, 1 foot 9 inches.

At the second frame, 2 feet.

At the third frame, 2 feet.

At the fourth frame, 2 feet.

At the fifth frame, 2 feet 1 in.

At the sixth frame, 2 feet 1 in.

At the seventh frame, 2 feet.

At the eighth frame, 2 feet above the height of breadth ribband, 2 feet 5 in. above the wing transom.

39. Make the Ribbands Around, Shores and Spalls

As mentioned previously, the shape of the upper works was determined with ribbands. Then the top timbers were fixed with shores and spalls, just like the futtocks earlier.

40. Make the hanging Knees, fit the ceiling in the Bilge, make the crutches and the Keelson, and the Mast Step.

(70 I 40) 21. *About the hanging knees.*

1. About the hanging knees in the hold, the hanging knees in the hold ⅓ of the thickness of the stem.

(75 I 33) 14. *About the hanging Knees in the hold.*

1. The hanging knees in the hold, thick 7½, or 8 inches.

2. The lower ends long 7, 8, 9, 10, 11 feet, according to their place forward, in the middle or aft.

3. The bills long 4, 5, 5½ feet, fit underneath the beam with a rabbet.

4. Let the hanging knees before and after the last 2 beams be cant knees: to every transom 2 knees, and to the broekstuk 4 knees.

(79 II 9) The crutches are thick, and wide 10 to 13 inches, and go as far as the turn of the bilge; the floor riders as mentioned: but the hindmost breasthook is deeper.

(77 I 16) the crutches as thick as the beams 10, 12 or 13 inches.

(267 I 23) The farther the breasthooks are set apart, the weaker the ship: which is why it should be minded.

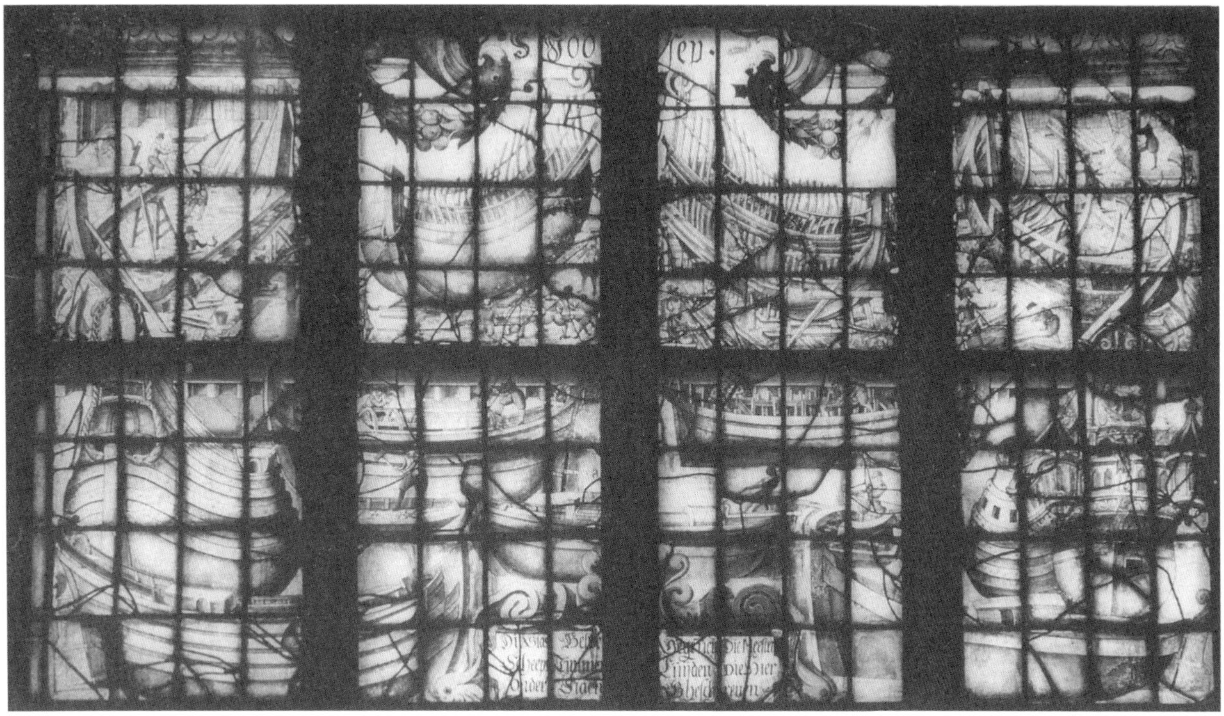

FIGURE 2.74. Detail of a stained glass window (1614) in Grote Kerk in Edam. The scenes in the top row illustrate the shell-first method in all its details. *Far left:* The keel with stem and stern; behind the sawyer is a large grindstone. *Far right:* The bilge, with the master ribband attached to some futtocks. *Second from right:* The top timbers are mounted, and the ship is "filled with wood." *Second from left:* The hull is planked up to the second wale; the ship is about to be launched. The bottom row shows various ship types; *left to right,* a fluyt and a wideship (*wijdschip*) or smallship (*smalschip*), a man-of-war, and a watership. (Courtesy Cees de Jonge, The Visual Art Box)

FIGURE 2.75. Plate XXII. Knee on the *broekstuk,* which connects the fashion pieces.

(273 II 21) The less ships are bound inside, whether with heavy floor riders, bulkheads or otherwise, the better they will make speed.

(75 II 14) 17. *About the rider floors, below and above the mast step.*

1. The rider floors broad 12 inches, thick 10 inches.
2. Below lie two and above four floor timbers: they are thick 9, broad 10 inches.
3. The ends long 7 feet.

(75 II 42) 19. *About the stern breasthooks and crutches.*

1. Aft there are four breasthooks, with crutches.
2. These breasthooks thick 7½ inches, broad 10 inches.
3. The ends long, 6, 7, 8 feet.

(74 II 17) The rising floor timbers are deep 1 foot, thick aft 6 inches: aft of the poort op de kiel [unclear] thick 10½ inches.

(79 II 14) The 2 rider floors at the main step, lie 3 feet 8 in. apart, and between the two, to each side of the keelson, lies a chock.

(54 II 20) The keelson [. . .], which can be called an inner, or second keel, is a heavy balk, lying along the entire ship below, or on the bottom, against the keel, to secure the frames or ribs of the ship, which are fastened to it.

. . .

(68 I 54) The keelson should be thick two thirds of the stem.

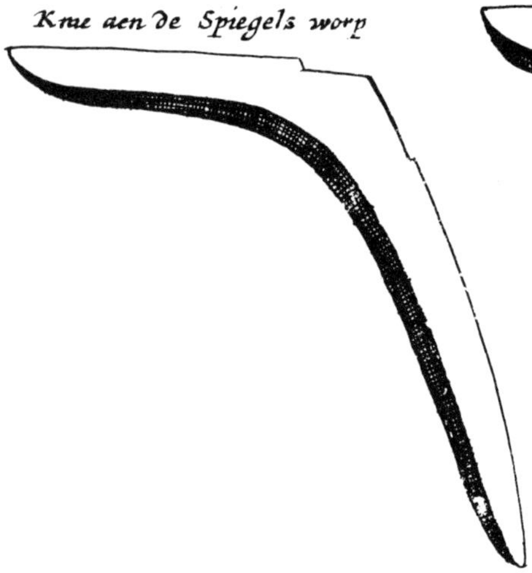

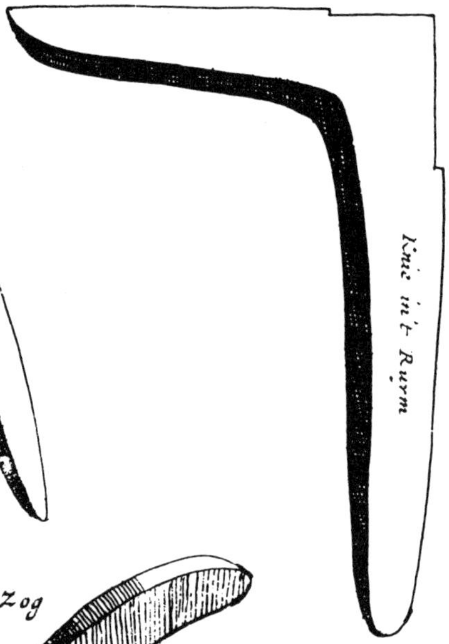

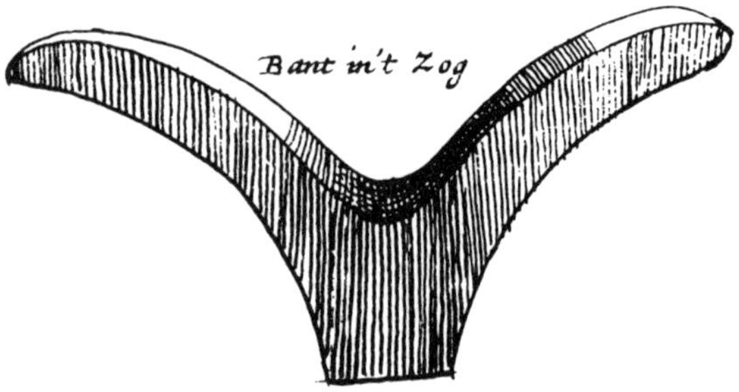

FIGURE 2.76. (*top left*) Plate XXIII. Knee on the transom.

FIGURE 2.77. (*top right*) Plate XXI. Knee in the hold.

FIGURE 2.78. (*bottom*) Plate XXIII. Crutch.

(267 | 52) The *Keelson* is broader than the keel, because the mast step is to be made thereon and the ceiling fastened to it. It is bolted onto the keel: serves to strengthen the entire ship, and may by right be called the inner keel. It often has a scarf forward.

. . .

(153 | 34) Meanwhile make *the Keelson* with a scarf, as shown at *a* on the plate *EE* [fig. 2.81]. Through every or every second floor timber a bolt with a clench is driven, through the Keelson, Floor, and also, but for one and a half inch through the keel.

(75 | 47) 15. *About the Keelson.*

1. The Keelson thick 9 inches.
2. Broad 2 feet 4 inches.
3. The Scarf long 5 feet.
4. The ends of the scarf thick 2½ inches, forward and aft, narrower and thinner.

(54 || 28) Mast steps are sturdy timbers, in the bottom of the ship, in which the masts stand with their ends.

(75 || 21) 18. *About the main Mast Step.*

This mast step is in the middle of the Keelson, and comes 6 feet aft of the middle of the ship.

1. Two rider floors surround the main mast step.
2. These are thick 12 inches, broad 14 inches.
3. They lie apart 3½ feet. The forward rider floor lies abaft the beam of the main hatch.
4. There four bilge riders are fitted, 2 forward, 2 aft: the bilge riders thick 10 inches: their breadth accords below with the rider floors: the lower ends long 8 feet, the upper ends long 7 feet, and 2 inches narrower than below as accordingly.

To each side of the keelson a chock is made, coming 4 inches above the keelson, thick 4 inches.

(152 || 35) And when the breasthook there is laid on the end of the Waterway, then place underneath the Foremast Step against the breasthook and across the ends of the ceilings a chock, called *Rising wood,* which is made as shown at *l* in *BB* [fig. 2.83].

FIGURE 2.79. (*top*) Plate XXII. Breasthook.

FIGURE 2.80. (*bottom*) Plate XXI. Keelson

(75 II 3) 16. *About the Foremast Step, and Chock.*

1. The Foremast Step broad 2 feet 5 inches.

2. Thick 2 inches.

3. The chock, underneath, is thick 10 inches. This comes with the fore end at a ninth part of the length, over the stem, meaning the fore face of the step, and the mast rakes 2½ inches, backward, from the partner: comes with a lip as high as the step itself. This step lies 4 feet 7 inches from the deck beams.

(79 II 18) The foremast step is broad 2½ feet, and thick 10 inches: underneath it lies a chock, that is broad 18 inches: the step lies 4¼ feet beneath the beams at the ends. The hole in the Step is broad at the front 1 foot, and aft 7½ inches, deep 6 inches, long 10 inches, deep 6 inches: it comes about 2 inches from the back.

(54 II 48) The futtock rider is a piece of timber, between the ceiling and skin: one makes these in ships, that carry guns, or have been found weak. [Riders are, of course, inside the ceiling.]

(75 II 48) 20. *About the Futtock Riders.*

1. The futtock riders, thick 10 inches.

2. The lowers ends come over the turn of the bilge.

3. And above to the upper deck.

4. The breadth, in proportion. They are about as thick as the hanging knees, but cut a bit stiffer at the waterway, and cut to each end of the waterway in the way the hanging knees have been cut at their bills: they continue up to the upper deck, and down across the bilge ceiling.

(74 II 11) The futtock riders thick 9 inches.

40. Make the Hanging Knees, Fit the Ceiling in the Bilge, Make the Crutches and the Keelson, and the Mast Step

In the meantime the workers in the hold prepared the hanging knees—one knee to either end of each deck beam. They were supposed to be two thirds of the thickness of the stem and lap-joined onto the beams. Timber for knees was expensive, so shipwrights would make two knees out of a thick timber, although they would become somewhat thinner, rather than use a thick piece or reduce it to the required size with much loss of wood. The hanging knees would go from the deck beams to the turn of the bilge inside the ceiling. Wherever there happened to be an open space between two frames, a chock could simply be inserted behind the ceiling to fasten the hanging knee securely. The outer planking had not yet been applied.

Like the sides, the bottom of the ship was now given ceiling planking—or for the most part. In the middle, right over the keel, the keelson was bolted down, with bolts straight through the floor timbers into the keel. The keelson was a lot thicker than the rest of the ceiling because it served as an inner keel, adding longitudinal strength to the vessel, and because it was to lodge the main mast step.

To either side between keelson and ceiling some space was left open to ensure the accessibility of the limber holes in case they became clogged. If desirable, both gutters could be covered with loose planks.

In the bow and stern a number of timbers were bolted across the ceiling, adding strength. In Witsen's pinas five breasthooks were placed in the bow and four in the stern. The ship's structure was reinforced by sleepers, which could stretch out fairly long. To each side of the main step rider floors were laid across the entire bilge and securely joined to a pair of bilge riders, which would be joined to the hanging knees or to the riders above (see fig. 2.238).

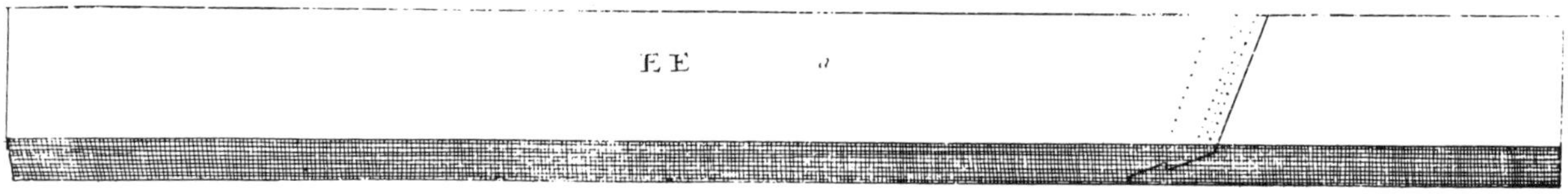

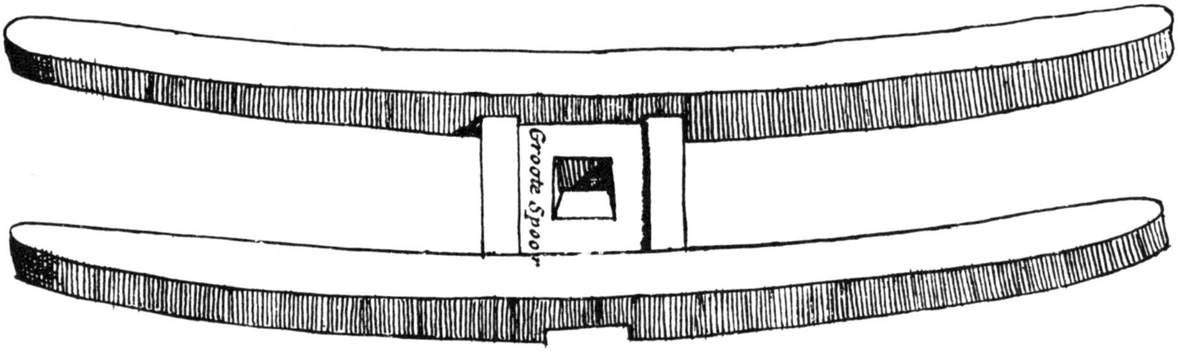

FIGURE 2.81. (*top*) Plate LIV (drawing EE)

FIGURE 2.82. (*bottom*) Plate XXII. Mainmast step

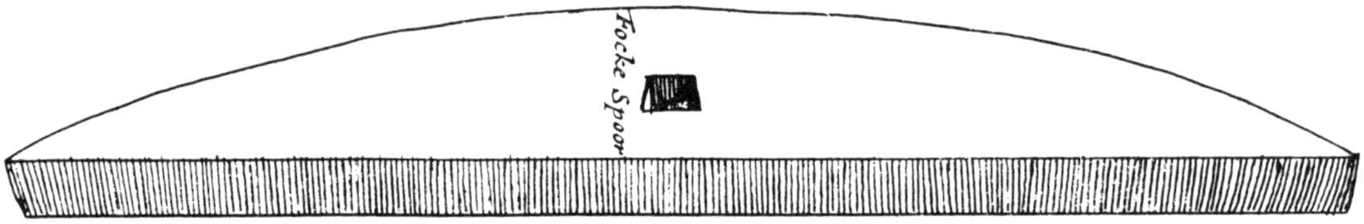

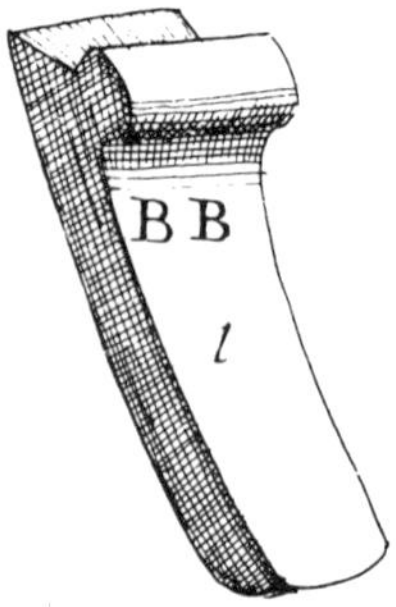

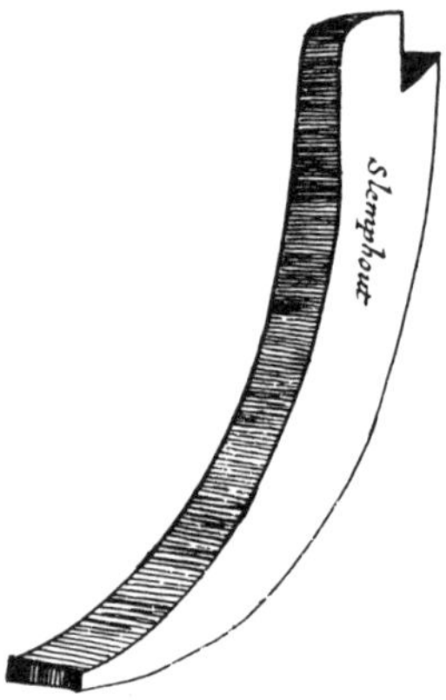

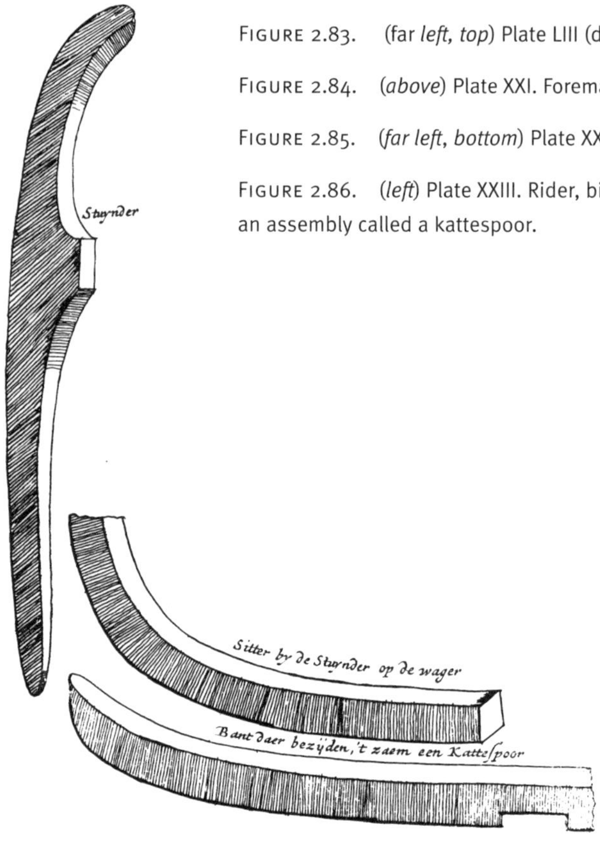

FIGURE 2.83. (far *left, top*) Plate LIII (drawing BB)

FIGURE 2.84. (*above*) Plate XXI. Foremast step.

FIGURE 2.85. (*far left, bottom*) Plate XXI. Chock.

FIGURE 2.86. (*left*) Plate XXIII. Rider, bilge rider, and rider floor, an assembly called a kattespoor.

The main step was placed six feet behind the middle of the ship and, as we saw before, secured between a rider floor forward and aft and on the sides by chocks (see fig. 2.87).

The foremast step lay all the way forward. It was a thick crescent-shaped plank, supported by a chock, which rested with its foot on the ceiling against a breasthook.

Witsen forgets to include the futtock riders in his enumeration of reinforcements. These large timbers went from the turn of the bilge past the main deck beams up to the height of the upper deck beams. They were as thick as the hanging knees and were bolted to the deck beams and futtocks with heavy bolts.

These riders were applied to compensate for the loss of constructional strength caused by the gunports and to strengthen the ship against the great strain caused by firing a broadside. In the pinas one was added between every two gunports. Thus, it could happen that a number of deck beams had a hanging knee to one side and a futtock rider to the other.

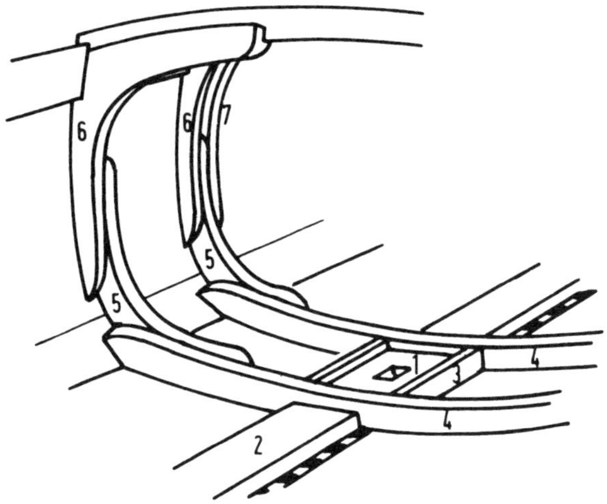

FIGURE 2.87. *Kattespoor.* The assembly consisted of the mainmast step (1), a hole hewn in the keelson (2), strengthened on both sides with a chock (3) and fore and aft with a rider floor. The rider floors were on both sides, placed against bilge riders, which were connected with a vertical futtock rider. (Drawing by A. J. Hoving)

41. *Level the upper* [should be "lower"] *Waterway.*

41. Level the Upper (Lower) Waterway

The waterway was a broad, thick plank lying on the deck beams against the ship's side, thus forming the edge of the deck (see also section 42, Lay down the Waterway). The thickness was one third, the breadth one and two thirds of the stem.

Witsen in fact means that the futtocks in the ship's side and the deck beams had to be leveled or faired to prepare them for the waterway. The waterway was let into the deck beams to level it with the other deck planks.

The inner side of the waterway was grooved, on which the ledges carrying the deck planks were to come. The depth of the groove corresponded with the thickness of the deck planks.

In the pinas the waterway was fitted around the futtock riders that went from the turn of the bilge to the upper deck. That was not a standard construction. In the restored *Wasa*, for instance, the waterways do not touch the ship's sides because of the riders. The space between these parts were first filled with filling pieces, and then the waterway was mounted.

42. *Lay down the Waterway.*

(54 I 50) Waterways are thick planks, that lie on the main deck, or on the upper deck, against the side, that have been bolted with bolts to or in the wales and through the hull planking. The ledges and beams here are tightened with heavy bolts, joined, and dovetailed.

. . .

(60 II 41) The waterway goes fore and aft on the side, on the beams, lying against the top timbers, and the futtocks.

. . .

(59 I 52) [. . .] lies on the ends of the beams, against the frames, on each side, up to the stem.

. . .

(68 II 17) *16. About the thickness of the Waterway.*

1. The Waterway, should be ⅓ of the thickness of the stem.

2. The breadth, to 1⅔ times the stem.

. . .

(153 I 29) Shown at the figure *DD* [fig. 2.89].

a the Waterway.

b The Beam, which is made as above, with a groove of ½ inch chiseled in the Waterway.

(76 I 47) *22. About the lower Waterway.*

1. The lower waterway broad 19 inches, thick 6 inches.

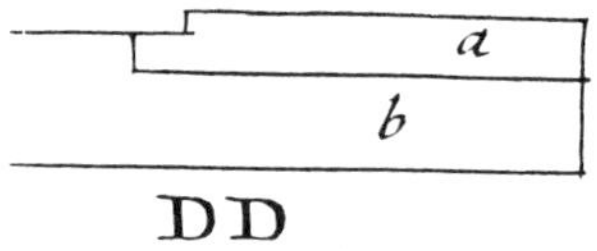

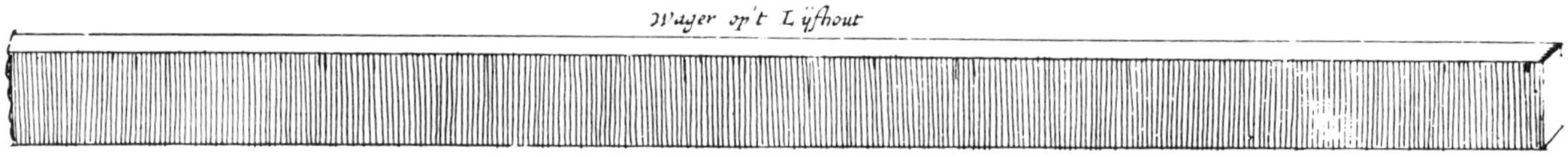

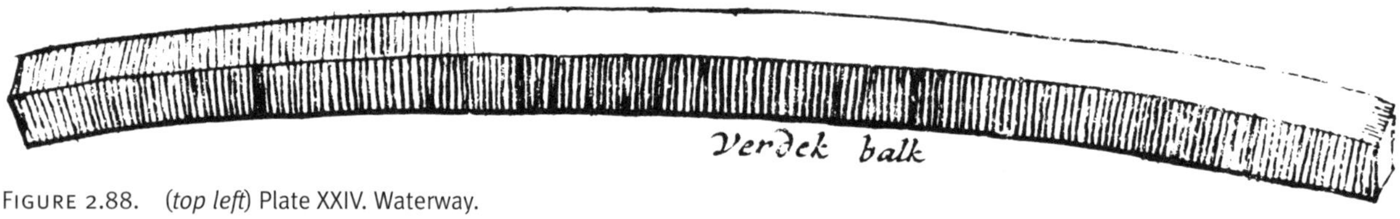

Figure 2.88. (*top left*) Plate XXIV. Waterway.

Figure 2.89. (*top right*) Plate LIII (drawng DD)

Figure 2.90. (*top*) Plate XXIV. Ceiling on the waterway.

Figure 2.91. (*bottom*) Plate XXIII. Upper deck beam.

2. Coming 2 inches above the beams. There is a knee on the rearmost beam across the waterway, fastened to the filling transom, the part coming against this, is fastened to the knee with a groove.

42. Lay Down the Waterway

The waterway could be fastened with bolts laterally, going straight through the ship's side and sometime through the wales. Such bolts were more than a yard long! This component of course strengthened the hull considerably.

43. *Then add a ceiling plank or two.*

(76 II 19) 24. *About the Ceiling on the Waterway, and others.*

1. The ceiling on the waterway be thick 3½ inches, broad 15 inches.

2. The ceiling on that one, thick 3 inches and 2½ inches, broad 16 inches.

43. Then Add a Ceiling Plank or Two

On the waterway, against the faired frames, a couple of spirketings were added, considerably adding to the interconnection between the futtocks and the top timbers. Witsen apparently knew of no methods for sealing the seam between waterway and ceilings other than with caulking. According to later sources, the waterways were shaped with a raised horizontal edge against the frames, raising the seam to be caulked above deck level.

44. *Fit the Beams of the Upper Deck, and make the Gun Room.*

(76 II 39) Their number is according to the strength the ship is to have and the number of guns she will carry: upper beams are lighter then lower beams, to each end of the beams a hanging knee is fitted.

18. About the beams of the upper deck.

1. For the thickness of the beams of the upper deck, ⅔ of the thickness of the stem.

2. For the camber ⅚ parts.

3. For the breadth ⅔ parts of the same.

(76 II 29) 26. *About the upper deck beams and knees.*

1. The upper deck beams, thick 8 inches, broad 9 inches.

2. Have a camber of 13 inches. Forward they lie 5 feet, 9½ inches high, in the middle 6 feet 2½ inches high, aft 6 feet 6 inches high. The aftermost upper deck beam lies 2 feet 4 inches away from the forwardmost beam of the gun room.

(77 I 13) The beams lie 3 feet 9 inches from one another.

(152 II 19) Then the Upper Deck beams are raised, as shown at *f* [in fig. 2.72] and which is first raised with the ends on chocks, and in the middle on props, and then the deck clamp is made, underneath it, as at *h* without dovetails, then further ceiling planking is added, down to the Ceiling on the Waterway, and then the hanging knees are applied; *i* is a scaffolding, on which the leveling is done, when placing the ceiling underneath the deck clamp.

At *k* the inner face of the Stem is shown: but the ends of the planks or ceilings have to meet one another there.

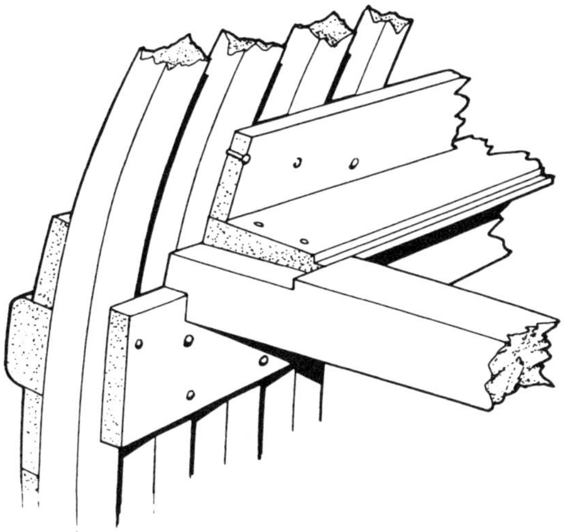

FIGURE 2.92. The waterway is partly let into the beam; the spirketing is fitted against the frames and on top of the waterway. (Courtesy G. A. de Weerdt)

(265 I 49) The *upper deck* sometimes drops one foot at the Gun room, which is for the guns to stand upright. In Merchant ships at the main mast the upper deck goes down, a step of half a foot, but the upper deck remains somewhat rising. So it is also with the main deck; but this need not drop as much, as it rises less: when this happens, and the main deck drops aft, the drop starts at the bulkhead of the Cabin.

The upper decks on North and East traders are made to rise fairly high, and such, that the crew may keep itself dry up there, because thus the water flows down and into the sea; which in the northern areas often sprays over the ship. On Men-of-War the guns need to lie parallel to the water, which is why the upper decks should not rise very high.

. . .

(276 I 6) But it must be avoided to build ships too high, because they are tender, and catch too much useless wind; nor should be built too low; because such are often run over by water.

(78 I 22) *About the deck Beams.*

2. The deck beams, thick 9 inches, broad 10 inches, have a curve of 7 inches. The waterways broad 2 feet, thick 4½ inches. These lie on the other beams of the main deck, separated 3½ feet, high fore and aft 4 feet. The forecastle sometimes goes one step down: its deck clamp is broad 1 foot and 8 inches, thick 2½ inches: forward it is to be 5 inches higher than aft: there are 7 beams, thick 5 inches, broad 8 inches.

(78 I 15) 30. *About matters in the cable tier.*

1. The deck clamp in the cable tier, forward high 4 feet, aft 4 feet 9 inches, the deck clamp, broad 18 inches, thick 3½ inches, the hanging knees long 4½ feet, their bills long 3 feet, and aft 4 feet 8 inches, thick, and broad 8 inches.

(78 I 34) 31. *About several matters, in the Gun room.*

1. The beams high 5 feet 9 inches, aft high 6 feet 5 inches.

2. The deck beams, thick 9 inches, broad 10 inches, curved 5½ inches. There lie 7 beams, one lying against the stern timber: is 1 foot 8 inches above the wing transom. The waterway here thick 1½ feet, thick 3½ in.

3. The room long 29 feet.

4. About the deck clamp on the Waterway.

   1. The deck clamp be broad 18 in.

   2. Thick 3 inches.

5. The hanging knees broad 8 inches, thick 7 inches. Aft there are 2 lodging knees. The bills long 3 feet, the lower ends long 5½ feet.

Between the cable tier and the Gun room lie 13 deck beams, below the forecastle lie 6 deck beams.

6. About the knees against the wing transom.

   1. The knees against the wing transom long 6 feet.

   2. In the side 6 feet.

   3 The knees thick 1 foot, broad 1 foot 2 inches.

44. Fit the Beams of the Upper Deck, and Make the Gun Room

When Witsen talks about the upper deck (*verdek*), he usually means only the central, uncovered part of the upper deck, not counting the forecastle, steering place, and cabin.

The beams of the upper deck usually were lighter than those of the main deck: thick and broad with two thirds the thickness of the stem. The camber was stronger: five sixths the inside stem. They generally lay straight above the beams of the main deck, joining the futtock riders well and making the hatches lie straight in line with each other.

The upper deck beams, however, were joined to the ship in a completely different way than the main deck beams. They were stuck between the frame timbers, temporarily supported with props in the middle and chocks in the sides. Only then was the deck clamp applied underneath. This method avoided a lot of measuring and marking, but it sounds to us like it took a lot of messing around to get a smooth run to the deck. The strong joinery with dovetails, as for the main deck beams, was apparently considered superfluous for the upper decks.

45. Then make the deck clamp.

(68 II 21) 17. *About the upper deck clamp.*

1. For the thickness of the upper deck clamp, 1/5 the thickness of the stem.

(76 II 25) 25. *About the upper deck clamp.*

1. The upper deck clamp be broad 19 inches.

2. Thick 3 inches.

(67 I 15) 2. The upper deck high 6 feet, and at the height of 8 feet it tumbles 3 feet: the other forward the same, depending on the height.

45. Then Make the Deck Clamp

The thickness of the upper deck clamp was only one fifth of the stem. As mentioned before, it was fitted when the beams were already in place.

The forward and aft parts of the upper deck of the pinas lay a foot or so closer to the lower deck than the middle—forward because the main deck needed less height there (that part of the main deck was used for the cable tier), while the upper deck could do with some more headroom in the forecastle as it was the crew's quarters.

More headroom was also needed in the stern for the steering place and the cabin above the gun room. So anyone going from the upper deck into the forecastle or the steering place had to descend a step.

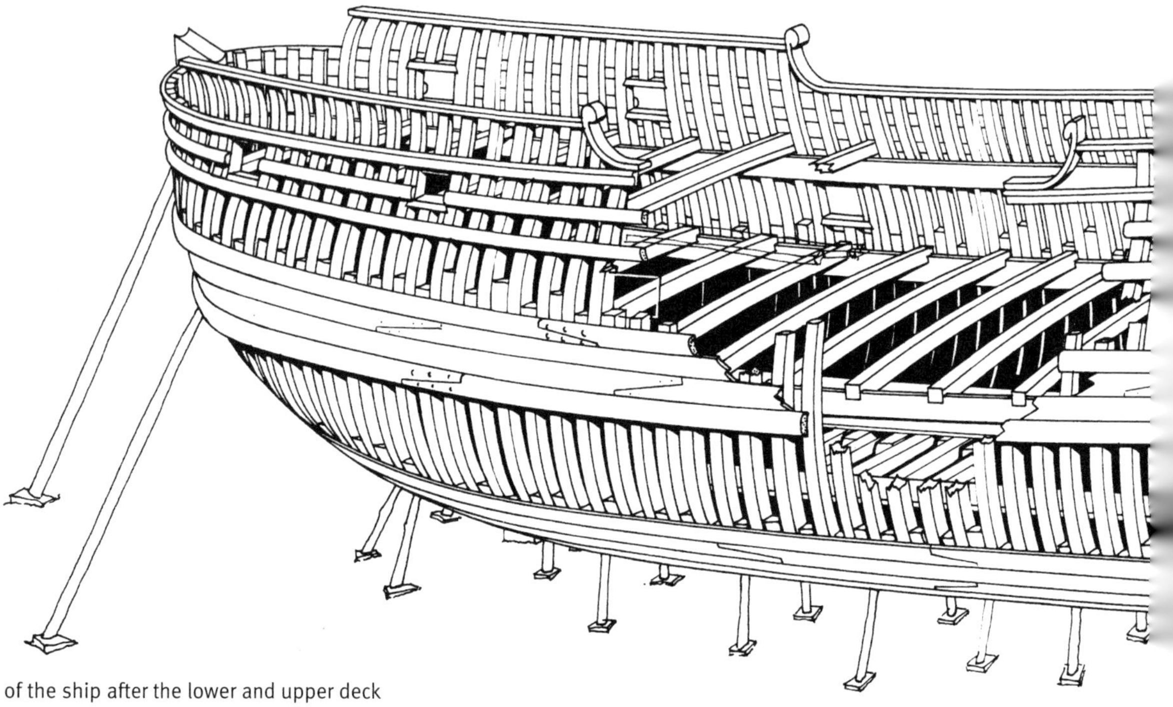

FIGURE 2.93. State of the ship after the lower and upper deck beams have been placed. (Courtesy G. A. de Weerdt)

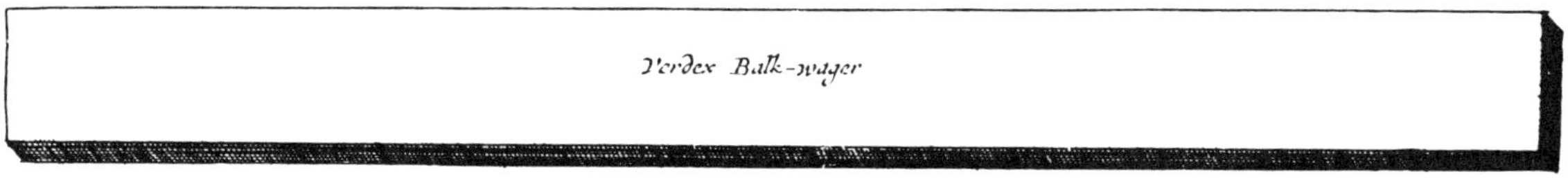

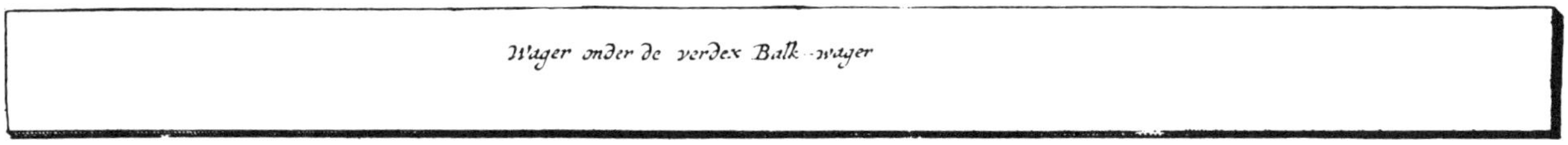

FIGURE 2.94. (*top*) Plate XXV. Upper deck clamp
FIGURE 2.95. (*bottom*) Plate XXV. Ceiling below the upper deck clamp

46. With the other ceiling underneath.

46. With the Other Ceiling Underneath

This other ceiling would not have been much thinner than the upper deck clamp. We should bear in mind that, although Witsen has not mentioned them yet, there were gunports here. Usually the shipwright would take these into account when placing the futtocks and top timbers, but Witsen postpones them until section 60 (Make the Chocks above the gunports around the outside).

47. Level the treenails in the Bilges.

(153 | 40) Once the Ship is lying on its side, the ends of nails are hammered into the treenails [Witsen mistakenly writes Dofjes, "bolt holes"]; they should be well and closely driven, so no water can seep into the ship.

47. Level the Treenails in the Bilges

The ceiling of the bilges was fastened to the floor timbers and bilge futtocks with treenails. To make sure the treenail would not become loose and cause leakage, a small

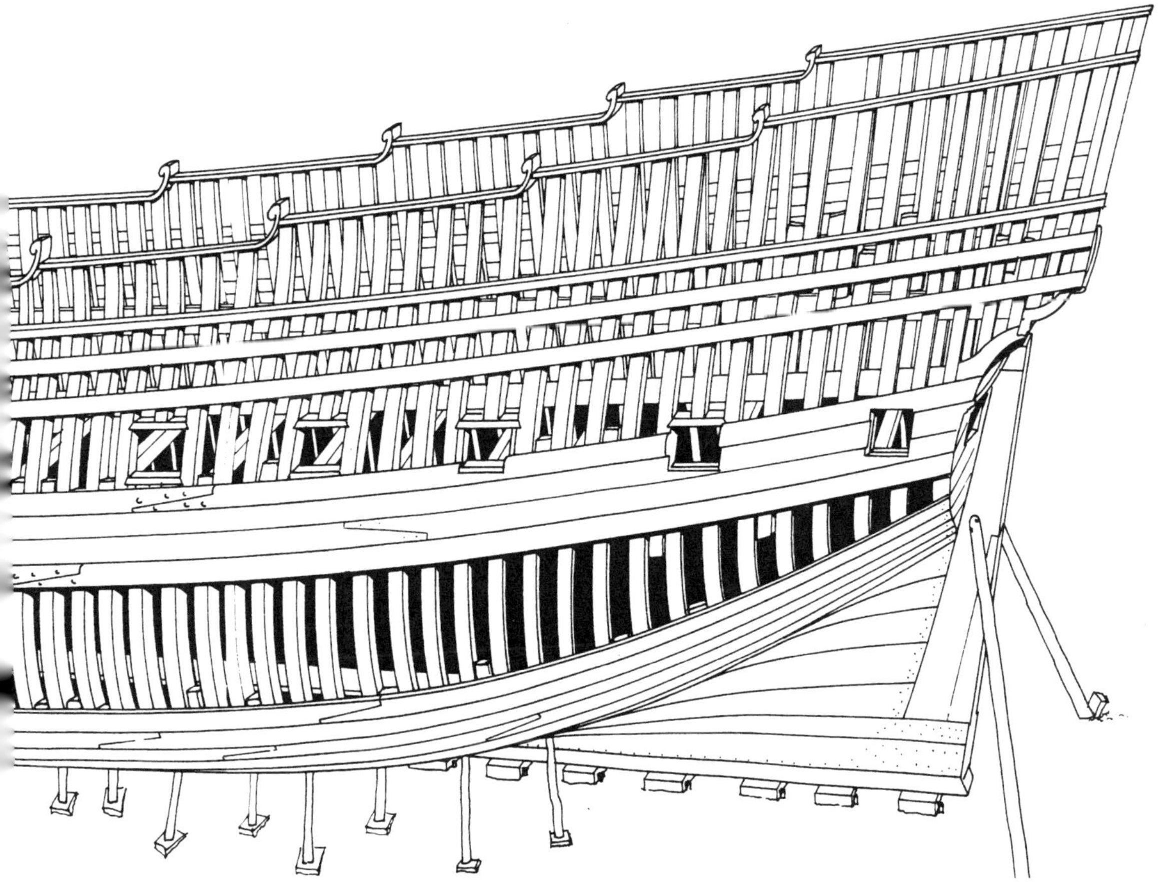

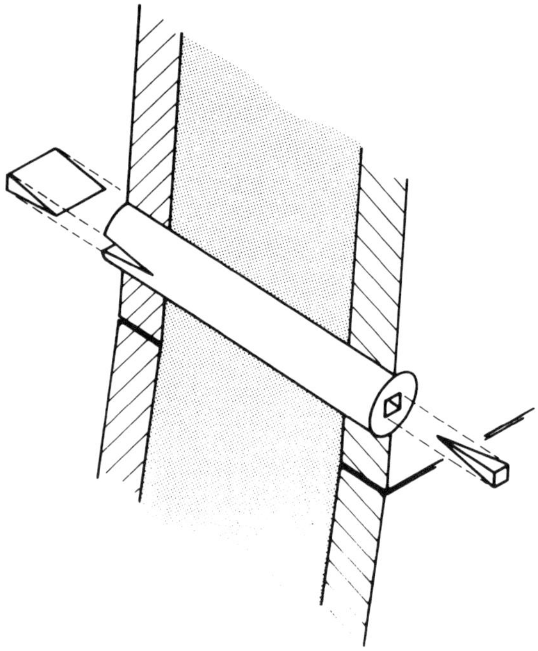

FIGURE 2.96. The treenail, a wooden dowel that attached the inside and outside planking against the frame parts. It was split on the inside to receive a wedge; on the outside nails or wooden pins (dottles) were hammered to close the open structure of the wood and keep water from seeping through. (Drawing by A. J. Hoving)

wedge was driven into it from the inside to fit in a small groove cut int the wedge; on the outside one or more dottles (square-sectioned wooden nails) were driven into the treenail. These now had to be smoothed off.

48. Make the Binding Strakes and the Tuck closed.

(54 II 3) The binding strakes, are thick, yet narrow planks, lying on the deck beams, those of the main deck with grooves, longitudinally: for the firmness of the main deck, which is laid on it. Some align with the width of the hatches, and close them in: the binding strakes aft of the mast, lie nearer to the ship's center, than those lying against the hatches.

(60 II 46) The binding strakes, hold the beams together; against toppling. The hold beam, holds the binding strake, supports, or props, against toppling.

. . .

(68 II 30) 19. *About the Binding Strakes, Carlings, and Ledges.*

1. ¼ of the thickness of the stem, for the thickness of the binding strake.

2. The breadth ¼ more then the thickness of the stem.

. . .

(152 II 43) *The Binding Strakes* must be joined to the beams with cogged laps, of half the thickness of the binding strake and an inch of the deck beam, from the side, and also from above.

a In the figure *CC* [fig. 2.98] is the upper face of a cog. And

b The lower face of the cogged lap joint.

(76 II 3) 23. *About the Binding Strakes.*

1. The binding strakes, broad 17 inches, thick 4½ inches, under the forecastle broad 14 inches, thick 3 inches; and on the forecastle broad 7 to 8 to 9 inches, thick 2 inches, the foremost is wide 2 feet, behind the mast 3 feet 2 inches. The one, going to the front of the main hatch, from behind, and the other, that goes from the partner forward, below the forecastle, they are somewhat narrower forward and aft, and are let half into the beams, and the beam is cut one inch in the side, but left uncut above, and this makes the cogged lap joint.

2. The rabbet is broad and deep 2 inches.

48. Make the Binding Strakes and the Tuck Closed

Again Witsen would not have won any prizes for clear formulation. It would have been better to say: Make the binding strakes and close the tuck. These two things, in fact, have nothing in common. The binding strakes were heavy, longitudinal planks, lying partly set into the deck beams, which also formed the sides of the hatches. On each side they had a groove or rabbet for the ledges and adjacent deck planks. The ledges, with their other ends laid in the groove of the waterway, supported the planking of the deck.

The binding strakes were farthest apart at the main hatch; forward and aft they would be laid closer together because all the other openings in the deck were smaller than the main hatch. It should be noted here that in the seventeenth century British and Dutch deck construction differed considerably. The Dutch *schaarstokken* were long pieces of wood, adding a remarkable longitudinal strength to the ship, while the English carlings were short pieces that connected only the beams.

That the tuck had to be planked was something altogether different, of course. This was done diagonally, from the corner of the wing transom and sternpost. Because

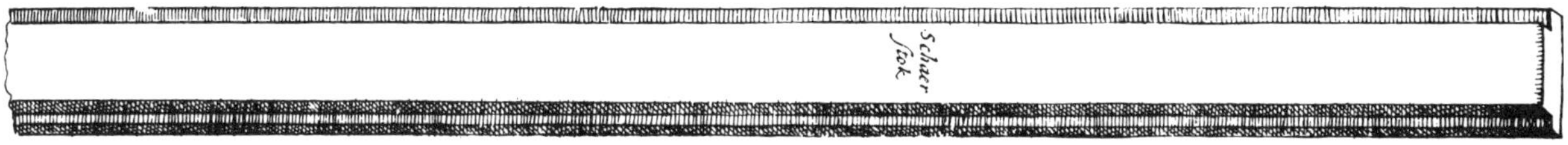

Figure 2.97. *(above)* Plate XXIV. Binding strake

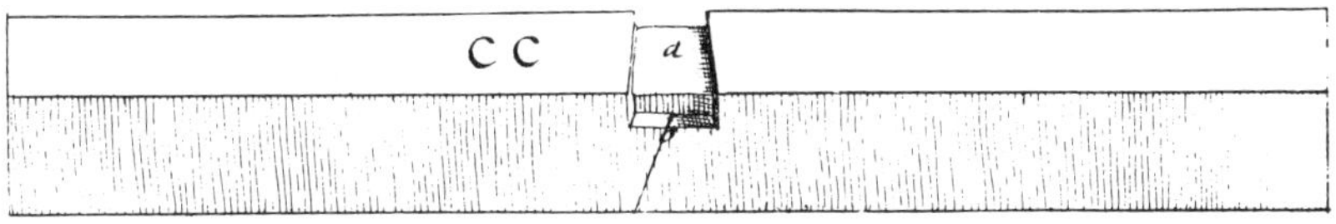

Figure 2.98. *(left)* Plate LIII (drawing CC)

the tuck was slightly convex, owing to the curves of the wing transom, transoms, and fashion pieces, the planks had to be bent either by scorching or sawn as such, so as not to have to mount them under tension.

49. Make the hanging knees, and have the Ledges sawn.

(76 II 51) When having light and heavy hanging knees, place them evenly on both sides against the deck beams.

(68 II 1) The ledges a quarter of the stem.

> (76 II 47) 3. The knees long 6 feet.
> 4. The bills long 3 feet.
> 5. Thick 7 inches.
> 6. Broad 8 inches, as for the Gun room.
> (78 II 1) The ledges thick $2\frac{1}{2}$ inches, broad $3\frac{1}{2}$ inches, between each deck beam 3 ledges.

49. Make the Hanging Knees and Have the Ledges Sawn

These hanging knees supported the upper deck beams. One was applied to each end of the beams. Like the beams themselves, they were lighter than the hanging knees of the main deck.

The ledges have been mentioned already. They supported the deck planking and spanned the space between the waterways and the binding strakes as well as the spaces between the binding strakes fore and aft the hatchways. They were one fourth the stem in thickness.

50. Go outside, and make the Wales under the gunports.

(55 I 10) Wales: are thick bands, which can be seen along the ship's side: they serve to hold the ship together.

. . .

(68 II 37) 20. *About the Wales.*

1. The Lower wale $\frac{1}{2}$ the thickness of the stem.
2. The breadth, as the stem is thick.
3. The upper wales less.

But for ships that are 170 or 180 feet long, then the wales 2 inches thinner.

> (77 I 44) 29. *About the Wales, and the Sheer rail.*
> 1. The lower wale, stands 1 foot 8 inches below the height of the scuppers, and is broad 13 inches, and thick $6\frac{1}{2}$ inches.
> 2. The second wale, 2 feet 5 inches higher, than the lower, and is broad 13 inches, thick 6 inches.

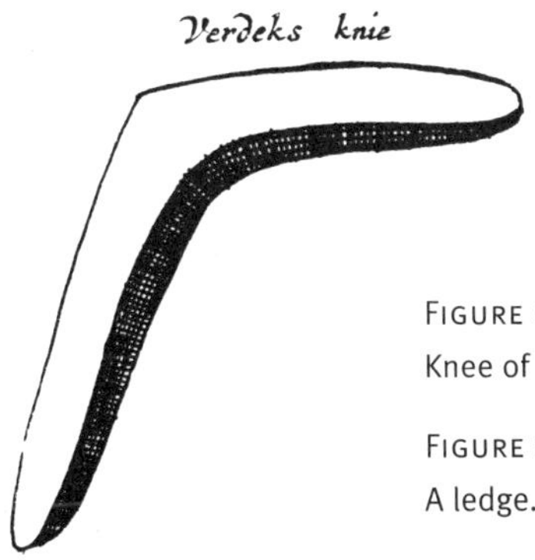

Figure 2.99. *(left)* Plate XXII. Knee of the upper deck beam.

Figure 2.100. *(below)* Plate XX. A ledge.

FIGURE 2.101. Plate XXV. Wale.

3. The planking at the gunports, broad 3 feet 1 inch, forward against the stem 2 feet 9 inches.

4. The Wale above this, broad 9½ inches, and thick 4½ inches, at the stem broad 9 inches, and at the main frame 10 inches, aft 1 inch.

(77 II 8) When a ship is 170 or 180 feet long, the wales are made 2 inches thinner, than half the stem. The number of wales is according to the size of the ships. Aft the wales are often placed lower, for the sake of the gunports.

(153 I 11) Then the futtocks and top timbers are faired, at the height where the first Wale is supposed to come, for this are used levers, ropes, and wedges, to wedge behind the levers, and save the wale from breaking. Between the futtocks holes for the ringbolts are drilled, to enter the levers, the ringbolts are secured with wedges, rings and fids, but the wale is fastened to the stem with nails. When it has been placed, it is first left to cool and harden, then the wood is removed, and half the levers are left in place, to make sure it has joined well, and if so, then continue fastening it.

50. Go Outside and Make the Wales under the Gunports

Wales (*barghouten, berghouten, berkhouten, barkhouten, barrighouten,* or *barckhouten*) are the thick strakes that emphasize the rise of the sheer to the bow and stern. Usually there were two below the gunports of the lower, main deck and two above. They were always paired, and the space in between was usually little more than the breadth of the wales themselves. The breadth was equal to the stem, the thickness half of it.

Between the lower wales are the scuppers, through which the ship discharges water from the main deck. On

The rising of the Wales.

Rising of the Wale at 7/100.

180 Foot	12 foot and 6	inches
175	12	2
170	11	9
165	11	6
160	11	2
155	10	9
150	10	5
145	10	1
140	9	8
135	9	4
1 0	9	
125	8	8
120	8	4
115	8	
110	7	7
105	7	3
100	7	
95	6	8
90	6	3
85	5	10
80	5	6
75	5	2
70	4	9
65 foot	4 foot and 6	inches
60	4	2
55	3	9
50	3	5
45	3	1
40	2	8

Rising of the Wale at 6 and ½ /100.

180 foot	11 foot and 7½ inches.
175	11	4
170	11	½
165	10	7½
160	10	4
155	10	½
150	9	8
145	9	4½
140	9	½
135	8	8½
130	8	4½
125	8	1
120	7	8½
115	7	5
110	7	½
105	6	9
100	6	5½
95	6	½
90 foot	5 foot and 9	inches.
85	5	4½
80	5	2
75	4	9½
70	4	6
65	4	2
60	3	9½
55	3	6
50	3	2½
45	2	10
40	2	6½

Rising of the Wale 6/100.

180 foot	10 foot and 8 inches.
175	10	5
170	10	2
165	9	9
160	9	6
155	9	3
150	8	10
145	8	7
140	8	4
135	8	1
130	7	8
125	7	5
120	7	2

FIGURE 2.102. Data for the rising of the wales (p. 69). Witsen's original table has been digitally altered to show the Dutch text in English. Scanned image modified by Emiel Hoving.

115 foot	6 foot	9 inches.
110	6	6
105	6	3
100	6	
95	5	7
90	5	4
85	5	1
80	4	8
75	4	5
70	4	2
65	3	9
60	3	6
55	3	3
50	3	
45	2	7
40	2	4

Rising of the Wale at 6 and ½ /100.

180 foot	9 foot	$9\frac{1}{2}$ inches.
175	9	$6\frac{1}{2}$
170	9	$3\frac{1}{2}$
165	9	$\frac{1}{2}$
160	8	$8\frac{1}{2}$
155	8	$5\frac{1}{2}$
150	8	$2\frac{1}{2}$
145	7	$10\frac{1}{2}$
140	7	$7\frac{1}{2}$

135 foot	7 foot and	$4\frac{1}{2}$ inches.
130	7	$1\frac{1}{2}$
125	6	$9\frac{1}{2}$
120	6	$6\frac{1}{2}$
115	6	$3\frac{1}{2}$
110	6	$\frac{1}{2}$
105	5	$8\frac{1}{2}$
100	5	$5\frac{1}{2}$
95	5	2
90	4	10
85	4	7
80	4	4
75	4	1
70	3	9
65	3	6
60	3	3
55	3	
50	2	8
45	2	5
40	2	2

Rising of the Wale at 5/100.

180 foot	9 foot	inches.
175	8	8
170	8	5
165 foot	8 foot	2 inches.
160	8	
155	7	8
150	7	5
145	7	2
140	7	
135	6	8
130	6	5
125	6	2
120	6	
115	5	8
110	5	5
105	5	2
100	5	
95	4	8
90	4	5
85	4	2
80	4	
75	3	8
70	3	5
65	3	2
60	3	
55	2	8
50	2	5
		for $10\frac{1}{2}$ inches.
45	2	2
40	2	

FIGURE 2.103. Data for the rising of the wales (p. 70). Witsen's original table has been digitally altered to show the Dutch text in English. Scanned image modified by Emiel Hoving.

flutes we sometimes see three wales at the height of the scuppers as well as on large ships of the first half of the seventeenth century, like the *Prins Willem* and *Wasa*.

The wales were intended to strengthen the ship longitudinally. This explains why shipbuilders had not cut these important constructional elements for the rear gunports before the mid-seventeenth century. Not cutting the wales was related to the flattening of the decks, which still followed the sheer in the early decades of the century, thus keeping all gunports between wales. Guns, of course, were better operated on a horizontal platform, which was also more comfortable. The decks were laid more and more level; yet the wales still followed the rise of the sheer. As a consequence, they crossed the gunports aft and had to be cut. This can first be observed on an engraving by Roelant Saverij of the *Aemilia*, Admiral Maarten Harpertsszoon Tromp's flagship in 1632, although admittedly there are three versions of the engraving, one of which shows the gunports neatly placed between the wales. When exactly the change came about remains uncertain.

The rise of the wales depended on the judgment of the shipwright and had direct consequences for the height of the vessel. Toward the end of the century the trend was to make the wales ever flatter, until in the eighteenth century the question of interrupting the wales no longer arose, as they ran parallel to the flat decks. However, this design did influence the construction. A flush ship has a greater tendency to hog than an arched ship, so it is to be expected that we find the introduction of diagonal framing systems in the eighteenth century to prevent this problem. With the strongly rising seventeenth-century ships this problem occurred less often.

The rise of the wales also depended on the purpose of the ship: men-of-war were flatter than merchant vessels. The shipwright could also express his personal preferences. Witsen supplies us with a couple of rather strange tables for calculating the rise of the wales (see figs. 2.102 and 2.103), varying from "$^7/_{100}$" (7 percent?) to "$^5/_{100}$" and for ships varying from 40 feet to 180 feet in length. What he meant exactly with his main categories—7, $6^1/_2$, 6, $5^1/_2$, and 5 (feet)—is unclear, but these tables do underscore the fact that the rise of the wales was open to choice.

Bending the wale planks, which were often 15 to 20 centimeters thick, by scorching must have demanded much insight and experience. Thus, it is not surprising that one of the most sharply curved wales on the *Wasa* shows a crack at the point of its greatest deflection. This bending process could go wrong even then.

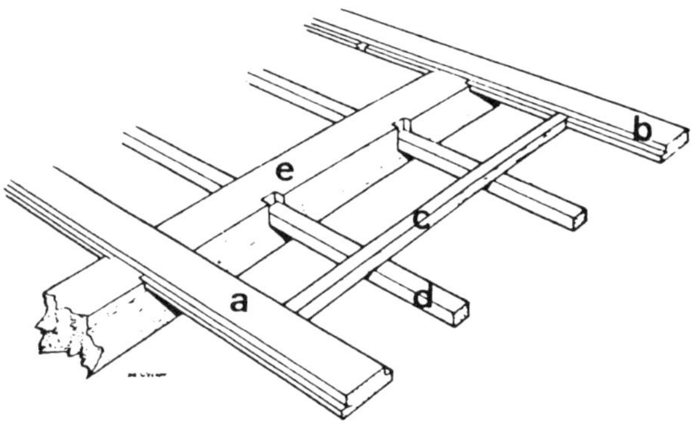

FIGURE 2.104. The binding strake (*a*) and the waterway (*b*) are both let into the deck beam. From their grooved sides the ledges (*c*) run, supported alongships by carlings (*d*), which are let into the deck beam (*e*). The weight of the guns was thus distributed over structural components in both directions. The length of both the binding strake and the waterway was important for strength. (Courtesy G. A. de Weerdt)

51. Caulk the Bilges and the garboard stern rabbet.

51. Caulk the Bilges and the Garboard Stern Rabbet

Caulking was done with unraveled rope called oakum, or with moss; this material was driven into the seams between the planks with the caulking iron (no. 43 in fig. 1.27), which looks like a blunt chisel with a groove instead of a cutting edge, and the caulking mallet (no. 41 in fig. 1.27). The seam was then sealed with pitch and, on some ships, even with a batten, fastened with *sintels,* small circular clips.

52. Make the side planking.

(153 | 46) On the plate *FF* [fig. 2.105] the side planking is shown at *a,* at *b* the Sternpost shores, and at *c* the bilge leveling shores.

And on the plate *GG* [fig. 2.106], the bottom *a* is shown, when the Ship lies on its side, where *b* is the side planking. Here the outer planks of the ship are applied from the top down, but in Rotterdam from the bottom up.

When having finished going around above, the *Scaffolding* is lowered, to close the side, and place it a little below the upper edge of the bilge planking, after which it is laid underneath the Wales to close immediately, and when it is closed, it is then treenailed, nailed and bolted; when it has been faired, it should be caulked to the bilges, but the bilges should be faired, nailed and caulked a little before.

52. Make the Side Planking

The open part between the bilges and the lower wale (*huiddicht,* here translated as "side planking") was planked next. The planks were joined with long scarfs in the same way as those of the bottom and bilges. It is hard to understand Witsen's formulation of planking methods. We know from Van Yk's treatise that in Rotterdam ships were planked from the wale downward. As Witsen indicates a difference between the Rotterdam and the Amsterdam method, he must have meant to say that in Amsterdam the planking was done from the bilges upward toward the wales.

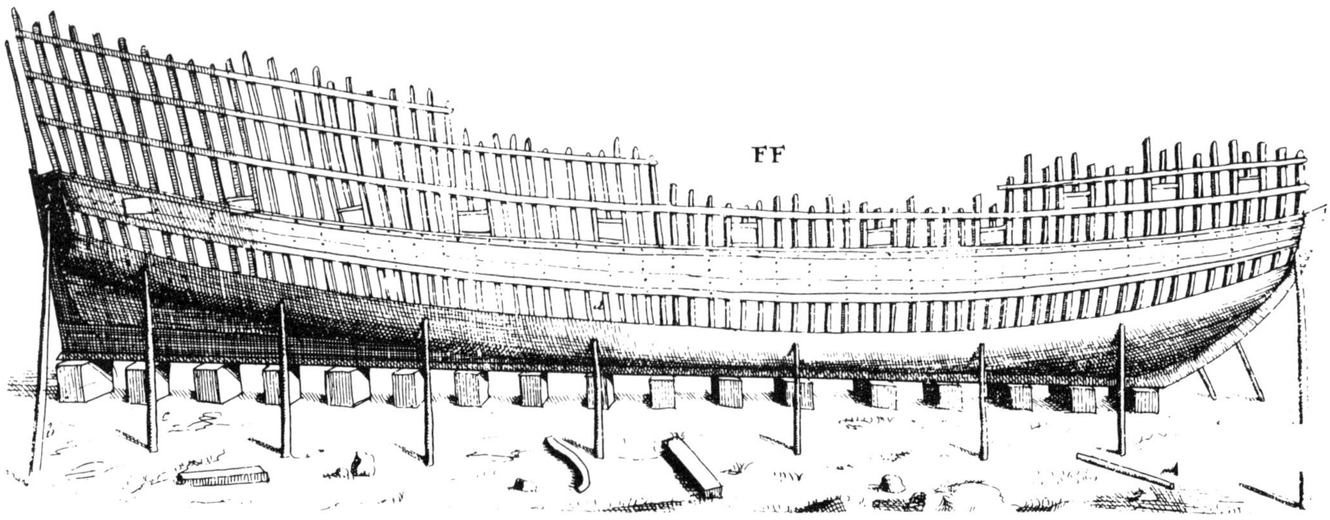

FIGURE 2.105. Plate LIV (drawing FF)

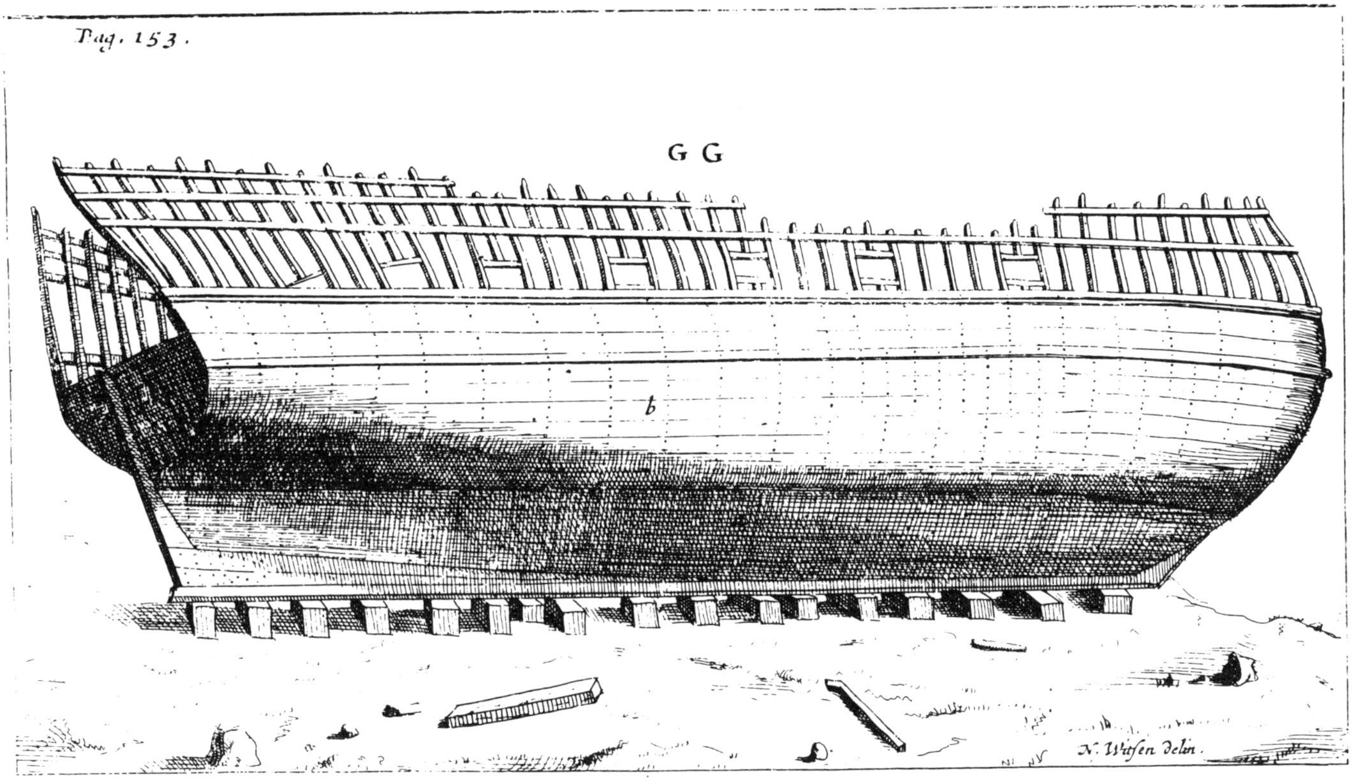

Figure 2.106. Plate LV (drawing GG)

53. Finish it further.

53. Finish It Further

Here Witsen must mean graving the caulked part (burning away oakum still sticking out), which could be a risky operation.

54. Let the Ship fall on its side.

(153 II 12) And then the Ship is made to fall on its side, to nail beneath the bilge, to fair and caulk, and the bottom is divided in as many parts as there are men working, or give each his share according to his strength, so that it will be done equally all together, and the Ship be hurt as little as possible: when it is done, the ship is raised, immediately, and then made to fall on its other side, and beneath the bilge is done as before.

. . .

(155 I 37) *When the Ship thus lies on its side, to raise it again,*

Then four or five Jackscrews are taken, and their lower end put on a plate, in which a hollow is made, in which the Jackscrew stands with its lower end, as in *a*, and the Yoke rests on the heads of the jacks, like *b* as shown in *SS* [fig. 2.107]. *b* [in fig. 2.108] Is the head of the Jack, and *c* the yoke, *d* the shore. A Chock is placed on top of the jacks in the stern because they are too short there to reach the wales and it is fastened to the jack on the other side to prevent it from slipping away during turning. The winding is first done with two Handspikes. When the Ship is raised, the Pans in which the Jackscrew stands must be brought somewhat nearer, or they might cause embarrassment. Many think that, turning Ships on their sides on land, hurts them, which is why they do not do it, and only do the same when the Ship lies in the water, when all parts can rest on the water. When the Jackscrews wind the ship from one side to the other, it must be heeded that the Ship be well supported on the side to which she is turning: But everything can be done without this turning over.

. . .

(155 II 30) The plate *RR* [fig. 2.109] shows how Ships are turned on their side. *a* Is the Jackscrew, *b* the head of the Jackscrew, *c* a Handspike, *d* the Pan in which the Jackscrew stands; this Pan and the Jackscrew should be well greased with soap.

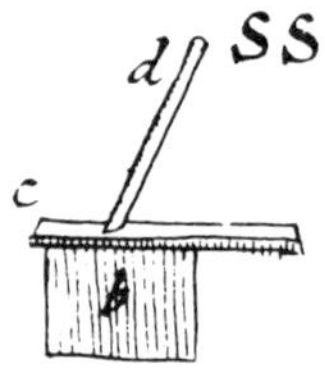

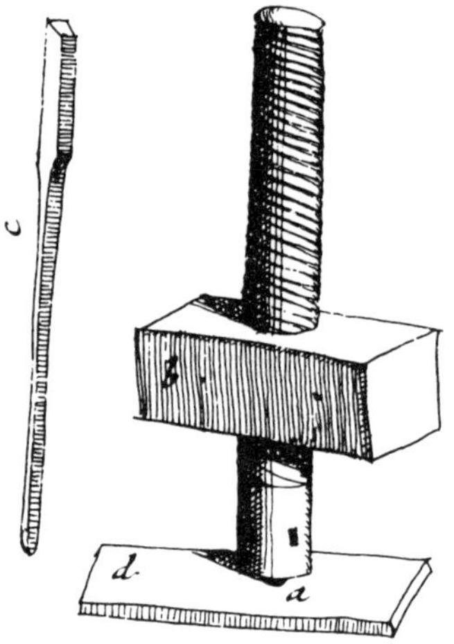

FIGURE 2.107. (*left*) Plate LVIII (drawing SS)

FIGURE 2.108. (*right*) Plate LVIII

54. Let the Ship Fall on Its Side

This sounds more spectacular than it was. Because of its flat bottom, the ship was never turned completely on its side: the bilge of the lower side rested on blocks, and the upturned side could be caulked and finished.

55. Raise it again.

(153 II 24) Then raise the Ship again, and catch it on both sides in a cradle, so it does not fall; so that the ship can stand fast, and finish it forward, ready for launching; when also the cradles are made under the keel, as also the bilge ways in the sides. The wedges are driven out from underneath the keel, by ramming or knocking with heavy beams, that got their name from the Ram's heads they used to have at their ends instead of the knobs; it is greased below to be slippery. When faltering during launching, Ships are rocked, which is done by letting them fall to and fro on shores, but this can hurt Ships, which is why it is not done unless necessary; ropes are also attached forward, which are then pulled; when it doesn't want to move forward, put jackscrews aft to each side of the sternpost, to prevent it from falling, and when then the rough blocks have been rammed away underneath, the shores that hold the Ship forward and in the sides are kicked away and thus it slides into the water.

On plate HH [fig. 2.110] a design is shown, of a ship standing on the stocks, ready to be launched.

55. Raise It Again

Launching a ship was a risky undertaking, as it could cause a lot of damage to the ship. Ships in Holland were launched headfirst and in an unfinished state. During the building process the ship stood more or less level. But before launching, it was given a greater tilt by lowering the bow and elevating the stern, pivoting the bottom on one of the stocks about one third of the ship's length forward of the stern. This stock was called the *dompblok.* After this operation, the ship was supported again until it was launched.

56. Make the Irons to the Post, and a Plate to the Keel.

56. Make the Irons to the Post, and a Plate to the Keel

The irons were the rudder irons of the sternpost, on which the rudder was to be hung. They were made from wrought iron and actually consisted of an eye, the gudgeon in which the pintle of the rudder fitted, and a pair of straps or tails encircling the sternpost and bolted to it.

The plate of the keel could be an hourglass-shaped

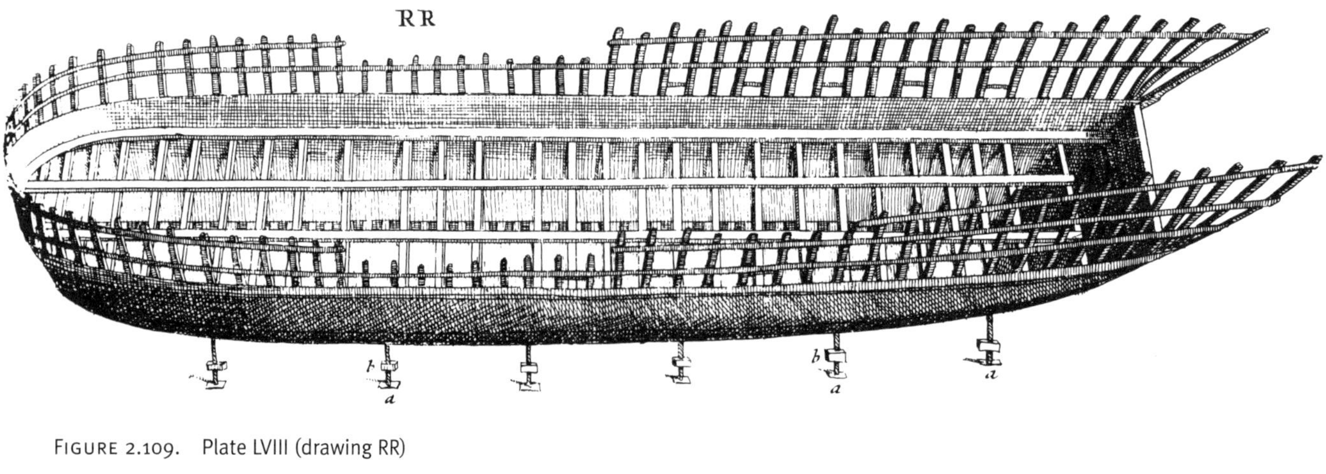

FIGURE 2.109. Plate LVIII (drawing RR)

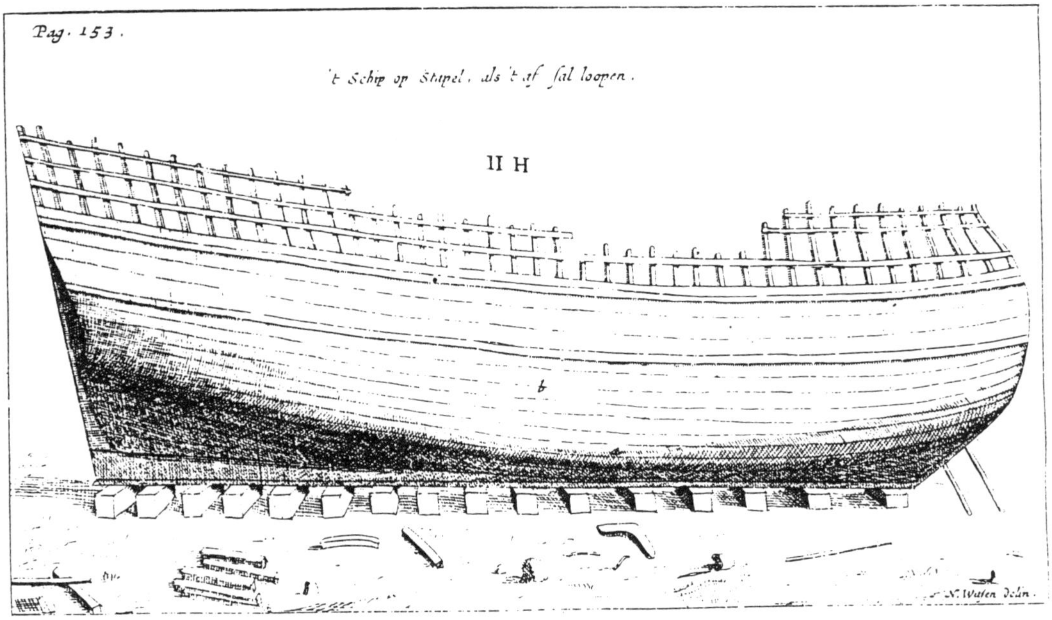

Figure 2.110. Plate LVI. The ship on the stocks when it will be launched.

iron plate set into the keel and sternpost, joining them in a double dovetail.

Furthermore, the foot measures indicating the draft were carved into the sternpost and stem. In archaeological finds and some models we see that roman numerals were used for this, in which the *V* was placed horizontally and the *I* was replaced by a dot.

57. Make the Mold for the Rudder.

57. Make the Mold for the Rudder

Every ship carried a rudder mold, indicating the exact position of the rudder irons. If the rudder was lost, a proper-fitting jury rudder could then be made.

58. Finish it all around, to prepare it for launching, and, when ready, launch it.

(153 II 54) The plate II [fig. 2.111] shows the *Model of a Ship being launched,* and how far Ships must be finished, when in our country they are made to slide from the land. In many places in Europe the Ships are finished completely on land before putting them to water, which in my opinion hurts them. Well-greased planks are put to each side of the Ship, on which it slides when gliding off the stocks, and which keep it from falling on its side. The men to be seen hauling the line, pull a rope that is attached aft to both sides, and thus help the momentum, when the ship does not want to go forward, because otherwise this is not necessary, with the stick, and the rope k which is to each side, the Ship is rocked to and fro, to make it loose, and slips into the water, but such is not done unless the ship is slow in leaving the stocks; *ll* are the wedges and beds, which have been kicked away from underneath the Ship, to raise the same; and to receive the kick of the rocking.

When the Ships are launched, a rope is run out aft, to steer it with, and hold it from running too far. When a Ship stands ready on the Yard, to be launched, and the shores have been removed, it should be launched as soon as possible, because without shores and standing on the keel alone for a long time, the ship could hurt; or if you want to wait, at least raise some shores against it outside. The beds, and the other wood, on which the ship slides, must be wetted, for it not to ignite, when the ship is launched. When the water in which the ship is to be launched, is short, ropes are attached to the stem forward, and it is turned, so that it doesn't hit into the other bank. Before the Ships leave the slipway, a chest is nailed to the side, holes made in the skin and the ship filled with water, to look for holes and cracks, and fill them.

. . .

(155 I 18) When a Ship has been launched with good fortune and without disaster, the workmen here are poured beer to be merry with, to which in former days for every hundred guilders the Ship's hull had cost, one was enjoyed.

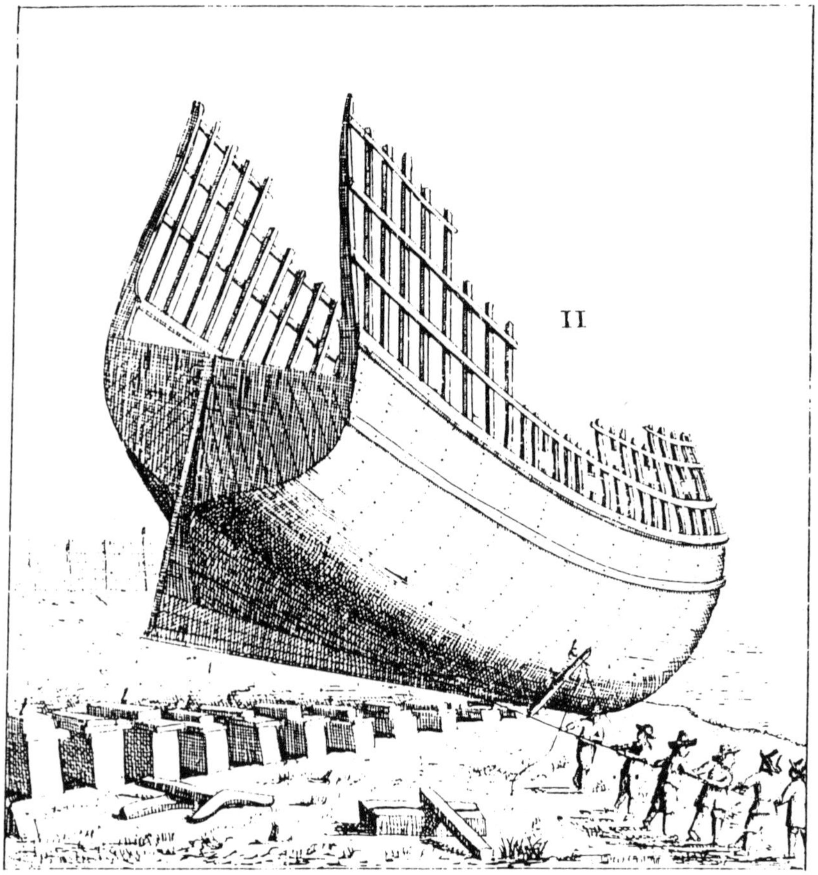

Figure 2.111. Plate LVII (drawing II)

58. Finish It All Around, to Prepare It for Launching, and, When Ready, Launch It

Witsen was not very generous in providing information on launching techniques, but they did not likely differ too much from the method described by Van Yk. First the ship was shored up with poles underneath the wales. The ship was held back with bolts, cheek knees, and a large tackle to the lower gudgeon to prevent a premature launch. The last shores, on which the ship's bottom and bilge still rested, were removed at this point. Then sliding planks were laid underneath both sides of the bilge, parallel to the keel, and smeared with soap or soot (not visible on Witsen's engraving). These bilge ways kept the ship from falling on its side while slipping into the water. The vessel was lifted somewhat with wedges rammed underneath the keel (nos. 25 and 26 in fig. 1.21), and then the open spaces between the keel blocks were filled with other blocks, ground ways, making a flat track toward the water. Heavy chocks alongside the keel kept the ship from deviating from the track. When the shores were finally removed, the tackle was cut and the ship usually glided off the stocks. It if did not, then the rope attached to the sternpost was pulled with great might while the ship was rocked with the aid of poles.

Thus the ship, planked only up to its second wale and not having even one closed deck, was launched, bow first. This early launching had three advantages: sagging was prevented, as the ship was supported best in the water, which gave prolonged support to the keel; the ship was still comparatively light, facilitating the launching; and the yard was now free, so construction of another ship could begin.

When the launching had been successfully achieved, the workers drank beer: the more expensive the ship, the more beer.

At this point, let us examine the interior of the ship.

In figure 2.114 deck beams of the forecastle are not yet planked, with the waterways to each side and the binding strakes in the middle. Forward, where the cable tier will be, we see the upper deck coming down to allow for more room in the forecastle.

In figure 2.115 we see the keel with the frame and the ceiling, on which the hanging knees of the main deck lock around the deck clamp, with their upper end next to the deck beam. *F* and *g* are riders; the combination of floor

FIGURE 2.112. Etching of a ship launch by Reinier Nooms, 1660. The artist's depiction fits Witsen's description in all respects except one: the scaffolding supports are already attached. See section 59 (When it has been launched, then make scaffolding outside and to the stern). Courtesy Rijksmuseum, Amsterdam.

riders and futtock riders on the ceiling were used in the pinas for the main mast step. Breasthooks can be observed. The fore step is not shown at all.

Figure 2.116 shows the gun room with the sternpost and the upper ends of riders coming up through the deck (*c*). In the foreground on either side we see the upper ends of a pair of futtock riders, which Witsen forgot to draw in the previous illustrations. It should be mentioned here that Witsen constantly mixes up upper and lower decks in his text. Throughout the translation this mistake has been corrected without comment.

In figure 2.117 we see the riders in the stern, with futtock riders, a rising floor rider—not the rising floor timber or crutch, as treated in section 26 (The Floor timbers, bilge Futtocks, Steekers, Crutches, and Transoms), which is part of the frame, but a rider to reinforce the ceiling—and the knees of the *broekstuk,* joining the fashion frames below the tuck (see section 7, The Broekstuk) and of the lowermost transom, just barely visible.

59. When it has been launched, then make scaffolding outside and to the stern.

(155 | 26) At *KK* [fig. 2.118] on the same plate, *a Ship is shown, that has just been launched,* to which the scaffolding is being made, ready before the Yard.

a Is the Putlog.
b Is the Putlog strut.
c Is the staging by which one enters the stern.
d The Mast, on which the scaffolding is made.
e Is the bank, on which the lower end of the scaffolding lies.

A notch is made in the lower face of the Putlogs, in which the strut is made to stand with its end, the strut is cut off with an angle, at the lower end coming against the ship.

At *MM* [fig. 2.119] *a Putlog* is shown with the lower face facing upward, and the notch in it, *a* is the hole of the putlog, and *b* the end toward the ship, it is fastened to the ship on the Wale, or whatever serves best.

a On the plate *NN* [fig. 2.120] is a Putlog, seen from below.
c The End of the Putlog, lying on the wale.
b The strut below the Putlog, the one end fastened to the ship, the other standing in the hole of the putlog, but nailed; and when the putlog is long, as they are forward and aft, then two struts are made to it.

FIGURE 2.113. The pinas, shortly before being launched. Clearly visible are the upper deck beams, which are not yet planked. The bottom is supported as well as possible after having been caulked (Courtesy G. A. de Weerdt)

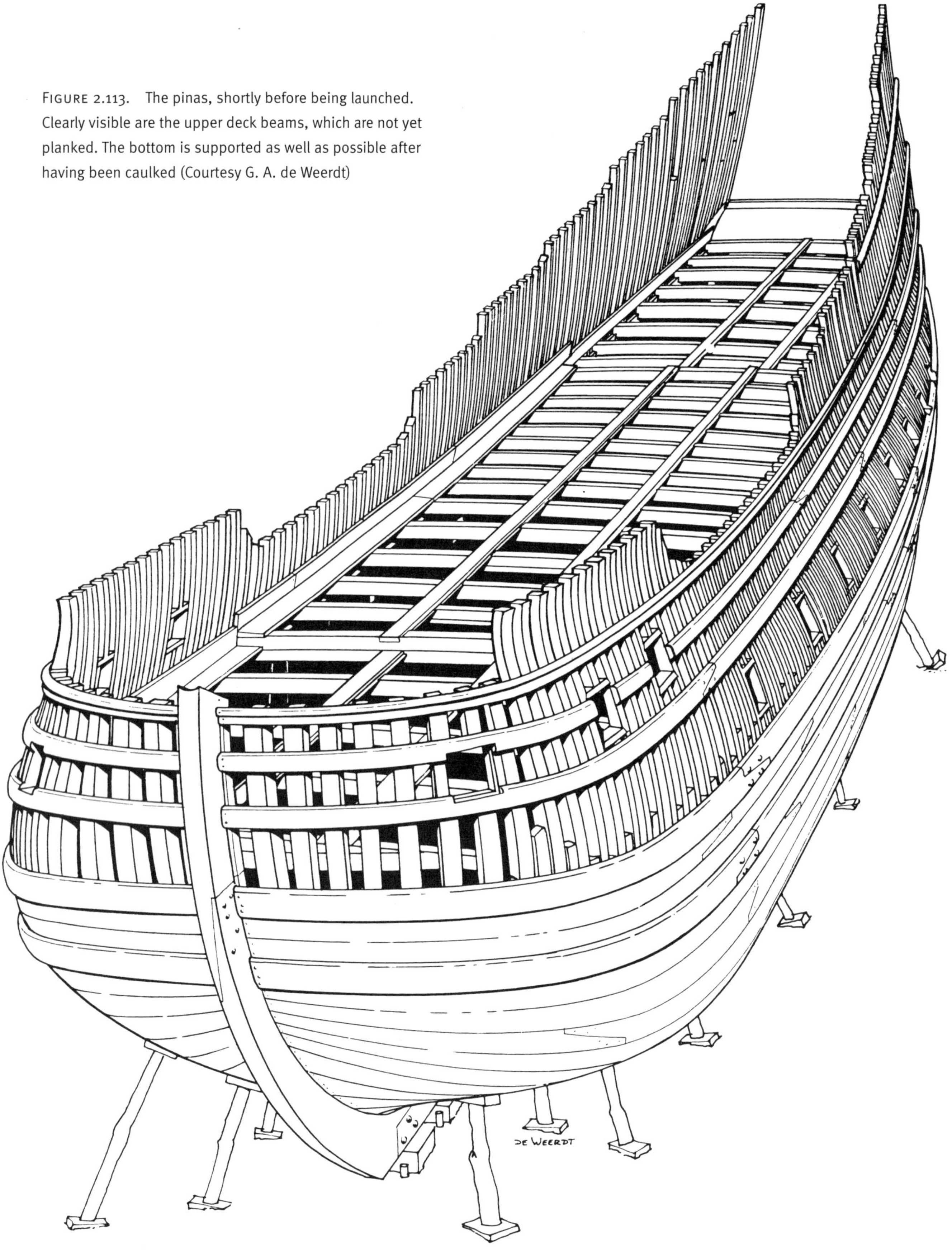

59. When It Has Been Launched, Then Make Scaffolding Outside and to the Stern

On the upper face of the wales putlogs were nailed, which were supported with struts, also nailed to the ship's side. Staging was laid across the putlogs, giving the workmen a place to stand. Again, this is a surprisingly simple kind of scaffolding.

60. Make the Chocks above the gunports around the outside.

(155 II 3) Then planks are laid on the putlogs, and when the scaffolding is finished, then the Chocks above the gunports are made, and the port sills at the base of the gunports, after which the upper wales and planking in between are made, and also the Sheer Rail, Kevel Rail and topsides: And inside, the upper Waterway, binding strakes and ledges; then plank the main deck, cabin and also the quarter gallery, make the capstan, also the knights, the crosspieces of the bitts, the cleats, the carlings, and railings, after which the masts are placed, and also the bowsprit.

On the plate at *OO* [fig. 2.121] *a* is *the Chock,* which is made with a dovetail at the lower edge, such that it comes to nil at the top; *b* is the Sill of the gunport,

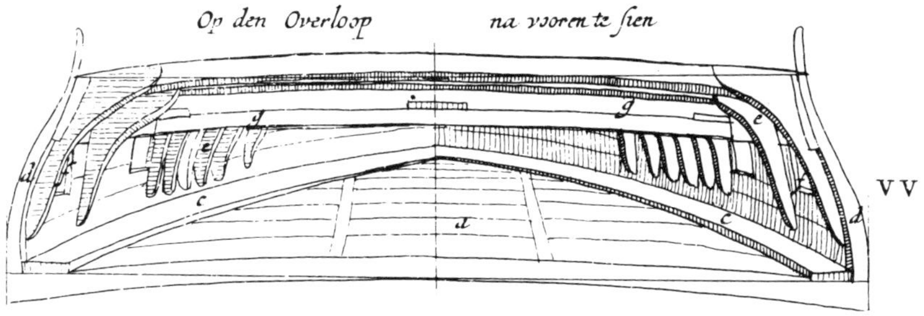

FIGURE 2.114. Plate LIX. On the lower deck, looking forward.
(156 I 53) VV Is a view inside the ship, when standing on the main deck, facing forward: where *a* is the Floor, *b* points out the Binding Strakes, *c* the Waterway, *d* the top timbers, *e* the hanging knees, and *f* a little of the gunports in the sides, *g* the upper deck beams, or beams of the second deck.

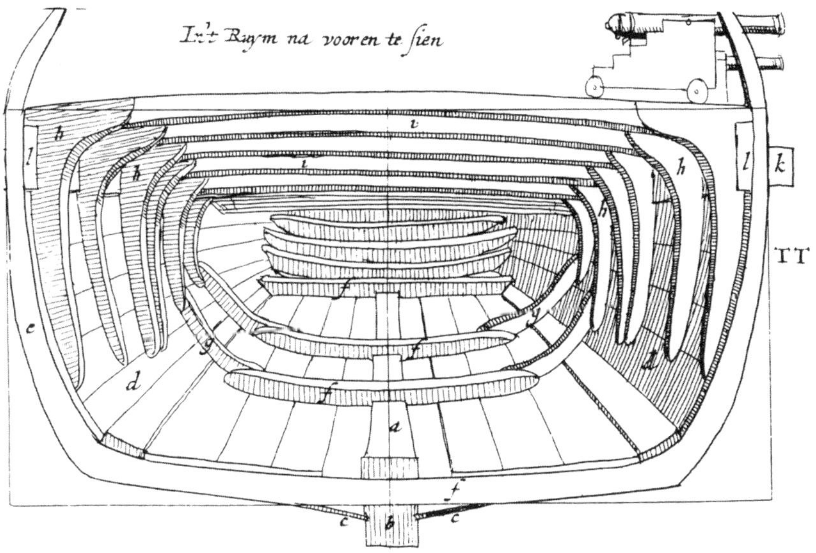

FIGURE 2.115. Plate LIX. In the hold, looking forward.
(156 I 42) On the plate at TT, is shown the interior of the Ship, when standing on the bottom, in the hold, and looking forward: Where *a* is the Keelson, *b* the Keel, *c* the garboard strake, *d* the ceiling, and *e* the Futtocks, *f* the Rider Floors, *g* the Bilge Riders, and *h* the hanging Knees, carrying the main deck, *i* the first main or lower deck beams, *k* the Wales, and *l* the Deck clamps.

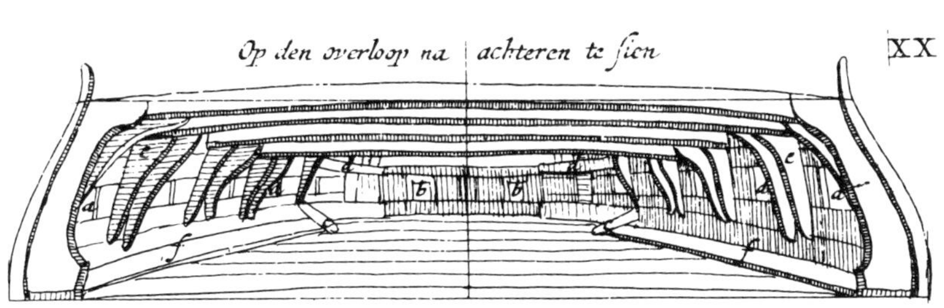

FIGURE 2.116. Plate LIX. On the lower deck, looking aft.
(156 II 15) In the fourth figure at *XX* the main deck facing aft, the bulkhead of the Gun Room removed, *a* are the gunports in the sides, *b* two gunports shooting aft, *c* riders going below, *d* are two transom knees, *e* the upper deck's hanging knees, *f* the Waterway on the lower deck.

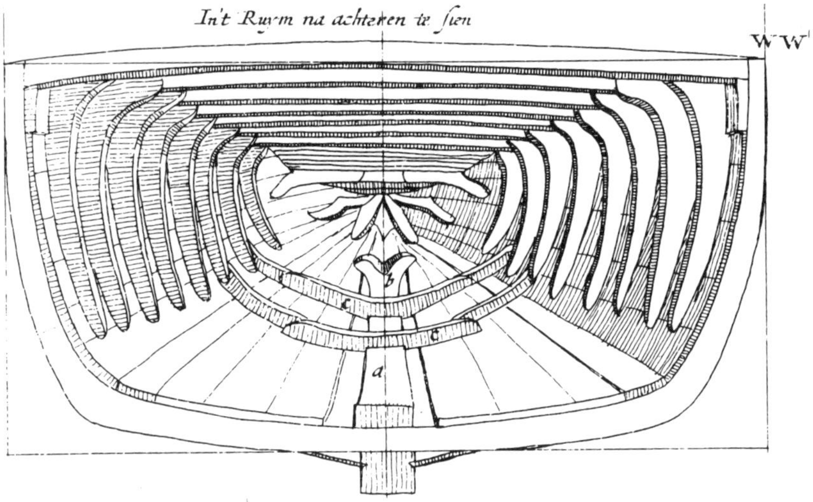

FIGURE 2.117. Plate LIX. In the hold, looking aft.

(156 II 7) The third picture on the same plate at *WW,* shows the Ship as seen, when standing in the hold on the bottom, looking aft, where *a* is the Keelson again, *b* a Crutch [or rising rider floor], *c* Rider Floors, *d* riders and crutches, etc.

FIGURE 2.118. Plate LVII (drawing KK)

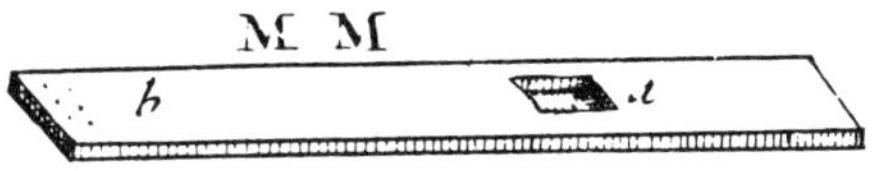

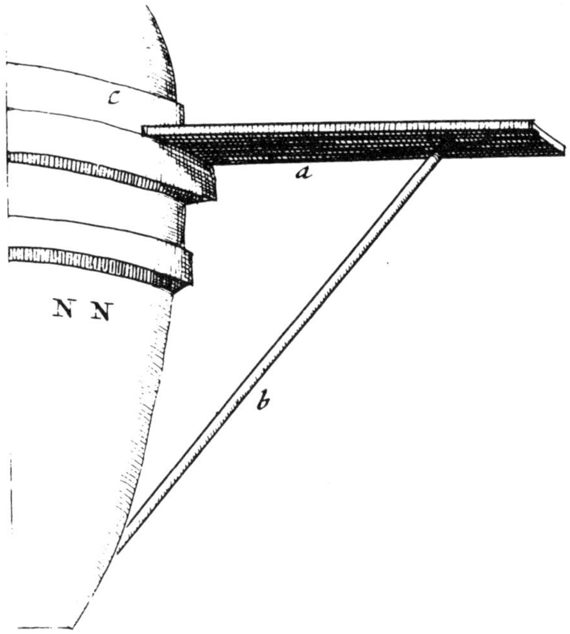

FIGURE 2.119. (*top*) Plate LIII (drawing MM)

FIGURE 2.120. (*bottom*) Plate LIII (drawing NN)

also made with a dovetail in the top timber, *c* are the top timbers. To this then the port lid is made, in the Chock the pegs of the port hinges are made, and they are shut from inside. The gunports are cut according to the rising and falling of the main deck and upper deck.

(55 | 41) Chocks are made to close something, and keep it secure.

. . .

(77 | 36) *About the chocks or lintels, above the gunports and sill.*

1. The chock, above the gunport, thick 4 inches: the sill, thick 4 inches, being equal with the top timbers inside and outside.

(77 || 46) The gunports come 2 feet 2 inches above the main deck, but sometimes also 2 inches wider: stand apart 8½ to 9 to 9½ to 10, and sometimes 11 feet.

(71 | 22) 31. *About the Gunports.*

1. At the main frame, the height of the upper deck is divided into three parts, one part below, one part above the gunports.

2. The sill of all the gunports after this height, aft and forward somewhat wider.

. . .

(77 | 41) The gunports are made wider today than before, to better aim the gun.

(78 | 1) The rearmost gunport, is made comfortably away from the stern timber and such 2½ feet, the other gunport 11 feet from the rear of the port, or 13 feet from the stern timber: the third gunport 10 feet from the rear, the fourth gunport 9 feet, the fifth gunport 10 feet from the rear, the sixth 10½ feet, the seventh 11½ feet the eighth 12½ the ninth gunport 12 feet from the rear, and so forth.

Forward in the forecastle, near the bow to measure, 6½ feet from the stem, comes the first gunport, at seventeen feet the second, and at 31 feet the third, etc.

60. Make the Chocks above the Gunports around the Outside

When the uppermost parts of the frames, the top timbers, were placed, the position of the gunports had already been taken into account. Usually they were arranged such that they fell between two frames, wasting as little wood as possible. The sills and lintels were joined with side scarfs between the frames.

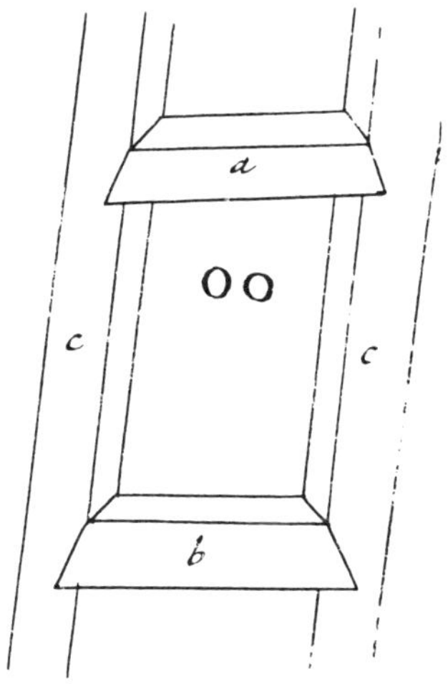

FIGURE 2.121. (*top*) Plate LIII (drawing OO)

FIGURE 2.122. *(bottom)* Plate XXXVII. Chock in the porthole, sill of the porthole.

The height between the decks was then divided into three equal parts, the middle part being the height of the gunport.

The width was generally slightly more than the height—by 2 inches according to Witsen, by one quarter of the height according to Van Yk. A formula relating the sizes of the gunports and the distances between them to the calibers of the guns can be found in a mid-seventeenth-century manuscript written for one of the admiralties, titled Evenredige Toerusting van Schepen ten Oorlog Bijder See (Proportional Equipage of Men-of-War). According to this source, for heavy ordnance the height of the port was six times the diameter of the cannonball, the width five times, and the distance between the center of the ports twenty times the diameter of the cannonball. In the eighteenth century the same sort of formula was used by, for instance, the Rotterdam admiralty shipbuilder Van Zwijndregt.[12]

There was a minimum and maximum distance between gunports (six and twelve feet), also based on the necessary space to handle the gun and the ship's construction, but Witsen is fairly careless in his calculations. The gunport distances of the pinas thus vary from 9 to 12½ feet, and the calibers of the guns do not seem to have been a factor whatsoever. Van Yk does not give another rule for the position of the gunports other than what would be practical, or what "Knees, Top timbers, Galley, Buttery, Dead-eye links, &c will allow" (Van Yk, p. 117).

For the pinas Witsen supplies several measurements for distances between his gunports. What is confusing is that his enumeration of the distances of the gunports on the main deck and in the forecastle both end with "and so forth," while later he clearly states that there were no more gunports than the ones he mentions here.

Witsen's text relating to the placement of the chesstree (section 114, Make the Chesstree) provides additional evidence that his gunport measurements were almost certainly taken from an existing ship and not merely the product of his fancy. The chesstree was to be placed at two fifth parts of the length of the ship taken from the bow, or straight above the second gunport, below the upper deck (90 I 37). It is clear that someone measuring the distances between the gunports, beginning in the stern, could not possibly be aware that the second gunport, seen from the bow, would be underneath the chesstree, which is at two fifths of the ship's length from the bow; Witsen or the person who gave him the data must have had a close look at the ship. More of such examples will follow.

61. And the Filling Strakes in the hold.

(54 II 15) The planks which are set in the ship's side are called Filling Strakes; gun filling is the flat wood, coming between the guns.

. . .

(80 I 42) They are called filling strakes, because they fill the place between the wales: they are made as broad as the ship allows. The lower filling strake must be broad enough for the scupper holes to be comfortably made in it; and the scuppers come out well underneath the second wale. The Gun filling is made wide according to the gunports. The filling strakes above the Gunports should be made according to the upper deck broad enough for the scuppers of the upper deck can comfortably drain through them. The uppermost filling strake, in an Armed ship, is also made broad after the height of the gunports. It is important to mind, that the filling strakes, going upward, should become thinner and thinner, and that from the Gun filling onward.

61. And the Filling Strakes in the Hold

Witsen's formulation again leads to confusion. There are no filling strakes in the hold. Filling strakes are the planks between the wales. Between the gunports they are called gun fillings. The scuppers would be led through the filling strakes, through which the water on the decks drained. They were half as thick as the wales and gradually became thinner than the ones below.

62. Hereafter it is done up outside, with Wales, Filling Strakes and Sheer rails, Washstrake, Vertuining and Railing.

(55 I 5) Sheer rail is the uppermost thick timber, which can be observed on ships, which covers the sides, and at the main frame, is the highest wood. This is shown in the sixth plate [fig. 2.124].

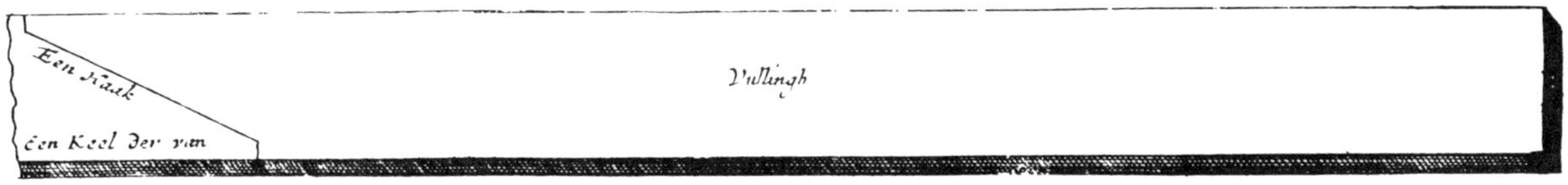

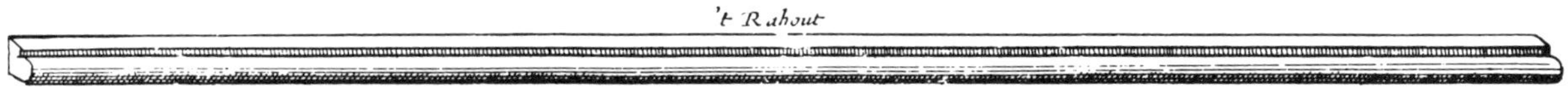

FIGURE 2.123. (*top*) Plate XXV. Filling plank.

FIGURE 2.124. (*bottom*) Plate XXV. Sheer rail.

(77 II 15) *About the Sheer rail.*

This is the uppermost wale, and is called as such [*ra-hout*] because the yard [*ra*] lies on it in some ships: and also has another form than the wales.

> (77 II 21) 1. The Sheer rail, 8 feet from the stem, broad 9 inches, thick 3½ inches.
>
> 2. The side broad 2½ inches, the other around the main frame wide 5 feet, aft one inch [mysterious formulation].

(54 I 38) Washstrake: is closed planking, which, at the main frame, where the ship's side is at its lowest, is made to remove and set at will: which is done to seal the ship better; on ships of war it has round holes for muskets, and small doors through which things are raised and entered.

> (77 II 26) 9 feet from the bow, or measured from the stem around the bow, there the washstrake begins, on the sheer rail: there it is broad, between the railing and the sheer rail, 19 inches, and above the first gunport broad 16 inches, farther aft at the end of the forecastle broad 18 inches, thick 1½ inches: the hole 3 feet 7 inches, past the aftermost beam of the forecastle: the hance comes forward: the washstrake is broad 20 inches, and let into the top-timbers above like the railing, broad 3½ inches, and thick 1½ inches: the molding 1 inch: the other railings wide 8½ inches, forward 9 inches, broad 5 inches, and thick 3 inches: the molding strip is 1½ inches, straight above and hollow below.
>
> The washstrake on the sheer rail, aft broad 18 inches, thick 1½ inches, forward 2 feet.

> (80 I 3) 37. *About the Washstrake.*
>
> 1. The washstrake on the sheer rail, forward broad 19 inches.
>
> 2. Aft broad 18 inches.
>
> 3. Thick 1½ inches.
>
> 4. Notched above, and comes to the outside equal to the top timbers.
>
> (80 I 18) 39. *About the Washstrake aft of the Mast.*
>
> 1. The washstrake, aft of the mast, broad 19 inches, thick ¾ inches.
>
> 2. The railing there broad 3½ inches, thick 1½ inches.

(54 I 32) Vertuining board, plank or wood: is wood, or are planks, that are placed on the Ship's sides, above the sheer rail, toward the stern, when ships are planked high.

Above the sheer rail, forward as well as aft, starting from the hances, it is vertuint [planked] as should be, and generally aft of the [main] mast 4 hances are made, or sometimes 3, depending on whether the stern is planked high or low; the lowermost strake, on the sheer rail, is always as broad forward as aft. Then the vertuining [planking above the sheer rail] is continued with wainscot boards, of a thickness as required. It should be minded that the vertuining boards aft against the large countertimbers are made a little broader, and coming in, so as not to have the vertuining look drooping. When it is vertuint properly, then the ship is left open between the rails and top planking, as wide as necessary; and on every hance the aftercastle side is made, one foot or a little less high, according to the size of the ship, but aft it is closed between the open strakes, and such around the cabin; that can be made so much higher.

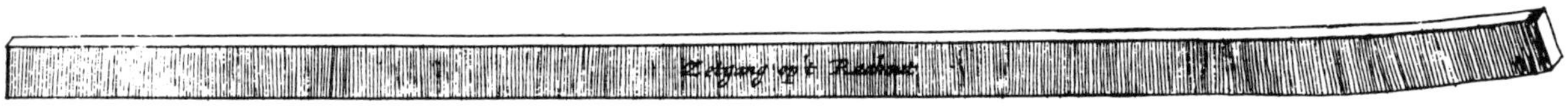

Figure 2.125. (*top*) Plate XX. Wash strake on the sheer rail

Figure 2.126. (*bottom*) Plate XX. *Vertuining* plank

(274 II 7) In heavy ships, the side planks, that rise aft, are made of wainscot, which for the sake of strength, lie with their edges overlapping, and for beauty's sake, rise with a curve inward.

(80 II 17) 41. *About the vertuining.*
1. The lowermost board broad 10 inches, forward 9 inches, thick 1 inch.
2. To the second railing from forward one board.
3. To the third two boards.
4. To the fourth three boards.
5. To the fifth five boards.
6. Thick 1 inch, broad 1 foot.
7. Notched above.
8. The fourth railing stands 5 inches apart.
9. The fifth hance is straight above, curved below.
10. The hance is thick and broad as the railing.

(55 I 1) Fillings: are pieces of wood, which are used to fill one place or the other, where wood is short.

. . .

(55 I 33) Railings, as depicted on the fifth plate [fig. 2.128], are broad battens, that are placed above and around the ship's side aft and forward.

(80 I 11) 38. *About the Railing forward.*
1. The railing here is broad 3 inches, thick 1½ inches.
2. The Molding broad, 1 inch.
3. The others, standing 8½ inches above it, are broad 5 inches, thick 3 inches.
4. The Molding is broad 1½ inches.

(80 I 23) 40. *About the uppermost railings.*
1. The uppermost railing at the main Mast stands 2 feet 8½ inches from the sheer rail.
2. The other, aft, 3 feet 6 inches, from the sheer rail.
3. The third 4 feet 5 inches.
4. The fourth 6 feet 1 inch.
5. The fifth 7 feet 1½ inches.
6. Aft straight up, 7 feet 9 inches.
7. The aftermost is long 18 feet 9 inches.
8. The other forward from there, 10 feet 5 inches.
9. The third forward, 13 feet 6 inches.
10. The fourth forward, 14 feet 5 inches.
11. The fifth forward, 14 feet 4 inches.

62. Hereafter It Is Done Up Outside, with Wales, Filling Strakes and Sheer Rails, Washstrake, *Vertuining,* and Railing

With this step the external planking of the hull was finished. The lower pair of wales had been placed before launching, and the second pair was now made above the gunports so that the scuppers of the upper deck would come between them.

Above this pair, in between which filling strakes closed the hull, was the area for the gunports of the upper deck, if need be. If these were not made, the space in between the second and third wale was often not much broader than that between the lower wales.

Figure 2.127. (*right*) Plate XXVI. Filling chock

Figure 2.128. (*below*) Plate XXIV. Railing

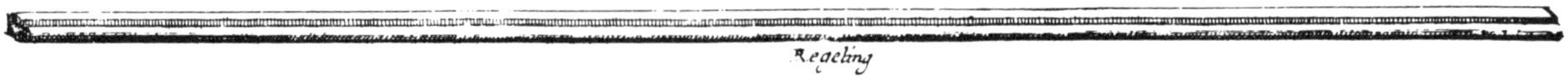

Above this the sheer rail or capping rail was made. Originally this was the rail onto which the main yard was lowered, but in Witsen's time ships were built higher. No ordinary wale, the sheer rail was distinguished by a decorative molded profile.

The washstrake was different amidships, forward and aft. Amidships the sheer rail was still the uppermost fixed part, on which a loose plank, the washboard, was set upright. The inboard side of the washboard was supported by small stanchions, the lower protruding ends of which sat in holes in the capping rail. The washboard could usually be removed, but sometimes it was fixed in place, supported by the tops of the top timbers. It often had a hinged section that served as a door for entering the ship.

Forward and aft of this midship section, the washstrake was always permanently fixed and served as the base of the upper works (the clinker-built upper part of the ship's side, called *vertuining*). The washstrake was often painted in color different from that of the sheer rail underneath it and the upper works above it.

The planks of the upper works (*vertuining*) were made of good wainscot oak and shaped slightly curved, broadening toward the stern, which was thought elegant. To keep the structure as light as possible, the planks were nailed in clinker fashion to the ship's side. Those of the pinas were only one inch thick (compared to the four inches of the bottom planks).

The top of the upper works was closed off with the so-called railings (*regelingen*) or capping rails that covered the upper ends of the top timbers. The uppermost plank was often left out, leaving the rail to look like banisters.

The transitions between the railings, which rose toward the stern in steps (and forward in the same way, though not as much), was made with carved hance pieces. Every hance marked the beginning of another strake of top-side planks.

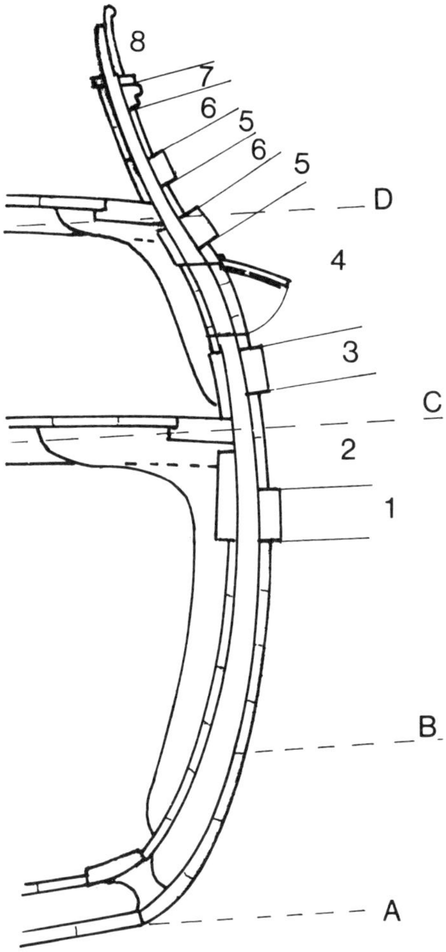

FIGURE 2.129. Parts of the ship's side. *A*, bottom; *B*, top of the bilges; *C*, the watershed; *D*, upper deck height; *1*, lower wale; *2*, filling strakes; *3*, second wale; *4*, filling strakes between the gunports; *5*, top wales; *6*, top filling strake; *7*, sheer rail; *8*, washboard, removable in the waist, fore and aft, where it forms the basis of the top side planking, the washstrake. (Drawing by A. J. Hoving)

63. Then the upper Waterways are made.

(77 | 19) The waterways are laid on the beams, close to the side, and made thick according to the size of the ship; often of the thickness of the main deck planking. In here the scuppers are made, as many as needed, which scuppers are made in pieces of square wood which are inserted in the waterway. These waterways are well fastened onto the beams, with bolts, and also from outside the ship with bolts, that go through the waterways.

(77 | 3) 27. *About the Waterway, on the Upper Deck.*

1. The Waterway, on the upper deck, is broad 16, or 19 inches, thick 4 inches.

The Waterway of the lower deck, broad 19 inches, thick 5 inches, in the forecastle broad 18 inches, on the cabin thick 3 inches, broad 16 inches, on the upper deck broad 17 inches, thick 3½ inches, on the forecastle broad 10 inches, thick 3½.

(77 | 31) The waterway lies about 2 inches above the height of breadth strake, and the beams are cut as much as the waterway is let in, about 2½ inches.

FIGURE 2.130. (*top*) Plate XX. Waterway of upper decks.

FIGURE 2.131. (*bottom*) Plate XX. Binding strake of upper decks.

63. Then the Upper Waterways Are Made (and the Binding Strakes)

On the upper deck the planking was prepared in the same way as for the lower deck: heavy waterways were laid against the sides, and in the middle, next to the locations reserved for the hatches, the binding strakes were laid. As on the main deck, the binding strakes did not run in one line from bow to stern. At the large skylight grating they lay farthest apart; forward and aft of that they lay nearer to one another, as the other hatches and openings were not as wide. The binding strakes and waterways both had grooves in which the ledges fitted (see 2.104).

There were different methods for joining the waterways and binding strakes to the deck beams. Sometimes a section of the beam was cut thinner to let them in, and sometimes a part of the binding strakes was cut to fit over the beam. With the pinas the beam was left untouched except for two small cuts at the edges. Holes were made in the waterway to fit in the *bos,* square wooden pieces with scupper holes drilled through them (see fig. 2.132).

64. With the Spirketing on it.

(77 I 12) 2. The spirketing on it, broad 10 inches, thick 2½ inches.

64. With the Spirketing on It

The spirketing was fitted against the ship's side on top of the waterway, as on the main deck. This ceiling or gunwale was continued all the way up to the rail. In the forecastle, steering stand, and cabin, however, there was only one ceiling plank, topped by the covering board, leaving the top timbers above the covering board exposed.

65. The Ledges on the Main Deck.

(90 II 17) The ledges thick 2½ inches, or 3 inches. There comes one on each end of the ledge [the carlings are meant].

65. The Ledges on the Main Deck

The ledges, as mentioned earlier, were small extra deck beams, lying in the grooves of the waterway and binding strake and giving extra support to the deck planking. They were as thick as the hull planking. Usually three ledges lay between a pair of beams, but there could be as many as five where the deck beams lay seven feet apart at the main hatch.

66. The Carlings underneath.

(56 II 48) the Carling comes longitudinally underneath the ends of the ledges.

. . .

(90 II 11) The carlings are let into the deck beams underneath the ledges, because the guns, would otherwise bend the main deck.

. . .

(68 II 35) [. . .] carlings ⅓.

(90 II 15) The carlings thick and broad 4 inches, the ledges 3½ inches, the carlings in the forecastle thick 3½ inches.

66. The Carlings Underneath

The ledges were in turn supported by longitudinal timbers let into the sides of the deck beams. Thus a web of supporting timber was laid underneath the deck planks. The

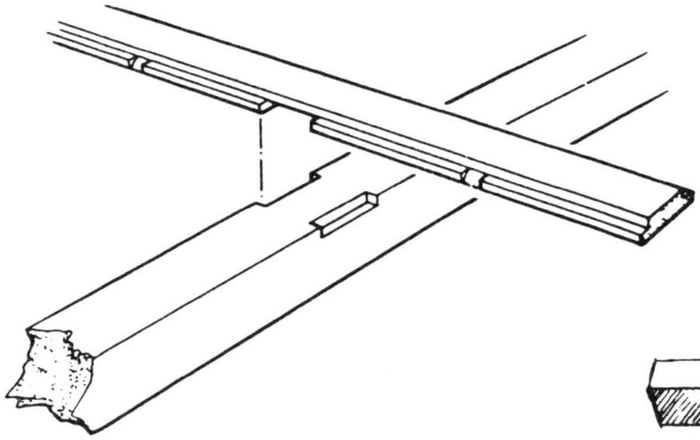

FIGURE 2.132. (*left*) There were several ways to connect binding strake and deck beam. Sometimes the binding strake was kept in one piece and the beam was adjusted. In this example the beam is only notched, like the binding strake, to weaken it the least. (Drawing by A. J. Hoving)

FIGURE 2.133. (*below*) Plate XXXVII. Carling.

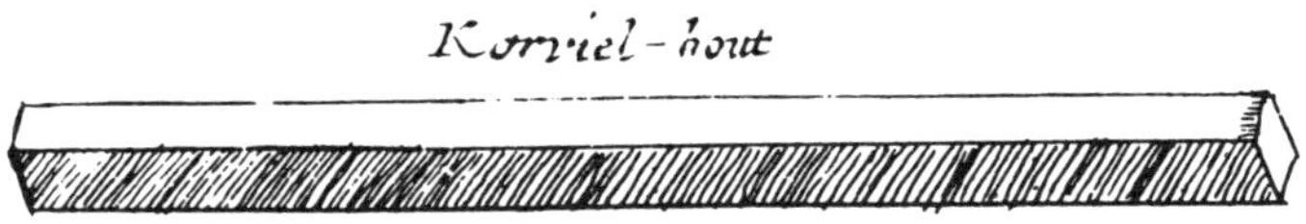

number of carlings depended on the length of the ledges. They were a little heavier than the ledges.

67. The step of the Capstan, and the Mizzenmast.

(79 I 41) 34. *About the Mizzen Step and the Partner.*

1. The Mizzen Step comes at ⅕ part of the ship's length.

2. The Step is thick 8 inches, broad 14 inches.

3. The Step rakes 8 inches.

4. The Hole in the step, long 8½ inches, deep 4 inches.

5. Aft wide 6½ inches.

6. Forward wide 5½ inches.

(79 II 7) the step long 4 inches thick 8 in., broad 17 inches: the hole wide 7 inches, deep 4 inches.

(79 II 16) The Back of the mizzen step, lies 5 feet 10 inches from the bulkhead.

67. The Step of the Capstan, and the Mizzenmast

These steps held the foot of the capstan and the mizzenmast and were on the lower deck, not in the hold, as for the main mast and foremast. They were heavy timbers with a nice profile forward and aft. They were located to receive maximum support from the underlying deck beams.

68. The lower Spur of the beakhead.

(55 I 30) The Spur of the Beakhead: is the timber, which sticks out forward: and makes the length of the beakhead, on top of it comes the kam.

(70 II 39) 1. The spur of the beakhead ⅕ of the length, [. . .]

2. Forward broad ¼.

3. The spur of the beakhead broad at the stem as the stem is thick inside.

4. Forward ½.

. . .

(86 II 25) The spur of the beakhead is to be fastened to the stem after the rise of the wales forward, and it is to be heeded that is does not droop, or appear to hang from the ship.

(86 II 4) 1. The spur of the beakhead, long 22 feet, aft thick 14 inches, broad 17 inches, has a curve of 17 inches.

2. Forward broad 12 inches, 9 feet aft, broad 11 inches at the end of the lion, broad 9 inches, 2 feet from forward, broad 5 inches, namely on the upper face.

68. The Lower Spur of the Beakhead

The construction of the beakhead differed from the English practice. Therefore the terms I use may not always be the best. Where no English term is available, the Dutch word will be used.

FIGURE 2.134. (*top*) Plate XXIX. Step of the capstan.

FIGURE 2.135. (*bottom*) Plate XXIX. Mizzenmast step.

This section should be placed with the rest of the beakhead in section 80, **The Beakhead**, but Witsen already has the lower spur made here as the basis of the beakhead. It would have been more logical to treat the rest of the beakhead subsequently, but Witsen doesn't. Only at section 80 does he return to it. We can assume that he was attempting to indicate that work on the ship was carried out at several places simultaneously.

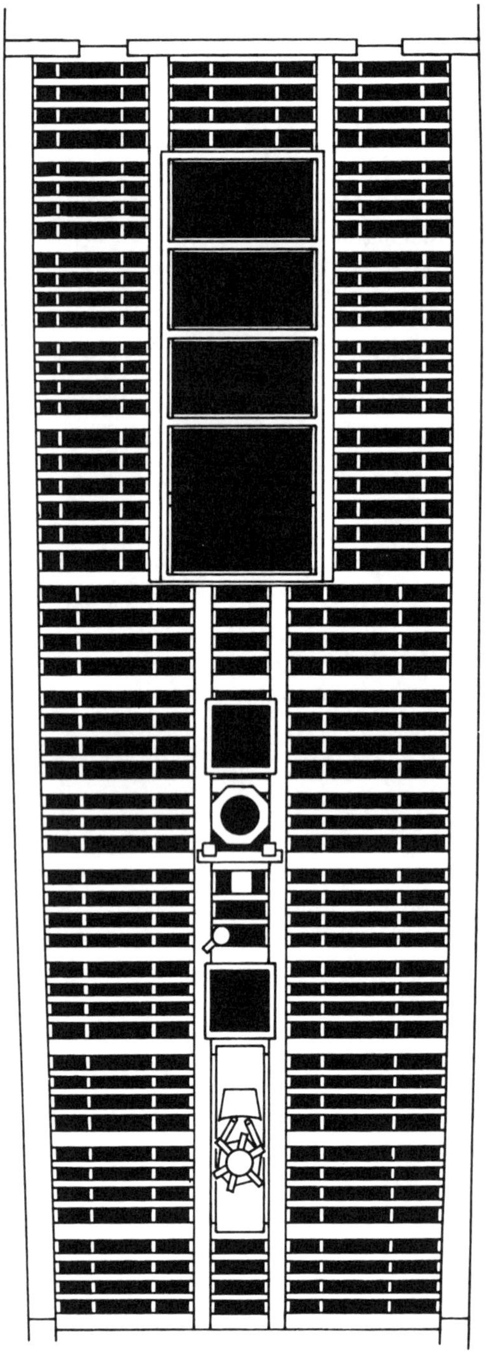

FIGURE 2.136. Bird's-eye view of the upper deck in its unplanked state. The function of the binding strakes is immediately apparent: they determine the width of everything between them, such as mast partners, hatches, partners, bitts, and so on (Drawing by A. J. Hoving)

The lower spur of the beakhead is a long, slightly curved timber, tapering toward the forward end and ending in a carved scroll. Its tail end was let into the stem and supported by the beakhead knee.

The angle of this part was very important for the line of the ship. Hanging too low, the beakhead would "droop" and disfigure the ship; coming up too high, it would place the beakhead in conflict with the bowsprit.

From the beginning of the seventeenth century until Van Yk's time, the length of the spur of the beakhead gradually decreased from one fifth to one eighth of the ship's length. At the stem it was as thick as the inside of the stem. As the front face of the stem was only three fifths of the inside, the lower spur of the beakhead would protrude a little on each side (2½ inches). This was no coincidence: when the cheek knees or cheeks were installed later on, bracing the spur of the beakhead laterally, these openings served as scuppers or drains for the water leaking in through the corners between hull, stem, and cheeks.

69. The coamings of the Gratings, on the upper deck.

(54 I 46) In some hatches, but mostly on Merchant ships, are covered holes, through which the cable goes.

. . .

(70 II 32) *27. About the Hatches.*

1. The coamings of the hatches, have ⅓ the thickness of the stem.
2. The inner frame, thick and broad ⅖.
3. The groove ⅛ wide.
4. The ledges ¼.
5. The hatch beam ⅓; the others less.

. . .

(55 II 18) Hatch beams: on the plate No. 10 [fig. 2.139], are pieces of wood, lying across the hatches, on which the hatch covers rest in the middle.

(86 I 26) The hatch beam is hollowed out a little in the middle, to each hatch is a ring, the hatch beams are athwartships: the main hatch, and the hatch behind the knight, have no hatch beam on one side.

FIGURE 2.137. Plate XXVI. Lower spur of the beakhead, which supports the kam.

(85 I 8) *About the Hatches.*

The *Main hatch,* long and wide 7 feet, the groove wide 1 inch, the inner frame thick and broad 2 inches, the hatch beam broad 4½ inches, the coamings 4 in. inside and out. The coamings of the *main grating* broad 9 inches, thick 5 inches, are wide and broad 7 feet, in the middle, the coaming broad 8 inches, thick 5½ inches. The laths broad 3½ inches, thick ½ inch. The small hatch before the mast 2 feet 5 inches, long and wide 3 feet. The coamings thick and broad 7 inches. The laths thick ½ inch, and broad 3½ inches, lie fore and aft on the ledges.

The *grating* for the main hatch, and on the forecastle, they are of even width, the coamings on the forecastle broad 8 inches, thick 4 inches: the ledges broad 1½ inches, thick 2 inches: the laths broad 3½ inches, thick ½ inch, the other on the upper deck as above. The hatch aft of the riding bitt, lying 3 feet away from it, long and wide 3 feet: the inner frame broad 2 inches, thick 1 inch: the ledges thick 2½ inches, broad 3 inches: the groove wide 1 inch, the outer frame, broad 8 inches, thick 2½ inches, inside and out ½ inch, having to either side a sloping plank, in the middle lie one inch deals. Hatches above the bitt standards are 2 feet wide and broad.

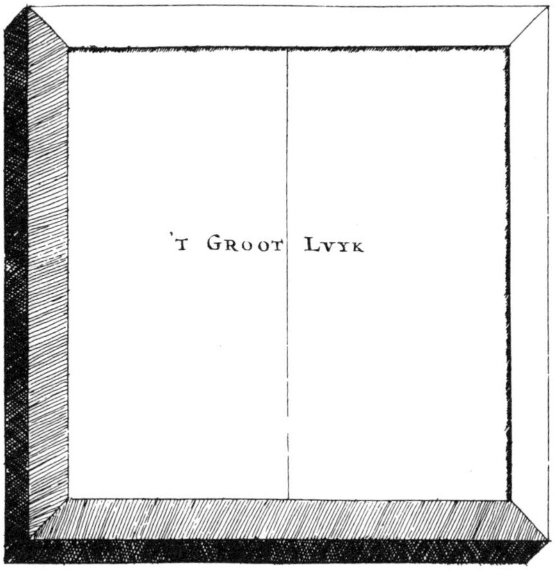

FIGURE 2.138. Plate XXIX. The main hatch.

69. The Coamings of the Gratings on the Upper Deck

The coamings of the hatches were placed on the planks of the main deck and on the upper deck. They were the profiled edges around the hatch openings. The hatch covers fitted in the square opening framed by the coamings and lay in a groove that had been planed into the coamings. If necessary, the gratings could either be covered with boards or with canvas, which allowed some light onto the lower deck.

At this time it is perhaps useful to note Witsen's division of the decks (see fig. 2.141). Apart from the two hatches mentioned above, the main deck also had a small hatch forward to the boatswain's storeroom (*hel*); aft, in the gun room, there were two small hatches, one leading to the peak and the other to the bread room (see section 101, After this make the Bordered Hatch Covers).

On the upper deck from bow to stern several hatches could be found. Behind the riding bitt (see sections 76, Place the Bitts; 84, The Bitt Knees; and 86, The Bolster) was the hatch to the cable tier, where the anchor cable was stored during the voyage. This hatch was different from the other hatches in that it had a sloping plank all around, probably for draining the water seeping out of the cable or so that kinks in the cable did not foul on the coaming.

In the forecastle, forward of the bulkhead, there was a hatch descending to the main deck. Aft of it lay the main skylight grating, admitting light and air to the main deck. The skylight was covered with a number of gratings (see section 100, And the Hatch Gratings). The aftermost part of the skylight grating was the main hatch, through which large pieces of cargo could be put into the hold. When hoisting, the last two gratings of the skylight and the hatch beam, an unconnected beam supporting the grating in the middle, were removed. There was a groove in the hatch beam through which water drained. Straight underneath was the main hatch of the main deck, of equal dimensions. It provided access to the hold.

Between the skylight grating and the mast there was a small hatch allowing the crew access to the lower deck; it must have had a companion ladder or flight of stairs.

Abaft the mast there was another grating, smaller than the other one and also straight above a similar hatch on the main deck. This was the access to the hold for smaller cargo. Behind the bulkhead of the steering stand was an-

FIGURE 2.139. Plate XXIX. *Above:* hatch beam; *below:* ledge.

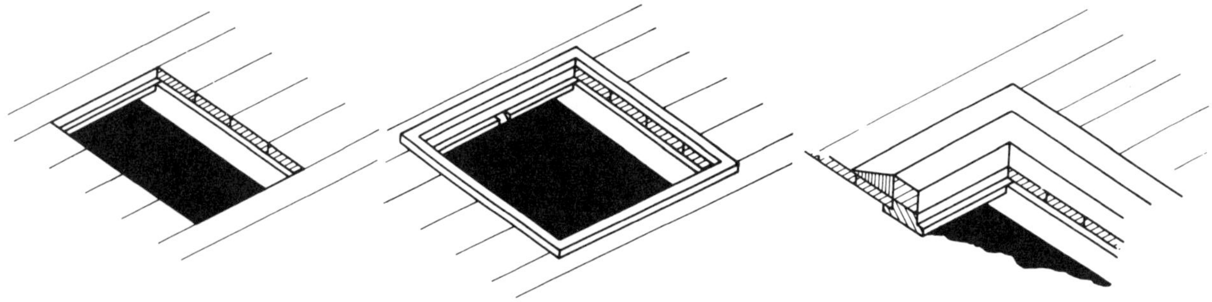

FIGURE 2.140. Hatches. *Left:* The simplest kind of hatch; the rabbet consists of the binding strake and a part of the beam. *Center:* Main hatch with coamings. *Right:* Cable tier hatch, bordered with a sloping plank (Drawing by A. J. Hoving)

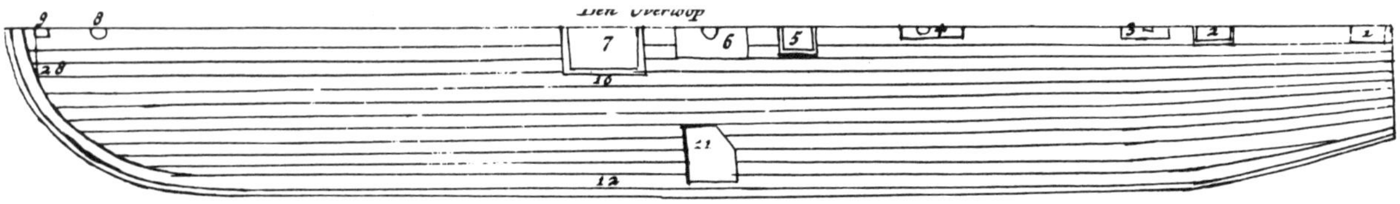

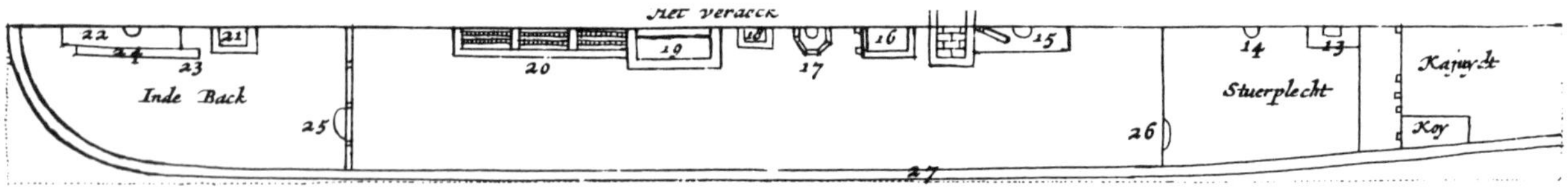

FIGURE 2.141. *(top)* Plate XLII. The main deck.

(59 I 35) On the main deck, the number 1 shows the small hatch abaft the bread room. 2 The hatch of the bread room. 3 The mizzen step. 4 The Step of the capstan. 5 A hatch abaft the mast. 6 The main partners. 7 The main hatch. 8 The hole for the foremast. 9 The small hatch in which shot or something else is stored, for which the hatch 1 is also used. 10 Are the binding strakes. 11 Is the buttery. The galley is at the same height as the buttery, but at the other side, which is why it cannot be shown here: on ships of war, it is placed on the Ship's bottom below, athwartships. 12 Is the waterway; lies on the ends of the deck beams, against the frames, on either side, up to the stem. It must be mentioned, that on the main deck, here only indicated with planks fore and aft, lie riders; deck beams across, and ledges, also waterways, and carlings, coming underneath the ends of the ledges. At 28 or about a breast knee lies against the bow. On ships carrying many people, the main deck is often raised a little to 7 or 10 feet from the stem; to have all dirt flow away more easily, and keep the ship clean forward.

FIGURE 2.142. *(bottom)* Plate XLII. The upper deck.

(59 II 10) At 13 on the plate of the upper deck, is the rowle of the whipstaff: in which the whipstaff goes and plays. At 14 is the mizzen partner and step. 15 Partner of the capstan. 16 The hatch behind the mast. 17 Small hatch around the mast. 18 Small hatch before the mast to descend. 19 The Main hatch. 20 Skylight grating with its coamings. 21 Small hatch behind the riding bitt. 22 Foremast partner. 23 A bitt standard. 24 Bitt knee [. . .]. 25 Is a companionway to descend to the main deck. 26 Likewise [another companionway]. 27 The waterway of the upper deck: as said above about the main deck.

other access descending to the main deck. Abaft the mizzen was the aperture for the whipstaff to the tiller under the upper deck beams (see section 110, The Rudder).

Witsen offers a description for the plan of the forecastle in figure 2.143. On the forecastle a large grating was made, just like on the upper deck, allowing light into the crew accommodations in the forecastle. There were also two small hatches above the bitt standards for better handling the anchor cables when they were belayed on the bitts.

Figure 2.144 shows a plan of the deck above the steering stand. In front of the mast we see a knight, and abaft it the aperture for the whipstaff (*ff*). The cabin, as can be observed, was divided into two separate rooms.

70. The Partner of the Mizzenmast and of the Capstan.

(54 I 27) On plate No. 1 [figs. 2.147 and 2.148], partners are shown, which are heavy plank-balks: lying in the upper deck, where the mast goes through, which hole is surrounded by a seam.

> (79 I 51) *About the Partner.*
>
> 7. The partner broad 2 feet 5 inches, thick 4½ inches.
>
> 8. The hole in the partner wide 15½ inches.
>
> (79 II 5) The partner of the capstan is long 8 feet, broad 2 feet 3 inches, thick 4 inches. The hole is wide 15½ inches.

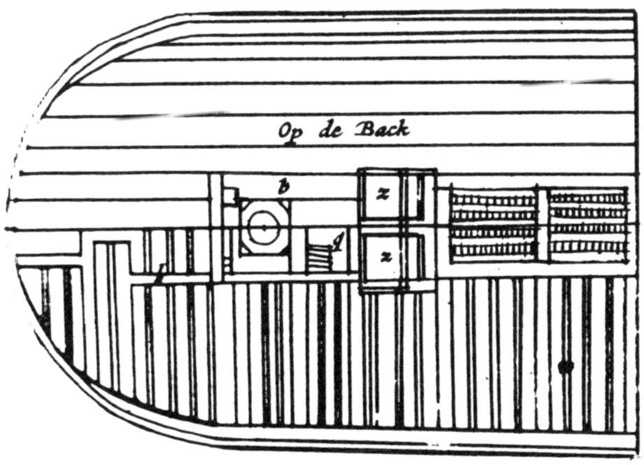

FIGURE 2.143. Plate XLIII. The forecastle above.
(59 II 30) In the plate which shows *the forecastle above,* B is the small hatch around the foremast. H the binding strakes. Z the small hatches above the bitt standards. Q the skylight grating abaft the bitt standards.

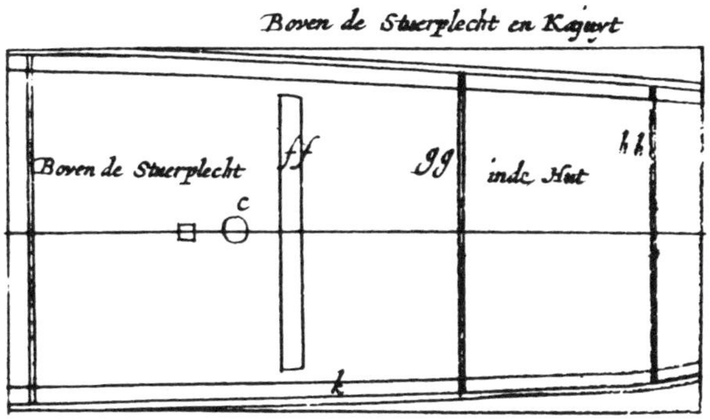

FIGURE 2.144. (*see p. 116*) Plate XLII. Above the steering stand and cabin.
(59 II 24) FF In the plate *above the Steering Stand and Cabin,* is the hole through which the whipstaff goes. GG is the door to the cabin, and the bulkhead for the same, HH the bulkhead inside the cabin. C the small hatch around the mizzenmast. K a waterway.

70. The Partner of the Mizzenmast and of the Capstan

This partner was an extra thick plank in the steering deck, lying one level above the step. The position of the partner determined the rake of the mizzenmast. Usually a partner went from deck beam to deck beam for maximum support, but sometimes it was longer and spanned several beams.

Sometimes the mizzenmast went through another deck, but this opening was no more than a simple hatch, as in the pinas, on the deck above the steering stand.

Like the coamings of the hatches, the partner lay on the deck planks. (Witsen continues the discussion of partners in section 93, Make the Partners ready.)

The partner of the capstan lay straight above the step, so the capstan could turn freely. The pawls of the capstan were fastened to the partner because it was stronger than the deck planks.

71. The beams of the Cabin, lies on chocks, as also in the Forecastle.

71. The Beams of the Cabin, Lies on Chocks, as Also in the Forecastle

Above the forecastle and the cabin, a deck was laid in the same fashion as the upper deck: first the deck beams were laid on chocks between the top timbers, and only then were the deck clamps fitted. These beams were half the thickness of the stem, the deck clamps only one fifth.

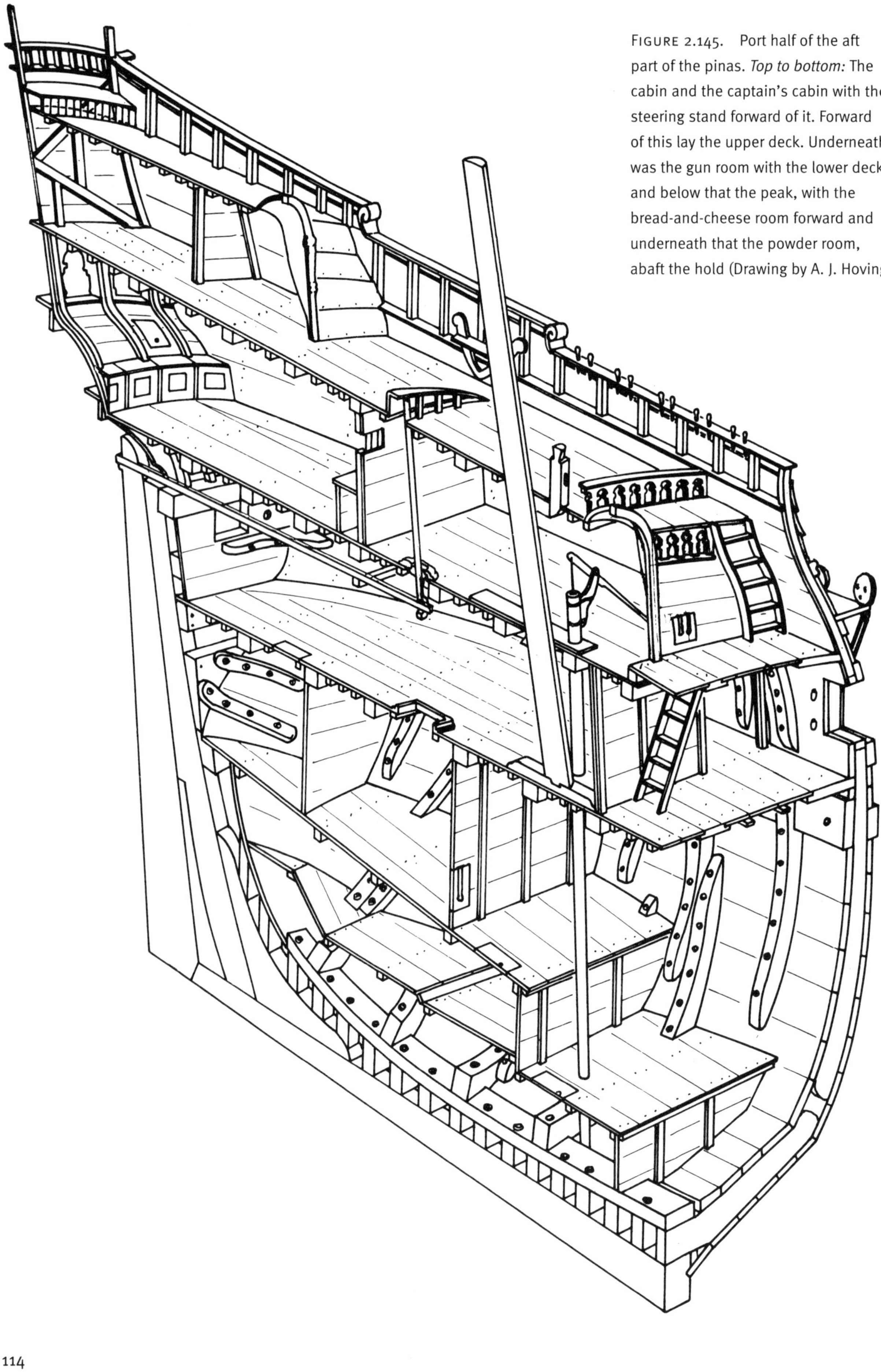

FIGURE 2.145. Port half of the aft part of the pinas. *Top to bottom:* The cabin and the captain's cabin with the steering stand forward of it. Forward of this lay the upper deck. Underneath was the gun room with the lower deck and below that the peak, with the bread-and-cheese room forward and underneath that the powder room, abaft the hold (Drawing by A. J. Hoving)

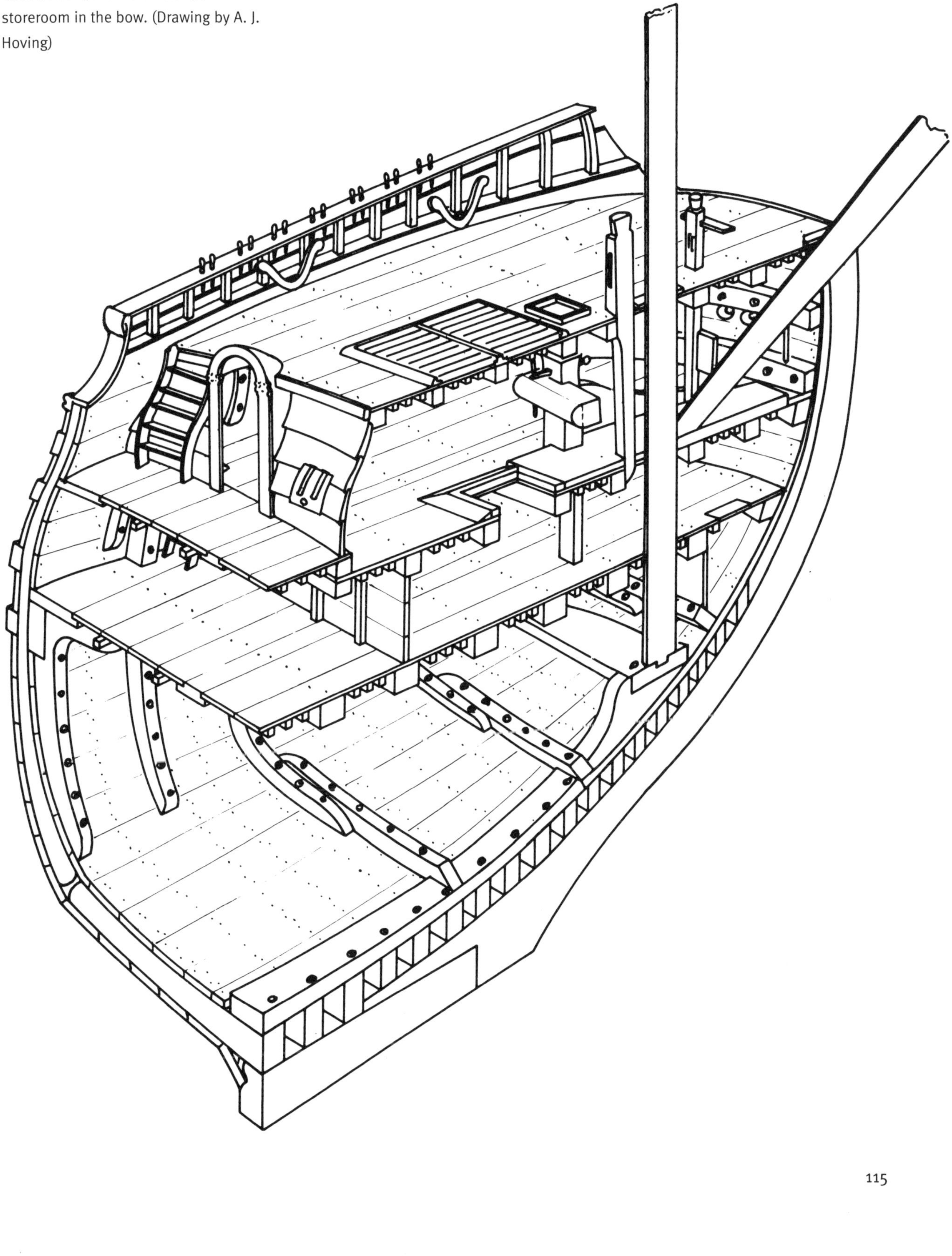

FIGURE 2.146. Port half of the fore part of the pinas. *Top to bottom:* The forecastle head, with the forecastle underneath and the upper deck abaft it, the cable tier with the lower deck abaft it, the hold below, and a storeroom in the bow. (Drawing by A. J. Hoving)

72. The Deck Clamp underneath it, and the other Ceiling.

(78 II 17) 32. *About several things in the Cabin.*

That always comes with a gallery aft, and on very large ships is divided in two. Sometimes a flight of stairs leads up to a stern hut, or upper cabin, as also 2 doors are made, although only the door to port is used; and that because when fighting, the guns can be hauled through there, and the opening allows more space for fighting. The cabin roof [or deckhead] is sometimes divided into compartments: which are painted after the fashion. In the gallery, which can have 3 turrets, of which the middle one is the highest; sometimes winding stairs are made going up. In the old days ships in this country were very ornate, and adorned with turrets, forward as well as aft, where the taffrail stands, as in the sides, at every hance, on the railings. On large ships round turrets were even placed in the middle of the main deck, and before the cabin, which served for the defense, because they carried light guns. Small windows, or rather shutters, were sometimes made in the cabin, to be able to look across the ship. Aft benches can be found, sometimes made strongly with cleats, done for the inclination of the ship.

. . .

(265 I 41) The *Cabin* is placed aft on the ship, because there the movement is the slightest, and the rest of the ship can best be overlooked from there, the rudder and sails attended: and also, because the ship can aft be built wide and high without problem.

. . .

(70 I 45) 22. *About things in the Cabin.*

1. The waterway 1/4 the thickness of the stem.
2. The deck clamp 1/5.
3. The deck beams 1/2.
4. The hanging knees as thick, as the top timbers are there.

. . .

23. *About the Waterways, deck beams, deck clamps and hanging knees in the forecastle.*

1. The Waterway 1/4.
2. The deck beams 1/2.
3. The deck clamps 1/5.
4. The hanging knees thick 1/2, of the thickness of the stem.

(78 II 51) 1. The deck beams are thick 7, and broad 8 inches, and have a camber of 10 inches.

2. Forward high 6 feet 3 inches.

3. Aft high 7 feet.

The foremost beam lies from the stem 6 feet 5 inches.

4. The hanging knees between the top timbers long 5 feet.

5. The bills long 12 1/2 feet.

6. The Waterway is broad 1 feet 5 inches, thick 3 inches.

7. The ledges are thick 2 1/2 inches, broad 3 1/2 inches.

8. Between each beam lie 3 ledges.

(79 I 25) 33. *About the Steering Stand.*

1. The steering stand high 6 feet, long 16 feet.

2. The hanging knees long 5 feet 16 inches, the bills long 3 feet, the knees thick 5 inches, broad 5 inches, the camber 12 inches.

3. The ledges thick 2 1/2 inches, broad 3 1/2 inches.

4. Between each beam lie 3 ledges.

5. The Waterway and deck clamp come from the cabin: the latter comes from the cabin, and is broad 9 inches, thick two inches, as the other parts; there are 7 deck beams, and the waterway is broad 16 inches, thick 3 inches.

(89 II 4) The foremost fixed deck beam in the forecastle lies 6 1/2 feet from the inside of the stem, high in the side, forward 5 feet 10 in. aft 5 feet 3 inches. The camber is 12 inches. There are 7 deck beams in this forecastle.

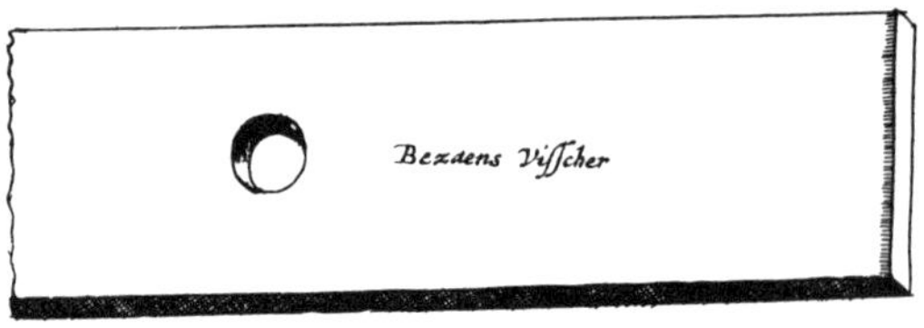

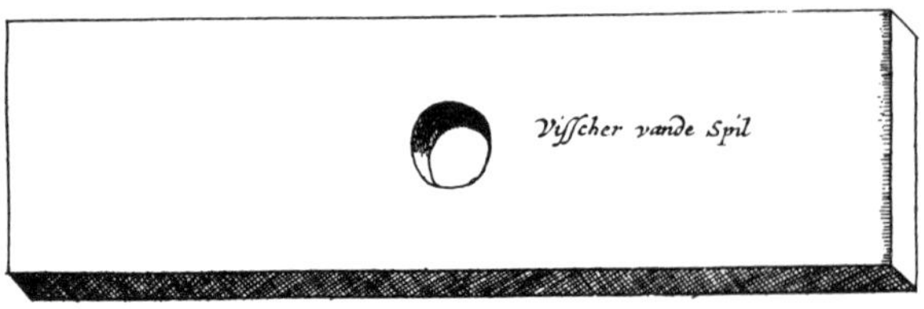

FIGURE 2.147. (*top*) Plate XX. Mizzen partner.

FIGURE 2.148. (*bottom*) Plate XX. Capstan partner.

The lower ends of the hanging knees long 4½ feet. The bills long 3 feet, thick 5 inches, broad 6 inches. The breasthook thick 9 inches, and also the futtock rider.

(75 I 44) The hanging knees, in the forecastle, have lower ends long 4 feet 4 inches, the bills long 3 feet, thick 5 inches, broad 6 inches.

72. The Deck Clamp Underneath It, and the Other Ceiling

When the deck beams were in place, the hanging knees were fixed—between the top timbers rather than on the ceiling to provide maximum space in the cabin. Then the deck clamps on which the deck beams rested were made. Finally, the rest of the ceiling was added underneath the deck clamp.

Witsen does not mention them, but the waterways and binding strakes of the cabin, steering stand, and forecastle were also made at this time. In general, it can be said that the upper structural elements became lighter so as not to threaten the stability of the ship.

73. The Ledges on the upper deck.

73. The Ledges on the Upper Deck

As for the main deck, three ledges were usually placed between each pair of deck beams of the upper deck. The ledges were one quarter of the width of the stem.

74. The knights ready in the Forecastle.

(92 II 34) The sheaves of these small fore knights are broad 12, thick 2 inches. The nail 2½ inches, and above the holes ["swallows" is the technical term] 5 inches, the head above that long 8 inches, thick 8 inches, broad at the base 8 inches, thick 3 inches, it stands a little more vertical than the mast, they stand against the binding strakes, and below fixed against the standard knees, with a bolt through the beam, and a small cross beam comes above the holes, thick and broad 5 inches, the heads long 8 inches.

(92 II 50) 70. *About the small Knights.*

1. The small knights are broad 1 foot, thick 10 inches.
2. The sheaves broad 12 inches, thick 1¾ inches.
3. The nail thick 1½ inches.
4. The sheaves below the head 2 inches.
5. The Head long 8 inches, above thick 7 inches.

(93 I 16) 71. *About the small belaying beam.*

1. The small Belaying beam, thick 5½ inches, broad 6 inches.
2. The heads long 8 inches.

74. The Knights (or Bitts) Ready in the Forecastle

In fact this section should be titled "The Knights *on* the Forecastle Head," for there were no knights in the forecastle. The lower ends of the small knights were in the forecastle, however, before the mast. Above, the ends were joined by a beam onto which lines were belayed. Heavier lines of the running rigging, like the topsail sheets, ran through sheaves in the knights or bitts.

Below, in the forecastle, where the ends were fixed to the bitt standards or standard knees (see section 84, The Bitt Knees), the bitts had a second purpose: they locked up the foot of the bowsprit. For this, heavy chocks were bolted onto the bitts, into which the bowsprit fitted closely. The heel of the bowsprit then lay on the partner of the foremast (see section 95, And the Bowsprit).

75. Cabin timbers in the Forecastle and underneath the Ledges.

(68 II 35) [. . .] carlings ⅓, the ledges ¼.

75. Cabin Timbers in the Forecastle and Underneath the Ledges

This is another mistake in the discourse. Witsen means the carlings underneath the ledges of the forecastle. These carlings were one third the thickness of the stem.

76. Place the Bitts.

(56 II 40) Bitts are heavy upright pillars that, together with the bolster and standard knee, make up the riding bitts.

. . .

(70 II 17) 1. The bitts ⅓ thicker than the stem.

(79 II 34) 35. *About the Bitts.*

1. The bitts thick 15 inches, broad 16 inches.
2. The heads long 15 inches.
3. Hollow one-half inch.
4. The lower ends thick 1 foot, broad 1 foot 1 inch.
5. The bitts stand 4 feet above the deck.

The bitts stand 23 inches apart.

76. Place the Bitts

The bitts or standards were the vertical timbers of the riding bitt, the heavy structure in the forecastle onto which the anchor cable was belayed. They were one third thicker than the stem and were partly let into the foremast partner. They bore against a deck beam forward, both on the main deck as well as on the upper deck.

It was actually "modern" to have the riding bitt in the forecastle. Before 1661 the practice was to stow the anchor cable in the hold, and the riding bitt would then stand on the main deck. This practice was adhered to in later years for men-of-war, but for merchantmen the arrangement described above was adapted because of an incident in 1661, described by Van Yk as follows:

> The [East India] Company had purchased a man-of-war, which it intended to use as an East Indiaman. While still in the Channel, the hold was found full of water and it was decided to return, fearing that the entire foreship had become undone. Back in the fatherland, however, it appeared that the water had entered the ship through the hawseholes, between frame and ceiling. The ship had never had that problem when a man-of-war, since it had never been loaded as deeply before. (Van Yk, p. 108)

The problem was solved by raising the stem, closing the hawseholes and making new ones one deck higher, as was done with the beakhead. The riding bitt was then placed in the forecastle, which head was prolonged to the stem (a closed forecastle). The additional advantage was that the hold remained free for cargo, the disadvantage that the fore side of the lower deck could no longer provide space for guns. Still, this arrangement was such a success that, according to Van Yk, all subsequent East Indiamen were built this way.

77. The main and Fore knight.

(55 II 44) With Knights the yards are hoisted:

. . .

(62 II 42) Stand below on the main deck; and are used to raise the sail with its yard, for the main mast as well as the foremast. Yet the knight for the mizzenmast is also used to haul and to pay. The small standing blocks or knights, which can be seen fitted on the main knight, are used to house the topmasts with the top rope, and to raise and lower the top.

(71 II 11) *37. About the Knight.*

8 Feet of the ship's length, gives 1 inch of breadth: being a little thinner, but the sheaves are this broad. On fluyts the sheaves are set in the ship's side, instead of knights.

> (55 II 45) the main knight is broad 1 foot 7 inches, thick 1 foot 6 inches, the sheaves long 1 foot 6 inches, thick 2 inches, the nail thick 2 inches, the holes of the head 8 inches, the head long 1 foot, made octagonal at the top: the small knights fixed there, thick 10 inches, broad 1 foot, the holes long 12½ inches, wide 2 inches, the head long 7 inches, above thick 7 inches, stand from the main 4 inches, above, below against it, here below often a crosspiece is made, fitted with a small bolt at each end; its heads are long 8 inches. The knights today are often doubled.

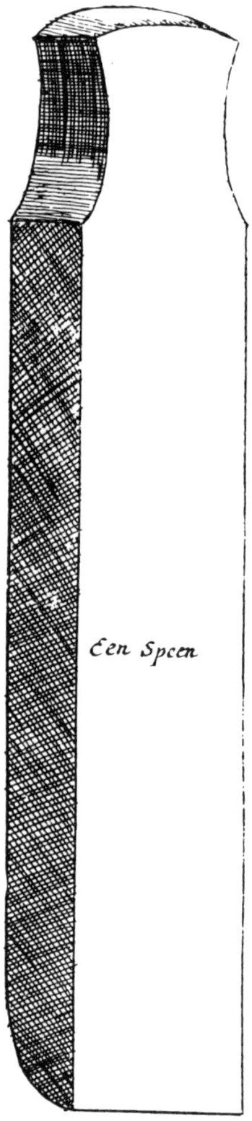

FIGURE 2.149. Plate XXXVII. A bitt.

(92 II 5) 69. *About the main Knight.*

1. The main knight, broad 18 inches, thick 17 inches.

2. The sheaves long 17 inches, thick 2 inches, the fourth on the outside thick 1½ in. the head long 1 foot, the nail thick 2 in. the holes of the head wide 9 inches, one is curved with a round edge of 3 inches, also two small knights are fitted to it at the side, below, which have to stand 4 inches from it, they are broad 1 foot, and thick 16 inches. The head long 7, and above thick 7 inches: above that is a bolster thick and broad ½ foot, the heads long 7½ inches, the sheaves broad 12½ inches, thick 2 inches, the nail thick ½ inch, with a small bolt fixed at each end.

The Fore Knight comes 13 inches behind the mast, and the inner sheave comes as the mast, stands 3 feet above the forecastle or deck, and there is thick 15 inches, broad 16 inches. The head is long 10 inches, above thick 9½ inches, below thick 8½ inches, and broad 1 foot and somewhat hollowed, for the rope. The 2 sheaves are broad 1 foot 6 inches, the third thick 1½ inches. The nail is thick 1½ inches, the 2 small fore knights are thick and broad 9 inches.

(93 I 9) *The mizzen knight* comes 1½ feet from the mast, and below even with the deck beam [*balck*; the text has *back*]: is long above 2 feet, broad 9 inches, and thick 8 inches: the head long 5½ inches, thick 6 inches, the holes long and broad 12 inches, thick or wide 1½ inches, the nail 1½ inches, the sheaves broad 8 inches, the small hatchet to this mast is of a shape as the others.

77. The Main and Fore Knight

These knights were used for all sorts of heavy hoisting besides raising the yards. Apart from the sheaves for the halyard, another sheave was often fitted in the knight for raising the topmasts and other objects, using the top line that ran to the capstan.

The knights stood abaft the masts, often slightly off center to avoid interfering with the stay, which came down to the mast at that point.

78. Plank the Main Deck.

(86 I 18) The planks of the main deck thick 2½ inches, also in the cabin, on the upper deck 1½ inches.

78. Plank the Main Deck (and the Upper Deck)

Only now was the planking of the main deck laid, probably along with that of the upper deck as well. The planks of the upper deck were thinner than for the main deck. These planks were adzed, which gave them a rougher surface than when planed and which made them less slippery. The planking had to be caulked to make the deck watertight.

79. Then the cross of the Cabin and the Bulkhead is made.

(57 II 34) The cross stands aft in the flat of the stern, it is faired, and the parts of the flat of the stern come onto it, the escutcheon stands against it.

. . .

(83 II 34) The cross is for strength, and to nail onto it parts of the flat of the stern, onto which ordinarily then the crest is placed. The counter above the rudder is called the curved counter, up to the flat of the stern.

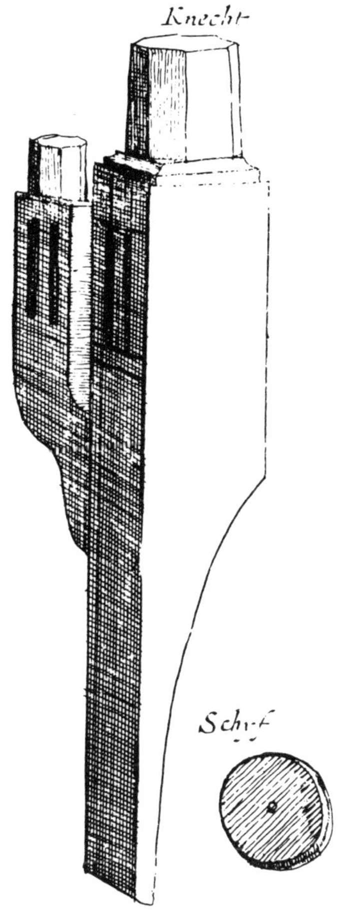

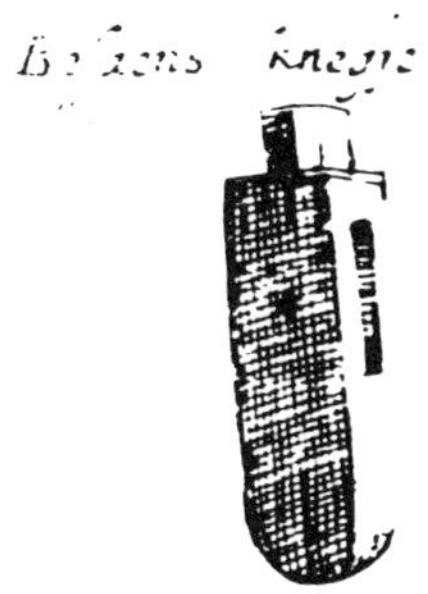

FIGURE 2.150. (*left and center*) Plate XXXI. Knight, sheave.

FIGURE 2.151. (*right*) Plate XXXI. Mizzen knight.

FIGURE 2.152. Plate XXIV. Lower deck plank.

(71 | 7) 3. The counter timbers thick ½ the stem, have as much curve as is deemed right.

. . .

(86 | 32) Counter timbers are made above to the sides of the flat of the stern for the sake of decoration, depending on the gallery, they rest on the transoms of the gallery.

(83 || 17) 5. *About the timbers of the flat of the stern.*

1. The timbers of the flat of the stern, rake one inch for every foot.

2. Are broad 5 inches, thick 4 inches.

3. There are 5 timbers, standing on the lower deck, the middle broad 6 inches, thick 4½ inches, the others 5 and 4 in. broad and thick: those in the sides stand. The lower end has a curve 2½ inches inward, it starts 3½ inches from the top of the second hackebord, it stands 3 inches from the side of the lower deck, the holes wide 2 feet 4 inches: the middle, and upper ends are attached to the small beam; and it is straight there, but to the beam comes a lap joint

(83 || 37) 6. *About the cross, in the flat of the stern.*

The cross serves for rigidity, and to nail the planks of the flat of the stern to, on which usually the crest is placed. The counter above the rudder, is called the hollow counter, up to the hackebord.

1. The cross stands 2 feet 3 inches above the second hackebord it is thick 2½ inches, broad 5 inches.

2. The upper end, stands 12 feet from the hackebord.

3. The small timbers in the cross, are thick 3½ inches, broad 4½ inches.

4. The counter and flat of the stern, is closed with one-inch deals.

5. The middle timber in the cross, onto which the flagpole will come, thick 6 inches, broad 7 inches.

The counter of the gallery is closed with one-inch deals. The middle timber, onto which the flag comes, is 6 inches thick, and a little broader than the others, and above somewhat thicker than below: the upper ends stand 1 foot 8 inches from the side: the lower stand 1 foot 5 inches from one another, are faired with one-inch deals.

(88 | 39) *About the Bulkhead forward of the Cabin.*

The doorposts, have sculptures on the outside, which are thick and broad 4 or 5 inches, the one standing above from the side 20 inches, has in the middle a ridge, the door above wide 3 feet ½ inch, below 3 feet 2½ inches: the door stands above from the side 20 inches, broad 4 inches, the door thick 1½ inches, the bulkhead 1½ inches.

The other battens broad 2½ inches, thick 2 inches, the cross broad 3½ inches, thick 3 inches, it comes to the fourth deck beam from forward, and lies below on a deck beam; the deals thick 1½ inches, paneled on the inside with sawn planks, the bunks against it wide 2 feet 2 in. above which lies a batten to keep it straight, the gunports are 4½ in. above the lower deck, wide and high 18 inches, there are to the lower hackebord 2 knees, which are thick and broad 6½ in. the end long 3½ feet, the other continues to the stern timber, the doorstep thick 2 inches, the cross bracing broad 7½ inches.

(56 | 25) Above the cabin door stands a small arch, and on the door lie three cross braces, it [the door] must be a little wider below than above: above 3 feet ½ inch, below 3 feet 2½ inches, the cross braces broad 9 inches, thick 1½ inches, the door is sawn, the part at the ship's side is wide 1 foot, the hinges have two flanges, the narrow part is with a bolt lock, the cabin doors as well as the upper cabin doors can often be seen on ships, protruding like porches, which seems to me to cause much discomfort, and therefore inadvisable. On the middle cross brace comes the lock. The grating for the planking, toward the upper deck, is often completely planked. The end of the arch above the door, should be somewhat meek [possibly meaning a somewhat flattened arch, elliptical rather than semicircular].

(58 | 46) The doorstep, at the inside of the cabin door, is thick 2 inches, and as broad as the door posts are.

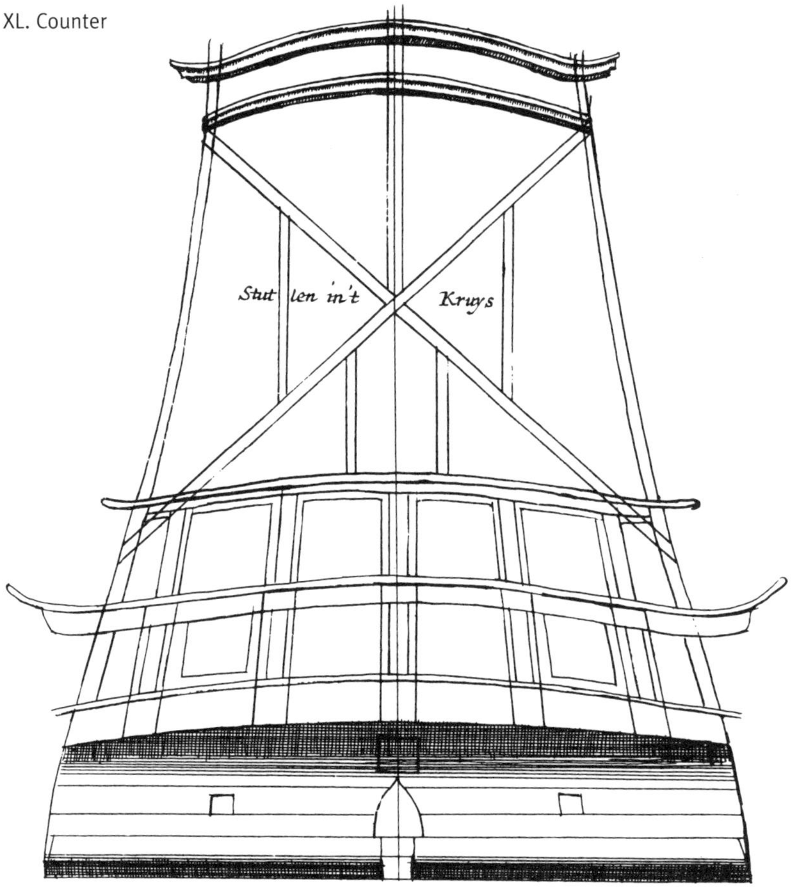

Figure 2.153. Plate XL. Counter timbers in the cross.

79. Then the Cross of the Cabin and the Bulkhead Is Made

Planks were nailed to the cross of the flat of the stern (or the upper transom), to which the escutcheon was attached. Before this could be done, a number of timbers had to be placed abaft the stern timbers, enabling the stern to project aft of the wing transom. This extra length of the ship had already been taken into account with the placement of the wales, filling strakes, sheer rail, and upper planking, and the last parts of the frames were undoubtedly placed at this stage (see section 62, Hereafter it is done up outside, with Wales, Filling Strakes and Sheer rails, Washstrake, Vertuining and Railing).

The opening between the wing transom and the floor of the cabin was closed with counter planks. Early on in the century this part was planked vertically, as the *Wasa* shows, but soon the counter was made to a hollow shape. Often a townscape was painted on this section when the ship was finished, or, in the case of ships of the Admiralty, a number of coats of arms. The counter and the flat of the stern above it were constructed with a number of counter timbers between the wing transom and the upper deck.

The closing of the stern will be dealt with in detail with the finishing of the *galdery* (section 81, The Stern Gallery).

The bulkhead of the cabin separated this area from the steering stand in front of it. Two doors were often made in it so that guns could more easily be brought into the cabin to use through the rear gunports underneath the stern windows. The pinas had no guns on the upper deck, so only one cabin door was necessary.

A trellis window was provided in the bulkhead, with a slide permitting the captain to keep an eye on the helmsman if he wished.

80. The Beakhead.

(60 | 41) Against the Ship, and the Beakhead at K [fig. 2.157], often a herm piece is made. Beakheads that are too long hinder ships because of their weight, which is why they are now made light and steep.

(286 | 18) Below the cathead at the Beakhead, stands a heavy herm piece, supporting it, when the anchor is raised with it.

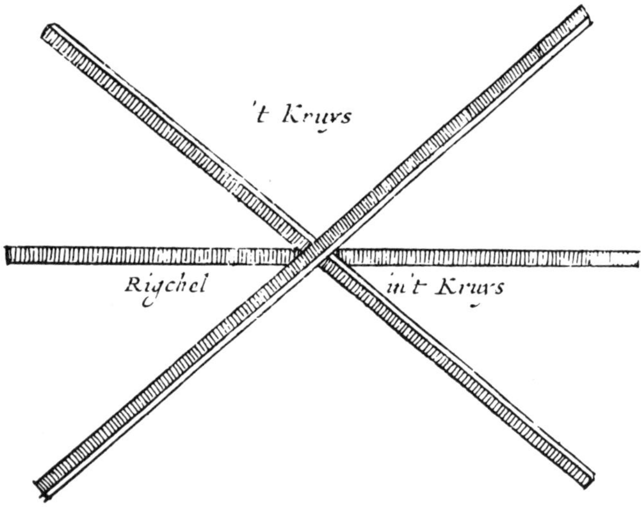

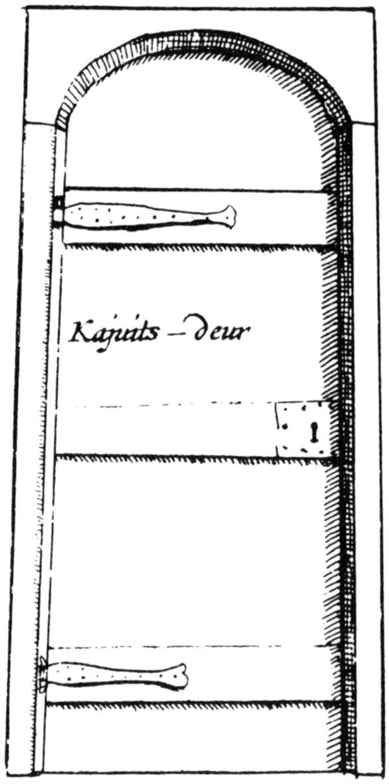

Figure 2.154. (*top left*) Plate XXXVIII. The cross, batten in the cross.

Figure 2.155. (*bottom left*) Plate XXX. Door post.

Figure 2.156. (*right*) Plate XXXIII. Cabin door.

(86 II 12) The support knee comes underneath the lower spur of the head, for the shape and security it is made about 4 inches hollow forward, between the feet of the lion and the water.

This is the pen [not clear; probably simply "the support"] in which the lion lies.

. . .

(70 II 40) [. . .] about the kam, aft, as many inches broad as the knee of the head is long.

. . .

(86 II 18) The kam is the plank that is carved, stands between both cheek knees of the head, and is made broad or narrow, as the filling strake between the two lower wales is broad or narrow: it is at its narrowest at the fore end. With its rear part the lion comes against the kam.

. . .

(54 II 33) Cheeks are knees outside, to stem and wale.

. . .

(55 I 36) Es: is the wood, lying on the lion's back.

. . .

(87 I 24) The upper head rail makes the shape of the beakhead; coming from the cant top timbers at one end, up to the es, where it is nailed underneath the scroll.

The others are faired according to the upper rail, and their one end is well nailed to the es, the other to the ship; a man's head is carved on the upper head rail for decoration.

. . .

(70 II 47) 5. the head rail aft broad ½.

6. Forward ¼ of the stem.
7. And half as thick as its breadth.

. . .

(87 I 44) The cheek knees are well fastened to the spur of the head, and the rails securely nailed to it.

Against the knees, on large ships, often open heads [toilets] are made.

The beakhead and bowsprit are bound together with ropes through the spur of the head.

. . .

(264 I 26) About the dimensions of the *Beakhead,* which is made to the bow of ships, there is no other

reason, but the pleasure of the shipwright: but the height of the knob, above the lion, never comes higher than the railings of the forecastle, if the ship has a forecastle: some years back the beakhead was made fairly long: today it is made short again, curved, and steep, which, in my opinion, is not to be recommended on large ships, because it gives too little space, for the required use, for which Beakheads are suitable. To ships though, made only to sailing performance, short beakheads fit best, and short bowsprits, because, the longer these are, the greater the angle of movement they give to the entire ship, when the sea happens to blow against them, which greatly impedes the headway of the ship: the lighter of wood they are, the less resistance the water, blowing against it, feels, and subsequently, the less force and momentum it passes on to the ship. When all too heavy and long Beakheads break down at sea, it is often necessary to saw them off: because they cannot be repaired because of their size and risk damaging the ship if left hanging loose and dangling. If made too heavy, they also make ships heel over forward too much, and dive into the water: on men-of-war the largest beakheads are required: they serve the quarters of the sailors, for cleaning, washing, urinating and for one's relief: as also to manage the rigging of the bowsprit from there; and especially for the adornment of the ships. On men-of-war the Beakhead is used for a prison, in which the fettered criminals are locked. The beakhead is gammoned with tight ropes for strength: it is made heavy or light, according to the ship's sturdiness or tenderness, and such to get the center of gravity in a suitable place: for this reason it can be seen that beakheads of ships are changed, made lighter or heavier, when ships have become tender, by placing this or that heavy thing on it, like masts etc. or also when it has become sturdy, by planking it from below, or in another fashion.

In these countries it is practice, to place a Lion on the fore end of the Beakhead, the coat of arms of the state: and aft, at the top, the coat of arms of the town, where the ship was built, and belongs.

. . .

(267 II 50) Forward on the Beakhead tarpaulins and pap-doeken [lit., "drenched cloths," perhaps a synonym for "tarpaulins"] are laid, across which drag the anchor cable and sheet.

The Beakhead and the Foremast are sawn off in an emergency, when the ship pitches, or is leaking forward.

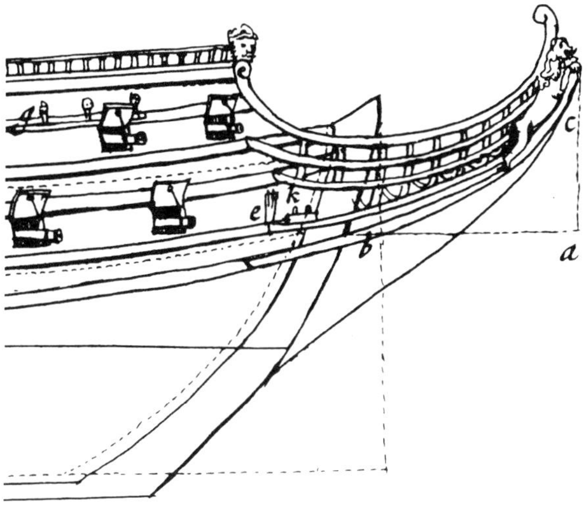

FIGURE 2.157. Plate XLIV

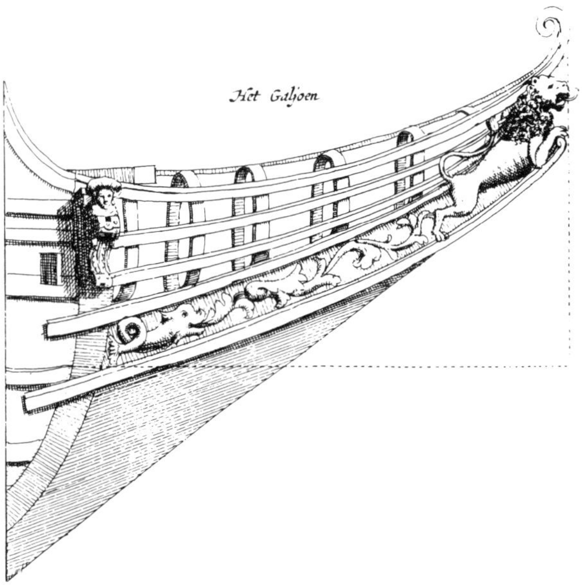

FIGURE 2.158. Plate XXXVI. The beakhead.

(60 I 1) The Beakhead rises, from A to C [fig. 2.157] 2½ feet less, than half of AB, measuring is done up to the underside of the lion's mouth, only halfway A to B, straight on. The hawseholes, shown at E, are 8 inches thick, the supporting knee to the spur of the head, which are against the front of the stem, broad 1 foot, thick 6½ inches, and long there 6 feet.

The lower cheek timber, comes 2 inches over the second wale. The upper, stands 13 inches, below the third wale, as to coincide with the pin.

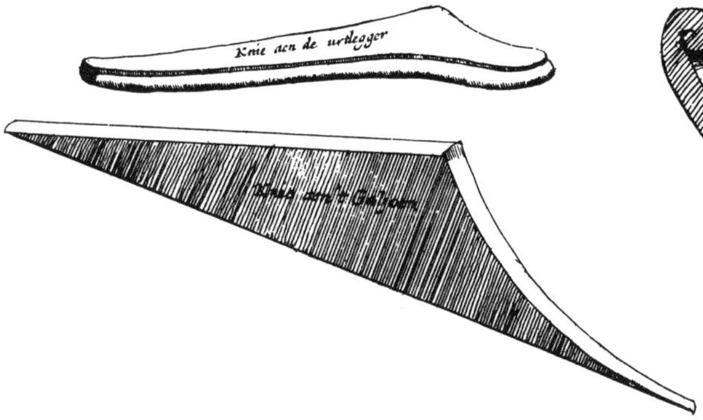

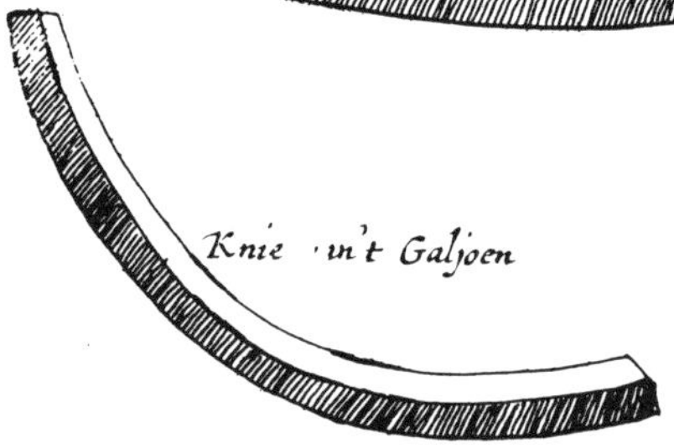

FIGURE 2.159. (*above*) Plate XXVI. Cheek, beakhead knee.

FIGURE 2.160. (*top right*) Plate XXVI. *Es* of the upper spur.

FIGURE 2.161. (*bottom right* Plate XXVI. Knee in the beakhead.

The upper head rails, aft are broad 8 inches, thick 4½ inches, and forward are broad 5 inches, thick 3½ inches, the middle ones broad 6 inches, thick 4½ inches, forward broad 4½ inches, thick 3½ inches. The lower ones broad 6½ inches, thick 4 inches, and forward 5 inches. The upper aft wide 1 foot 3½ inches, forward at the lion wide 6½ inches. The lower wide 1 foot 1 inch, aft: and forward wide 5 inches.

The first knee stands to the stem. The other 2 feet 1 inch farther, thick 5 inches, broad 7½ inches. The third wide 2 feet, thick and broad as above. The others wide 2 feet 2 inches, thick 4 and broad 5 inches. The little deck forward thick 4 inches, wide and long 10 inches, the two crosspieces thick and broad 5 inches, with holes, and a curve of 3 inches, the laths thick 1 inch, broad 1½ inches, the ridges thick 2½ inches, broad 3 inches. At the end, straight across, lies a ledge, thick 3 inches, broad 4 inches. The cathead sticking 2 feet over the head rail, 10 inches broad at the top timber on which it lies, thick 4 inches. The cathead knee thick 10 inches, broad 1 foot.

(86 II 41) 3. The upper knee of the head, forward broad 6 inches, aft broad 1 foot, forward thick 9 inches, aft thick 12 inches.

(86 II 35) 4. *About the Kam.*

1. The kam, forward broad 7 inches.

2. The kam, aft broad 30 inches, thick 3 inches, let in 1½ inches. The kam always, forward ½ inch broader, than one third part from behind.

(86 II 31) 3. The lion, long 9 feet, thick 19 inches, the end comes 14 feet forward, the end rises 2 feet 7 inches, it is one foot thick aft.

(86 II 44) 5. *About the Es.*

This is a timber, lying on the lion's back, and with a scroll.

1. The Es lies 2 feet 2 inches above the lion, above broad 14 inches, thick 10 inches.

2. Below coming 3 feet past the lion.

(87 I 1) 6. *About the Knees.*

They come onto the ends of the spurs, to hold onto the ship.

1. The Knee forward on the stem, broad 12 inches, thick 6 inches, long 6 feet, the other 5½ feet, without tongue [probably the bill, the long leg of the knee].

2. The other knee [on top of the upper spur], broad 9½ inches, thick 4½ inches, forward broad 8 inches, thick 4 inches.

(87 I 9) 7. *About the Head Rails.*

These make the shape of the beakhead, run up to the es, and are to both sides 3 in general.

1. The upper rail, aft broad 8 inches, thick 4½ inches, forward broad 5 inches, thick 3½ inches.

2. The other, broad 6 inches, thick 4 inches, forward broad 4½ inches, thick 3½ inches.

3. The lower, broad 6½ inches, thick 4 inches, forward broad 5 inches, thick 4 inches.

8. *About the width of the Head Rails.*

1. The upper head rail, aft wide 14½ inches, forward wide 6½ inches.

2. The lower, aft wide 12 inches, forward wide 5 inches.

(87 I 35) 9. *About the Ribs in the Beakhead.*

The First rib stands against the stem.

The Other 2 feet farther, and are thick 5 inches, broad 7½ inches.

The third wide 2 feet, thick 6, and broad 1 inch.

The fourth, wide 2 feet 2 inches, thick 4½ inches, broad 6½ inches.

(87 II 3) 10. *About the Little Deck* [in the beakhead].

Forward the little deck is thick 4 inches, broad 10 inches.

(87 II 6) 11. *About the beakhead Pinrails.*

1. There are 2 pinrails, thick 5 inches, broad 5½ inches.

2. Having a curve 3½ inches.

3. Made with 4 belaying pins.

(87 II 11) 12. *About the Battens.*

1. The battens, thick 1 inch, broad 1½ inches.

2. Lying apart 2 inches.

3. They lie 1 inch underneath the upper face of the second head rail.

4. There are 6 beams, thick 2½ inches, broad 3 inches.

5. At the end they are athwart the beams, thick 3 inches, broad 4 inches.

80. The Beakhead

We saw how the lower spur of the beakhead was placed in section 68 (The lower Spur of the beakhead). It was secured by knees from below as well as from the sides. A groove was cut into the top, in which the kam fitted. This was a carved plank, sculpted with floral motifs and other embellishments. The upper spur of the head rested on top of the kam. It was slightly less massive than the lower spur and secured by cheeks at the sides as well as a knee above.

The lion rested on the lower spur of the beakhead, its rear bordering the kam. The *es*, the extension of the upper spur of the head, terminated with a nice scroll that protruded above the lion's head.

Between the *es* and the ship's hull were the head rails, which were held in place by ribs. The ribs were connected by beams, on which the battens forming the beakhead deck were nailed. These were set a little apart so that the deck was always drained.

Forward in the beakhead a pair of crosspieces were

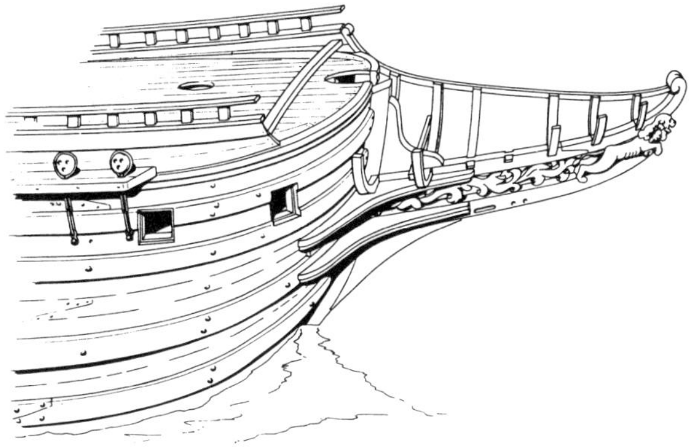

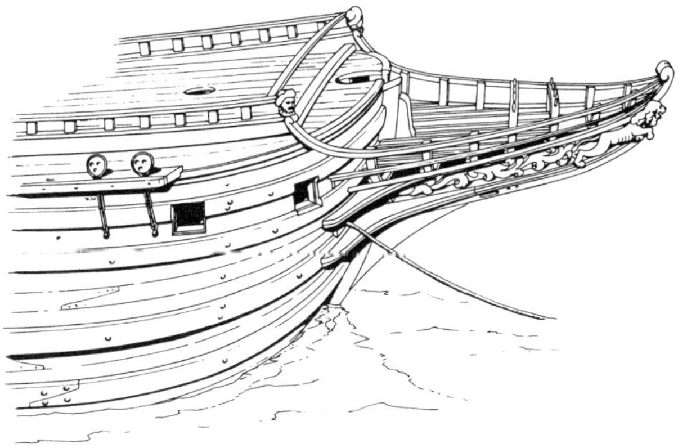

Figure 2.162. (*top*) The beakhead in the first stage of construction. The lower spur is supported from underneath by a knee and laterally by the cheeks. Abaft the lion we see the kam with the upper spur on top. (Courtesy G. A. de Weerdt)

Figure 2.163. (*center*) The upper spur is secured laterally by cheeks. The top headrail has been fitted, connecting the *es* with the ship. (Courtesy G. A. de Weerdt)

Figure 2.164. (*bottom*) The headrails and ribs have been completed. The battens of the grating deck are placed on top of the beams, and the pinrails connect the headrails of both sides. (Courtesy G. A. de Weerdt)

Figure 2.165. The beakhead on the pinas model. (Courtesy Cees de Jonge, The Visual Art Box)

Figure 2.166. General representation, based on Witsen's construction sequence, of how the pinas should look upon the completion of the beakhead. (Courtesy G. A. de Weerdt)

mounted; these were the "ranges" with belaying pins for the rigging. The heads for the crew were also located here.

On men-of-war, lawbreakers were locked underneath the beakhead grating, where they sat out an uncomfortable detention. On the model of the Dutch two-decker man-of-war known as the Hohenzollern model, a part of the beakhead floor could be removed, apparently for this purpose.[13]

The eye of the shipwright determined the way the beakhead joined the rest of the ship.

The order of construction described by Witsen is less compelling from this point on than it was in the beginning. As pointed out earlier, Witsen sometimes uses the sequence to solve the literary problem of describing simultaneous processes. For instance, although placement of the lower spur of the beakhead is described in section 68 (The lower Spur of the beakhead), the rest of the beakhead construction is not dealt with until this section. Figure 2.166 shows that the craftsman has also taken some liberties with the sequence. Bulwarks, decks, and channels are already in place, but the stern decorations are still missing.

81. The Stern Gallery.

(56 | 10) No. 13 [fig. 2.167] shows a Stern, or ship seen from astern, cut in half, because both sides are the same, but the coat of arms, which is the ship's distinguishing mark, was shown completely because half would have been unrecognizable.

. . .

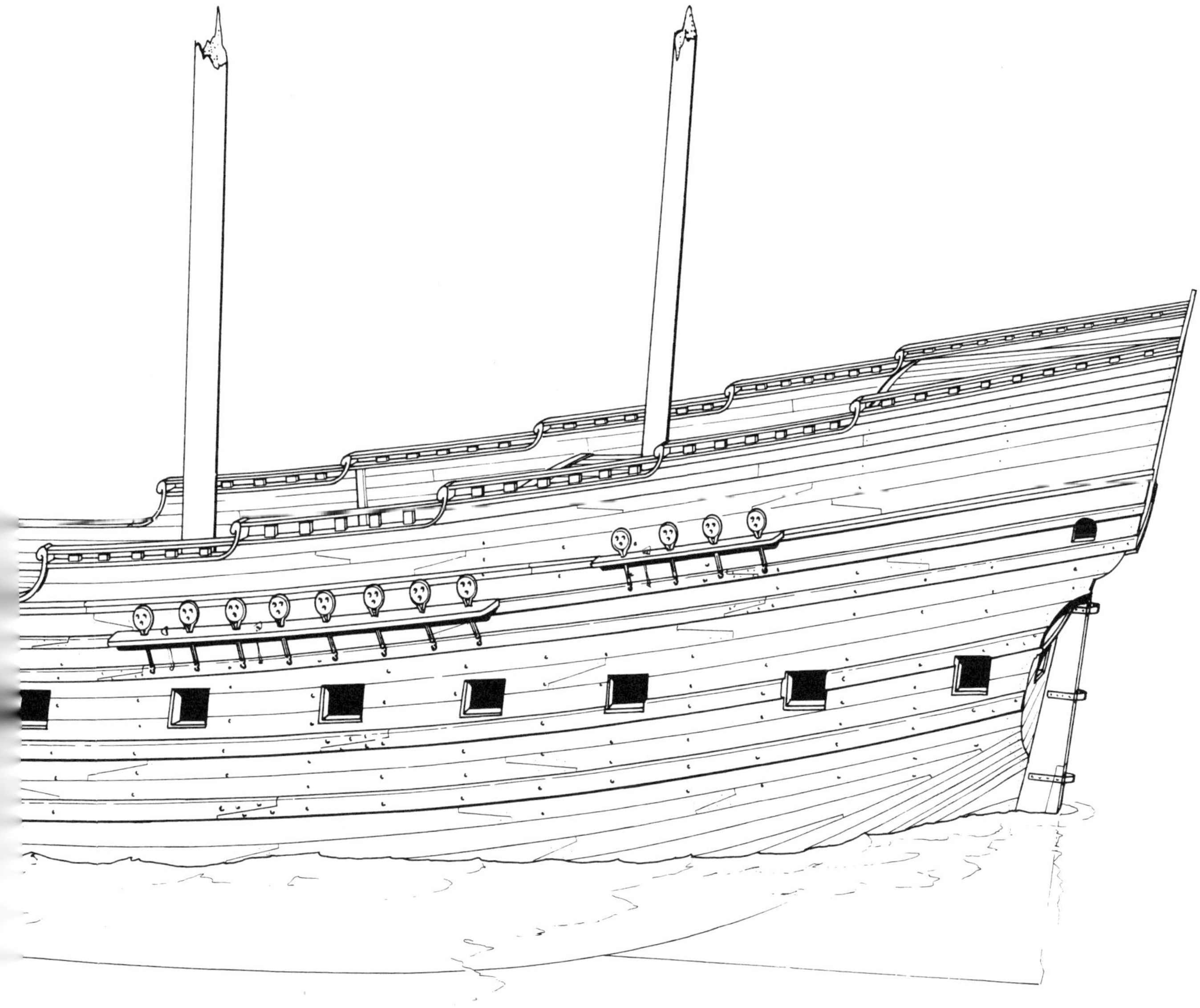

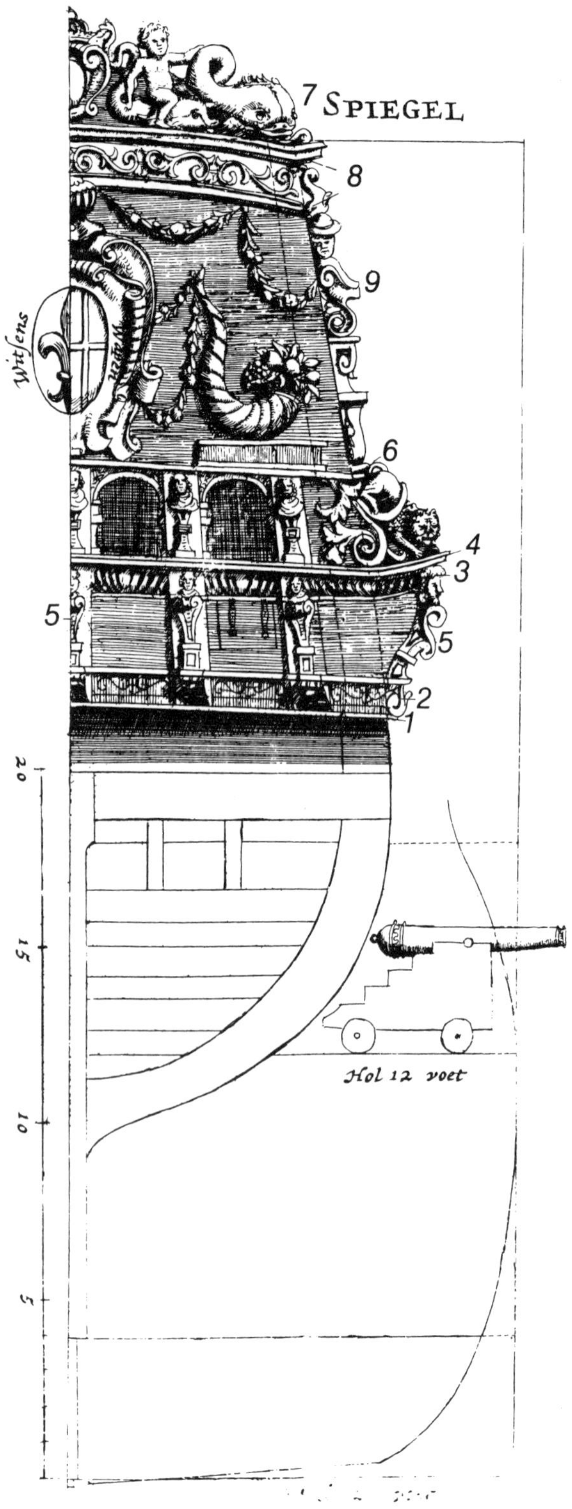

FIGURE 2.167. Plate XXXII. Stern, Witsen's crest; depth 16 feet, beam 27 feet.

The stern consisted of a number of components: *1*, first *hackebord; 2*, first cover; *3*, second *hackebord; 4*, second cover; *5*, quarter consoles; *6*, third cover; *7*, upper *hackebord; 8*, cornices (*vaezen*); *9*, "statues in the side."

(71 | 1) *29. About the Stern Gallery.*

1. The Lower hackebord, in the middle is as broad as the stem and ⅛ its thickness.

2. The cover on top of it, as broad, as the hackebord.

4. The other hackebord, and cover, has less curve, and with the counter timbers, and other carvings, the stern gallery is decorated.

(83 | 10) 47. *About the Stern Gallery.*

1. The upper side, of the lower hackebord, comes 12 inches above the upper side of the second wale, at the ends broad 1 foot, in the middle broad 13 inches, thick 4 inches, the underside has a curve of 5 inches, and above as much more as it is broader in the middle.

2. It stands 2 feet 2 inches aft of the sternpost, half joined into the filling strakes, they are made with little turrets or without, but more often without.

2. *About the cover thereon.*

1. The Cover thick 2 inches, in the middle broad 13 inches, at the ends broad 1 foot ½ inch.

2. At the ends 7 inches longer, than the hackebord.

3. *About the footing or base of the quarter gallery.*

1. The Foot in the side, long 5 feet 6 inches.

2. Aft broad 7 inches.

3. Forward broad 5 inches.

4. Thick 2 inches.

5. Corresponds with the narrowness of the wale.

(57 || 13) The first cover, shown on plate No. 20 [fig. 2.168] is thick 2 inches, in the middle at c broad 1 foot 2 inches, at the ends a 1 foot 1½ inches, goes 7 inches past the filling, the foot in the side long 5 feet 5½ inches, broad 7 inches, thick 2 inches, it has some curve downward, and the underside stands, as far outside the wale, the little post at the end broad 6 inches; lies on the second hackebord; is that on which one leans looking aft, thick 5 inches, stands 1 foot from the wale, at the height of 1 foot 4 inches, from the foot, and has some curvature in between, 3½ inches below the foot, the quarter gallery architraves are broad 9 inches, the upper side high 2 feet 2 inches, aft broad 1 foot, thick 1½ inches, curve above 2 inches, and aft 1½ feet outside the wale on the inside. The arch comes 8½ inches before the post, the post heels 5½ inches backward, the arch comes 1½ inches on the wale.

From the errata section, page 40, of Witsen's appendix: The Cover thick in the middle at *C* [in fig. 2.168]: 2 inches, broad 13 inches at the ends at *A*.

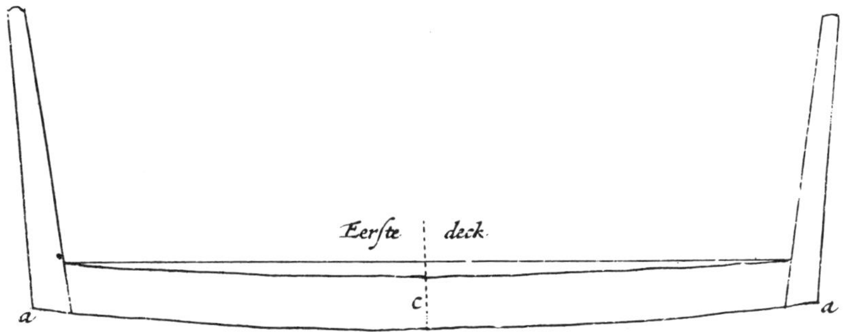

FIGURE 2.168. (top) Plate XLI. First cover.

FIGURE 2.169. (bottom) Plate XXX. Cymatium.

(55 II 33) Underneath the lowest hackebord, on which the carved console figures stand, lies the *Architrave*.

And the *Cymatium* stands to the lowermost cover.

(83 I 41) 4. *About the second hackebord.*

1. The second hackebord, is 2 feet 7 inches higher, than the first hackebord, stands 19 inches farther aft, it is 1 foot broad in the sides, but in the middle broad 14 inches, thick 2 inches.

The hackebord is the base and foot of the stern gallery, to which the bottom ends of the carved figures are attached. The cover lies on the second hackebord, and is that on which one leans when looking out. The Foot in the side, is fixed to the hackebord, on the side to make the eye of the quarter gallery. The counter timbers hold the gallery across the stern.

The lowest hackebord, stands with its underside, 12 inches from the upper side of the wale, and it is broad 1 foot at the ends, in the middle 13½ inches, and thick 4 inches, has a curve of 5 inches toward aft, as from below a curve of 5 inches, it stands 2 feet 2 inches back from the stern post.

2. The corners project 3½ inches, and have a curve downward of 7 inches.

3. This hackebord has a curve of 4 inches aft.

(84 I 8) 7. *About the Quarter Gallery in the side.*

This is laid, according to the line of the ship, not to look misshapen, which is why below and above it slants against the ship, is often sheathed with copper or plated lead, serves as special abode, for the skipper or captain, to stow something in it, and often as privy, as well as lookout; often they are made open, and only for decoration's sake, and also with small spires, for show.

. . .

(57 II 38) In the Gallery privies are made, cases, or lockers and lookouts, they serve for the decoration of the ship. A Gallery, with a window in it, is shown in plate No. 19 [fig. 2.172].

(84 I 31) 8. *The aft second cover, on the Wale.*

1. The aft cover, broad 8½ inches, has a curve backward 4½ inches, the end broad 13 inches, thick 4/5 inch, then end comes 2 feet past the ship.

2. The small Knee in the meeting of the master ribband, and the second hackebord, long 2 feet, broad 5 inches, thick 4 inches.

9. *About the cover in the side.*

1. Aft this is broad 9 inches, forward on the post broad 6½ inches, thick barely 2 inches, on the arch broad 7 in.

2. Aft with a corner console projecting in and out.

(84 I 21) 1. The Foot long 6 feet, forward broad 5 inches, aft broad 7½ inches.

2. The Upper end, 1 foot from the ship.

3. The quarter gallery architrave to it, broad 9 inches, thick 2 inches.

4. The Arch 8 inches past the gallery framing timbers.

5. The frame timber rakes 5 inches aft, and is broad 4½ inches, thick 3½ inches.

(57 I 37) The second cover, has a curve aft of 4 inches, and is 1 foot 1 inch broad there in the middle, on the wale B [in fig. 2.170] broad 8½ inches, and at the end 2 feet from the ship, broad 1 foot 1½ inches, thick hardly

2 inches, in the side at N, broad 9 inches, on the post C the cover broad 6½ inches.

The corners jut up and out. Forward on the gallery frame timber the quarter gallery architrave stands 1 foot from the wale.

Aft against the hackebord, and above the wale 1 foot 5½ inches, the small knee long 2 feet forward, the side thick 4 inches, broad 5 inches. Many shipwrights, have their ships jut aft, and make it high there, to their satisfaction, and as custom demands, without rule.

(55 II 38) The *Thorus* to the second cover, to which the carved figures are attached with their heads.

(87 II 37) The ***windows aft,*** they are wide inside 18 inches, the outside 16 inches, because the pillars come onto them there, and these are broad 3 inches, thick 3 inches: the pillars stand on a small tuigeltjen [no English equivalent], against water seeping in, thick 1½ inches. The shutters made from sawn deals, also the cross braces. The windows in the cabin wide 1 foot, high 1 foot 3 inches, on the outside there are frames around the windows, thick 3 to 4 inches, and also as broad.

(87 II 21) 49. ***About the window in the Gallery in the side.***

1. The Window, high 1 foot, 5 inches, wide 16 inches.

2. The lower sill, broad 5 inches, has an angle in the middle, and is paneled with wainscot 1 inch thick in the side planking.

3. Coming from behind 16 inches.

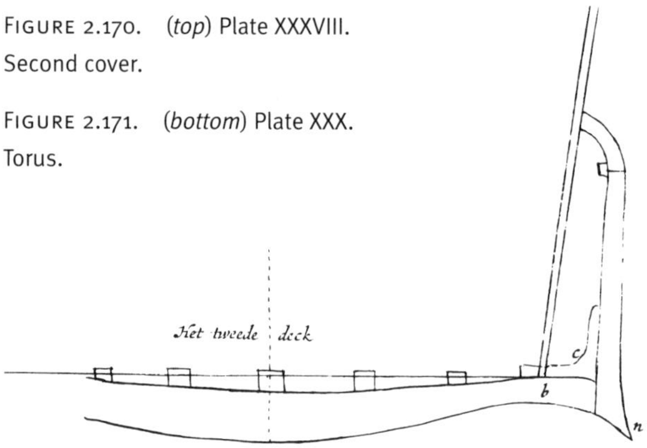

FIGURE 2.170. (*top*) Plate XXXVIII. Second cover.

FIGURE 2.171. (*bottom*) Plate XXX. Torus.

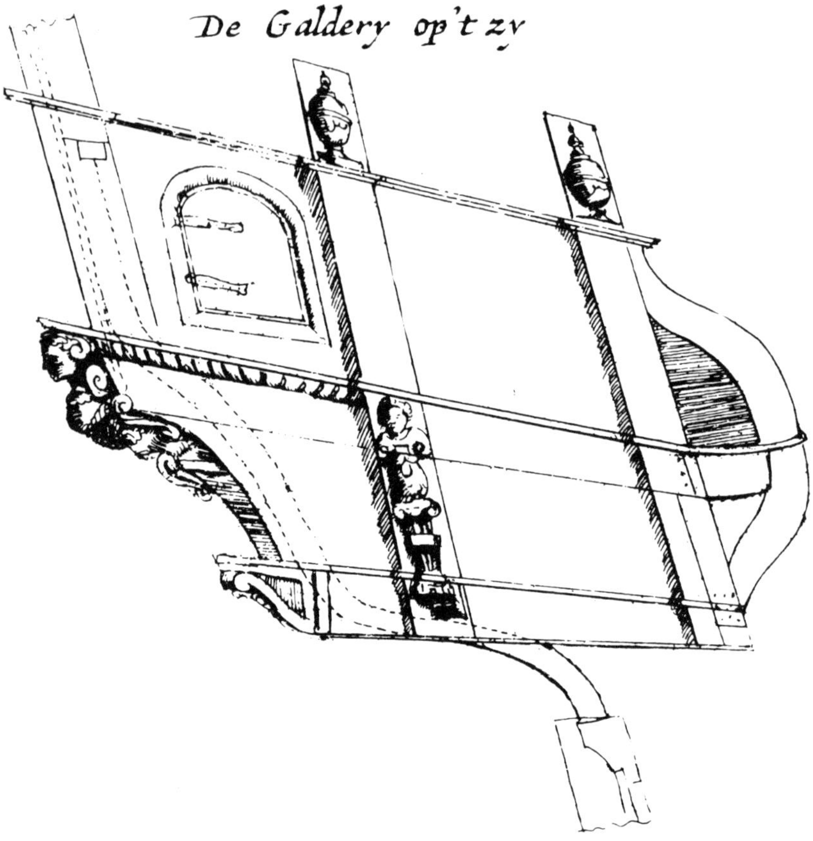

FIGURE 2.172. (*left*) Plate XXXIX. The quarter gallery.

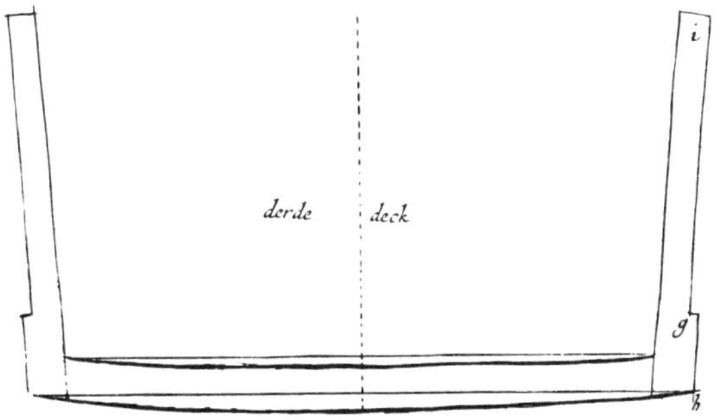

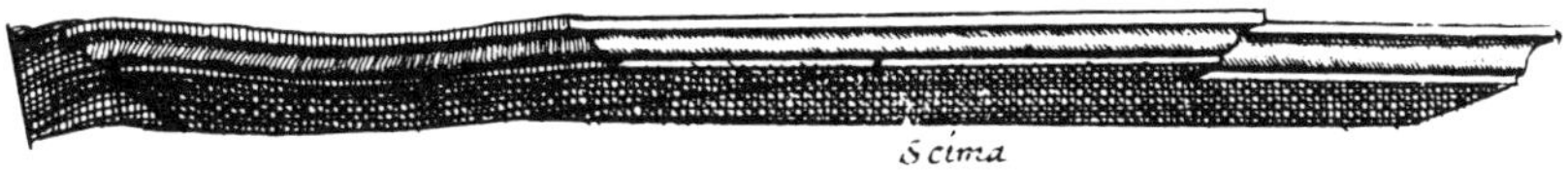

Figure 2.173. (*top*) Plate XXX. Decorations around the cabin door.

Figure 2.174. (*center*) Plate XXXVIII. Third cover.

Figure 2.175. (*bottom*) Plate XXX. Cyma.

The upper chock must be as broad as to accord with the curve, above it is square, and is standing from the flat of the stern 1 foot 4 inches.

4. Above the arch is a chock, which is made square, hewn out above and below, forward slight, aft hewn short.

(57 II 1) The third cover stands, in the middle from the second, 2 feet 4½ inches, at the ends measured at the flat of stern timber 2 feet 10 inches, the same is broad 9 inches in the middle, and the ends broad 8 inches, the side thick 2 inches; placed on the wale, the broadest from H to G [fig. 2.174], is long 1 feet 8 inches, and broad there 1 foot, from I to G, is broad 8 inches, thick 2 inches in the middle, the deck from I to H, long 8 feet 5½ inches, the outside is planed with a cyma.

(55 II 42) The Cyma stands to the third cover.

(84 II 33) 13. *About the third cover.*

1. This cover stands 2 feet 5 inches from the other cover, to be measured in the middle at the counter timber. In the middle it stands from the second, 2 feet 4 inches, at the end it is wide, 2 feet 10 inches, in the middle broad 9 inches, at the end broad 8 inches, thick 2 inches.

2. Aft on the side, it is broad 1 foot, long 1 foot, and rebated 9 inches, forward broad 8 inches, long 8 feet 6 inches, is joined into the wale a little aft, and has a Cyma.

(84 II 46) 14. *About the herm piece standing on the thorus.*

1. The herm piece broad 17 inches on the chest, thick 10 inches, below even with the outer edge of the cover.

2. The Hance Piece 3 feet above the railing: the first hance broad 14 in., the other stands at a distance of 2 feet 5 inches below. Above wide 2 feet 6 inches: the next wide above and below 2 feet 4 inches, thick 9 inches, coinciding with the quarter piece below.

(86 I 37) 15. *About the uppermost hackebord, &c.*

1. This hackebord, is broad 3 feet 6 inches, thick 5 inches.

2. Stands 10 inches above the cornice.

3. The baluster [probably the balusters between the two *vaezen*, or top cornices], broad 4½ inches, thick 2 inches.

4. The balusters stand 5 inches apart, and fixed in the lower cornice.

5. The top cornice broad 6½ inches, thick 2 inches.

6. The curtains, thick 2 inches.

7. The pilasters in the windows, broad 3 inches, thick 1½ inches.

8. The arches above them, thick 1½ inches, the breadth comes against the deck above, nicely cut.

Above on the hackebord, over which one can lean, either an elegant arch is made, or sculpted dolphins or lions, or something of the sort.

(84 | 46) 10. *About the Quarter Consoles.*

1. The quarter console, broad 9½ inches, thick 8 inches.

2. The middle console, broad 8½ inches, thick 7½ inches.

3. The other four consoles broad 8 inches, thick 7 inches; the consoles broad 12 inches on the chest.

4. Below they stand on the architrave, the architrave broad 2½ inches, which goes straight on, thick ⅘ inch.

11. *About the carved figures in the side.*

1. The carved figure on the post, broad 9 inches, on the chest broad 10 inches, thick 8 inches, below broad 8 inches.

2. Leave a stump at the end, broad above 8 inches, thick 7 inches.

12. *About the top Cornices.*

1. The top cornices between the figures, broad 5½ inches, thick 1½ inches, these cornices come below the uppermost cover.

2. The lower cornice, broad 6 inches, thick 1½ inches.

3. The cornices on the side, broad 4 inches, thick 1½ inches: at the end the cornices are broad 3 inches.

4. The cornice beneath the hackebord, against the flat of the stern, come even with the lower one, and are broad 8 inches, thick 3 inches.

5. At the end broad 7 inches, thick 3 inches, and has a curve of 8 inches.

6. The Hackebord has 8½ inches curvature, the curve at the end 1 inch.

(81 | 36) In the old days sometimes two galleries were made aft around the outside, for the cabin as well as for the gun room, fixed with iron arches above, and open.

. . .

(267 II 48) The curve or the counter, hanging backward, is decorated or undecorated, as is deemed fit.

FIGURE 2.176. Plate XXXIX. Herm pieces.

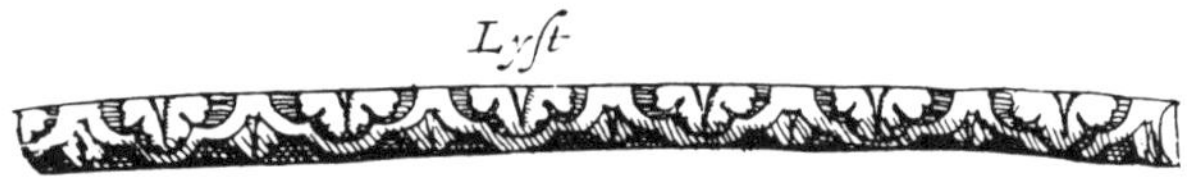

FIGURE 2.177. Plate XXX. Molding.

FIGURE 2.178. Plate XXXI. Decorative moldings.

81. The Stern Gallery

Terminology is complicated here because we do not have real English equivalents. Even in Dutch many of the terms used by Witsen have disappeared. The stern gallery was a complicated architectural whole (see fig. 2.167). As was customary in baroque decoration, the main parts were separated by horizontal curved cornices and vertical carved consoles.

The first cornice, the lower or first *hackebord* (lit., "carved plank"), served as the aftermost support of the deck of the captain's cabin. This piece of wood was often carved as an ornamental molding, the architrave, and planed into a cymatium, called an ovolo if it had an egg-and-dart motif (see fig. 2.169). Other patterns were also used (see fig. 2.178).

The first *hackebord* was covered by the first cover, which was as broad as the height of the *hackebord*. The cover concealed the cross-grained ends of the deck planking and had the same curve. It formed the basis for the consoles that were installed underneath the windows for statues to rest on. On each side the plank protruded to become the foot of the quarter gallery.

A set of counter timbers was raised from this base, forming a bulbous or S-shaped counter called *het ronde wulf* (lit., "the round counter") and continuing as the posts between the windows and higher up as vertical timbers of the flat of the stern, to which the coat of arms was attached at a later stage.

The round counter was planked up to the windows. Below the windows was the second *hackebord*, carved as well and covered by the second cover. This cover was quite complex in shape, high in the middle, sloping downward toward the sides and then rising again at the end, where the plank was twisted to align with the rail of the quarter gallery. It was therefore also called the *slingerlijst*, or twisted cornice; it was planed into a torus, a convex molding (see fig. 2.171).

Between both *hackebord* covers were the herm pieces and quarter consoles, nailed to the counter timbers straight through the counter planking. At the quarter galleries carved consoles were also applied.

The windows between the second and third cover opened into the captain's cabin, and their posts were decorated with figurines, mostly putti (angels in the shape of cherubs). In the quarter galleries these carvings were often dolphins, while the forward end was closed off with a small arch or buttress. Left and right of the windows were ornamental decorations, filling the gap between the ship and the statues in the side. The quarter galleries of the pinas were thus nothing more than an open, cagelike construction, encasing the window in the cabin's side.

The third cover, placed above the windows flush with the planking of the upper cabin deck, joined the upper ledges of the quarter galleries across the stern. It was planed into a cyma (2.175).

The section above the windows and the third cover was called *het rechte wulf* (the flat counter) and was closed off at the top by the uppermost *hackebord*. Straight underneath this *hackebord* were what Witsen calls *vaezen*, profiled cornices, in between which the planking could be left off; or this narrow section could be planked and decorated with carved floral motifs or the year of construction and sometimes the name of the ship. If the space between the two cornices was left open, it was often filled with balusters, shaped as vases, which might be the explanation for Witsen's confusing use of the word *vaezen*. From the text it is clear that he is actually referring to the railings, or maybe we should call them "top cornices" or "covers," like the ones below. He even specifies the amount of camber they have.

On the flat of the stern or upper transom, the escutcheon was applied. Until about 1640 this was almost invariably the arms of the town of origin; afterward the name of the ship was symbolically depicted. Witsen uses his own coat of arms for his illustration, so it seemed to me a good idea to give my test model the same coat of arms.

The flat of the stern is high enough to give ample room for the cabin above the main cabin and for a bench on the uppermost deck with its back to the taffrail. Underneath the bench, of which the front side was closed off by small bars, the captain often kept his chickens.

Flanking this flat of the stern were "statues in the side," which continued downward between the second and third cover.

Finally, at the top, above the top cornices, was the third or upper *hackebord*. Often this section was decorated with carved dolphins or lions and other figurines. Almost all carvings were fashioned from rejected masts and spars, so they were mostly made of pine, though sometimes also lime and, in exceptional cases, leftover oak. See also section 91 (Then plank it there, and make the Channels) and figure 2.187.

---

82. [erroneously numbered "92"] *The Gunports.*

(56 II 53) The Pins of the port lids come onto the chocks, and the port lid below closes on the port sill, the port lid is lined on the inside, and nailed with lozenges. It has ropes to it, to pull it open and shut.

. . .

FIGURE 2.179. Stern and quarter gallery of the pinas model. (Courtesy Cees de Jonge, The Visual Art Box)

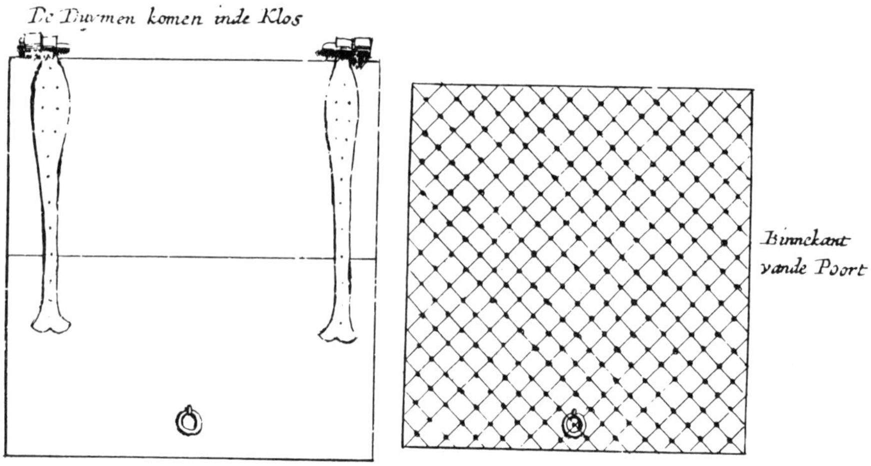

FIGURE 2.180. Plate XXXVII. The pins of the port lids come onto the chocks; inside of the port lid.

(275 I 19) Simulated gunports and gun plugs which are white, can often be seen painted on merchantmen, to appear more defensible.

> (77 II 50) the hinges broad 2½ inches, and thick ¼ inch: the pins thick one inch: the iron work of the port lids is ¼ inch: the lining somewhat thinner than the port lid.

82. The Gunports (Port Lids)

Because the gunports have already been discussed in section 60 (**Make the Chocks above the gunports around the outside**), it is probable that the port lids were intended here. They consisted of two planks, the one slightly thinner than the other, nailed to each other with the grain of the wood at right angles. For this a lozenge pattern was gouged on the inside, the nails coming on each crossing. Supplied with hinges and rings, they were

Figure 2.181. Plate XXVI. Hawsehole.

hung to the hooks of the hinges that had been nailed to the chocks above the gunports.

83. Hawseholes.

(55 I 26) Hawsehole: a hole, in a thick plank, standing in the side, against the stem, through which the cable runs, with a fairing made around it on the outside.

. . .

(276 I 1) Forward against the bow, on the upper deck, one can see a trough, which is called the pis-bak: it serves to catch the water coming in through the hawseholes in bad weather, named such because of it similarity to that trough on the main deck, which the sailors use as their urinal.

> (79 II 50) 36. *About the Hawseholes.*
> 1. The first hawsehole is wide 12 inches, the second 10 inches, the third 9 inches.
> 2. To starboard are 2 hawseholes, one of 1 foot, and one 10 inches wide.

83. Hawseholes

Curiously enough, this pinas had three hawseholes to port and two to starboard of varying size.

The proper placement of the hawseholes was a problem in this ship. Usually inside, underneath the hawse, there was a trough to catch the water coming in through the hawseholes in bad weather and dripping from the anchor cables when they are hauled up. It was called the *pisbak* (urinal), which explains its second function.

Although Witsen mentions the manger, he does not include one in his description of the pinas. That leaves two possibilities: either the hawseholes were placed at the usual height and Witsen just forgot to mention the manger (and the scupper belonging to it); or the hawseholes were so low above the deck that manger and scupper became unnecessary, as the drippings could drain through the hawseholes themselves. (It is also possible that the curious fifth hawsehole served as the scupper when four cables ran through the hawseholes). For the model I chose this design, as it also worked better on the outside: the anchor cable thus risks damaging the beakhead much less than when entering the ship at a higher level.

The size of the hawseholes depended on the thickness of the anchor hawse or cable, which in turn was derived from the size of the ship. Witsen's illustration (fig. 2.181) shows a crescent-shaped fairing under the hawsehole, but other iconography shows a thick "boxing" around the hawseholes on the outside of the hull.

84. [erroneously numbered "83"] *The Bitt Knees.*

> (79 II 45) 7. The bitt knees, long 10 feet, thick 10 inches. The bitt knees at the bitts posts broad 2 feet 2 inches, forward 12 inches.

84. The Bitt Knees

The bitts were fitted in section 76 (Place the Bitts) and are reinforced in this section with two large knees forward. At the same time, they provided an important support for the bowsprit because they held the two knights or standards in between which the foot of the bowsprit rested.

85. The Covering Board.

(56 I 55) The Covering Board, also called Plank Sheer: is the wood, which covers the Ship's side, and closes it off, so that the water cannot come between the sides, and to the frames, the posts which are depicted in the plate, are the ends of the frames.

> (77 I 15) *About the Covering Board.*
> The Covering Board, thick 1½ inches,

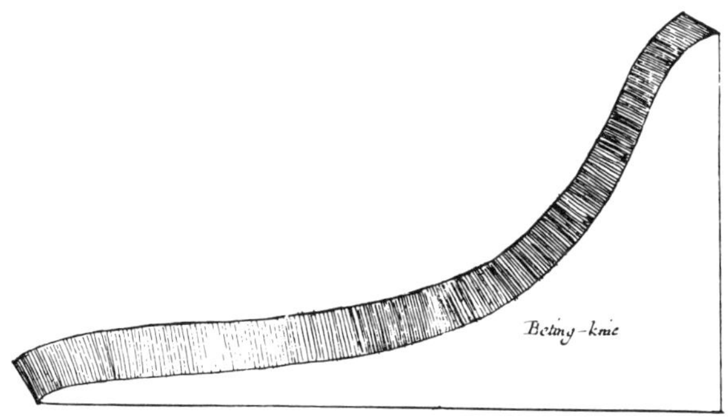

Figure 2.182. *(right)* Plate XXXVII. Bitt knee.

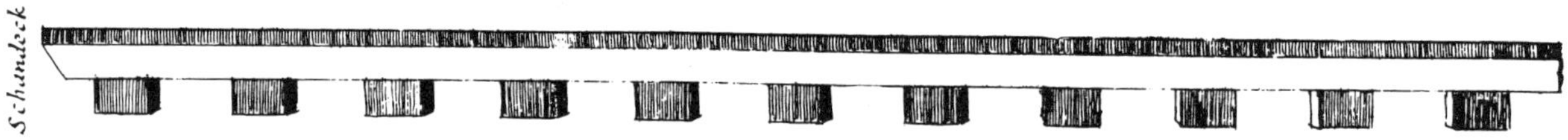

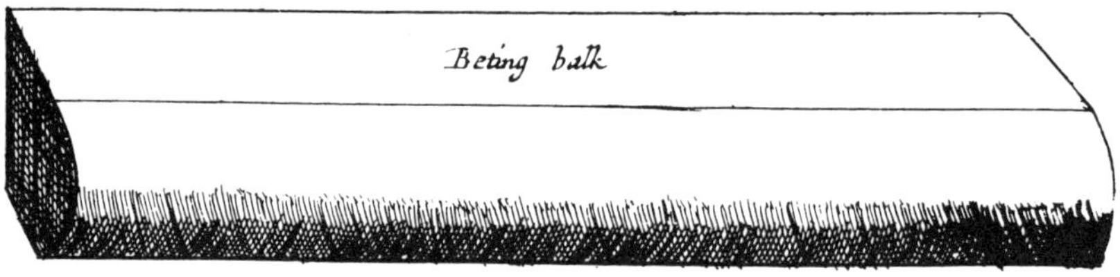

Figure 2.183. (*above*) Plate XXXIV. Covering board.

Figure 2.184. (*right*) Plate XXXVII. Bolster.

Figure 2.185. (*below*) Plate XXXII. Cabin beam.

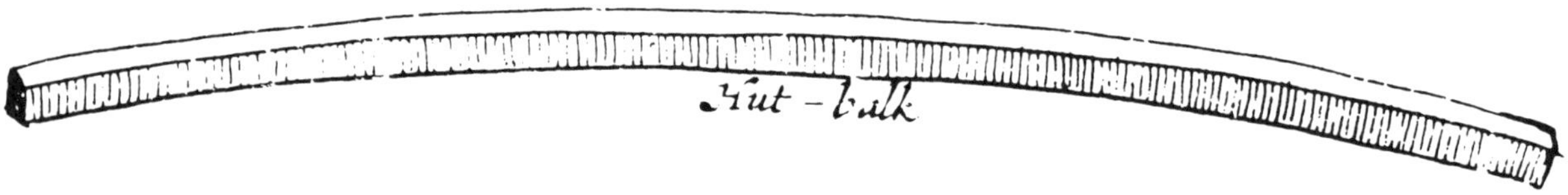

85. The Covering Board

The covering board closed off the top of the top timbers, preventing water from entering between the frames from above. At the beginning of the century, for instance, on the *Wasa* (1629) and even on the *Prins Willem* (1649), the covering board at the top of the gunwale lay very low above the deck, with the tops of the top timbers still visible at the inside of the gunwale. Later the inwale was raised, lining the inside of the gunwale up to the covering board everywhere. This can be seen on the Hohenzollern model of the Dutch two-decker, for instance. Van Yk was not very taken by this idea and devotes a lengthy section to the low gunwale, which in his opinion is much easier to maintain (p. 115).

The plank that formed the covering board was half as thick as the ceiling; it was mortised to fit over each top timber and caulked after positioning.

86. The Bolster.

(55 I 14) Bolsters: serve to belay ropes, and cables onto, to make bitt anchors. The large riding bitt also helps to strengthen the ship in the bow.

. . .

(70 II 18) 2. The bolster ⅓ thicker than the stem.

3. The cheek of deal, on the aft side half as thick as the stem.

(79 II 43) 6. The heads of the bolster, long 15 inches.

86. The Bolster

The bolster was the last part of the riding bitts, which consisted of the bitt posts and bitt knees before them (see sections 76, Place the Bitts, and 84, The Bitt Knees).

This horizontal beam was provided with a deal or pine fish. It was here that the wear was the greatest. It was much cheaper to occasionally change such a piece of soft wood than the oak crosspiece itself. Furthermore, the anchor cables were subjected to much less chafing from the softer wood.

87. Plank the forecastle.

(86 I 18) [the planks] in the forecastle 2 inches

87. Plank the Forecastle

This section refers to laying the deck planks on the forecastle head. These filled the spaces between the waterways and the binding strakes at each side and between the binding strakes in the middle.

88. Make the Waterboard on the Forecastle.

(83 I 32) The waterboard on the forecastle, 3½ inches, broad 10 inches.

88. Make the Waterboard on the Forecastle

Waterboard was another word for "waterway" and was the plank lying flat on the deck beams in the sides. Apparently this word was used only for the uppermost decks, and Witsen is very consistent in his use of the term. If we can judge from his illustration of it (see fig. 2.186) in section 89 (As also the Waterboard in the Upper Cabin, lay the Deck Beams against it, with the Ledges), there was no difference in shape between the waterway and the waterboard; yet it was called this only for the forecastle, the steering stand, and on top of the upper cabin.

89. As also the Waterboard in the Upper Cabin, lay the Deck Beams against it, with the Ledges.

(56 | 22) The ledges of the Upper Cabin are as long as the deck beams, The planks of the upper cabin are as thick as the waterboard.

(81 | 24) 1. The Waterboard is long 17 feet, broad 6 inches, thick 2 inches.

2. Forward in the side high 5 feet, aft 6½ feet.

3. The deck beams have a camber of 13 inches, are thick 3 inches, broad 4 inches.

4. The ledges lie between each deck beam, and are thick 1½ inches, broad 2 inches.

5. Faired with ¼ inch planks.

6. The knees, long 3½ feet. The bills, long 2 feet 3 inches, broad 5 inches, thick 4 inches.

Aft, for light on board, square deck lights, or round open holes are made, and above also a grating.

7. Aft the ceiling planks are pierced for muskets

8. And the ceiling is made of ¼-inch planks.

9. The bulkhead, next to the door, is also pierced for muskets.

10. The door wide 3 feet 6 inches. The posts, broad 3½ inches, thick 3 inches.

11. The arch, high 13 inches, broad 3½ inches, thick 3 inches.

12. The gunports, wide 14½ inches, high 13½ inches.

(88 II 13) The waterboard is broad 6 inches, thick 2 inches, the deck beams lie against it from below, the foremost deck beam lies 5 feet 5 inches high in the side, 7 feet and 8 in. farther aft, it is high 6 feet 2 inches in the side, and there a small beam lies between two large ones, to the back of the bulkhead lies a ledge. The deck beams have a camber of 15½ in. on a length of 14½ feet, the deck beams thick 3 in., broad 4 in., the foremost high 5 feet in the side, the knees broad 5 inches, the deck clamp thick 1½ inches. The ledges thick and broad 2 inches, between each deck beam 3 ledges, the schildje [probably a small grating to allow air and light into the upper cabin] is broad 8 inches, thick 3 inches, in which are 4 holes, wide 3 inches, and it stands behind the third deck beam, 13 inches.

89. As Also the Waterboard in the Upper Cabin, Lay the Deck Beams against It, with the Ledges

When laying the lower deck beams, first the deck clamp was made, then the deck beams, and lastly the waterway. When laying the deck beams of the upper deck, the forecastle, and the steering stand and cabin, the deck beams were laid first, followed by the deck clamp and the waterway.

According to Witsen's text, first the waterboard was placed, then the deck beams set against it from below, and finally the deck clamp. The reason for the difference in procedure with the lower decks remains a complete mystery and sounds almost impossible.

The deck beams of the upper cabin were one third of the thickness of the stem and had a much more pronounced camber than the deck beams of the lower deck. The general rule was that the deck had more camber and was thus more convex the higher it lay.

90. Make the Deck Clamp and other Ceiling.

90. Make the Deck Clamp and Other Ceiling

After the deck beams had been placed against the waterboard on chocks, the deck clamps and the ceiling underneath were placed and nailed. The deck clamps were one sixth of the thickness of the stem.

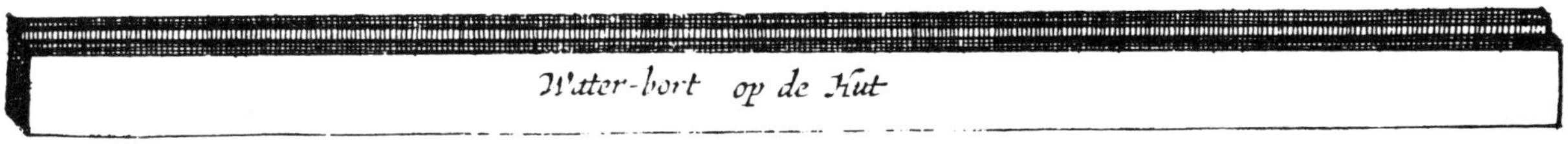

Figure 2.186. Plate XXXIII. Waterboard on the cabin.

91. Then plank it there, and make the Channels.

(81 I 6) 42. *Several matters about the Upper Cabin.*

Which on large ships is divided in two, and has a third cabin above it. Before its door a porch is often made, supported by small pillars, and at the end at half the upper deck, banisters, made differently, high or low, heavy or light, elegant with carvings, drolleries, or without. Above the upper cabin against the flat of the stern, often a bench is made, and underneath it a chicken roost or dovecote. The bulkheads of the upper cabin, the cabin and the forecastle, which stand in the air, are made slanting or bulging, outwardly, such for drainage, and also for the sake of decoration.

. . .

(70 II 5) 24. *About the Upper Cabin, and matters in the Upper Cabin.*

1. The beams of the upper cabin thick 1/3.
2. The deck clamps thick 1/6.
3. The deck beams have a camber of 10 inches.
4. The Waterboard thick 1/6 of the inside of the stem.

In front of the upper cabin, on very large ships, sometime loose decks are placed, that can be removed.

(88 I 26) The foremost binding strakes thick 2 inches, broad 7 inches, the others behind them thick 2 inches, broad 9 inches: the fourth and fifth as above, the sixth wide 4 feet, 8 inches, thick 1 inch, broad 8 inches.

(86 I 20) [the planks] on the forecastle and the upper cabin, 1¼ inch.

(70 II 22) 26. *About the Channels.*

1. The main channel long 1/8 of the length of the ship.
2. The thickness 1/3 of the stem.
3. The breadth, as the stem is thick.
4. The fore channel a little less long, thick and broad.
5. The mizzen channel 1/3 of the main channel, namely, its length and breadth, but its thickness as the fore channel.

(81 I 51) 43. *About the main Channel.*

1. The main channel, long 27½ feet, broad 17 inches, thick 3½ inches.
2. The fore end comes equal with the main mast with its fore side, and with the uppermost wale with its upper side.
3. The lath lying on it outside, broad 3½ inches, thick 2½ inches.
4. The foremost chain plate, equal with the aft face of the main mast.
5. There are seven chain plates.
6. The chain plate bolts, thick 2 inches.

(82 II 33) 44. *About the fore Channel.*

1. The fore channel, long 22 feet 3 inches, broad 16 inches, thick 3 inches.
2. The Fore end, 6 inches, past the foremast, lying on the sheer rail.
3. The lath on it, broad 3 inches, thick 2 inches.
4. The foremost chain plate equal with the fore face of the foremast.
5. There are 6 chain plates.
6. The chain plate bolts, thick 2 inches.

(82 II 44) 45. *About the mizzen Channel.*

1. The mizzen channel, long 10 feet, broad 9 inches, thick 2½ inches.
2. The Fore end, comes equal with the aft face of the mizzen mast, lying on the sheer rail.
3. The lath, broad 2½ inches, thick 2 inches.
4. The foremost chain plate, coincides with the aft face of the mizzen mast.
5. There are 4 chain plates.
6. The chain plate bolts, thick 1½ inches.

91. Then Plank It There and Make the Channels

When the deck of the upper cabin had been planked as well, all decks were finished. The difference in construction of this deck, compared to the other decks, was that it had no binding strakes. The ledges were as long as the deck beams, which meant they ran from side to side. As with the other decks, three ledges were placed between each pair of deck beams.

Channels

Channels were the heavy timbers lying with their edge against the ship's sides to hold the shrouds and deadeyes clear of the sides. On the outside the channels had indentations into which the chain plates fitted, the irons holding the deadeyes. This was then covered with a lath over the entire length of the channel, sealing off the indentations.

In the case of the pinas the main channel was let into the uppermost wale. Ordinarily the fore channel would have been fixed in the same way and at the same height, but on Witsen's pinas the fore and mizzen channels have been fixed onto the sheer rail, thus distinguishing this ship from many others in the seventeenth century. One can only speculate about the reason for this deviation: for one thing, this particular ship had guns in the forecastle, which between the chain plates certainly had more room to shoot than between the deadeyes above the channel.

92. Panel the Cabin, and make the Berths in there.

(56 II 52) At No. 18 [fig. 2.187] a Cabin can be seen from the inside. AA Are posts on which pilaster are to come. bbb Are shelves. C Arches. D Small lockers. E is a small wine cellar. F A door of a cupboard, to store something inside. G A pass-through. At H the cabin door would be; here left out for lack of space. The posts are broad 3 inches, thick 1½ inches. The cabin door, is wide 2 feet 3 inches, and the large wine cellar door wide 3 feet 6 inches, the lower shelves of the treasury, come from the floor 4 feet, bb is high 3 feet 7½ inches, the middle b is high 2 feet 7 inches, the pillars, or pilaster lying on the posts, are broad 3 inches, thick 1½ inches, met gestreeken fermetten [no English equivalent]. The door of the wine cellar, through which the wine is drawn, wide 9 inches, high 1 foot, the door F, wide 1 foot 8 inches, all stands from the cabin bulkhead 7 inches.

Above the door of the cabin, which is not shown here, lies an architrave, to close the door to.

And above all this paneling, comes a cornice, to fill the roundness of the roof.

Above these cupboards, often straight shelves are made against the bulkheads, with pillars, to store glasses and plates. All around in the cabin cleats are made, from which to hang arms.

(83 I 6) *46. About the windows in the Cabin.*

1. The Window in the cabin, long 1 foot 3 inches, wide 1 foot, comes 3 feet above the floor.

(58 I 39) The window in the berth, coming from the bulkhead of the cabin 5 feet, and from the deck: it is high 1 foot 3 inches, wide 10½ inches, the berths in the cabin are sometimes made athwartships, and such above the letters bbb on the plate No. 18 [fig. 2.187].

(57 II 43) The bench standing in the cabin, against the stern of the ship, is high 1 foot 6 inches, the legs are thick and broad 2 inches, it is 1 foot 8 inches broad above, at the ends it is broad 1 foot 6 inches, the lid is broad 1 foot 20 inches, thick 2 inches, shown in plate No. 21 [fig. 2.189]. To starboard at C there is a bench, long 1 foot 8 inches, and the middle bench at letter A, long 1 foot 7 inches, through which passes the rudder, when it is being placed, and when shooting through the back, the bench at b, to both sides, can be lifted, in which the gun carriages go; often a privy pipe is made in the bench at d, 6 inches from the side of the little stern timber, and 1 foot from the ship's side, below wide 4½ inches, above wide 16 inches.

Often hatches can be seen in the floor of the cabin, through which the rudder is being placed, instead of having these benches, but in such a case, the ships overhang aft considerably.

Inside the cabin, in the porch, there are often two treasuries, with a drawer in the upper one.

The framing of the berth, fore and aft, to port, they are thick 1½ inches, and also the legs, the berth is long 5 feet 7 inches, the foremost leg stands 1 foot 5 inches from the ship's side above, below 3 feet 5 inches, the aftermost from the side 1 foot 3 inches above, and below 3 feet 1½ inches. The batten, which are at the foot of the berth; through which go the ropes, are high 2 feet 2 inches, the ends above that; are high 3 feet 6 inches,

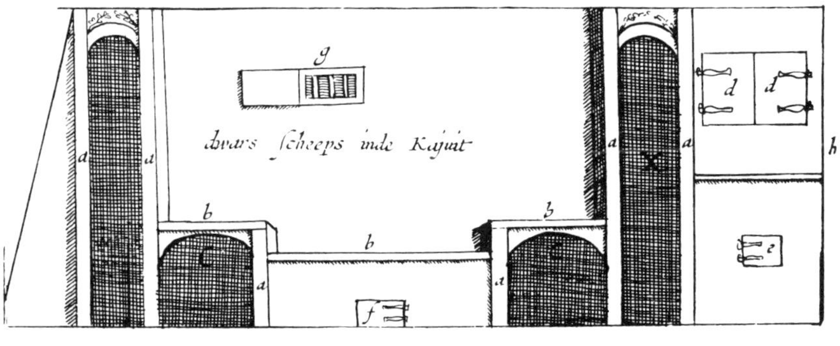

Figure 2.187. Plate XXXVIII. Inside the cabin athwartships.

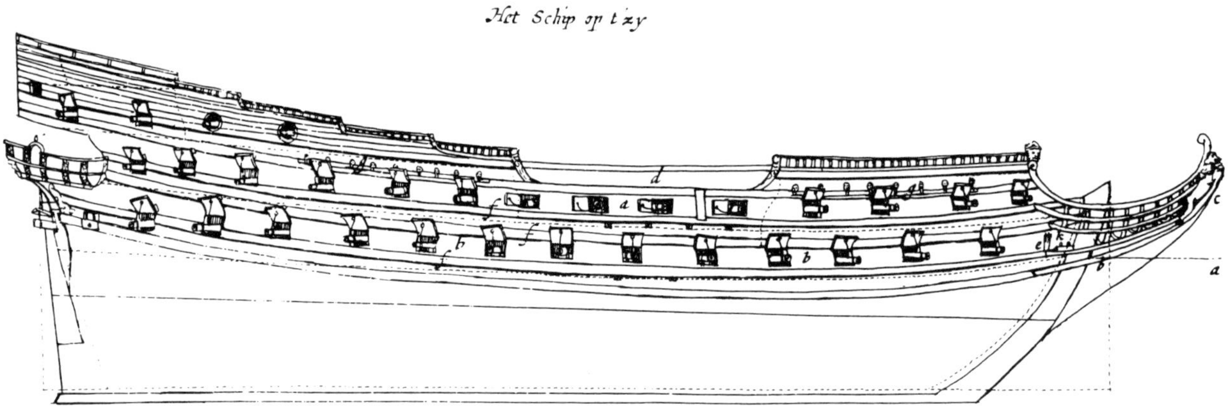

FIGURE 2.188. Plate XLIV. The ship from the side.

(59 II 35) On the plate showing *the Ship from the side:* is clearly visible, the placement and shape of the Quarter gallery, the window in the Cabin, the gunports, how most of them open upward, but others in the waist as at letter A, the shape of the rudder, and the stem: as at B between the wales F, the railings at C, and the washstrake at D, besides the hawseholes E and channels G. [. . .]

Against the Ship, and the Beakhead at K, a console figure is often made.

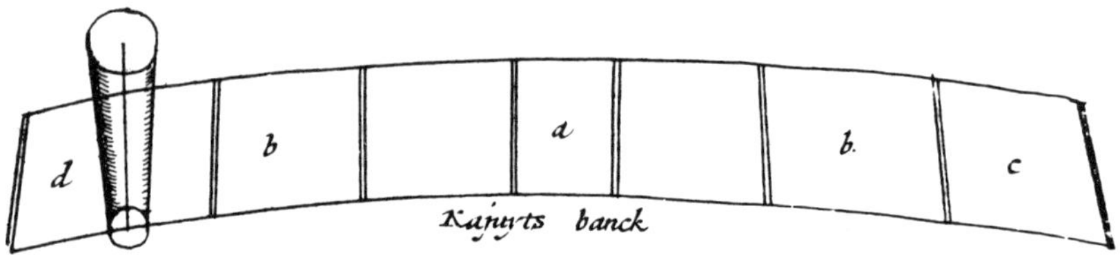

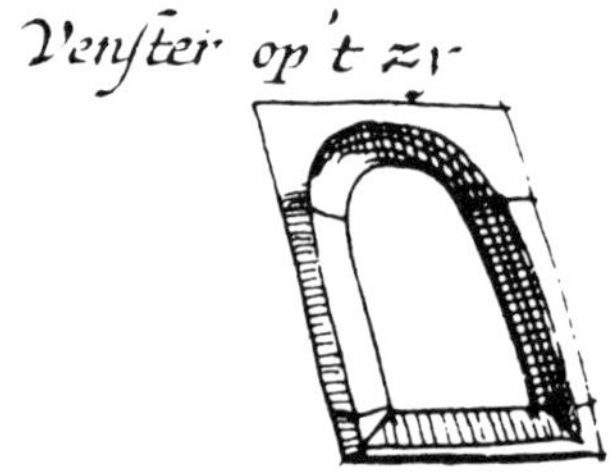

FIGURE 2.189. (*left*) Plate XLI. Cabin bench.
FIGURE 2.190. (*right*) Plate XXX. Window in the side.

> the legs are broad 2 inches, and the batten making the foot of the berth, is broad 3 inches; thick 2 inches, wide between the aftermost leg 10 inches, between the hanger wide 3½ inches, and between the other wide 12 inches, The post underneath the foot, wide 6 inches.
>
> (89 I 34) 51. *About the bench, aft in the Stern.*
>
> 1. The bench in the stern, high 1 foot, 6 inches, broad 1 foot, 8 inches.
> 2. The posts, thick 2 inches, broad 2½ inches.
> 3. The Lid, broad 1 foot 9 inches, thick 2 inches.
> 4. Forward closed with one-inch planks.

92. Panel the Cabin, and Make the Berths in There

Witsen was pretty expansive about the paneling of the cabin of this pinas, which is fortunate, as a number of characteristics also figured on other ships.

Thus here, as on other ships, the bench built against the stern was partly removable, providing space for guns firing aft. A privy was made in the bench to one side, where the captain could relieve himself. The lead soil pipe ended between the decorations of the counter.

The windows were quite high, and a person could look through them only when standing. Wall cupboards for household effects were installed against the bulkhead between cabin and steering stand

Berths were sometimes made in the sides and sometimes against the bulkhead; in the fore-and-aft berth there was a small window in the ship's side. Another window, farther aft, opened into the quarter gallery. On large ships this would be a door that opened into the quarter gallery, where cupboards and privies were located.

93. Make the Partners ready.

(95 I 4) To obtain the rake of the main mast, measure the depth from the keel to the underside of the main deck, or the upper side of the hold beams, and as many

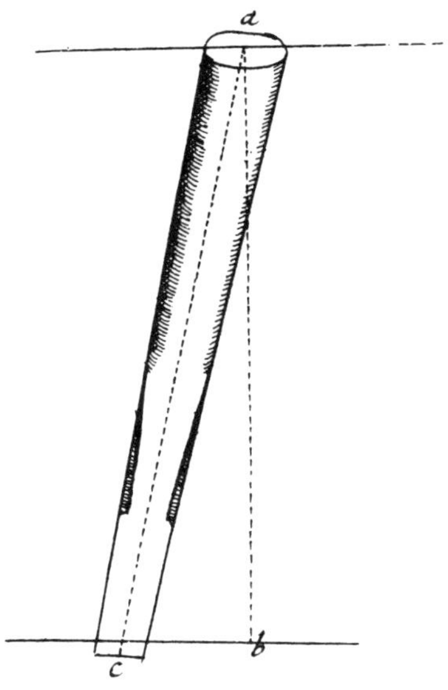

Figure 2.191. Plate XLI

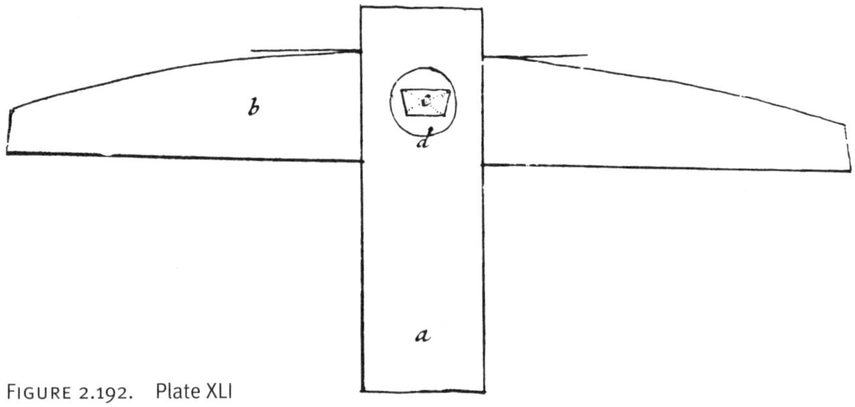

Figure 2.192. Plate XLI

feet as that is high, as many inches thou shalt take, for the rake of the main mast. The step below, make five and a half feet aft of the middle. When placing the masts, they are raised with sheer hulks or hoists, and are made to descend downward along a perpendicular, as fit. A [in fig. 2.191] is the main deck [. . .] B the keelson. C the main step. A is also the middle of the partner, and from there a plumb line is lowered onto B, and then again from B to C as many inches are measured, as it is high in feet from B to A. And this is taken for the rake of the mast. The mizzenmast is made to rake equally. The fore step, make for the entire height between the two, two inches forward. A is the fore partner. B is the fore step. In the round circle, the mast will stand, and in the edged C, the lower end of the mast stands, and this is called the step. D is called the hole of the partner.

. . .

(71 | 30) *32. About the Steps.*

1. The Ship having been divided into 5½, from behind, so place the mizzenmast on the first part, and on the third part the main mast, and on the fifth part the foremast.

2. In the hold the main mast should rake 1 foot 8 inches for every 10 feet depth.

3. The foremast rakes 2 inches to 10 feet or stands right up.

4. The mizzenmast rakes 1 inch to 10 feet.

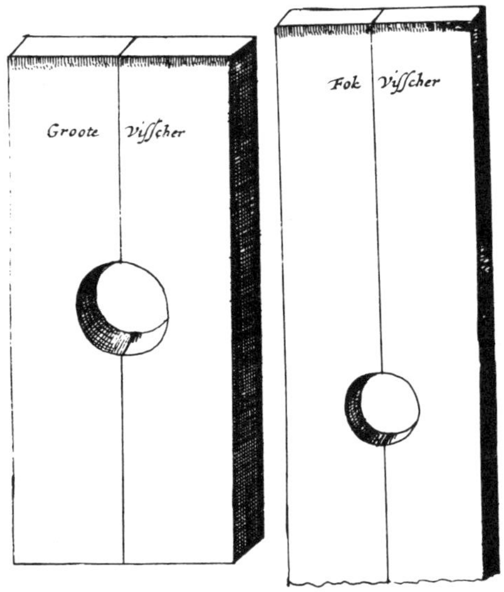

Figure 2.193. (above) Plate XXII. Mainmast partner, foremast partner.

(79 || 3) The main partner is long 5 feet 3 inches, broad 4 feet 7 inches, thick 6 inches.

(79 || 27) The partner up on the forecastle or in the forecastle, long 15½ feet, broad 3½, thick 4½ inches: the fore side of the hole lies 10 feet from the breasthook: the hole is wide 1 foot 8½ inches, and rakes backward 3 inches in the partner.

93. Make the Partners Ready

The partners of the main and foremast were much broader than those of the capstan and the mizzenmast. Each consisted of two broad planks with a hole in the middle, into which the mast fitted.

The fore partner was situated in the forecastle between the bitt knees, and the bowsprit heel rested on it. This

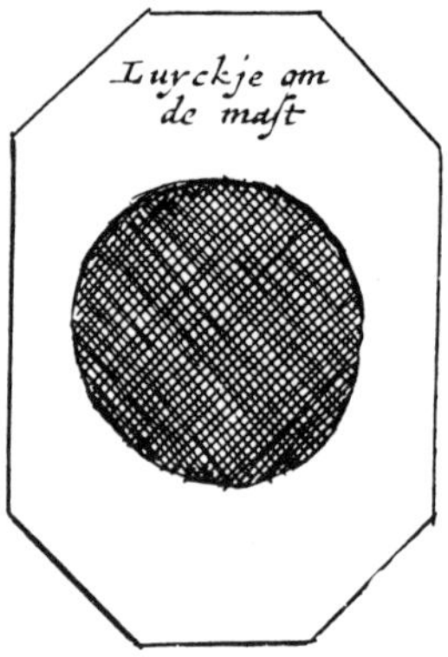

FIGURE 2.194. (left) Plate XXIX. Little hatch around the mast.

heavy oak plank, almost four and a half meters long, contributed considerably to the strength of the fore ship together with the riding bitts.

The main partner was much smaller—not even one and a half meters. It lay on the main deck between the binding strakes.

On Witsen's pinas all three masts had to go through another deck above the partner. In these, as mentioned before, a small partner was fitted; often the aperture was octagonal.

94. *Step the Masts.*

(156 II 36) When the Ship has been launched, the masts are then brought into it, which on small ships is done with a crane (as said before); with which the masts are hoisted: but on large Ships, which cannot be brought under the cranes, sheers are made of heavy masts, on the ship's main deck, against the sides, which sheer legs lock onto each other at the top, and are joined with ropes; from this then blocks are hung, with which the masts are raised and stepped. But as this also hurts ships somewhat, often sheer Hulks or Barges are used, ships of which the shape is shown below: these then are laid alongside the ships, if it is a heavy vessel, which by consequence needs a very large mast, two of these hoists, or sheer hulks, and secure them to each side of the Ship, on which stands a very heavy and thick mast, or rather post, with which the mast is hoisted with sheaves, and set in the ship in its appropriate place: In this winding either on one side or to both sides of the Ship, cleats are nailed onto the masts, to which ropes and lines are tied, when winding; care is to be taken that the sheer hulks are not too light for the weight of the mast that is to be stepped, and the same shall not break out, which I have seen at times.

. . .

(62 II 38) The main mast standing in the middle of the ship, deserves the first place. After which comes the foremast, that is the fore straight-up mast, and the bowsprit pointed lower and the mizzen, or rear mast: and also the main topmast, or second upper-middle mast: and the fore topmast. The main yard, or cross-mast from which the sail hangs. And the foreyard. Lateen yard, or cross mast; from which hangs the mizzen. Spritsail yard, from which hangs the foremost and lowest sail. Main topsail yard, or second middle cross-mast, from which the highest but one sail hangs. Fore topsail yard. The main-topgallant yard, or third and highest middle cross-mast, from which hangs the highest sail, the fore-topgallant mast. The mizzen topsail yard, or aft cross-mast, above the mizzen. Besides the mizzen topmast. The spritsail topmast, or little mast, coming up from the bowsprit. With its spritsail topsail yard. Mizzen and main-topgallant masts. Fore-topgallant mast.

. . .

(94 I 27) On short ships fit long masts and short yards; because as the wind blows parallel to the water, long masts on long ships would plunge the bow too much into the water, which on short ships cannot happen, because the angle which the bow makes with the water cannot be as large. The longer the lines, the larger the arcs they describe, when they are moved.

. . .

(94 I 40) *Proportions of the Masts.*

To arrange the masts in proportion, so take the depth and width twice, as for example, the width is 29, the depth 12.

29 41
12 41
41 82 of the height of the mast. The thickness of the mast is derived from the depth of the ship, for as many 6 feet depth, one foot thickness.

To get the measurement of the end, the thickness of the mast in the partner, ¾ for the end: Example,

18 palms [1 palm = about 10 centimeters]

13½ for the thickness of the end.

For the thickness of the topmast, this proportion is, namely ⅚ parts of the end of the mast, is the thickness of the topmast:

Example, 13½ thickness of the end of the mast,

11½ for the thickness of the topmast, which is in the mast cap.

The length of the yard, is according to the width of the ship: Example:

	28½ width
twice	2
	56 [should be 57]
depth once	13 depth
comes to	69. [should be 70]

The foreyard is one 7th shorter than the main yard, which in this case is 60 ft. and a little less than ¾ of a foot.

The thickness of the yard is ¼ part less than the mast.

The lateen yard a foot or two longer, than the mizzen mast.

The spritsail yard, ¼ shorter than the length of the bowsprit: for Example, the bowsprit is long 60 feet, the length of the spritsail yard comes to 45.

The main topsail yard, is nearly half shorter, than the main yard is long.

The fore topsail yard, hardly half as short as the foreyard.

The spritsail topsail yard, is shorter by ⅔ of the bowsprit: for example, the bowsprit long 60 feet, comes to 20 feet for the length of the spritsail topsail yard.

The spritsail topmast, is one foot or two longer than ⅓ of the bowsprit.

The mizzen and main-topgallant masts, have a length, ⅓ of the main topmast, or a little more.

The fore-topgallant is ⅑ part shorter, than the main-topgallant mast:

Example	18
comes to	16.

At the end the masts are a little thicker, for ropes to pass through and 8-sided.

. . .

(132 I 15) *The Ship's masts are made, thus*

For a Ship of 135 feet length, wide 34: found the mast to be long 85 feet, thick 21½ palms.

A Ship of 130 feet length, and 30 feet wide: the mast long 83 feet, thick 20 palms.

A Ship of 100 feet, the mast long 74 feet, thick 15½ palms.

A buss of 74 feet, the mast long 55 feet, being a square sail, thick 11 palms.

A boyer of 86 feet, the mast thick 19 palms, and long at the shrouds 58 feet, the top 19 feet.

A galliot of 88 feet, the mast thick 20 palms, and long at the shrouds 59 feet, the top 19 feet.

A large yacht, the mast thick 16½ palms, long on the shrouds 54 feet, the top 17 feet.

A schietschuit [?], the mast long 50 feet, thick 11 palms.

. . .

(133 I 46) *Proportionate measure of the thickness of the Masts.*

When a mast is thick 16 palms at the foot, then it should be 12 palms thick at the platform and 11 palms at the top.

When the topmast is 10 palms thick at the foot, as will be appropriate for a mast such as above, then at the top it shall be 6½ palms.

The crosstrees come 6 inches beneath the octagon.

The Upper end of the mizzenmast, is thick two thirds of the lower end.

As goes for the topgallant masts, mizzen and spritsail topmast. A yard is made thick at the ends, at two fifths from where it is thick in the middle.

To arrive at the thickness of the main mast, take one foot thickness at the partner for every six feet of the depth.

Masts on East Indiamen have a considerable rake, and are shrouded forward, but the topmast is held backward with the backstays: such that the mast is bent in misshape. [The iconography—i.e., the work of Nooms—suggests that Witsen has this reversed: lower masts were raked aft while topmasts were bowsed forward with the stays.]

The lower foot of the masts, where they stand in the bottom, and also underneath the mast-top, is planed in eight or six sides, where often cheeks are placed, for strength, below the mast top chocks are nailed to the mast, on which it comes to lie.

(94 II 49) *Necessary observations on the Main and Fore mast.*

The mainmast is set at about the middle of the ship. The foremast at one fifth: and the mizzen at the first part from the stern, when the entire Ship is divided in five and a half.

. . .

(133 I 36) The masts are tarred at the top, about where the yards and the tops are, for durability; as also below and everywhere, where they suffer most.

. . .

(3133 II 54) On very large Ships, masts are made from several parts, and well bound together with rope.

. . .

(274 II 16) Below sometimes iron hoops are laid around the mast.

. . .

(133 II 51) The masts are moved into the ships with cranes, lighters and sheer hulks.

(93 II 31) 73. *About the Masts of our Ship.*

Which is long between perpendiculars 134 feet, wide 29 feet, deep 13 feet, above 6 feet, with a forecastle of 5½ feet, and a cable tier of 4½ feet.

The main mast be long 80 feet, thick 18 palms, the top 7 feet long.

The foremast long 70 feet, thick 16 palms, the top 6½ feet long.

The bowsprit long 60 feet, thick 18 palms.

The mizzenmast long 60 feet, thick 11½ palms, the top 4½ feet long.

The main topmast long 51 feet, thick 11½ palms, the top 4½ feet long.

The fore topmast long 41 feet, thick 9½ palms, the top 3½ feet long.

The main-topgallant mast, long 22 feet, thick 4½ palms.

The fore-topgallant mast long 18 feet, thick 4 palms.

The mizzen topmast long 26 feet, thick 4½ palms.

The spritsail topmast long 20 feet, thick 4 palms.

The main yard thick 13 palms, long 69 feet, 26 ells [1 ell = 69 centimeters] without yardarms.

The foreyard, thick 11 palms, long 60 feet 6 inches, 22 ells without yardarms.

The mizzen yard, thick 7½ palms, long 62 feet, 24½ ells without yardarms.

The spritsail yard, thick 7 palms, long 45 feet, 17½ ells without yardarms.

The main topsail yard thick 7 palms, long 36 feet, 13½ ells.

The fore topsail yard thick 6 palms, long 32 feet, 11½ ells.

The spritsail topsail yard, thick 4½ palms, long 20 feet, 8 ells.

The main-topgallant yard thick 4½ palms, long 22 feet.

The fore-topgallant yard, thick 4 palms, long 16 feet.

The fore-topgallant mast thick 4½ palms, long 18 feet.

The mizzen topsail yard thick 4 palms, long 19 feet.

The topmast long 26 feet.

(134 I 15) The head of the main mast 7 feet, of the Foremast 6 feet, of the Mizzenmast 4¼ feet: the head of the main topmast 4 feet, of the fore topmast 3 feet, as for the mizzen topmast and main-topgallant mast, the spritsail topmast scarcely 2 feet. The head of the fore-topgallant mast 1¼ feet long.

(268 I 14) The *Trestletrees*, standing on the Top-gallant masts, serve to hold the flagpole: which is why they only have one hole.

(134 I 5) The trestletrees of the main mast of our Ship are 9½ feet long, of the Foremast 8½ feet, of the Mizzenmast 4½ feet, of the bowsprit 4½ feet, as also those of the main topmast; the trestletrees of the fore topmast long 3½ feet, of the mizzen topmast 2 feet: as also of the main-topgallant mast, and the spritsail top-mast, the trestletrees of the fore-topgallant mast 1¼ feet.

(95 II 20) The trestletrees are as long as the opening in the floor of the mast-top is wide: the main trestletrees thick 4½ in. broad 7½ inches: the others are thinner, flatter, and narrower, of which those below broader.

(133 II 36) On every mast a heavy block, hollowed out in a half circle, is laid, called the mast cap, this half-round hole is closed with a brace, in which the top masts, and the flagpole, and the top gallant masts stand. The top masts, including the poles, all slide through the forward holes of the crosstrees, and such because heeling over backward, they will stand more firmly, and can be easily hauled downward.

. . .

(274 II 29) The Mast caps are pierced, through which holes go ropes holding the yards firmly.

(95 II 10) The main mast cap, is long 3 feet, 1½ inches, and broad 2 feet, high 14 inches. The foremast cap, long 2½ feet, broad 21½ inches, high 12½ inches. That of the mizzen, main topmast, and bowsprit, long 16 inches, broad 12 inches, thick 7 inches; that on the fore topmast and main-topgallant mast, long 14 inches, broad 1 foot, high 6½ inches; on the flagpole 12 inches, on the spritsail topsail 9 inches.

(95 I 36) *About the Tops.*

Our main Top is 10 feet in diameter. The foretop 9 feet. The other mast tops 4½ feet, on the decking, and thick 2 inches, broad 26 inches. The knees are broad 4 inches, thick 2½ inches, 14 knees lie on the floor. The capping, broad 5 inches, also rises as much above the floor, thick 2½ inches. The edges broad 3½ inches, thick ¼ inch. The decking 3½ inches thick, the deadeyes to the mast top are 11 inches, also 7 and 8 inches.

(258 II 4) As for the *Tops,* they are to hold the masts steady, for which the shrouds are fastened to it at either side; and also to provide room for the seamen, on which to stand, when they have something to do up there.

The tops are covered with lambskins, for the chafing of sail and ropes.

. . .

(62 II 8) Tops are the round garlands that lie above around the masts, and on which one can stand; made to attach ropes to, and save sails from, etc.

. . .

(274 II 24) Underneath the bowsprit top there are clamps, well studded with iron, to support the top.

Around the tops hoops are drawn, bands, serving for the stability of the sailors, who have to perform much work on the mast tops; and without these, would easily fall down. Forward they are the lowest, to prevent the chafing of the sail and ropes; for which it is also hung with servings.

. . .

(95 II 1) The shape of a top is shown in the plate No. 20 [fig. 2.196]. Most ships have a top to one mast only, but some have two; and that when the topgallant masts are raised, for which the top is also carried on board; around tops of large ships often a railing can be seen, half a man's height.

. . .

(95 I 47) The top has a number of holes, through which ropes pass, which expediently are sheathed with iron. But, to arrive at the width of the main top, for every 10 feet of the length of the ship 9 inches are taken.

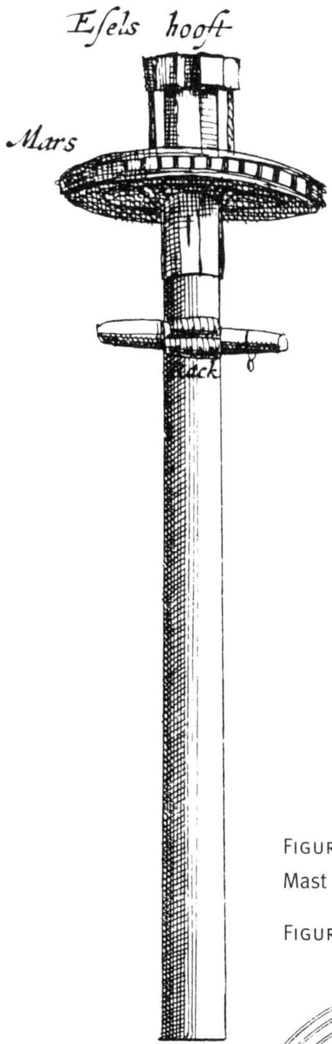

FIGURE 2.195. (*left*) Plate XXXVI. Mast cap, top, parrel.

FIGURE 2.196. (*below*) Plate XLI

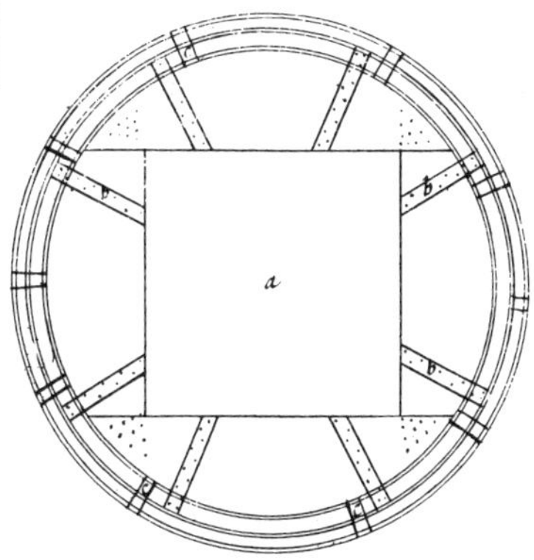

94. Step the Masts

Large masts were generally composed of several lengths of pinewood from Riga, called *gekuipte masten* (coopered masts). To keep the parts together the mast was wound with rope or iron bands at several places, the same as with the cooperage of casks.

The mast was planed into an octagon at the foot and also at the hounds, the upper part of the mast to which the trestletrees were nailed. Crosstrees were let into the trestletrees, and the top was bolted on top of them. To strengthen the hounds, cheeks were applied.

The length of the main mast was twice the total of the depth plus the width. For the pinas this should result in 84 feet, but Witsen indicates 80 feet. The length of the masts was apparently also a characteristic that the shipbuilder could vary according to his wishes.

The thickness of the mast was one foot for every 6 feet of depth. For the pinas, this comes to $2^1/_6$ feet (59 centimeters). This measurement is the diameter of the mast, but in his lists Witsen used circumferences only in the measurements for the masts and spars of the pinas. This may imply that he took the measurements from the ship itself, which is another indication that the ship actually existed and was not simply "built in our thoughts" (53 I 28), as Witsen says himself.

The head of the mast is the part above the hounds. The thickness there is still three fourths of the total thickness because the mast became thinner toward the head. The head was often cut to an octagon as well, or rather a square with beveled corners.

The dimensions of all the other spars were arrived at in this way, although it must be said that Witsen does not supply all the formulas. (For the formulas used by Witsen as well as Van Yk, see table 1 in the appendix.) According to Van Yk, the length of the head of the mast was one tenth of the total length, and the length of the head of a topmast one twelfth (p. 170). Witsen is thriftier: 7 instead of 8 feet for the 80-foot main mast. Perhaps this represents a gradual change. Over time mastheads became longer.

Mast length was measured in feet and thickness in palms, giving the circumference. A palm was one third of an Amsterdam foot (95.4968 millimeters). To arrive at the diameter, divide the circumference by 3.14 (pi).[14]

The raising or stepping of the masts was done with cranes at some yards and sometimes with floating sheers. Another technique employed improvised sheer legs stepped on deck, but this was mainly done in case of damage underway.

Trestletrees and Crosstrees

The trestletrees and crosstrees were fixed on the hounds. They were longitudinal and were crossed by the crosstrees, surrounding the mast closely and supporting the top. When no top was made, they served only to separate the mast from the topmast, to support the fid (which held up the topmast), and to spread the topmast shrouds.

According to Van Yk, the length of the trestletrees was one third of the ship's width; they were five inches thick for every six feet of its length, while their breadth was three quarters of the thickness (Van Yk, p. 177). The crosstrees were a bit shorter.

For the trestletrees of the foremast Van Yk counted one sixth less; those of the mizzenmast, bowsprit, and main topmast were half the size of the main trestletrees. The others were even smaller (see table 1 in the appendix).

Mast Caps

A heavy, semicircular block of wood, the mast cap, was placed on the head of the mast; it held the topmast and was grooved for the halyard ties of the yard. As with many other parts, the size of the mast cap depended on its place in the rigging: the main mast had the largest mast cap, the rest in diminishing sizes.

The mast cap drawn by Witsen (see fig. 2.195) is one of a shape unknown to me.

Tops

The top was the platform lying on the crosstrees and trestletrees, onto which the deadeyes of the topmast shrouds were fixed. It was a working platform for taking in the topsails and as a place for snipers during battle.

In the seventeenth century the mast tops were round, becoming square with a rounded front in the eighteenth century.

Hoops were laid around the edge, fixed on low knees lying radially on the top.

95. And the Bowsprit.

(256 II 34) The foot of the Bowsprit stands against the foot of the Foremast, to support one another, and supply firmness, without which it makes no difference whether the bowsprit comes far or short into the ship.

. . .

(133 II 30) On the bowsprit, mainly on very large Ships, often laths are nailed, to be used as steps.

The bowsprit rests on the stem or on a breasthook, beside the stem: often comes between the riding bitts, against the masts. The masts are sometimes painted below.

. . .

(133 II 6) At the stay the bowsprit is one fourth part thinner than at the heel.

The stay comes at 14 or 15 feet, from the fore end of the bowsprit.

. . .

(256 I 44) The bowsprit is wound, for strength, and this winding again is covered with a skin, against wear and tear.

. . .

(265 I 21) [. . .] it stands thus far above the Lion, that the sail just touches it, to which the proportion supplied by me conforms. It would be of no use for it to be longer, because it would only serve to catch the wind sneaking underneath, and between the sails, and always feel the side winds. Which may be done, however, if it stands low, and the lower it stands, the better it is managed.

(93 II 12) The bowsprit lies on the partner: against the mast lies a cleat, notched in the front of the knights, and straight on the bowsprit: is thick 4 inches, broad 12 in. there at the end a brace lies across, which is fastened on the knees, through a beam: the brace is broad 4 inches, and thick ¼ inch: the bolts thick 2 inches.

95. And the Bowsprit

What has been said about the masts in general applies to the bowsprit as well. It was thickest at the stem and thinned down to three fourths of that size at the head.

Witsen does not supply us with a formula for the length. According to Van Yk, it was three sevenths of the ship's length, which, even in his day, was diminished to three eighths, because too many accidents happened. When the bowsprit broke, this usually led to the loss of the entire foremast, the main topmast, and the main-topgallant mast.

In contrast to the masts, the bowsprit had neither a partner nor a step, but its foot rested between the posts of the knights forward against the foremast. Farther up, it rested on the stem or, when lying next to the stem post, as, for instance, on fluyts, on a breasthook (a heavy knee in the bow). The knee at the top, supporting the spritsail topmast, was one foot high for every ten feet of the bowsprit's length (Van Yk, p. 147).

96. With the Capstan.

(71 I 15) 30. *About the Capstan.*

1. The upper end of the capstan thick 2½ inches, for every 10 feet of the length of the ship.

2. The whelps half as long as the head of the capstan.

3. The holes wide ⅙ of the thickness.

. . .

(267 I 11) On some large merchantmen the capstans stand on the upper deck and are turned with four bars above and two below; to wind all the more powerfully. Before the capstan, forward, often a crossbar is nailed to the main deck; for the cable to wind more easily around the capstan; and not hitch below.

(93 I 28) *About the Capstan.*

The capstan comes in the middle, between the steering stand and the front of the knight, or midships 15 feet abaft the mast: is 5 feet, 3 in. long above the partner the hole in the partner wide 1 foot, 4½ inches, fore and aft, thick above square 21 inches, above underneath the score 20½ inches, underneath that again 15 inches, and at the lower end 14 inches, it has 6 whelps, long 2 feet 7 inches: thick 5½ inches, let in ½ inch, below broad 8½ inches, and at the score 4 inches, and the score from below 20½ inches, and thus have 3 sides, the intermediate whelps below 5 inches, thick 3½ inches. The capstan has four holes, wide 4 inches, the lower end long 7½ feet, the scheenen [probably a small chock of fatwood to serve as a bearing] underneath the partner 1½ inches, the pin below long 4 inches, thick 7 inches, turning on a point standing on an iron plate. Above it also has a brace around it, which is broad 3 inches, and thick 3 inches. The hoop in the partner is thick ½ inch, broad 3 inches, 3 tenons with 16 shins [?] long 9 inches, broad 1½ inches, thick ¼ inch, the pawls broad 6 inches, thick 5 inches, long 2 feet 5 inches. The pawls stand 1 foot 5 inches from the capstan, the rear end fixed with a bolt, and come to the front of the capstan.

(93 II 20) the beam of the capstan broad 1 foot, thick 2½ inches, in the middle wide 2¼ feet: the ends come even with the front of the gunports.

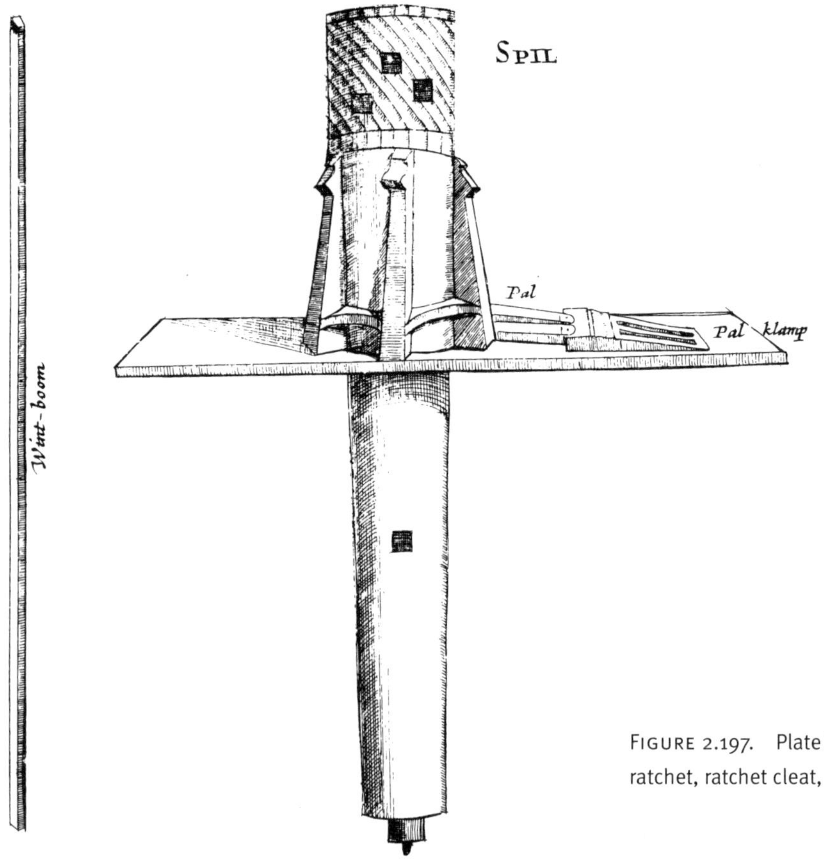

FIGURE 2.197. Plate XXXV. Capstan, ratchet, ratchet cleat, bar.

96. With the Capstan

The capstan was the most important device aboard for hoisting heavy loads, such as the yards, the anchor, or parts of the cargo; it could even be used to shift the ship—that is, kedge it using a kedge anchor (kedging). Large ships also had a second small capstan, not resting with its foot in a partner on the lower deck but in a kind of brace, mounted against the underside of the deck.

Where the capstan turned in the partner, it was furnished with small, easily replaceable pine chocks to facilitate rotation and to minimize wear on the expensive partner or pivot. The capstan was turned with ash bars, which were removable to save room.

To prevent the capstan from turning back, pivoting pawls with which it could be secured were mounted onto the deck forward or aft of the capstan.

When the anchor was being heaved, it was not the anchor cable that was wound around the capstan; the cable was simply too thick. Instead, the so-called messenger was used, a loop of line running around the capstan and a heavy vertical bar or roller in the bow next to the hawseholes. This line had mousings (thickenings). The anchor cable was tied to the messenger with short lengths of rope (nippers) at short intervals as it entered the ship, and the mousings prevented the nippers from slipping. The cable was drawn to the hatch of the cable tier with the capstan, where the nippers were untied and the cable stowed below. Ropes (halyards) for hoisting yards were always led to the capstan through a sheave in one of the large knights.

97. Fasten the Chain Plates.

(275 II 53) Chain plates are irons or ropes, which are attached underneath to the wale and the ship, well fixed with bolts. At the main channels one can observe two longs staves of iron, also called chain plates, which come through and are mobile: with the one the shrouds are hauled taut, and the other is used to hoist boats in and out of ships. Likewise one can be found in the small channels, to which tackles with blocks are attached, used to take heavy loads in and out.

. . .

(268 I 21) At every channel an iron *Chain Plate* can be seen, which is smooth and of one piece, and goes through. Thereto a line is attached, going over four sheaves, with which the masts are held to the side and hauled taut. Another is loose and mobile, which is used to hook a tackle onto, and take on loads. The futtock

shrouds underneath the mast tops are fairly long, holding the topmast shrouds: but those of the Foresail and Mizzen are the longest, because there the shrouds do not reach the edge of the mast top. The topmast shrouds are hauled taut with Deadeyes. The top of the bowsprit has no futtock shrouds.

97. Fasten the Chain Plates

Chain plates were the iron fastenings of the deadeyes. They usually consisted of four parts: a band with an eye around the deadeye; a long link underneath the channel going to the ship's side; a short link; and a plate in the shape of a figure eight. The chain plate was fastened to the ship's side with two bolts, one of which went through both the short link and the plate, the other only through the plate. For Witsen's remarks on chain plates specific to the pinas, see section 91 (Then plank it there, and make the Channels).

A 100-foot ship had six pairs of deadeyes in each side (and thus chain plates) for the main mast, one less for the foremast, and half as many for the mizzenmast. For every 15 or 16 feet of the length of the ship, a deadeye and shroud was added, so a ship of 134 feet had to have eight deadeyes.

Apart from these, a number of chain plates used for tackles in the shrouds were installed in the wale. They were not composed of links but were made of one rod or plate. The whole was fastened to the ship's side with two bolts. Furthermore, smaller chain plates to which the backstays were attached were often fixed on the wales.

98. Also the bulkhead before the Steering Stand.

(55 II 27) The Schilt [shield] is an ornament, stands in front of the steering stand, as the coat of arms of the Owner of the ship, is often applied forward against the forecastle. Sometimes two schilden can be found forward in heavy ships.

(88 II 48) 50. *About the Bulkhead before the Steering Stand.*

1. The door of the steering stand, wide 4 feet, above wide 3 feet 9 inches.

2. Door posts, broad 4 inches, the one post comes 2 feet from the ship's side, is thick 3 inches.

3. The planks thick 1 inch, and between both shot-free.

4. The grating thick 3 inches, the bulkhead rakes aft 2 inches, broad 4 inches, long 14 inches, below with a concealed scarf.

Inside the steering stand against the bulkhead, there are apertures through which one descends to the main deck, 2 feet square.

Next to the portholes in the upper deck, comes a post, broad 5 inches, the portholes wide 1 foot 9 in. the paneled woodwork is pierced for muskets. The border underneath the arch above the grating, broad 8 inches: also the carved plank, above which the grating comes another carved plank equal with the outside, in which the arches are made, the porthole comes 10 inches from the door, and 1 inch from the deck, the cross brace broad 12 inches, thick 12 inches.

5. The arch, high 15 inches, broad 14 inches, thick 3 inches.

6. The portholes wide 15 inches, high 14 inches.

7. The covering board thick 1½ inches.

8. The border around the grating, thick 1 inch, broad 8 inches.

9. The border below thick 2 inches, broad 3 inches.

Above the cornice broad 7 inches, thick 1½ inches.

The steering stand long 14 to 16 feet, high 6 to 7 feet.

(58 I 51) *The arch above the bulwark of the steering stand,* is broad 7 inches, and above thick 4 inches, below thick 1½ inches, in this way, as is shown in plate No. 20 [2.198] at letter L [*L* is missing]: below lies a batten, on which the small pillars stand, broad 2 inches, thick 2 inches, in the way as shown on the same plate at letter M [*M* is missing], and it lies 9 inches from the bulkhead: the small pillars are long one foot, thick 2 inches, broad 5½ inches, and also lie 5½ inches apart: in the middle lies or stands a head, on which the arch lies: this is thick 10 inches, broad 1 foot.

(90 II 44) Above the steering stand, often stands an arch, on the edge of the upper deck, as shown at letter N on plate No. 20 [2.198], raised one foot, and on the planks lies a batten, thick 1½ inches, broad 3 inches, it lies 9 inches farther aft, (which front can be seen at letter O:) than the ends of the planks come, the arch is 7½ inches broad, above thick 3 inches, and below 1 inch, the small pillars underneath the arch are broad 5½ inches, thick 2 inches, the holes between the same 5½ inches, in the bulkhead of the steering stand, are openings letting in light, decorated with frames all around, which pillars have hidden scarfs to the back: shown with letter P.

FIGURE 2.198. Plate XL. Bulwark of the steering stand.

98. Also the Bulkhead before the Steering Stand

The bulkhead in front of the steering stand was made to bulge outward, and it was covered with weatherboarding or clinker planking. It had a door in the middle with a round arch, so it had a little more height. As we have seen, the deck of the steering stand was a step lower than the main deck.

To the left and the right of the door carved arches made the bulkhead *schootsvrij* (lit., "shot-free"). This somewhat obscure word did not mean bulletproof, but rather that the openings offered the opportunity to fire through them with muskets, which was convenient if the ship was in danger of boarding. Below these apertures to either side there was a hatch, through which light could be admitted behind the bulkhead. There a ladder descended onto the main deck.

On the exterior of the bulkhead, next to the apertures and close to the ship's sides, stairs led up to the deck above the steering stand (see section 103, The Bulkhead before the Upper Cabin). On that deck there was a rail in the same style as the apertures in the bulkhead. This rail marked out the deck above the steering stand and was mostly ornamental. The arch was actually a low decorative rail and consisted of small pillars that were placed between a batten nailed onto the deck planks and a cornice placed one foot higher. (The labels *L* and *M* mentioned in passage 58 I 51 are nowhere to be found on Witsen's plate—they correspond to the letters *N* and *O* referred to in passage 90 II 44.)

99. With the bulkhead before the Forecastle.

(89 II 12) The bulkhead forward, will be bulging, the hollow side on top, in which one or two doors are made, with portholes to the sides. Also forward and above. On the forecastle there are banisters, the heaviest and most elegant forward, because they are the most exhibited.

Apart from the portholes also loopholes are made fore and aft into it, on men-of-war nowadays a musket-shot-free bulwark is made above all around the forecastle. The forecastle often goes down one step, to gain height, especially on ships which have their galleys, and berths there.

(89 I 42) 52. *About the Bulkhead at the Forecastle.*

1. The posts of the bulkhead at the forecastle, thick 5 inches, broad 6 inches, with a curve of 8 inches, above hanging forward, 16 inches.

2. There are two doors, each wide 2 feet 6 inches, the arches high 14 inches, thick 5 inches, broad 6 inches.

3. The portholes wide 21 inches, high 18 inches.

4. The bulkhead, with the door shot-free.

5. The frame broad 5½ inches, thick 5 inches.

6. The planks inside and out, thick 1 inch, the deck above the forecastle is flat or roundish.

Figure 2.199. The quarterdeck of the pinas model. The bulwarks for the steering stand and cabin are visible, as is the arch for the whipstaff (see section 119, Arch above the Whipstaff, if necessary). At the aft end of the poop deck we see the bench, which also serves as a chicken coop. (Courtesy Cees de Jonge, The Visual Art Box)

Figure 2.200. The forecastle deck of the pinas model. In the forecastle bulwark the ports are visible, through which the messenger runs to the capstan. The boat that is normally stowed on the gratings has been removed for the photograph (cf. fig. 2.219). (Courtesy Cees de Jonge, The Visual Art Box)

99. With the Bulkhead before the Forecastle

The forecastle in Witsen's pinas had a bulkhead only at the aft end. Forward it was closed off by the stem. Ships having an open forecastle also had a bulkhead forward, with doors leading onto the beakhead. What Witsen refers to as the bulkhead before the forecastle is actually the bulkhead aft of the forecastle. It had two doors with arches; next to these doors there were also two small hatches, giving access to the main deck below forward of the bulkhead.

Between the doors the bulkhead had two portholes, through which the messenger was led to the capstan and back.

Between the doors and the side of the ship were also ladders. There are no indications that the forecastle head had a banister on the aft edge, but it did have a cornice, probably ornamented.

100. And the Hatch Gratings.

100. And the Hatch Gratings

The dimensions of these gratings were treated at section 69 (The coamings of the Gratings, on the upper deck). From Witsen's description it appears that these were not the complex gratings found mainly on English models, with the parts countersunk onto each other, but simple athwartship ledges and longitudinal rafters nailed to one another.

101. After this make the Bordered Hatch Covers.

> (86 I 21) The small hatch aft near the transoms, is 20 inches wide, and as long: made of one plank.
>
> The small hatch behind the mast, has a bordered hatch cover.

101. After This Make the Bordered Hatch Covers

Bordered hatch covers had edges that closed around the coamings. These were almost always the small hatches, like the ones giving access to the cheese and bread room. Other hatch covers were sunk into the deck, like the ones of the *hel* (the forward storeroom) and the peak aft. The gratings of the main hatches were sunk into the coamings, as we have seen.

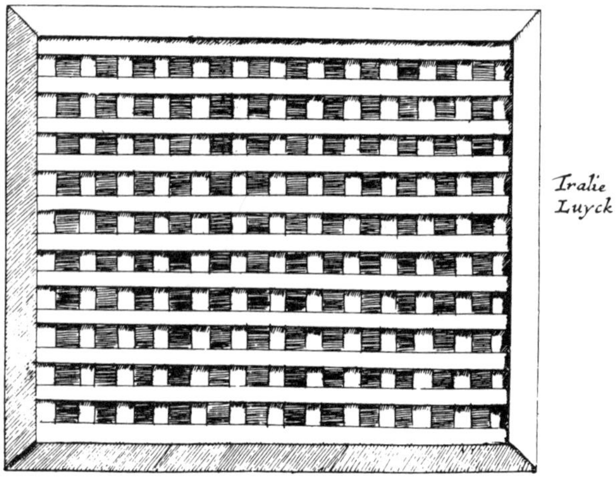

102. The Bushes, Pumps, and the Privy.

(56 I 48) Bushes, are tubes, through which the water goes into the sea.

. . .

(60 II 43) The bush, is the wood into which comes the hole for drainage, or scuppers, on the main deck, and the upper deck.

> (93 I 20) 72. *About the bushes on the upper deck and main deck.*
>
> The bushes, on the main deck, broad 6 inches, thick 5 inches, the holes wide 3 inches.
>
> 1. The bushes, on the upper deck, thick 4 in. broad 4 inches.
>
> 2. The holes wide 2 inches, the curve is of 14 inches.

(56 I 50) In large ships two pumps are made well served.

(56 I 45) The Scupper Hose to the pump, is made of cloths, or canvas: the water goes through this.

. . .

(269 II 5) In large ships two pumps are to be found before the steering stand, and one near the main mast, well fitted, served and surrounded by casings in the hold.

> (85 II 19) *About the Pumps.*
>
> Abaft the main knight, stands a pump, which is 1½ feet before the hatch, for drainage.
>
> Both pumps aft are on the steering stand, and 5 inches from the mizzenmast, and the rear of the pump, comes about in the middle of the mast, the holes are wide 12 inches: the pumps stand with their lower ends in

Figure 2.201. (top *left*) Plate XXXV. Grating.

Figure 2.202. (*bottom left*) Plate XXIX. Bordered hatch cover.

Figure 2.203. (*below*) Plate XXXIII. Bush.

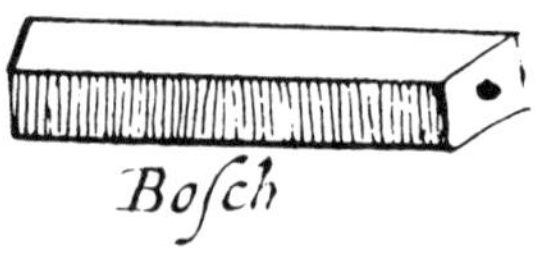

Figure 2.204. Plate XXXIII. Pump, rod, piston, valve, scupper hose, copper kettle, pivot and handle.

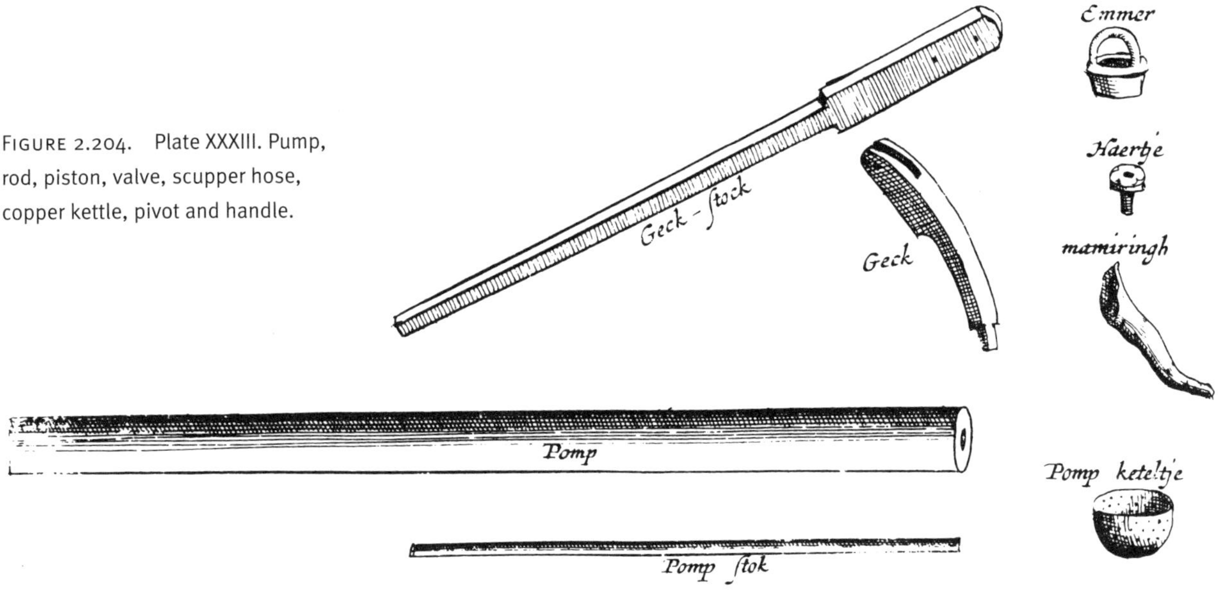

a copper kettle which is full of holes, which rests down on the planking, and to the side of the keel. The hole that is in the pump, is 1½ inches wide, broad 1½ inches, thick ¼ inches: the eye to which comes the pivot is 4 inches long, wide 4½ inches, the pivot is 1½ feet long above, thick and broad 7½ inches, the lower end long 14 inches, below around the pump is a round border, on the main deck and the steering stand, the pump dale comes 6 inches abaft the mast, and it is wide 8 inches, long 14 inches, the lip comes against the steering stand below: the pipe is wide 4 inches, on each side, and goes through the ship, and is paneled with planks, of 1 inch thickness: the pipe is of 2 pieces on each side, the piece that is fixed to the kettle, is not attached to the side: the other piece is fixed to the side, so that it will not leak at the kettle. The one at the side is fixed around the scupper, and this is broad 8½ inches, thick 3 inches. Above around the lead cistern or kettle, there is a border, which is 2 inches high and thick, the lid on it, that is thick 2 inches, and it has been let in one inch, and the other above that, which stands over it, and there is 1 hole or 2, in which the scupper hose hangs, the scupper hose is 1½ feet high on the pump, the hole in the pump above is wide 5 inches, the handle in the pump 9 feet long, the handle in the pivot long 8½ feet, and above the square part long 2 feet, and there thick 3 inches, where it is broad 5 inches, the lower end thick and broad 2 inches, the bolts have a small plate to each end, the pivot stands with a pin in a small enclosure, with its lower end.

(58 | 50) The pipes of the privies are made of lead, and the cover is planed, with a parrot beak.

102. The Bushes, Pumps, and the Privy

The translation of the Dutch word *bossen* into "bushes" may not be correct. *Bossen* were square lengths of wood in which a hole had been drilled lengthwise and which were placed in apertures in the ship's side. The waterways had been cut away where a bush began. The bushes would stick out a little outside the planking so the scupper hose could be nailed onto it. This was a leather or canvas hose that prevented the water from coming up the scuppers and bushes when the ship was heeling.

The bushes always protruded between two wales and were placed so that the water could not accidentally enter the ship again via a gunport below.

The pumps were made of elm. They were bored out accurately through the heartwood, then tarred and served

Figure 2.205. Plate XXXI. Privy pipe, lid.

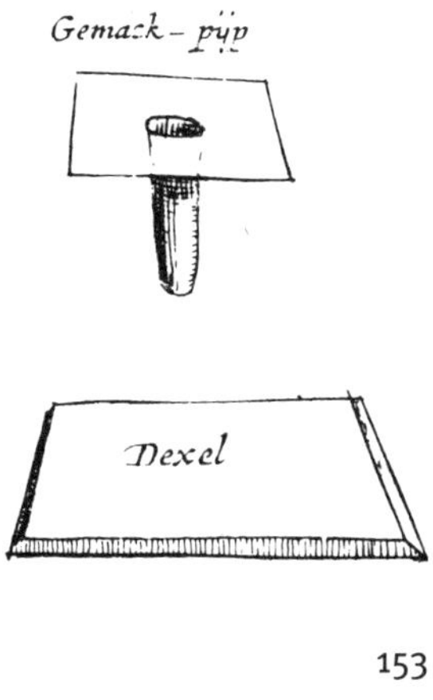

with canvas and rope to prevent cleavages due to drying out.

The foot of the pump stood below decks in a semicircular copper basket, which prevented the pump from clogging.

The handle, fixed into the pivot, was connected to the pump rod. The "bucket" and, within it, the leather valve or "clapper"was attached to the rod. Inside the pipe was another valve, the *haertje,* which kept water from flowing back into the ship.

On the steering stand there were two pumps near the mizzenmast and one near the main mast.

To work the pump, the pipe first had to be filled with water. Above, in the steering stand, the pipe had a hole to which a hose was attached, which took the water from the pump to the pump dale; from there it drained through pipes to the sides of the ship.

Down in the hold casings were made around the pumps to prevent damage to them by parts of the cargo.

The heads for the crew, as we have seen, were on the beakhead. The privy of the captain was in the bench inside his cabin. The latter had a lid with a rounded border, called the parrot beak.

103. The Bulkhead before the Upper Cabin.

(266 II 22) The upper cabin is sometimes divided in two: and in that case the entrance is sometimes made with two doors, of which one leads into a special room: the dome of the door being slightly raised above the deckhead, to make access all the easier.

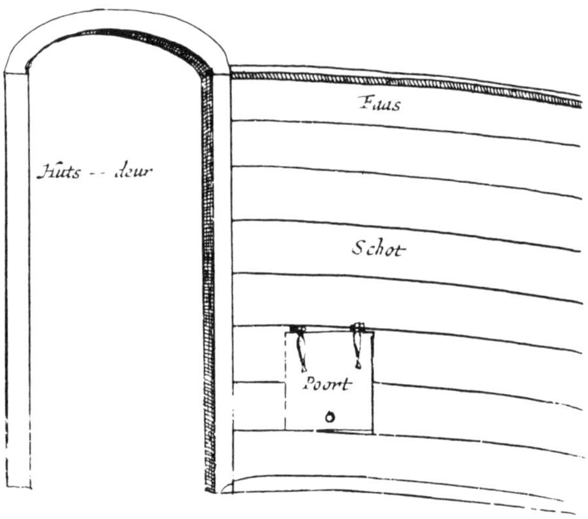

FIGURE 2.206. Plate XXXIII. Cabin door, cornice, bulwark, porthole.

(56 I 17) In No. 14 [fig. 2.206] the door of the upper cabin, and the bulkhead of the upper cabin are shown.

Above the bulkhead stands a cornice, and also below, against the seams of the planks.

(88 II 34) the posts are thick 3½ inches, and as broad, the portholes they are 1 foot from the door, they are high and wide, from the deck, 1½ feet. The bulkhead stands upright, the door wide 2 feet 9 inches, the posts thick and broad 3½ inches, paneled inside and out wide one-inch planks, and the upper cabin is also roofed with these, the porthole in the upper cabin comes 2 feet 2 inches from the bulkhead, and from the deck 1 foot 8 inches, high and wide 1 foot 6 inches. On the aft side of the bulkhead there are 2 knees, thick and broad 5 in. and the ends are long 3 feet. To the schiltje [unclear; probably the front bulkhead of the forecastle] there is a bulkhead of sawn planks, the door is as above, the ridges of 2 inches.

(87 II 48) The *window in the upper cabin* is wide 1 foot, high 14 inches, it comes almost as high as the ledges, outside, a frame is made thick 2 inches, broad 3 inches. Around the window in the gallery the ridge frame is thick 3 inches, broad 4½ inches. Around the window of the cabin it is broad 4 inches, thick 2½ inches.

103. The Bulkhead before the Upper Cabin

In the middle of this clinker-built bulkhead there was a door with a porthole to the left and right, admitting light and air into the room that was used as the officers' quarters abaft it. Above, a cornice marked the edge of the deck above the upper cabin. The upper cabin was often divided into forward and aft rooms.

104. Pinrails.

(65 II 7) The Pinrail, comes onto the top timbers, against the gunwale; for the sake of strength cut with hidden scarfs.

. . .

(56 I 52) In plate No.15 [fig. 2.208], is the Pinrail, from another perspective than shown before.

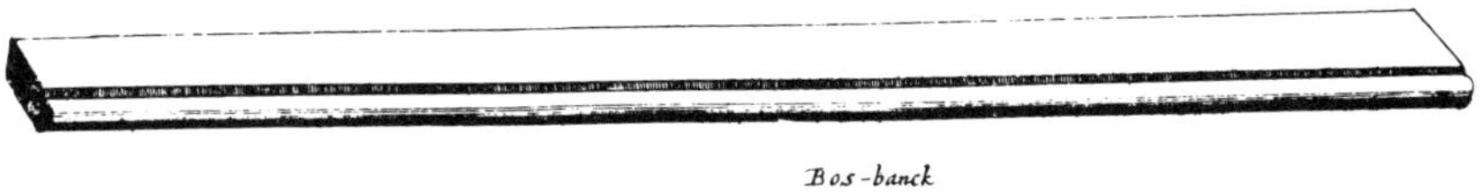

Figure 2.207. *(top)* Plate XXXIV. Pinrail.

Figure 2.208. *(bottom)* Plate XXXIV. Pinrail and post to the pinrail (incorrectly called “kevel rail” and “post to the kevel rail”).

(90 I 24) *55. About the Pinrail.*

The pinrail lies on the upper ends of the top timbers, joined with a scarf. In it some holes are drilled, to put the belaying pins in, onto which the ropes are belayed.

1. The pinrail, broad 8 inches.
2. Thick 4½ inches.

104. Pinrails

The pinrail covered the upper ends of the top timbers in the forecastle and above the top planking. Outboard were the rails, which had been fastened onto the top of the top timbers (see section 62, Hereafter it is done up outside, with Wales, Filling Strakes and Sheer rails, Washstrake, Vertuining and Railing). Holes were made in the pinrail for the belaying pins, onto which ropes were belayed. The horizontal parts of the smaller bitts were also called pinrails.

Witsen mixes up the terms *pinrail* and *kevel rail* (see section 112, Apply the Kevel Rail, if necessary).

105. Kevels.

(90 II 24) *60. About the Kevels, behind the Foremast.*

1. There are two kevels on each side, thick 3 inches, broad 4½ inches, broad above 8 inches, broad below seven inches, long 1 foot.
2. The kevels go through the kevel rail, into a chock, thick 3 inches, broad 4 inches.
3. The kevels, above wide 3 feet 8 inches, the lower ends on the kevel rail, wide 7 inches, the lower ends long 15 inches.

(89 II 37) *54. About the Kevels.*

1. Here are four kevels, broad above 1 foot, broad below 9 inches, thick 3 inches, above wide 18 inches, below 9 inches.
1. The belaying pins, long 8 inches, broad 6 inches, thick 2½ inches.
2. The kevels, long 10 inches, broad 7 inches, thick 3 inches.

(90 II 37) *61. About the kevels behind the main Mast.*

1. These kevels, between the heels, wide 18 inches, broad 6 inches, thick 2 inches, long 7 inches, below long 1 foot, standing crookedly in a chock, broad 2½ inches, thick 1½ inches.

(91 I 31) *62. About the kevels behind the mizzenmast.*

1. The kevel between the heels, wide 15 inches, broad 6 inches, thick 2 inches.
2. Long 7 inches.
3. Below long 1 foot, and curved, standing in a chock, broad 2½ inches, thick 1½ inches.

(71 II 18) 38. *The kevels,* 8 feet 1 inch.

105. The Kevels

Kevels were belaying points placed in the ship’s side and used for heavy parts of the rigging, such as the sheets, braces, and tacks. On the forecastle of the pinas there were two to each side, one pair for the tacks of the foresail and one for the sheets of the spritsail (these could also be placed amidships). Amidships there were another two pairs, one for the sheets of the foresail and one for the tacks of the mainsail, and abaft the mainmast another

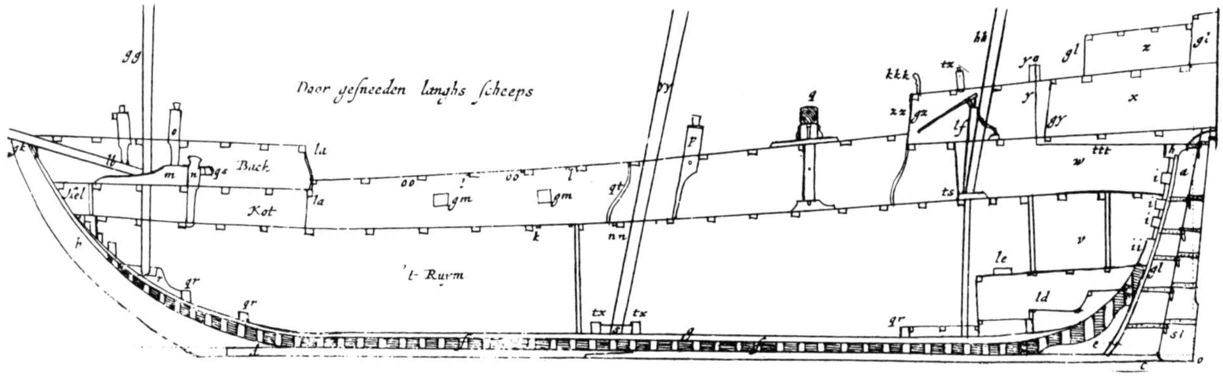

Figure 2.209. Plate XLII. The ship's longitudinal section.

(58 II 24) In the *Ship's longitudinal section,* A is the sternpost, B the stem, on which or beside which on a chock or fender the bowsprit lies, C the heel of the keel, on which the rudder rests, D the kinback ["boxing scarf" or "boxing of the stem" in English, but it could be called the "chin," as it is in Dutch, being the bottom of the head of the ship] of the keel; which is a scarf to the stem and the keel, E a knee on the keel, holding the sternpost to the keel, F the floor timbers, lying on the keel and bilge, G the keelson: this lies above the keel, on the floor timbers. H is a wing transom, fastened onto the sternpost with a shoulder and dovetail. I are the transoms, holding the fashion pieces; ii the broekstuck, coming across the ends of the fashion pieces, on the sternpost. K the beams, in the hold. L the beams of the main deck [actually the upper deck]. M riding bitt knee; the cables are belayed onto the bolster of the bitt and the bitts. N is a bitt, to which the bolster is fastened. O fore knight. P main knight; with it the topmasts and yards are hoisted. Q capstan, of which the foot often rests on an arch. R the fore step: in which stands the foremast, and next to which a chock. S the main step. V bread and cheese room. W gun room. X cabin. Y Steering stand. YO hutch, or aperture into which the whipstaff goes; namely, when the upper deck is too low to use the whipstaff. Z the upper cabin. ST the rudder, pintles, and gudgeons of the rudder, made of iron. TS mizzen step. TX riders, lying before and aft of the [main] step. TZ the mizzen knight. CO the rake of the sternpost. OM height of the sternpost, GL the stern rabbet. KK rising floor timbers. GK the chock, in which the bowsprit lies. GM gunports. GY the bulkhead before the cabin. GZ bulkhead or trellis for the steering stand. LA the bulkhead for the forecastle and cable tier, against which stairways are standing. YY mainmast. HH mizzenmast. GG foremast. LB bowsprit. LG the door, and bulkhead of the upper cabin: against which stairways are standing. NN scuppers of the lower deck. OO scuppers of the upper deck. QR are breasthooks. From d to f is the raking of the stem. QS is the bolster. At QT a stairway goes down. TTT points out the sweep and tiller. KKK the arch above the steering stand: in which the trellises come. ZZ the stairway before the steering stand. LD the powder room. LE are hatches. LF the pump. Often a pump was also installed near the mainmast, which was called the brake pump. LG the aft cabin.

Sometimes cabins or, instead, henhouses are made on top of the cabin.

---

pair for the sheets of the mainsail. Behind the mizzenmast there was another pair on which the braces of the mainsail were belayed.

106. The Bulkhead before the Gun Room.

(78 II 2) The bulkhead is made against the aft side of the mizzenmast, between the second and the third beam: the door is to port, and is 4 feet from the ship's side, wide below 4 feet, above 3 feet 9 inches. The posts are thick 3, and broad 4 inches. The ledges thick 1½ inches, broad 2 inches. The deals thick 1 inch. The batten below broad 2½ inches.

(79 I 12) 9. The bulkhead is made against the mizzenmast.

10. The door, to port, stands 4 feet from the ship's side, and is wide 4 feet below, but 3 feet 9 inches above: in Fluyts it mostly stands to starboard.

11. The stanchions thick 3 inches, broad 4 inches.

12. The ledges thick 1½ inches, broad 2 inches.

13. The deals thick 1 inch.

14. The batten below broad 2½ inches, thick 2 inches.

(90 I 12) *About the hole through which the painter of the boat goes.*

Inside, above the first transom, between the stern timber and the chock of the gunport, stands a length of plank on its side, into and against which, comes this hole; is thick 3 inches, broad 16 inches, 1 foot removed from the transom, wide 3½ inches, and outside on the tuck a Cardinaels hoet [lit., "cardinal's hat"] is placed around the hole, thick 3 inches, broad 6 inches.

2. The cleats of the kevels, broad 4 inches, thick 2 in. long 20 in.

106. The Bulkhead before the Gun Room

On the lower deck, forward of the gun room, a bulkhead was raised; it coincided, as we saw in section 33 (Fasten the Deck Clamp), with an inflection of the deck: the deck of the gun room was made flatter than the rest of the lower deck. A door was made to port in the bulkhead, which had to be wide enough to admit a gun. Often racks for storing cartridge bags were mounted against the inner face of the bulkhead.

Since we now are in the gun room, let's also take a look toward the stern where a hole is made through the tuck for the towing line or painter of the boat. On the outside the hole is ornamented with a round rim, a "cardinal's hat." The origin of this name is obvious. The towing line was belayed onto a kevel next to the hole.

107. *The Butlery.*

(59 I 45) The galley stands at the same level as the butlery, but at the other side, which is why it cannot be shown here. In men-of-war it was mostly lowered down to the hold, athwartships.

(91 II 51) *66. About the Butlery to starboard.*

The forward side comes 19 inches before the mast, 4 feet and 8 inches from the ship's side, but in men-of-war, it is placed in the hold, against the galley, in which there is a hole, through which the food is handed.

1. The butlery wide 4½ feet, long 5 feet.

2. The corner posts thick 3 inches, broad 4 inches.

3. The ledges and deals as with the galley, the door braces broad 7 in., in which there are ledges, but only serving to nail something to it.

107. The Butlery (or Pantry)

The butlery was the small room to starboard near the main mast across the galley. This is where the cook kept a working stock and where he and the coxswains often had their berths.

108. *The Galley.*

(91 II 5) *65. About the Galley.*

Which, on Merchantmen, stands on the lower deck, but in men-of-war in the hold: it is put amidships, to this or the other side, or also close to the cabin, with the hearth toward the bulkhead.

The galley should be well covered with copper plates, and the hearth brick laid: 7 square feet can be brick-laid with 1000 Leiden bricks, and for every 3000 stones a hoet [probably a cubic measure] of mortar.

. . .

(91 II 49) The galley is often sheathed with copper all around.

(91 II 18) 1. The posts broad 3½ inches, thick 3 inches.

2. The galley, wide 6 feet, long four feet.

3. The planks thick 1½ inches.

4. The foot thick 3 inches, broad 15 inches.

It comes 8 inches before the mainmast, and below is 6 feet from the ship's side.

(91 II 40) 5. The lower sill broad 3½ inches, thick 3 inches, is half round underneath the sill as a *cymatium.* The planks come 8 inches past the posts.

6. The upper sill broad 5 inches, thick 3 inches.

7. The chimney long 5 feet, above wide 12 inches, below wide 16 inches, the cover long 15 inches.

108. The Galley

The galley was placed to port and covered inside with copper plates. Then the hearth was laid inside with fireproof bricks, reinforced with wrought iron strips. The galley chimney covered an opening through the deck above the hearth. The top cover of the chimney could be removed and turned away from the wind.

More unexpected evidence of the authenticity of Witsen's data is the fact that the cover of the chimney comes to exactly the right height between the railings, although

Figure 2.210. The galley on the port side of the lower deck. (Drawing by A. J. Hoving)

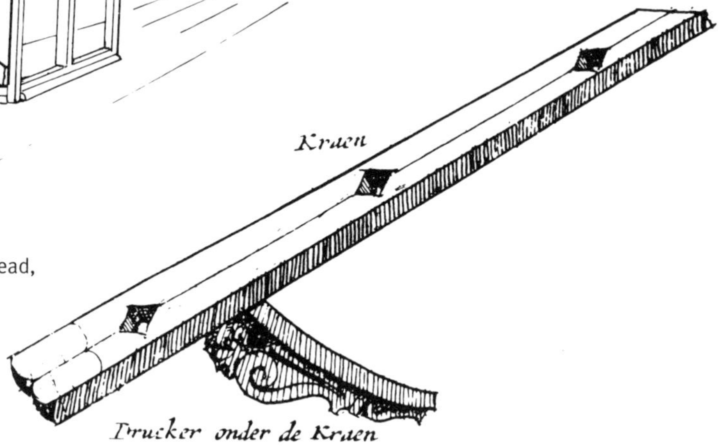

Figure 2.211. Plate XXVI. Cathead, cathead support (*drukker*).

the height of the rail at that spot is not mentioned explicitly anywhere.

In men-of-war the galley was placed in the hold so as not to stand in the way of the guns. In smaller merchantmen it was also placed under the forecastle.

109. The Catheads.

(55 | 19) Cathead: is a timber, which forward sticks out of the beakhead over the water: when wound up, the anchor cable goes through it, either through one, or through two sheaves. Near it is another timber, called the Cathead Knee, which supports the same.

(90 | 41) *57. About the cathead.*

The one end of the cathead lies at 4 feet from the mast, the other end straight on the forward top timber.

1. The cathead lies on the forecastle, 2 feet across the railing, and is thick 8 inches, broad 10 inches.

2. At the other end it is broad 8 inches, thick 4 inches.

109. The Catheads

The cathead was a beam across the railings of the beakhead, from which it protruded on each side, for catting the anchor. "Catting" was pulling the anchor out of the water with a sheaved block hooked to the anchor ring in order to secure it to the ship's side. Contrary to Witsen's statement, the anchor cable itself did not run through the cathead.

The so-called fish davit was a loose beam used in the operation to stow the anchor against the ship's side. It was shoved across or between the railings, for which a special aperture was sometimes available.

The cathead was secured on top of or below the forecastle deck and was supported at the beakhead by a cathead knee, a carved truss going from the beakhead railings back against the ship's side.

110. The Rudder.

(71 | 43) *33. About the Rudder.*

1. Every 12 feet length of the ship, gives 4 inches breadth of the rudder.

Example 100 feet length,	12	
	8	8
	6	4
		32

2. Comes to 33 inches [should be 32½ inches], the breadth of the rudder.

. . .

(262 I 19) Against the lifting of the rudder, bars are hammered through the rudder irons; for which end a preventer rope also goes through the rudder, which holds the ship,

The tiller is secured with an iron pin aft against the rudder itself.

. . .

(274 I 19) A groove, made in the bottom of the rudder, in the middle of the ship or keel, is thought to benefit the course of the ship, and to steady the rudder itself.

. . .

(274 II 17) On the lower deck, underneath the whipstaff, is a sheave, through which in bad weather goes a line, to facilitate the handling of the rudder; and it is manned by two men, the more the rudder is turned, the heavier it is to handle.

. . .

(275 I 24) The hole, through which the tiller enters the ship is filled with sailcloth when at sea, to avert the water; on some ships the water is admitted through there; but then drained again at the sides.

When hanging the rudder to ships with a stern gallery, with the cabin overhanging aft, a hatch is cut in the cabin, through which the same can be admitted, which, after the rudder is hung, is closed again.

. . .

(275 II 12) The whipstaff goes through the tiller, and then upward, so as it is possible to steer from the steering stand.

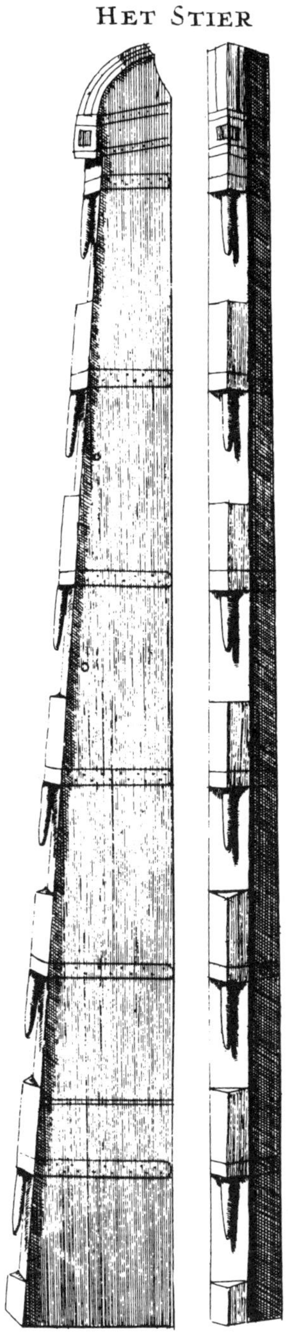

Figure 2.212. Plate XXXIV. Rudder.

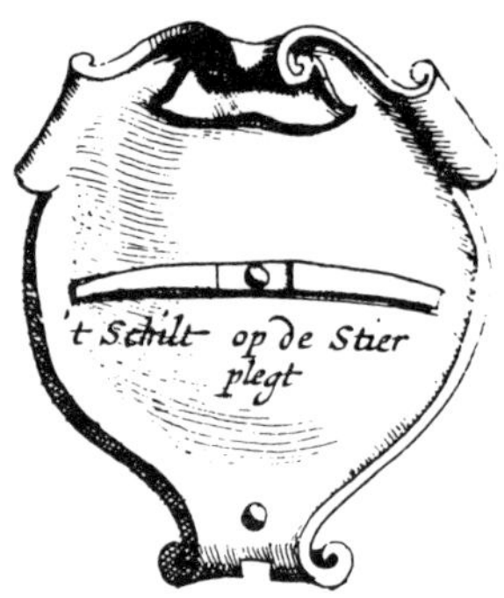

Figure 2.213. *(above)* Plate XLI. Rowle.

Figure 2.214. *(right)* Plate XXX. *Schilt* in the steering stand.

(268 | 38) In ships, which have low steering platforms, the place where the man stands to the rudder can often be seen to be raised, for him to be able to do his work with ease, and well.

(88 | 34) In the plate No. 20 [fig. 2.213] letter K, the moving chock going in the deck can be seen, otherwise called the rowle, through which goes the whipstaff.

(92 | 31) 68. *About the Rudder.*

1. The Rudder, long 26 feet 7 inches, wide below 3 feet 6 inches, wide above 1 foot 6 inches, thick at the front 12 inches, thick at the back 8 inches.

2. The Hole, long 8 inches, the lowermost hole is wide 3 inches, the uppermost 2 inches: the lowermost comes 6 inches from every side, the uppermost 5 inches, it is doubled with one-inch planks, like the ship itself, the front sheathed with copper, if doubled. To the rear of the sternpost it is doubled with one-inch nails, the lozenges wide 1½ inches.

3. The shins, broad 1½ inches, thick ½ inch.

4. The strap around the rudder head, broad 2 inches, thick ½ inch.

5. The pin of the pintles, long 17 inches, there are 6 pintles: broad 3 inches, thick ½ inch.

6. The tiller, thick 9 inches, forward thick 6 inches, long 23 feet, hewn square.

7. The whipstaff long 9 feet, thick 4 inches, the upper end round or octagonal.

(88 | 13) *The Sweep* lies 4 feet 9 inches high at the ship's side, and it is thick and broad 4½ inches, lies 4 feet 4 inches behind the hole, through which the whipstaff goes, the pin lies on 2 chocks, is long 14 inches: the hole comes 2 feet from the bulkhead, and is wide 4 inches. The schilt [a huge plate of timber carved as an ornamental shield] is 4 feet long, and broad 3 feet; thick 3½ inches. The tiller is thick and broad, forward and aft, 6 inches, the whipstaff long 8½ feet, thick 14 inches, and at the bottom end broad 1 foot, the tiller long 19 feet.

110. The Rudder

The width of the rudder depended on the length of the ship: 4 inches for every 12 feet (according to Van Yk, 3 inches for every 10 feet). Above the water the width was reduced to half. Its thickness was approximately equal to the aft face of the sternpost. The forward face of the rudder was hewn to an angle. Below, the rudder was made as thick as the sternpost together with the garboard strake (see section 15, Then the garboard stern rabbet is made, with the keel rabbet).

Often the rudder was composed of several parts. With hooks (the pintles) it was hung into eyes (the gudgeons) fastened to the sternpost. At the top, the tiller went through a hole that was one third of the thickness of the head. On the aft side the tiller was secured with a wedge.

The tiller entered the gun room through the helm port. There it rested on the sweep, a beam athwartships lying on chocks in the ship's sides. Often the tiller lay on the sweep with a roller, facilitating its movement.

The forward end of the tiller was attached to the whipstaff, which came up vertically through the deck into the steering stand. A pivoting point was made in the deck, or rather in the *schilt* (a piece of wood in which the rowle pivoted), which lay on the deck. A chock with a fulcrum bearing, called the rowle, could turn in it. The whipstaff could move up and down in the rowle, enabling a maximum reach of the tiller (in this case, 23 degrees to each side). The whipstaff therefore had to be fairly long to be pushed as far as possible through the rowle. Because of its length, an opening was made in the upper deck, allowing for the movement of the top of the whipstaff. Stories about helmsmen of this period sticking their heads through this aperture to see how to steer the ship seem pure fantasy to me. The helmsman could somehow see the luff of the mainsail, or fore topsail if the mainsail was furled, and could thus sail full and by when beating to windward. To some extent, he could also see stars at night and therefore steer a course, occasionally checking his heading by the compass. He also received his orders from the officer on deck and sailed on his compass.

111. The Sheet Blocks.

111. The Sheet Blocks

The sheet blocks were mostly attached to rings in the ship's side, but as can be seen on the *Wasa,* they could

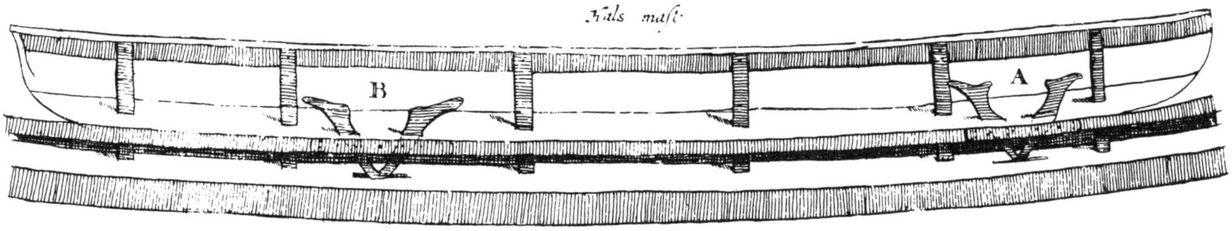

FIGURE 2.215. Plate XXXI. Kevel rail.

also be made part of the construction. If so, they were bolted onto the top of the railing above the mizzen channels and had a sheave over which the sheets came inboard.

112. Apply the Kevel Rail, if necessary.

(56 I 6) The kevel rails are for belaying ropes, A and B, in plate 12 [fig. 2.215] are kevels for it, their place in the ship is at the main frame.

(89 II 28) 53. *About the kevel rail.*

1. The kevel rail, broad 5 inches, thick 3 inches.
2. The posts broad 4 inches, are numbered 56 to one side, thick 2½ inches.
3. The kevel rail, stands 15 inches from the sheer rail, and its upper face coincides with the top of the lowermost railings.

112. Apply the Kevel Rail, If Necessary

The kevel rail was the plank in the bulwark of the waist, onto which both midship kevels were mounted. The "posts" that Witsen mentions are probably the tops of the top timbers, to which the pinrails were attached (see section 104, Pinrails).

113. And also the Railings above the Pinrails if required.

(89 II 42) There is a small *railing* above the pinrail, and is secured on stanchions; these stanchions are thick 2 inches, broad 3 inches, its foot standing on the waterboard, long 12 inches. There are 7 stanchions; the railing is broad 3 inches, thick 1½ inches, and in height leaving 9 inches in between them. At the end of the forecastle, the railing comes in 1 foot, toward the rear. Below there is also one, and one above in the steering stand, where there are 5 stanchions, broad 2½ inches, thick 2 inches, the railings 14 inches between them, broad 3 inches, thick 1½ inches, the topside coinciding with the other railings, at the top, but this hance is 1 foot shorter than the other, on which it stands, and these are made in the way the cleat rail is made; the cleat rail is bolted to the stanchions; the others are nailed.

(91 I 40) 63. *About the loose Railing, above the forecastle.*

1. The Railing, above the forecastle, broad 3 inches, thick 1½ inches, standing 9 inches above the pinrail.
2. The stanchions broad 3 inches, thick 2 inches.

(91 I 47) 64. *About the Railing, above the Steering Stand.*

1. This Railing long 12 feet, thick 1½ inches, broad 3 inches.
2. Here are 5 stanchions, each broad 3 inches, thick 2 inches.
3. The railing stands 9 inches above the fife rail.

113. And Also the Railings above the Pinrails If Required

These railings stand athwartships and apparently could be left off (if we may interpret Witsen's section title in this way). There are two of these railings: the first was placed forward on the forecastle, between the ends of the uppermost headrails, the other above the steering stand. Witsen's formulation is so vague in this respect that guesswork is inevitable.

114. Make the Chesstree.

(55 II 22) Chesstree: is outside against the ship, through which goes the rope, named the tack: of which the shape is shown in the plate No. 11 [fig. 2.217].

FIGURE 2.216. Bow of the pinas model. The athwartship rail on the forecastle is peculiar: to reach the beakhead, the crew would have to climb over it. The text leaves sufficient room for doubt (see section 117, **The Pinrail on the forecastle.** (Courtesy Cees de Jonge, The Visual Art Box)

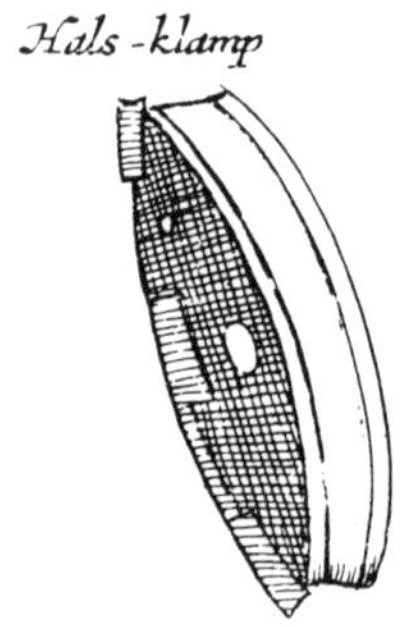

FIGURE 2.217. Plate XXX. Chesstree.

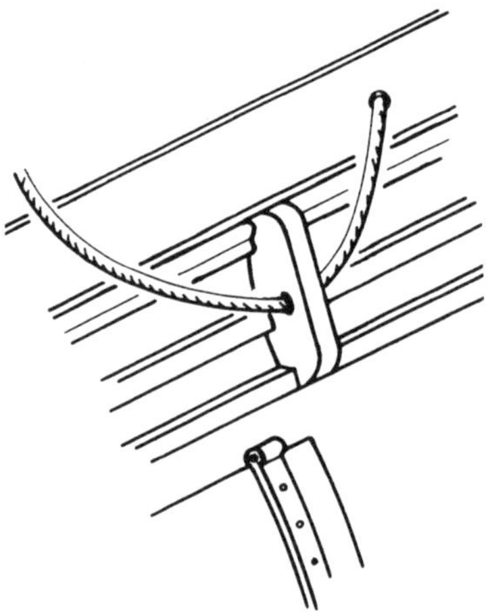

FIGURE 2.218. Witsen's drawing of the chesstree for the pinas (fig. 2.217) is not very helpful for understanding how it works. The main tack is led through the fairlead and then through an aperture in the ship's side. Sometimes chesstrees were beautifully carved as grotesques, with the tack led through the mouth. (Drawing by A. J. Hoving)

FIGURE 2.219. The waist of the pinas model. The capstan, the knight heads, the boat on the gratings, and the chesstree are clearly visible. (Courtesy Cees de Jonge, The Visual Art Box)

(90 | 32) *56. About the Chesstree.*

1. The chesstree broad 8 inches, thick 7½ inches.

2. The Large hole wide 3½ inches.

3. The Upper hole wide 2 inches.

4. The chesstree stands at two fifth parts of the length of the ship taken from the bow, or straight above the second gunport, below the upper deck.

114. Make the Chesstree

The chesstree was a chock or fairlead on the outside of the ship through which the tack of the mainsail passed. It was placed two fifths of the ship's length from the bow. On expensive ships this chock was extensively carved, often as a mask with bulging cheeks, a beard, and a mouth as the hole for the tack.

On the pinas the chesstree was a simple chock, standing between the sheer rail and the third wale. The rope went through a hole in the chesstree and then through a hole in the ship's side forward of the chesstree.

During the last stage of preparing this English translation of Witsen's book, the exact location of the chesstree was the subject of an interesting discussion. Van Yk gives a formula that, if applied to the pinas, gives the same location for the chesstree as Witsen did. Nick Burningham, who was involved in the research for the building of the Dutch yacht *Duyfken* in Fremantle, Australia (and who also sailed the ship), noticed that the location for the chesstree was too much aft to be effective while sailing close-hauled. Olof Pipping, working on the rigging of the *Wasa,* estimated the true location as about one gunport forward. Perhaps the main tack was taken out of the chesstree if close-hauled and led to the snatch block at the foot of the foremast. Or perhaps sailing with full bellying sails, as is often depicted, called for a tighter bowline, not a farther forward hauled tack.

115. The Pumps.

(59 || 22) Near the main mast often a pump is placed, which is named the brake pump.

115. The Pumps

The pumps were discussed extensively in section 102 (The Bushes, Pumps, and the Privy).

116. The Bread Room and the Boatswain's Storeroom.

(274 || 13) The bread room is sheathed with tinplate, to conserve the bread, and the hatch of the powder room covered with lead.

. . .

(275 || 17) Boatswain's Storeroom and Peak are corners forward and aft below, beneath the rooms, used only for storing rags, and sometimes shot.

(85 | 44) *About the Bread Room.*

The forward beam, lies 9 feet from the lower deck, the aftermost 8 feet, the foremost lies 1 foot behind the pump, it is thick and broad 5 inches, the second lies 2 feet behind it, and is broad 10 inches, thick 6 inches, the third lies 2½ feet from there, broad 7 inches, and thick 5 inches: there is a hatch, wide 17 inches, long 2 feet: the binding strakes thick and broad 4 inches. The bread room itself long 9 feet, and as deep: the laths underneath the planks thick 1 inch, broad 2 inches, lying 1 foot apart: the deals thick one inch, and are connected by tongue and groove. Before the powder room there are 6 posts, and they are thick and broad 3½ inches, in the middle there is a hatch, to enter the well. Before the bread room there are also 6, and these are thick 3½ inches, broad 4 inches: two of them are next to the hatch. To the fore end of the deck clamps, also 6 posts are standing, and 4 at the well around the pump, thick and broad 3½ inches. Near the front of the hold is the boatswain's storeroom at the foremost beam, of which the posts as above.

116. The Bread Room and the Boatswain's Storeroom

The bread was stored in the bread room, along with biscuits, flour, and cheese. This compartment was entirely covered with tinplate to keep out insects and rats. The storeroom was below the gun room and could be entered through a cap scuttle. It was situated on top of the powder room and closed off with a bulkhead forward and aft. Behind this compartment was the peak, a room where the master gunner kept spare parts, and to the fore was the part of the hold where the victuals were kept. In front of the powder room was the well, in which the bottom ends of the pumps stood. The rooms could be entered via hatches.

The boatswain's storeroom (*hel* in Dutch) was in the bow before the foremast. It was used for the storage of miscellaneous supplies and also as a detention room. It was entered through a hatch in the cable tier.

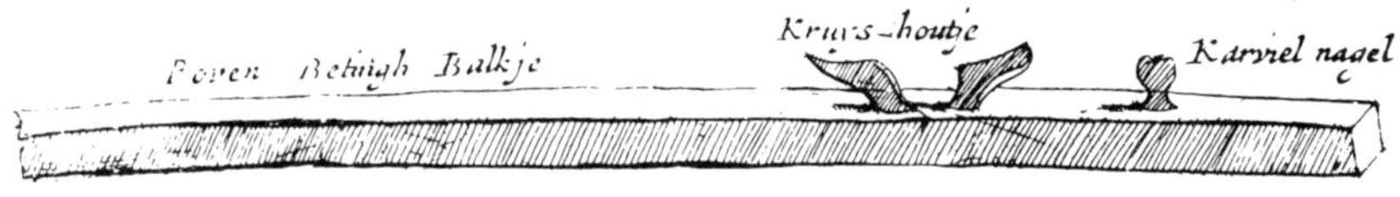

FIGURE 2.220. Plate XXVI. Forecastle crosspiece, kevel, belaying pin.

117. The Pinrail on the forecastle.

(90 II 1) 58. *About the pinrail, before the Foremast on the forecastle.*

1. This pinrail, thick 8 inches, broad 9 inches.
2. Lying 8 inches above the forecastle.
3. At each end there is a knee, in the side long 3 feet, at the pinrail long 2½ feet, broad 8 inches, thick 6 inches.

117. The Pinrail on the Forecastle

Across the catheads and below the transverse railing there was a beam into which belaying pins were set, along with kevels for the cathead stopper, which was used when catting the anchor. On the *Wasa,* which strictly speaking had a closed forecastle, the same solution can be observed.

118. Kevel and belaying pin.

(71 II 19) 39. *Belaying pins,* ¾ inch.

118. Kevel and Belaying Pin

The kevels were treated in section 112 (Apply the Kevel Rail, if necessary), and for belaying pins Witsen here uses the word *karvielhout* (carling) instead of *karvielnagel* or the more modern equivalent, *korvijnagel* ("belaying pin" or "jack pin"). The jack pins had a diameter of ¾ inch and were placed in the pinrails.

119. Arch above the Whipstaff, if necessary.

(88 I 6) *About the arch above the whipstaff.*

This comes 18 inches from the bulkhead, and is 15 inches wide, and high in the middle 19 inches, it fades to naught at 1½ feet from the ship's side. The sides consist of a scant plank, and on the top, it is covered with deals of 1 in.

119. Arch above the Whipstaff, If Necessary

In section 110 (The Rudder), we saw why the whipstaff was made of some length. It could therefore protrude above the quarterdeck, and an opening was provided for it to move freely. Often a hutch was built over the opening, going from waterway to waterway and closed on the aft side. The front was left open for the helmsman to have enough visibility to see the lower sails and to communicate with the quartermaster when the latter was standing on deck. The helmsman would also steer by the compass, which was placed in the binnacle, a small wooden cabinet with one or more compasses and an oil lamp for the night.

120. Then commence finishing off the outside.

120. Then Commence Finishing Off the Outside

In the title of this section Witsen uses the verb *verzoeken,* which has no clear meaning here. It probably means finishing off the outside—that is, leveling the nails to make them flush, caulking, removing uneven patches, and so on.

121. And sheathe if necessary.

121. And Sheathe If Necessary

Wood sheathing was applied on ships destined for the tropics because of the risk of shipworm (*Teredo navalis*), a type of mollusk that could damage ships so badly that they would fall apart. A second skin of deal planking was applied below the waterline to prevent the shipworm from eating into the ship. Between the main planking and the wood sheathing a layer of tar and cattle hair was applied to form an additional barrier.

Before the wood sheathing could be applied, it was necessary to hove down the ship (i.e., to haul it onto its side in the water).

122. The Sternpost and Rudder sheathed with copper.

122. The Sternpost and Rudder Sheathed with Copper

The sternpost was sheathed with copper plates to protect this important timber from decay. Yet ironically the combination of copper with the iron of the rudder irons resulted

Figure 2.221. Side view of the pinas model (Courtesy Cees de Jonge, The Visual Art Box)

in galvanic corrosion in the saltwater, not a blessing for the ship.

Thus, with this section, Witsen has reached the last of his 122 steps in the building of a ship while acknowledging that he has omitted certain details:

(61 | 12) And so the most important parts of the ship, which show on a ship, are pointed out; the others are too small, or not important enough, to be depicted on the engraving, like bolts, nails, cramps, belaying pins, stop waters, copper for the rudder etc.

These missing elements, such as the ladders, iron mountings, anchors, ropes, blocks, sails, flags, and other articles of equipment, will be treated in the following sections, drawing on the information provided by Witsen.

123. Ladders

(91 | 23) The ladders, forward of the steering stand, are often made with handrails on the upper deck, and the steps themselves quite broad, and coming down in a curve: and also, on ships having quarterdecks, going from the steering stand upward, but this only seldom, as on men-of-war.

. . .

(267 | 20) In the hold of the ship, sometimes a pillar can be seen standing straight up, and hewn as a rack, it serves for a ladder, near which a rope hangs down.

(92 | 17) 67. *About the Ladders.*

1. The side rail broad 5 inches, thick 2 inches.

2. The steps wide 9 inches, broad 8 inches, thick 1½ inches.

3. The ladder wide 2 feet 2 inches, also the ladders abaft the forecastle, before the mast, and before the steering stand, below and above.

4. Both stairways for the upper cabin, wide 17 inches, the steps, wide 10 inches, thick 1½ inches, broad 8 inches.

5. The side rails broad 6 inches, thick two inches. The chocks broad 6 inches, thick 1 inch.

123. Ladders

The pinas had five ladders on the upper deck: two against the forecastle bulkhead, another two before the bulkhead of the steering stand, and one for the upper cabin. These all had an S-shaped curve, like the bulkhead against which they were placed.

To descend onto the lower deck there were also five ladders: two abaft the steering stand bulkhead, two inside the forecastle, and one before the main mast.

Access into the hold was by means of a pillar notched with steps at the main hatch as well as steps fastened against the bulkheads of the boatswain's store, peak, and bread room.

124. Bolts and Nails

(71 II 1) *34. About the Bolts.*

1. The bolts, in a ship of 120 feet, are 1 inch thick, more or less.

35. About the Nails.

1. The nails for this ship, are 1¼ inch. 100 feet of the ship's length gives 1 inch thickness.

. . .

(279 I 22) Holland nails made from Swedish iron, are considered the best, those from Liège follow in quality. But to make known, what weight of iron is used in ships, here follows a short notice on the iron, on ships of specific size.

In a ship of 150 feet length, wide 38½ feet, deep 15 feet, there are 80000 pounds of ironwork.
And about 15000 pounds of nails.
In a ship of 140 feet length, wide 37, deep 13½ feet, there are 70000 pounds of ironwork.
And about 13000 pounds of nails.
In a ship of 136 feet length, wide 34 feet, deep 13 feet, there are 66000 pounds of ironwork.
And about 11000 pounds of nails.

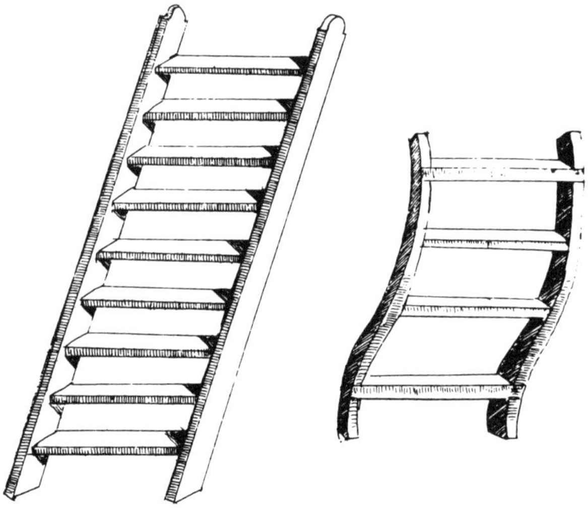

FIGURE 2.222. Plate XXXV

In a ship of 130 feet length, wide 32 feet, deep 13 feet, there are 55000 pounds of ironwork.
And about 9000 pounds of nails.
In a ship of 126 feet length, wide 31 feet, and deep 12 feet: there are 50000 pounds of ironwork.
And about 7500 pounds of nails.
In a ship of 121 feet length, wide 30 feet, and deep 12 feet, there are 40000 pounds of ironwork.
And about 5000 pounds of nails.
In a ship of 103 feet length, wide 25 feet, and deep 10 feet, there are 34000 pounds of ironwork.
And about 4300 pounds of nails.
In a ship of 100 feet length, wide 24 feet, and deep 10 feet: there are 30000 pounds of ironwork.
And about 4000 pounds of nails.

(93 II 24) The bolts are an ample inch thick, the nails 5 inches thick.

The nails, where they are applied, must be half as long as the wood is thick.

(59 II 45) the bolts in the heart of the ship, are 1 inch thick. The nails thick 1¼ inches. The nails 7, 8 to 9 inches. The bolts thick ¼ inch at the top.

(279 I 3) To fully build the ship I propose would be needed:

21000	pounds thumbnails.
13800	pounds staples or cramps.
8000	pounds double medium-size nails and drielingen [nails that were 6 inches long].
6000	pounds single medium-size nails and drielingen.
7000	pounds scarf irons.
13000	pounds bulkhead nails [nails for bulkheads, about 4 centimeters long and with a flat head; 1,000 weighed 7 pounds] and round-headers.
32000	pounds countersunk nails.
22000	pounds lead and slate nails and ley-nagels [no English equivalent].

Furthermore some stelspijkers [double-headed nails for easy removal], bolts with heads, pivots and rings: bolts with turning rings, ringed and rag bolts. Of which to tell where and for what purpose they are used, would be too circumstantial a story. The Master carpenters are fully acquainted with it.

Figure 2.223. Plate XXVI.
Ring bolt, hook bolt.

124. Bolts and Nails

These details speak for themselves. It is strange, however, that the total amount of iron that Witsen gives for his 134-foot pinas (122,800 pounds) does not at all comply with figures he supplies for the amount of iron needed for ships of various sizes. For a 136-foot ship he calculates 77,000 pounds, for a 130-foot ship only 66,000 pounds, which is about half of what the pinas has!

Van Yk supplies two formulas for the amount of iron in a ship (pp. 49, 50): for the nails the length, breadth, and depth of a ship are multiplied, with one pound of nails for every six cubic feet; for the pinas the weight of the nails would come to 8,420 pounds (134 × 29 × 13 ÷ 6). Apart from the nails, there was other iron in the ship in the form of bolts, chain plates, rudder irons, "piecework" (i.e., all iron fittings weighing less than one pound), and so on. For this category, known as "coarse iron," the formula was half a pound for every cubic foot (so for the pinas, 25,259 pounds). For a light vessel three quarters of this weight was reckoned, and for a ship with an orlop deck, five quarters.

So according to Van Yk, only 33,679 pounds of iron would go into a ship the size of the pinas, which in no way can be reconciled with Witsen's statements. Yet Van Yk concludes: "But remember that the one master Shipwright will build his ship with much more Iron, and so heavier than the other, and so no fixed Rule can be given for this" (p. 50).

125. Anchors

(65 I 14) *The Ship's Anchors are these.*

The *Bower;* which is used in heavy weather, when the Sheet Anchor is too light.

The *Sheet anchor;* which is used together with the Bower.

The *Kedge;* is brought ahead, and is used to kedge the ship, in a calm.

The *Grapnel,* is an anchor with four arms; with which the launch and the longboat are anchored.

. . .

(117 I 24) *About Anchors in general; and first, their decree or order.*

Take twice the thickness with calipers, and for the length of the shank take as many feet as this comes to in inches.

For instance, thick 6 inches, 6
2
12 feet,

Then take as many inches, as there are feet, which adds to 13 feet 1 inch, for the length of the shank. Take the thickness three times, for the weight of the anchor, two zeros added to the outcome, assuming

For instance, thick 6 inches 6
3
1800
pounds for the weight.

Beneath 1000 pounds, then take 2 inches, where above one inch was taken. Below five hundred pounds, take the thickness three times for the length: for instance, thick 2½ inches, comes to a length of 7½ feet, as
2½
3
7½

Also take the length, and halve it, for instance, half 7½, is 3¾ feet, and for every foot of the half, as 3¾ feet, 100 pounds, coming to 375 pounds for weight. Also take three double thicknesses for the length: for instance, thick 2½ inches, comes to the length of 7½ feet. Then take the length and halve it: as, half 7½, is 3¾, every foot 100 pounds, comes to 375 pounds for the weight, as above.

. . .

(118 I 11) It is also done to take a thousand pounds of weight for every hundred feet of length of the Ship: or ten feet of length, one hundred pounds of weight. And to roughly estimate the thickness, when the length is known, as many inches thickness are reckoned as there are feet in length, measured with a rope around.

The arms together are 1½ feet shorter than the shank: when traced on the outside they make a segment of a circle.

The crown wide two thirds.

The flukes or palms half as long as the inside face of the arms, and broad two thirds.

. . .

(117 II 24) All our anchors have two arms at least. It is true, that it is possible to make anchors with one

FIGURE 2.224. Plate LVIII

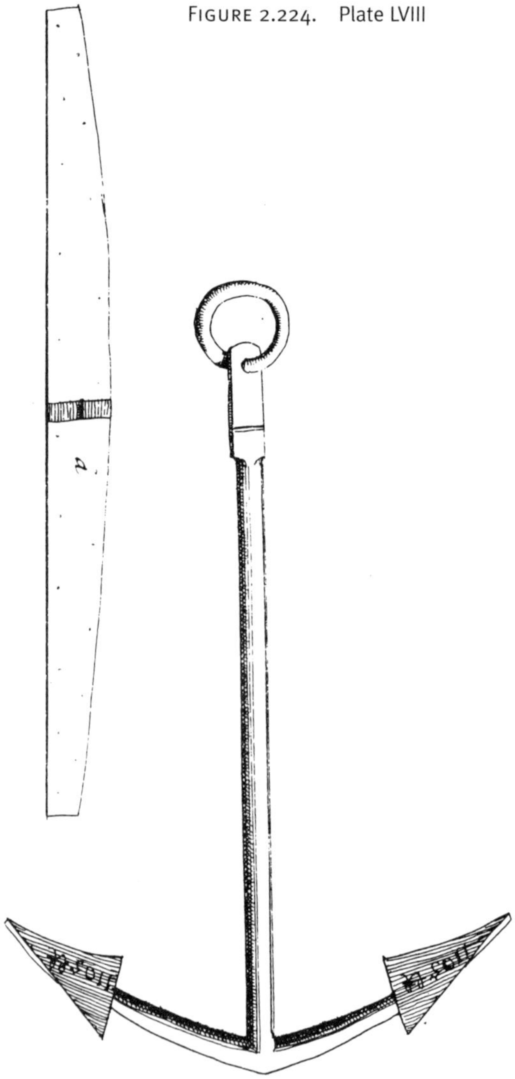

arm, which would be lighter, and hold as well in calm weather; but because there is also a need for weight, I think little of this invention, and leave it aside, more so because more are needed, to make it hold, tackles which foul up easily: matters of no benefit to the ship.

Some want the fluke of the anchor to be half the length of the arm, and the shank to be one and a half times as long as the arm.

(117 I 16) *Anchors of our Ship.*

Its bower is to be 1800 pounds heavy. The Sheet anchor 1600 pounds. The Stream anchor 1500. The Kedge 450 besides a light kedge, and one or two grapnels, of no specific weight.

(119 I 15) Grapnels are anchors with four arms, of which the weight is not measured according to the size of the ship, but to the boats, for which they are used: and while these vary little in size, so do the grapnels.

. . .

(117 II 40) *A small Notice, on the Anchor's length, thickness, and weight.*

13½ feet long	6⅔. inches thick	2000 pounds of weight.
13¼	6½	1900
13	6	1800
12¾	5¾	1700
12½	5½	1600
12	5	1500
11¾	4¾	1400
11½	4½	1300
11¼	4¼	1200
11	4	1100
10¾	3¾	1000
10½	3½	900
10¼	3¼	800
10	3⅛	700
9¾	3	600
9	2¾	500
8	2½	400
7	2¼	300
6	2	200
5	1¾	100

. . .

(118 I 14) *On the length, thickness, and weight of the Anchors, as follows.*

Weight	Length	Thickness	
2000 pounds	13½ feet	20 inches	6⅔ inches
1900	13¼	19	6½
1800	13	18½	6
1700	12½	18	5¾
1600	12	17¾	5¾
1500	11¾	17¼	5½
1400	11½	17	5¼
1300	11¼	16½	5
1200	11	15¾	4¾
1100	11½	15	4½
1000	10	14	4
900	9½	13	3¾
800	9	12	3½
700	8½	11	3¼
600	8	10	3

500	7½	9	2¾
400	7	8½	2½
300	6½	8	2¼
200	6	7	2
100	5	7	1¾
Weight	Length	Rope thickness	Caliper thickness

. . .

(118 I 36) An anchor then, heavy 1100 pounds, long 11½ feet. Is square at the top 4½ inches, and at the bottom end 5¼ inches, measured with the calipers. The arms have 5 inches curve, and are 5 feet long. The flukes long 2½ feet, and broad 2 feet, flat on the inside, but on the outside like the shank. The arms are bent 2 feet and 9 inches, the square is long 1¼ feet, the notch on it 1 foot, thick 1 inch. The hole for the ring is wide 2½ inches, the ring wide 16 inches. The anchor stock has to be as long as the anchor with the ring and all, which is 13 feet. Every foot length of the stock one inch for the breadth, but the ends half of this, the bottom side is made curved, or arched, for it to be stronger, and better able to bend.

Anchor flukes are taken as wished, yet fitting, according to the size of the anchors; and the unfitness of the ground, on which they have to hold, which is why there are no rules for this.

. . .

(119 I 15) Grapnels are anchors with four arms, of which the weight is not derived from the size of the Ships, but of the boats, because that is what they are used for: and because these differ little in size, so do the grapnels.

. . .

(118 II 51) A fluyt of 200 lasts, carries a bower of 900 pounds. A sheet anchor of 800 pounds. Stream anchor of 700 pounds. Kedge of 200 pounds.

A ship of 600 lasts, has a bower, 3000 pounds of weight. A sheet anchor, 2800 pounds of weight. Stream anchor of 2600 pounds. And kedge of 1800 pounds.

A ship of one hundred lasts, has a bower of 800 pounds. Sheer anchor of 700 pounds. Stream anchor, of 600 pounds. Kedge, of 500 pounds.

A fluyt of 220 lasts, carries a bower of 290 pounds [should probably be "1290 pounds" compared with the first ship listed above].

A Merchantman, long 140 feet often carries anchors, heavy 1700 pounds.

The anchor on an East Indiaman long 170 feet, I have found to weigh 3300 pounds.

125. Anchors

A ship carried several anchors: the bower, which was the largest; the sheet anchor, somewhat smaller; the streamer, smaller again; and the kedge, the smallest. The longboat and launch both had a grapnel, an anchor with four arms.

Anchors were severely tested before approved to be taken aboard. Small wonder, considering that they were often used in dangerous situations and the safety of the ship depended on their quality. To test its strength, the anchor was dropped on the barrel of a gun to check whether it would bend or crack.

Witsen's tables speak for themselves. Dimensions of thickness refer to diameters, except "caliper thickness," which indicates the cross section. The diameter of the anchor cable of the pinas was about eleven centimeters.

126. Anchor Cables

(120 II 9) The thickness of the working bower cable, is proportionate to the length of the Ship: for instance; let the Ship be 100 feet long: then for every ten feet of the length, one inch is taken for the thickness for the cable.

The bower usually has the same cables. In distress, two ropes are stitched together, to ride the sea more easily.

. . .

(127 I 13) *The weight of the cable laid on one hundred fathoms.*

A cable of 4 inches,	weighs 325 pounds
4½	400
	90
5½	580
6	680
6½	800
7	930
7½	1060
8	1200
8½	1340
9	1490
9½	1640
10	1800
10½	1970
11	2150
11½	2325

Cable of	serves for a	anchor of	Cen.	Quart.	Pont.
21		bower	60	0	3
20 inches,		sheet	56	0	1
14 serves		stream	55	0	0
10 for a		kedge	25	0	0
9		light kedge	7	0	2

Cable of	anchor of	c.	q.	p.
20	bower	43	0	0
19 inches,	sheet	40	0	5
13 serves	stream	39	0	3
9 for a	kedge	20	1	2
8	light kedge	7	2	0

Cable of	anchor of	c.	q.	p.
17	bower	35	3	3
16 inches,	sheet	34	0	2
12 serves	stream	31	2	7
8 for a	kedge	11	3	5

Cable of	anchor of	c.	q.	p.
17 inches,	bower	32	0	0
16 serves	sheet	30	2	2
11 for a	stream	27	0	0

Cable of	anchor of	c.	q.	p.
	bower	29	0	0
15 inches,	sheet	25	0	0
14 serves	stream	23	3	2
9 for a	kedge	9	0	0
	light kedge	2	2	7

Cable of	anchor of	c.
13 inches, 12 serves for a	bower	28
	sheet	27

Cable of	anchor of	c.	q.	p.
10 inches, 9 serves for a	bower	22	0	2
	sheet	11	0	0

Cable of	anchor of	c.	q.	p.
16 inches,	bower	27		
15 serves	sheet	23	3	5
10 for a	stream	23		

Cable of	anchor of	c.	q.	p.
14	bower	18	0	0
13 inches,	sheet	17	0	0
12 serves	stream	16	0	2
12 for a	kedge	4	2	0
8	light kedge	2	2	0

Cable of	anchor of	c.
12 inches, 11 serves for a	bower	11
	sheet	10

Cable of	anchor of	c.	q.
8 inches,	bower	7	
7½ serves	sheet	6	
6 for a	stream	5	2

Cable of	anchor of	c.	q.	p.
8 inches,	bower	5	0	2
7 serves	sheet	4	0	0
6 for a	stream	3	2	3

FIGURE 2.225. *(at left)* Cable sizes by anchor weight (pp. 126 [mistakenly numbered "112" in the original] and 127). Witsen's original table has been digitally altered to show the Dutch text in English. The column headings for anchor weights are not retouched and reflect the English system of hundredweight ("Cen."), quarters ("Quart."), and pounds ("Pont.") (Scanned image modified by Emiel Hoving)

12	750
12½	2750
13	3000
13½	3250
14	3500
14½	3800
15	4900
16	5332
17	5900
18	6600
19	7000
20	8000
21	9000

For the largest Ships, which are equipped for War, cables of 120 fathoms are laid, to better ride at anchor, and these are thick 20 or 22 inches: they are laid from three ropes, each made of three strands, which are composed of about 900 yarns each, so that the entire rope consists of 18000 yarns, at a thickness of 20 inches, such a cable weighs 9½ thousand pounds, when tarred.

126. Anchor Cables

There is something peculiar about Witsen's table of cable sizes according to anchor weight (fig. 2.225). It seems to be a list of English origin, as the anchor weights are shown in hundredweight ("Cen."), quarters ("Quart."), and pounds ("Pont."). One hundredweight is 112 pounds. There are twenty hundredweight to a ton. A quarter is 28 pounds or two stone.

As these units of measurement were not in use in Holland, it appears that Witsen borrowed this information from lists in Edward Hayward's *The Sizes and Lengths of Rigging for All the States Ships and Frigats* (1655). Hayward was clerk of the survey at the Chatham Dockyard, one of several Royal Navy dockyards. Other lists that Witsen appears to have borrowed are the ones on rope thicknesses relative to their masts (see figs. 2.226 through 2.230). Although Hayward even gives rope lengths and does not refer to mast thicknesses, the sizes of ropework for his largest ship (*Sovereign*) correspond precisely to the dimensions Witsen gives for a 34-inch mast. The following order of the ropes and the apparently laboriously translated terms are identical. The numbers for rope sizes in both tables are the same (with few exceptions), both vertically and horizontally. Witsen's printer must have rather freely inserted a few columns (8, 9, 10, and 11) in the wrong places. These are the four columns that Hayward did not include in his tables, presumably for lack of space, and presented in a separate list.[15]

Witsen usually gives rope thicknesses in circumference. An anchor cable of Witsen's pinas weighed 800 pounds, according to the table (fig. 2.225), and was 100 fathoms (170 meters) long and about 11 centimeters thick in cross section. Usually four of these cables were carried.

127. Ropes for the Rigging

(120 II 19) The Ship's main ropes, proportionate to the Ship's length, can be observed in the next tables [figs. 2.226 through 2.230]. That does not mean that all sizes follow the rule, while it is not unknown to me that most of them are usually made at a guess and according to custom: but if made to order they will be suited and according to requirements.

When the strength of the rope is to be increased, then thin yarns are twisted between the turns of the rope, and tarred, together with the rope itself.

At many places, like the yards and elsewhere, the ropes are clad with leather; against wear and tear, for which the same are also served, at many places, particularly the forward stays, because the yard rubs against them, as also the tacks. Ropes going through holes, as the ones hoisting the yards, have to be laid smooth, so as not to jam.

. . .

(134 II 39) The lines aboard are tied below in the Ship, as deemed fit, without rule, to Mast, railing, knightheads, to cleats, inboard and outboard, to pins, nails etc.

To describe where they go, their course, would be of little use, as such could not be well examined without seeing, which is why the reader is referred to the plate [fig. 2.231] for this purpose, inserted here [. . .]: in which all lines and ropes are shown, on their appropriate places, in the ship. The lines to the top end of the Mizzen, are split, and spread themselves widely to the same, to prevent it from swaying, and to stretch the sail more firmly and better.

. . .

(129 I 37) The heavy ropes of 17 inches, are to be found 100 fathoms long. A cable of 8 inches thickness, long 100 fathoms. A cable of 6 inches thickness, long 110 fathoms. A hawser of 4½ inches.

(275 II 33) The shrouds are made of thick ropes, or served poles splinted athwart ships, for the sake of strength: above the shrouds are laid around the top. As all the heavy ropes they are clad with leather above.

When hauling the shrouds taut, they need to be untied and served anew.

(119 II 45) *State of the Ropes, on my proposed Ship.*

There are four cables, thick 13 inches, long 100 fathoms, and one of 12 inches, together weighing 14560 pounds. The largest [main] stay thick 13 inches, long 24 fathoms. The forestay thick 9½ inches. The main shrouds thick 6¼ inches. The fore shrouds thick 5½ inches, weighing 740 pounds. The Mizzen shrouds thick 4 inches, weighing 408 pounds: 6 yzeren troszen [strong hawsers], weighing 1290 pounds: two wiel troszen [no English equivalent; a *wiel tros* is a thin line consisting of 27 strands and made with the use of a wheel instead of a ropewalk] of 122 pounds: 34 peerde lijnen [*paardelijnen*: cables of small dimensions, used for mooring, towing, and working with the anchors] of 377 pounds, 300 pounds spun yarn: 12 marlines of 25 pounds: 36 bundles of houseline and 36 bundles of marling; 2 wiel-troszen of 218 pounds.

Of the Main Topmast, the shrouds thick 4 inches: the fore topmast 3½ inches, together weighing 690 pounds. The topmast stay, and both top ropes, thick 5½ inches. 2 pairs of topsail sheets of 1062 pounds; thick 4¾ inches. The ties thick 4 or 4½ inches. The cathead stopper thick 5½ inches. The main course halyard tie, thick 6¾, long 26 fathoms. The fore course halyard tie 6 inches, long 24 fathoms; weighing 461 pounds. The mizzen and main topmast halyard tie, thick 4 inches: the fore 3½ inches, weighing 112 pounds. 5 Different strong hawsers of 883 pounds. 8 Wiel-troszen long 100 fathoms, of 775 pounds. 300 Pounds spun yarn: 18 Hawsers, of 200 pounds. 24 Bundles of housing, and 24 bundles of marline. Two pairs of sheets of 521 pounds, thick 4½ or 5 inches, long 14½ fathoms, of 296 pounds. Two iron straps, 312 pounds. 7 Wiel-troszen, thick 10, 12, 15, 18, and 20 inches, weighing 512 pounds: another 20 tarred lines. And this is the rope, which the proposed Ship would need, to be fitted out according to requirements.

Table, in which can be seen, the thickness of all the main Ship ropes, proportionate or ranged after the thickness of the masts. And first, the ropework of the Mainmast, which has its

	Mast of 34 inches.	Mast of 32 inches.	Mast of 30 inches.	Mast of 29 inches.	Mast of 28 inches.	Mast of 27 inches.	Mast of 26 inches.	Mast of 24 inches.	Mast of 23 inches.	Mast of 19 inches.	Mast of 13 inches.	Mast of 12 inches.	
	1	2	3	4	5	6	7	8	9	10	11	12	
	in.	in.	in.	in.	in.	in.	in.	in.	in.	in.	in.	in.	
Schenckels	8½	8	7	6½	6	5½	5	7	6	5	4	4	Pennant of tackles
Standers of Taeckels	6	5½	5	5	4½	4	3½	5	5	4½	3½	3½	Runners of tackles
Vallen of Taeckels	4	4	3½	3½	3	3	3	3½	3½	2½	2½		Falls of tackles
Hooft of Want-touwen	8½	8	7½	7	6½	5	5	7	6	5	4	4	Shrouds
Taeli-reeps	4½	4	4	3½	3	3	3	4	3½	3¼	2½	2½	Laniards
Servingen om 't want te zwichten	8½	8	7½	7	6½	5	5	5½	5	4½	3¾		Standing back-staies
Taeli-reeps, of ſtag wand	4½	4	4	3½	3½	3	2½	3½	3	3	2½		Laniards
Stagh	17	16	15	14½	14	10	8	12	11½	9½	6½	6	Stay
Kraeg hals-banden, of ſtroppen om	16	15	13	12	11	9	8	10	10	8	6		Collar of the stay
Stagh Tali-reeps (de ſteven	6	5½	5½	5	4	4	3½	4	4	3½	2½	3	Laniard of the stay
Toppenants	4½	4	3½	3½	3	3	2½	3	3	2½	2½	2	Lifts
Smeet	9½	9	8½	8	6½	6	5	6½	6	5½	4	4	Tacks
Schooten	6½	6½	6	6	5	4½	4	4½	4½	3½	2½	3	Sheats
Boelijns	5½	5	4½	4½	4	4	3	3	3	2½	2½		Bowlines
Spriet-touwen	4½	4½	4	4	3½	3½	3	3	3	2½	2	2½	Bridles
Bras ſchenckels	4	4	3¼	3½	3	3	2¾	3	3	2½	2	2½	Pennants of braces
Braszen	3	3	3	3	2½	2½	2	2½	2½	2	1½	2	Falls of braces
Knie-touwen	4	3½	3	3	2½	2½	2	3	2½	2½	1½	2	Clugarnets
Cordeelen of Valletjes	8½	8	7	6	5½	5		6	4½	4			Tye
Rack	6	6	5	5	4½	4	3	4½	4	3½	3	3	Parrel-ropes
Borſtreep	8	7	6	6	5								Brest-ropes
Buick-gordingh	2½	2½	2½	2½	2	2							Bunt-lines
Nock-gordings	3	2¼	2¼	2	2	2		2½	2	2	1½		Leech-lines
Knie-ſchenckels	8½	8	7½	7	6	5	4½	6	5½	5	5	4	Pennant of the garnet
Reepen	6	5½	5	5	4½	4	3½	3½	3½	3	3	3	Guy
Groote kardeel	4½	4	4	4	3½	3½	3	3	3	2¼	2	2	Fall of the garnet

The Ropework of the main Topmast, has its

	1	2	3	4	5	6	7	8	9	10	11	12	
	in.	in.	in.	in.	in.	in.	in.	in.	in.	in.	in.	in.	
Schenckels van	5½	5	4½	4	3½	3	2¾	3½	3½	2½		2	Pennants of tackles
Vallen of Taeckels	2½	2½	2½	2½	2	2	1½	2	2	1½		1½	Falls of tackles
Want-touwen	5	5	4¼	4	4	3	3	4	3½	3	2¼	2½	Shrouds
Taeli-reeps	2½	2½	2¼	2	2	2	1½	2	2	1½	1	1	Laniards
Back-ſtaggen	5	5	5	4	3½	3	2½	4	4	3½	2	2½	Standing back-staies
Stagh want	3	3	3	2½	2½	2	2	2½	2	1½	1	1	Laniards
Stagh	8	7	6	5	5	4½	4	5	4½	3½	3	2½	Stay
Taeli-reeps	4	4	3½	3½	3	3	2¾	3	2½	2	1½	2	Lanyard
Toppenants	3½	3	2½	2½	2	2	1¾	2	2	1½	1	1	Lifts

FIGURE 2.226. Rope thicknesses by rigging category (p. 121). Witsen's original table has been digitally altered to show Hayward's terminology in the source table (p. 7). The column heading "Mast of 27 inches" corrects a misprint ("Mast of 20 inches") in the original text. (Scanned image modified by Emiel Hoving)

FIGURE 2.227. (*facing page*) Rope thicknesses by rigging category (p. 122). Witsen's original table has been digitally altered to show Hayward's terminology in the source table (p. 3). The column heading "Mast of 27 inches" corrects a misprint ("Mast of 20 inches") in the original text. (Scanned image modified by Emiel Hoving)

Thickness of the Ropework for the Foremast, for every thickness of the Mast as above

	Mast of 34 inches.	Mast of 32 inches.	Mast of 30 inches.	Mast of 29 inches.	Mast of 28 inches.	Mast of 27 inches.	Mast of 26 inches.	Mast of 24 inches.	Mast of 23 inches.	Mast of 19 inches.	Mast of 13 inches.	Mast of 12 inches.	
	1 in.	2 in.	3 in.	4 in.	5 in.	6 in.	7 in.	8 in.	9 in.	10 in.	11 in.	12 in.	
Schenckels	8	7½	7	6	5	5	5	6¼	5½	4½	3½	6	Pennant for tackles
Standers of Takels	5½	5	5	5	4	4	4	5	5	4½	3	2½	Runners for tackles
Vallen	4	3½	3½	3½	3	3	3	3½	3	2½	2		Falls for them
Want-touwen	8	7½	6½	6	5½	5	4½	6	5½	4½	3½	3½	Shrouds
Tali-reeps	4	4	4	4	3½	2½	2	3½	3½	3	2	2	Laniards
Puttings	8	7½	7	6	5½	5	4½						Standing back-staies
Tali-reeps	4	4	4	4	3½	3½	3						Laniards
Stach	15	13	12	11	9	7	6½	10½	8½	7½	5	4	Stay
Taeli-reeps	5	5	5	4½	4	3½	3	4	3½	3	2	2	Laniard to the stay
Reepen	8	7½	7	7	6	6	5	5½	5	5	3½	4	Tye
Valletje	6	6	6	5	4½	3½	3	3	3½	3	2	2	Halliards
Kordeelen	6½	6	6	6	4	4	5½						Jeers
Toppenants	4	3½	3½	3	3	2½	2	3	3	2½		2	Lifts
Rack	5	5	4½	4½	4	4	3	3	3	2½	2	2½	Parrel ropes
Borst-reeps	8	7	6½	6	5								Brest ropes
Schoots	6	6	5	5	4½	4	3½	4½	4	3½	2½	2½	Sheats
Smeets	8½	8	7	7	6	5	4½	6	6	5	5	3½	Tacks
Boelijns	4½	4½	4	3½	3½	3	3	3¼	3	2½	2	2½	Bow-lines
Spriet-touwen	4½	4	3½	3	3	3	2	3½	3	2½	2	2	Bridles
Bras schenkels	4	3½	3	3	2½	2½	2½	3	3	2½	1½	2	Pennants of braces
Braszen	3	3	2½	2½	2	2	2	2½	2½	2	1¼	1½	Falls to them
Knie-touwen	3½	3	3	2½	2	2	2	3	2½	2	1½	1½	Clu-garnets
Bint-touwen	3	3	2½	2½	2	2	2	2	2	2	1½	1½	Bunt-lines
Buick-gording	3	2½	2½	2½	2								Bunt-line
Nock-gording	2½	2½	2½	2½	2	2½	2	2	2				Leech-line

The Ropework of the Topmast.

	1 in.	2 in.	3 in.	4 in.	5 in.	6 in.	7 in.	8 in.	9 in.	10 in.	11 in.	12 in.	
Schenckels	4½	4	3½	3	2½	2½	3½	3½	3	2½			Pennants for tackles
Scheerlijns	2½	2	2	2	1½	1½	2	2	2	1½			Falls for them
Want-touwen	4½	4	3½	3½	3	3	2½	4	3	3	2	2	Shrouds
Tali-reeps	2	2	2	2	1½	1½	2	2	1½	1½	1	1	Laniards
Want puttings	5	4½	4½	4	3½	3	2¾	3	3	2½	1½		Puttocks
Stagh	5	5	5	4½	4	3	3	4½	4	3½	2	2½	Stay
Talie-schenckels	4	3½	3½	3	2½	2½	1¼					1½	Pennants of the laniard
Val van de taeli	3½	3½	3	2½	2	2	2	3	2½	2	1½	1½	Fall of the Laniard
Reep	7	7	6½	6½	5½	5	5	6	5	3½	2		Tye
Standers	5	5	4¾	4¾	4	4	4	4½	4	3½			Runner of the Halliards
Valletje	5	4½	4	4	3½	3½	2¾	3	2½	2	1½		The fall to them
Boelijns	4	3½	3	3	2½	2½	2½	2	2	1½	1¼		Bowlines
Schoots	3½	3½	3	3	2½	2	1½	2	1½	1½	1½		Sheats
Knie-touwen	4¼	4	4	3½	3	2½	2	3	2½	1½	1½		Clugarnets
Raks-touw	4	4	4	3¾	3½	3	2	3½	3	2½	2		Parrel rope

Thicknesses of the ropes of the Bowsprit.

	Mast of 34 inches.	Mast of 32 inches.	Mast of 30 inches.	Mast of 29 inches.	Mast of 28 inches.	Mast of 27 inches.	Mast of 26 inches.	Mast of 24 inches.	Mast of 23 inches.	Mast of 19 inches.	Mast of 13 inches.	Mast of 12 inches.	
	1 in.	2 in.	3 in.	4 in.	5 in.	6 in.	7 in.	8 in.	9 in.	10 in.	11 in.	12 in.	
De schenckels van de schoot	6	6	5½	5½	4½	4	3½	5	4½	4	3	2½	Pennants for sheats
Schoot	4½	4½	4	3	3	2½	2½	4	3½	3¼	3	1½	Falls for the same
Knie-touwen	3	3	3	2½	2	2	2	3	2½	2	1½	1½	Clulines
Blinde-takel	3½	3	3	3	2½	2½	2	3	3	2½	1¾		Garnets
Bras-takels	4	4	3	3	2½	2½	2½	3	2½	2	1¾	1½	Pennants for braces
Brassen	2½	2½	2½	2½	2	2	2	2½	2	1¾	1¾	1½	Falls for the same
Takel	4½	4	3½	3	3½	3	2½	3½	3	2¾	2	2	Halliards
Reeps	7	6	6	6	5	4½	4	6	5	5	4½	2	Tye
Buik en Nock-gordens	3	3	2½	2½	2	2	1½	2	2½	2	1½		Bunt and leech lines
Groot blinde schoot	6	6	5½	5¼	5	4	3	3	3	3	2½	2½	Spritsail sheat
Blinde toppenants	4	3	3	3	2½	2	2	3	3	2½	2	1½	Spritsail lifts

Thickness of the Ropework of the spritsail topmast.

	1 in.	2 in.	3 in.	4 in.	5 in.	6 in.	7 in.	8 in.	9 in.	10 in.	11 in.	12 in.	
Puttingh-touwen	3	3	2½	2½	2	2	1½	2½	2¼	2	1¼	1	Shrouds
Taeli-reeps	2	3¾	1¾	1½	1	1	1	2	1½	1	1	1	Laniards
Bras schenckels	2½	2½	2	2	1½	1½	1½	2	1¾	1½	1½	1	Pennants of braces
Brassen	2	1½	1½	1½	1½	1	1	1½	1	1	1	1	Falls for them
Rack	3	3	3	2½	2	2	1½	2	1½	1½	1¼	1	Tye
Boven blinde val	2	2	2	2	1½	1½	1½	2	1½	1¼	1	1	Halliards
Knie-touwen	2½	2	2	2	1½	1½	1	2	1½	1½	1	1	Clulines
Schenckels	3	3	2¼	2	1½	1½	¾	½			¾		Pennants for the backstaies
Scheerlijn	2	2	1¾	1½	1	1	1						Falls for the same
Reeps	2	2	1½	1½	1½	1½	1	1½	1½	1	1	1	Lifts
Want-puttens	3	3	2½	2	2	2	2	2½	2	2	2	1	Puttocks
Parel racktouw	2	2	1¾	1½	1½	1½	1	1¾	1½	1½	1	1	Parrel line

FIGURE 2.228. Rope thicknesses by rigging category (p. 123). Witsen's original table has been digitally altered to show Hayward's terminology in the source table (p. 1). The column heading "Mast of 27 inches" corrects a misprint ("Mast of 20 inches") in the original text. (Scanned image modified by Emiel Hoving)

FIGURE 2.229. (*facing page*) Rope thicknesses by rigging category (p. 124). Witsen's original table has been digitally altered to show Hayward's terminology in the source table (p. 11). The column heading "Mast of 27 inches" corrects a misprint ("Mast of 20 inches") in the original text. (Scanned image modified by Emiel Hoving)

Mizzen Ropework.

	Mast of 34 inches.	Mast of 32 inches.	Mast of 30 inches.	Mast of 29 inches.	Mast of 28 inches.	Mast of 27 inches.	Mast of 26 inches.	Mast of 24 inches.	Mast of 23 inches.	Mast of 19 inches.	Mast of 13 inches.	Mast of 12 inches.	
	1	2	3	4	5	6	7	8	9	10	11	12	
	in.	in.	in.	in.	in.	in.	in.	in.	in.	in.	in.	in.	
Schenckels	5½	5	4½	4									Pennants of tackles
Standers	4	3½	3½	3									Runners
Vallen	3	3	2½	2½									Falls of tackles
Want-touwen	5½	5	4¾	4½	3½	3	2½	4	4	3	2	2½	Shrouds
Tali-reeps	3	2½	2½	2½	2	2	1½	2	2	1½	1½	1½	Laniards
Rack	7	6½	6	5	5	4½	4						Tye
Val	5	4½	3½	3	2½	2½	2	3½	3	2½	2		Halliards
Stach	6	5	4	3½	3½	3½	3	4	4	3	2½	2½	Stay
Tali-reeps	3½	3	2½	2½	2	2	1½	2½	2	1½	1	1½	Laniard
Schoot	4	3½	3	2½	2½	2½	2	3	3	2½	2	1½	Sheat
Valletje	5½	5	4	3½	3½								Jeer
Scheerlijn	3½	3	2½	2½	2	2	1	2½	2	2	1½		Trufs
Toppenands	4	3	2	2	2	2	1¾	2½	2	1½	1½	1½	Bowlines
Knie-touwen	2½	2	2	2	2	2	1½	2	1¾	1½	1½	1	Brails
Raksperel-tou	5½	5	4	3	3	3	2½	3½	3	2½	2	2	Parrel ropes

Ropework of the Crossjack.

	1	2	3	4	5	6	7	8	9	10	11	12	
Toppenants	4	3½	2½	2½	2	1½	1	2½	2½	2	1½	1	Lifts
Braszen	2½	2½	2	2	1½	1	1	2	1¾	1½	1	1	Braces
Schenckels	3½	3	2½	2¼	2	1½	1¼	2	1½	1¼	1	1	Pennants
Val	4	3½	3	3	2½	2	1½	2	1¾	1½	1½	1½	Halliards

Thickness of the Ropework for the Mizzen Topmast.

	1	2	3	4	5	6	7	8	9	10	11	12	
Schenckels	3½	3	2½	2½									Pennants of tackles
Vallen	2½	2¼	2	2									Falls of tackles
Wand	3	3	2½	2½	2	1½	1	2	2	2			Shrouds
Taeli-reeps	2	1½	1½	1½	1½	1	1½	1	¾				Laniards
Want-puttens	3½	3	2½	2½	2	1½	2	1½	1½	1			Puttocks
Schenkel braszen	2½	2½	2	1½	1½	1½	1	2	1½	1¾			Pennants of Braces
Braszen	2	2	1¾	1½	1	1	1	1½	¼				Braces
Boelijns	2½	2	1¾	1½	1¼	1	1	1½	1	¾			Bowlines
Kruis-ſprie-touwen	2	2	1½	1¼	1	1	1	1	1	¾			Bridles
Kruis-knie-touwen	2	2	1¾	1½	1½	1½	1	2	1½	1½	1	1	Clulines
Rack	3	3	3	2½	2½	2½	2	3	2	2	2	1½	Tye
Valletje	2½	2½	2	2	1½	1½	1¼	2	1½	1½	1½	1	Halliard
Toppenants	2	2	1¾	1½	1½	1½	1	1	1	1	¾	1	Lifts
Parel rack-touwen	3	2½	2	1¾	1¾	1½	1½	2	1½	1½	1	1	Parrel ropes
Staghlooper	2½	2½	2										Runner
Schinckel op de ſtag	3	3	1¼	2	1½								Pennant on the stay
Kruis-zeil reep	4	3½	3½	2½	2	2	1½						Top-rope
Backſtagh ſchenkel	3	2½	2½										Pennant for Back-staies
Een Garnatje	2½	2	1¼										Falls
Voort kruiszeils wand.													

Thickness of the Ropework to the main Topgallant Mast.

	Mast of 34 inches. 1 in.	Mast of 32 inches. 2 in.	Mast of 30 inches. 3 in.	Mast of 29 inches. 4 in.	Mast of 28 inches. 5 in.	Mast of 27 inches. 6 in.	Mast of 26 inches. 7 in.	Mast of 24 inches. 8 in.	Mast of 23 inches. 9 in.	Mast of 19 inches. 10 in.	Mast of 13 inches. 11 in.	Mast of 12 inches. 12 in.	
Schenckels	3	3	2½	2	2	2	2					1½	Pennants of tackles
Vallen of tacckels	2½	2	3¾	3½	3	3	1¾					1½	Falls of tackles
Wand	3	3	3¾	3½	3	3	2¾	2	2	1¾	1½	1	Shrouds
Taelireeps	1¾	1½	1½	1½	1½	1½	1	1	1	1	1	1	Laniards
Bram puttens	3½	3	2½	2	2	2	1½	1½	1½	1	1	1	Puttocks
Bakſtagh ſchenckels	3	3	2½	2	2	2							Pennants of Backstaies
Scheerlijn	2	2	1½	1½	1	1							Falls of back staies
Stagh	3	3	3	2½	2	2	1½	2	1½	1	1	1	Stay
Taelie	2½	2½	2½	2	2	2	1½	1¼	1¼	1	1	1	Laniard of the Stay
Braszen	2	1½	1¼	1¼	1¼	1	1	1	1	¾	¾	¾	Braces
Schenckel braszen	2½	2½	2	2	1½	1½	1	1	1	1	1	1	Pennants for braces
Boelijns	2½	2	1¾	1½	1½	1½	1	1	1	¾	¾	¾	Bow-lines
Spriet-touwen	2	1½	1¼	1¼	1	1	1	1	1	¾	¾		Bridles
Reeps-touw	4½	4	3½	2½	2	2							Top-rope
Paerel racks-touw	2½	2	2	2	1¾	1½	1	1½	1	1	1	1	Parrel-ropes
Rack	3	3	3	2½	2½	2	1¾	2½	2	2	1¾	1¼	Halliards
Val	3	2½	2	2	1¾	1½	1	1½	1½	1½	1¼	1¼	Tye
Toppenans	2½	2	1½	1½	1	1	1	1	1	1	1	1	Lifts
Vlag ſtaf-ſtagh	2½	2	1½	1½	1	1							Flag-staff stay
Knie-touwen	2	2	1¾	1½	1½	1	1	1½	1½	1¼	1	1	Clulines

Thickness of the Ropework of the Fore Topgallant Mast.

	1	2	3	4	5	6	7	8	9	10	11	12	
Rack	2½	2½	1¾	1½	1½	1½	1¼	2½	2	1½	1¼	1¼	Parrel
Valletje	2½	1½	1½	1	1	1	1	1½	1½	1	1	1	Halliard
Boelijns	2	2	1½	1½	1¼	1¼	1	1	1	¾			Bow-lines
Spriet-touwen	1¾	1½	1½	1½	1½	1¼	1	1	1	¾			Bridles
Stagh	2½	2½	2	2	1¾	1½	1	1½	1½	1			Stay
Want	2½	2½	2	2	1½	1½	1	2	1½	1			Shrouds
Taeli-reeps	1½	1½	1½	1	1	1	1	1¼	1	¾			Laniards
Parel rak-touw	2	2	1½	1½	1	1	1	1½	1½	1			Parrel brace line
Knie touwen	2	1¼	1½	1½	1¼	1¼	1	1½	1½	1			Clugarnets
Braszen	2	1½	1¼	1¼	1¼	1	1	1	1	¾			Braces
Toppenants	2	1¼	1½	1½	1¼	1	1	1	1	1			Lifts
Reeptouw	3½	3	3	2½									Top-rope
Schenckels	3	2½	1½										Pennants
Vallen	3¼	2	2										Laniards
Puttens	3	2½	2½	2	1¼	1½	1½	1	1	1	1		Puttocks
Back-ſtaggen	3½	2½	2										Backstays

127. Ropes for the Rigging

As stated in the previous section, the tables for the sizes of rigging parts (figs. 2.226 through 2.230) have been "borrowed" from Edward Hayward. This is confusing because there were many differences between English and Dutch rigging practices. For example, Dutch yards were hoisted using a ramshead, English yards were hoisted with jeers. Witsen must have been aware of this, and sometimes he changed the terms (and sometimes not), which explains why he did not adopt some of Hayward's rope sizes. Furthermore, Witsen's tables are somewhat odd because it was highly unusual for Dutch shipbuilders to work from tables. Rather, they stuck to their formulas, as tables might make their work subject to interference by commissioners and, in the case of the Dutch Admiralties, by high-ranking officers. In contrast to English practices, the Dutch shipbuilders were kings of their profession, and they allowed no interference in their work. If a ship turned out to be a bad ship, they simply stated that it had "fallen off the axe that way"—and there was no changing that.

As Witsen borrowed his list of ropes from Hayward, it seemed appropriate to show his translation of the English terms along with Hayward's terminology (in its original spelling), which has been added to the original tables in a right-hand column. I realize, of course, that the two systems were not completely identical and that comparing them may lead to wrong assumptions.

In Witsen's tables the dimension in the middle column heading, "Mast of 27 inches," was originally "20 inches"—which violated the incremental progression of the other mast dimensions. Although Witsen may have meant to place this column between the columns for 19-inch and 23-inch masts, the data seems to fit best in this location, so "20" is likely a misprint; it has been changed here to "27" (a logical progression from 26 to 28).

The drawing in figure 2.231, which is not complete, indicates the position of most of the ropes. At first glance, the rigging seems a chaotic whole. Yet it was a tightly organized system in which few fundamental changes occurred between the seventeenth and nineteenth centuries. One exception is the disappearance in the 1620s of most of the complicated and decorative crowfeet and bridles, which are so apparent in Hendrik Cornelisz. Vroom's paintings.

Rope thicknesses were often given in circumference, measured in inches. The diameter is obtained by dividing the circumference by 3.14 (pi). The length was given in fathoms (1.698 meters).

In the following overview of the rigging for Witsen's pinas, the parenthetical numbers refer to my rigging plan for the pinas model (see drawing 1 in appendix).

The ship's rig has three masts and a bowsprit. The main mast (205) and foremast (219) are lengthened with the main topmast (206) and the fore topmast (220), and again with the main-topgallant mast (209) and the fore-topgallant mast (223). Above these flagpoles are carried (248 and 249). The mizzenmast (236) was lengthened with the mizzen topmast (244), topped with a flagpole (258). The bowsprit (226) had a small vertical mast at the end, called the spritsail topmast (233), with a flagpole (251) on top.

The connections between masts and topmasts, and between topmasts and topgallant masts, consisted of the trestletrees (207 and 221) and the mast caps on the mastheads (204 and 211). These enabled the topmasts to be struck or housed (i.e., to be lowered when circumstances demanded).

The yards were named after the mast to which they were attached and the sail they carried. The main yard (252) was hung on the main mast, the main topsail yard (253) on the main topmast, the main-topgallant yard (254) on the main-topgallant mast. At the foremast we find the foreyard (255), at the fore topmast the fore topsail yard (256), and at the fore-topgallant mast the fore-topgallant yard (257). The mizzenmast carried the lateen yard (258; the mizzen was a lateen sail) and the crossjack yard (259), which had no sail but was used to spread the clews of the mizzen topsail, which hung from the mizzen topsail yard (260). Below the bowsprit was the spritsail yard (261), and the spritsail topmast carried the spritsail topsail yard (262).

The rigging, as stated before, had two functions: to stay or hold up the masts and to handle the yards and sails.

The "standing" part of the rigging stayed the masts: shrouds, stays, backstays, and pendants for the tackles. These ropes were usually tarred. Furthermore, to protect them from chafing and moisture, they were often served (i.e., wound with rope, spun yarn, or houseline). Before serving, the rope had to be wormed and parceled: worming consisted of filling in between the strands of the rope with worming thread or houseline to obtain a more even surface; parceling involved wrapping the wormed rope

Figure 2.230. (*facing page*) Rope thicknesses by rigging category (p. 125). Witsen's original table has been digitally altered to show Hayward's terminology in the source table (p. 9). The column heading "Mast of 27 inches" corrects a misprint ("Mast of 20 inches") in the original text. (Scanned image modified by Emiel Hoving)

FIGURE 2.231. Plate XLV

(61 | 20) Here follows *the proposed Ship, on the page plate as a whole,* which above is shown in its parts, the *Ropes* pointed out with letters, the standing as well as the running rigging, names of the masts, blocks, sails, and their appurtenances.

a Are the mizzen shrouds, b the main shrouds, c the fore shrouds, [d is missing in Witsen's text] e the mizzen topmast shrouds, f the main topmast shrouds, g the fore topmast shrouds, h the sprit topmast shrouds.

i Mizzen stay, K main stay, 1 forestay, 11 lanyards.

On large ships, a rope with knots can be found stretched forward on the bowsprit, which goes through the top and is tied to the bow, along which the Sailors climb up.

m Crossjack lift, n main lift, o fore lift, p the mizzen topmast lift or mizzen topsail yard lift, q main topsail yard lift, r fore topsail yard lift, s spritsail topsail yard lift, 31 the spritsail yard brace.

The crossjack yard is the yard underneath the mizzen top: Called begijn, because it is of as little use and as little service, as the womenfolk, found among the Roman Catholics, which are called Beguines, because no sail is bent to this yard, for which it is actually designed, but it is only applied to keep the rigging above out of the way.

Lifts are that with which the yards are leveled, or cockbilled, when sailing close-hauled.

t Is the mizzen topsail brace. v the main topsail brace, w the fore topsail brace, x the spritsail topsail brace, y the crossjack brace, z the main brace, ef the fore brace. Braces are the ropes with which the sails are turned, when lying on the mast.

1 the mizzen yard lift, 2 the main [topmast] backstay, 3 the fore topmast backstay, the backstay prevents the topmast from breaking, when running before the wind, instead of the stay.

4 the spritsail topsail crowsfoot, 44 the mizzen topmast crowsfoot.

5 the Mizzen topsail bowline, 6 the main topsail bowline, 7 the fore topsail bowline, with its bridles, also called the skinny man. With bowlines the sail's sides are trimmed, when close-hauled.

8 The Mizzen topsail sheet, 9 the main topsail sheet, 10 the fore topsail sheet, 11 the spritsail topsail sheet, 50 the spritsail sheet, 27 the foresail sheet, 23 the main sheet. The main [topsail] sheet goes from the ends of the main yard, through the topsail sheet block and through the single sheet block, it is belayed to a knee or pin, and used to haul the sails, and to ease, like the other sheets.

Spritsail yard braces are lines, with which the spritsail is raised, when hauling close.

T is the puddening, or the serving around the anchor ring, making it easier to handle, and to secure.

25 Is the main tack, 28 the foresail tack.

20 The mizzen clew lines [actually brails], 21 the main topsail clew lines, 24 the mainsail clew garnet, 26 the foresail clew garnet. These are on the inside, to clew up the sails.

32 the Main topmast stay, 33 the fore topmast stay. At the forward side of the foresail is a buntline, to enable taking in the sail.

11 is the mizzen halyard, 13 the main topsail halyard, 14 the fore topsail halyard, 15 the spritsail topsail halyard, 17 the mainsail halyard, 18 the foresail halyard: these are used to haul up the sails, and to drop them again.

30 spritsail preventer braces

51 The standing lift, which does the same work as the spritsail lifts, but is attached to the yard, and the bowsprit.

19 Is the spritsail halyard, 45 the anchor stock. 29 The mizzen sheet. 46 The flagpole. 43 The main sheet block. 42 A ring around the pump dale. 38 Mizzen deadeyes. 39 Mainmast deadeyes. 40 Fore deadeyes. [*continued on facing page*]

35 Mizzen chain plates, 36 mainmast chain plates, 37 fore chain plates.

41 The chesstree. 34 Gammoning around the bowsprit.

52 Spritsail brace. 53 Foresail sheet block 54 The lizard; this leads forward to the bowsprit, as do both spritsail sheet blocks; with these the spritsail is boarded, thick about 5 inches, goes with the other end through a sheet block, attached to the spritsail sheet or pendant. Is belayed in the shrouds, or below the channels.

55 Is the cathead; with which the anchor is catted, after it has been wound up, with the cable.

47 Mainstay block, 48 forestay block [these are hearts or dumb sheaves rather than blocks].

Gaskets are used to furl the sails. At the top of the topsail there are spilling lines.

with strips of tarred sailcloth. Sometimes parts of served rope were covered with leather.

The masts were stayed with stays forward; the stays were named after the masts to which they were attached: main stay (1), forestay (2), mizzen stay (3), main topmast stay (4), fore topmast stay (5), mizzen topmast stay (6), main-topgallant stay (7), fore-topgallant stay (8), and, in exceptional cases, the backstay of the spritsail topmast (9), staying the spritsail topmast to the forestay. The stays were set up taut with lanyards in the hearts, or dumb sheaves (blocks without sheaves), similar to the deadeyes, at the ends of the shrouds.

To the sides and aft the lower masts were stayed with shrouds (10, 13, and 16), which had ratlines for going aloft. The topmasts had shrouds with ratlines as well (11, 14, and 17), fastened through the tops to the lower mast shrouds. They also had backstays (21, 22, and 23) going aft to the ship's topsides, where, like the lower shrouds, they were set up with lanyards and deadeyes to the channels and chain plates. The topgallant masts also had shrouds (12 and 15), spread by the crosstrees and made fast to the topmast shrouds via the futtock shrouds, but generally no ratlines. Sometimes topgallant masts also had backstays.

Although standing rigging refers to the ropes with which the masts were stayed, the pendants of the pendant tackles (24, 25, and 26) are also included in this category. These were found underneath the mastheads. The lower pendant tackle blocks were hooked to eyebolts in the channels when the pendant tackles (or burtons) were not being used for any other purpose. In general there were two pendant tackles to each side of the main and foremast: one with a fiddle block (the sheaves one above the other) and one with a double block (sheaves lying next to each other). The mizzenmast usually had only one pendant tackle to each side. Tackles could also be stropped to stays and the arms of the main and foreyards and the lateen yard.

The running part of the rigging, including the line rove through the blocks of the pendant tackles, was mainly used to handle the sails. It consisted of halyards, halyard ties, lifts, clew lines, brails, sheets, braces, tacks, bowlines, and other gear. These ropes remained untarred and so had the color of hemp fiber. They were rove through blocks (over sheaves) to gain mechanical advantage. The falls were belayed to cleats, belaying pins, and knightheads on deck or in the tops.

The running rigging can be divided into two main groups of lines: those attached to the yards and those attached to the sails.

The yard was raised with a halyard. The halyards of the main yard and foreyard had a huge block, called the ramshead, which had several sheaves. The halyard was rove through this block and a knight, which protruded through the deck. From the block both ends of a rope called the tie (27 and 47) went up over the mast cap and were bent to the yard. Smaller halyards, such as the ones of the topgallant yards (55 and 61), had a single halyard falling aft of the topgallant mast and through a sheave in the top of the topgallant mast. The spritsail had a halyard rove through the forward end of the bowsprit.

The yard was held to the mast by the parrel. The parrel consisted of lines rove through trucks and ribs, the ends of which were bent to blocks (28 and 48); the parrel fall rove through these blocks belayed at the foot of the mast.

The lifts (29, 42, 49, 62, 74, 86, and 92) are used to hold the yards level when they are lowered and the sails furled, and they can be used to cockbill the yards. The lower sails (mainsail and foresail) had their lifts rove through their own sheaves in the same fiddle block as the topsail sheets. Often the topsails did not have lifts because the sheets of the topgallant sails could be used for that end.

The braces (30, 36, 43, 50, 56, 63, 70, 75, 87, and 93) were used to turn the yards so the sails would fill with the wind. The brace block was usually attached to the yardarm with a pendant. Braces were belayed on deck, where they were passed via a number of blocks and fairleads.

The sails were bent to the yards with robands. When taken in, the sails were furled with gaskets, which were tied to eyes fixed onto the yard with cramps. The clews (lower corners of the sails) were hauled up to the yards by the clew lines (32, 38, 45, 52, 58, 65, 72, 77, 81, and 89), which hung behind the sail. Buntlines (33, 39, 53, and 59) were used for the same end, but they ran on the forward side of the sails and were attached to footropes. Leech lines, spilling lines, or throat brails were made fast to the leech ropes.

The sheets (31, 37, 44, 51, 57, 64, 71, 76, 82, and 88) were used to set the sails by pulling the lower corners (clews) of the sails out to the yardarms of the yards below them, or aft in the case of the fore and mainsail. The tacks (94) were used to pull the clews of the foresail and mainsail forward on the weather side. The bowlines pulled the weather leech forward. The bowlines were bent to bridles, which were made fast to the leech ropes of the sails (34, 40, 46, 54, 60, 66, and 90).

The mizzen had a couple of special lines: the mizzen yard lift (80), attached to the upper end of the lateen yard with a running bridle, and the mizzen heel tackle (84), attached to the heel of the mizzen yard, with which the yard could be hauled to either side. The mizzen sheet (82) was a tackle attached to the knee of the flagpole on the poop.

The spritsail yard had a standing lift (69) on both sides, consisting of a lanyard between two deadeyes, one attached to the bowsprit and the other to the yard. It also had spritsail lifts, serving as spritsail topsail sheets at the same time, and braces, with which the yard could be braced when sailing close-hauled. The yard had a more or less vertical position on those occasions.

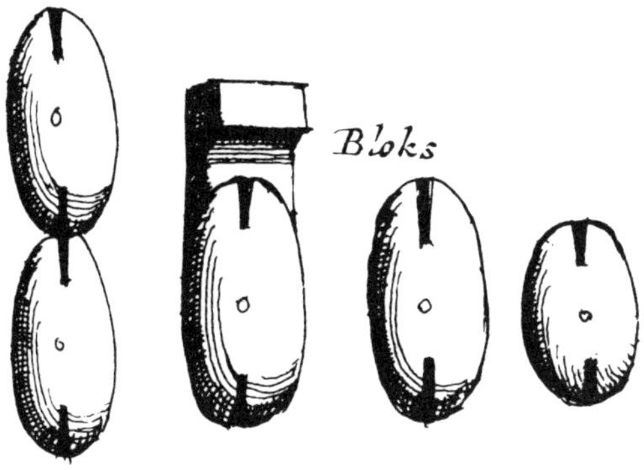

Figure 2.232. Plate XXVI. Blocks.

128. *Blocks*

(64 II 22) These blocks are made in varying shapes [see fig. 2.232], and each one according to its use. Some are round, others elongated, in some one sees 2, 3, to 4 sheaves next to one another. Others have two sheaves one above the other, making a double block that way. Others are running. Others are made fast. Most of them are attached, at the top or the bottom, at both ends sometimes, and the smallest are suspended with a score, which is cut in its sides. At the top and the bottom, on each side, one can see grooves on the outside, in which the rope that holds the block is laid. Some are also affixed with hooks. The sheaves are made of the heaviest and strongest wood, and sometimes of copper.

. . .

(97 I 9) I think it is of no use to describe the right places, in which the blocks are to come on the ship, because it is better shown, than described: for this the reader is advised to turn his eyes to the plate [fig. 2.231] inserted here before, on which this is shown: as they are placed in a true ship. Some blocks are round, others elongated, deadeyes are flat, and egg-shaped.

(62 II 14) *About the Blocks of our Ship.*

At the Mainmast there are:

32 Main deadeyes. These come to each side against the side of the ship: are attached to the channel, half in iron, half in rope, and are used to tighten the shrouds, and when they are slack, to lash them tightly, new shrouds generally slacken. Therefore they are to be hauled taut. [Witsen forgets to mention the main ramshead here.]

2 pendant blocks with 3 sheaves.

2 ——— With 2 sheaves.

2 Double tackle blocks with 2 sheaves. These are used to load and unload the ordnance, and are affixed above the upper deck. The main and fore sheets go through the topmast sheet blocks, hanging from the main yardarms. These deadeyes are broad 1 foot.

2 Topsail sheet blocks on the yardarm.

2 Below the yard.

4 Bowline blocks without sheaves.

2 With sheaves.

2 Small brace blocks.

4 Small clew-line blocks.

2 Small flat yard blocks.

4 Small blocks to the main stay.

2 Standing blocks or knights. Standing down on the upper deck; and are used to hoist the sail with the yard, for the mainmast as well as the foremast. But the knight at the mizzenmast, is also used to hoist and to veer out. The small standing blocks or knights one can see affixed to the main knight, are used for striking the topmasts with the top rope, and to haul the topsail halyard in and out.

2 Main blocks.

1 Parrel with 3 rows of trucks.

3 Reeving blocks with 3 sheaves.

3 ——— With 1 sheave.

1 top rope block.

1 Deadeye for the parrel.

48 Shroud bull's-eye fairleads. These are hanging in the main shrouds or fore shrouds, mostly for show, ropes go through them, like clew lines, braces, brails &c.

1 Snatch block.

8 Blocks for the crowfoot lines.

(63 | 14) *To the Foremast.*

28 Deadeyes.

2 Stay blocks.

1 Ramshead. These are hanging one below each top. Through these go the halyards, which also go through the knights, to hoist sail and yard.

2 pendant blocks with 3 sheaves.

2 ——— With 2 sheaves.

2 Tackle blocks with 2 sheaves.

2 ——— With 1 sheave.

2 Topsail sheet blocks on the yardarm.

2 ——— Below the yard.

1 Sling block.

1 top rope block.

1 Parrel with 3 rows of trucks.

3 Reaving blocks with 2 sheaves. These are found in the shrouds of both masts, and to the side of the mizzenmast, and are used, to transfer light wares.

3 ——— With 1 sheave.

1 For the topmast stay.

2 Lift blocks.

4 Brace blocks.

4 Brail blocks.

2 Flat on the yard. Are fixed to the yard, and are used to haul the yard backward, and to veer it out. [This is doubtful. The braces haul the yards aft (except mizzen braces); flat cheek blocks fastened to the yard would probably be for buntlines.]

4 Bowline blocks with 1 sheave.

2 Without sheaves.

2 Foresail sheet blocks.

6 Blocks for the crowfoot lines.

To the Mizzenmast.

16 Deadeyes.

1 pendant block for the parrel. Pendant blocks are hanging in the main or fore shrouds; or lie in the ship, as long as they are used, to hoist aboard the guns, or other heavy loads.

1 Block with 2 sheaves.

1 Ditto with 1 sheave.

4 Garnet blocks.

8 For the Clew lines. These are hanging from the yards, and are used, to take in the sails, when making to much speed, or when reefing.

4 For the braces.

4 For the lifts.

2 For the lifts on the yard.

1 For the sling.

1 For the top rope. There is one hanging from each mast cap, to house the topmasts.

2 Deadeyes to the stay.

A snatch block is fixed on the ship to the foremast; a hawser is rove into it, with which everything is unloaded from the ship, and with which also the shrouds are hauled taught, some also call them cat blocks, because it can be used here and there on the ship.

To the Bowsprit.

4 Deadeyes for the bowsprit horse.

4 Deadeyes for the standing lift.

1 Parrel.

1 Block for this with 2 sheaves.

1 With 1 sheave.

4 Clew-line blocks.

4 Spritsail yard lift blocks with 2 sheaves. A single hawser goes through this, and is there for splendor.

2 ——— With 1 sheave.

1 ——— With 1 sheave.

1 With 2 sheaves for the bowlines.

2 To the topmast stay.

2 Spritsail sheet blocks.

2 Spritsail sheet blocks, alias lizards.

1 Spritsail halyard block with 2 sheaves.

1 ——— With 1 sheave.

Blocks of the main topmast.

16 Deadeyes.

1 Parrel.

4 Clew-line blocks.

4 Pendant tackle blocks, are of the same use, as the swifter blocks. [This is remarkable, as topmasts don't have pendant tackle.]

(64 | 1) 4 For the braces.

2 For the buntlines.

6 For the bowlines without sheaves.

2 ——— With 2 sheaves.

1 Topsail halyard block with 2 sheaves.

1 ——— With 1 sheave.

4 For the backstays with 2 sheaves.

1 For pressing: this one stands on the second deck near the guns, in the middle of the ship, of great length, it has three sheaves, and when pressing, the capstan is used, pressing is also called wringing, and in such a way it is possible to stow once as much of a soft cargo, such as wool, cloth or yarn, as otherwise.

Blocks for the Fore topmast.

16 Deadeyes.

4 ——— For the backstays.

1 Parrel.

1 Topsail halyard block with 2 sheaves.

1 ——— With 1 sheave.

4 Lift blocks.

4 For the garnets.

4 For the braces.

2 To the topmast stay with 2 sheaves.

4 For the bowlines with 1 sheave.

4 Without sheaves.

4 Clew-line blocks.

4 Buntlines.

[Blocks for the fore-topgallant mast:]

2 For the tackle of the topmast stay with 2 sheaves.

4 For the bowlines with 1 sheave.

4 Without sheaves.

4 For the clew lines.

2 For the buntlines.

1 For the tackle of the topmast stay.

1 For the topgallant stay.

Blocks for the Main-Topgallant mast.

12 Deadeyes.

4 Lift blocks.

4 For the braces.

2 For the clew lines.

4 For the bowlines without sheaves.

4 ——— With sheaves.

2 Small blocks.

1 Parrel.

For the Mizzen topmast.

12 Deadeyes.

1 Top rope block.

1 Parrel.

4 Lift blocks.

4 For the braces.

4 For tackles without sheaves. With these the sails are hoisted in the middle, to slow the speed, and these are also called spruit blocks.

2 With 2 sheaves.

2 Halyard blocks.

2 For the mizzen yard lift.

6 Deadeyes for the bridles.

Blocks for the topmast on the Bowsprit.

12 Deadeyes.

1 Parrel.

3 Halyard blocks.

3 Small dumb sheaves for the stay, and 6 heads for the flagpole.

4 Lift blocks.

6 For the braces.

4 For the clew lines.

128. Blocks

The rigging plan (see drawing 4 in appendix) was reconstructed as much as possible with the help of Witsen's lists of blocks. From the number of blocks it was often possible to deduce the exact course of a rope, but in other cases this was not so easy, especially because the fashion changed several times during the seventeenth century.

I should point out that in Witsen's treatise the blocks of the fore-topgallant mast are listed as a continuation of the blocks of the fore topmast; I have separated them for clarity.

129. Sails

(135 II 8) *About the Sails in general.*

Concerning the dimensions of the sails, these are different, their width is conformed to the length of the Yards, and the height to the length of the Masts, and such in accordance with the master's judgment, some hanging loose, others taut and stretched. Fixed and invariable laws for their dimensions I have not found.

. . .

(64 II 42) The Sails are.

The mainsail. The foresail to the foremast. The Bonnet; attached below to the sails, to catch more wind. The main topsail: or the second sail, on the mainmast, counting from below. The fore topsail, or second sail on the foremast. The spritsail, the sail forward, underneath the bowsprit. The mizzen or sail aft, with its bonnet. Spritsail topsail, the sail on the bowsprit. The mizzen topsail; or the sail above the mizzen. The main-topgallant sail, or the topmost sail of the mainmast. The fore-topgallant sail, being the topmost sail of the foremast. Staysails: studding sails: fatzen en geiken [*fatsen en gijken*: small additional sails, mostly to fill the gaps between larger sails]: the last of which are sails which, in exceptional cases, either when in a hurry, or in a dead calm, or when chasing, are bent to the other sails.

The vinnetje, or jachtenduivel [no English equivalents] is a small sail, which can be hoisted in the poop flagstaff.

. . .

(139 | 29) Sailcloth, today, is spun from finely combed hemp. The dimensions of the sails are in some ways adapted to the use for which the ships are intended.

On Yachts, or ships which are out to chase others at sea, many sails are made.

. . .

(139 | 50) The sails should not be attached too tightly to the Yards, but in such a way that, the wind blowing into them, gives them a certain roundness; thus the wind is better caught in the Sail, or can pass more movement to the cloth, and the Ship will have more speed. Unless when sailing very close to the wind.

(135 | 10) *Content of Sails of my Ship.*

The Mainsail is broad 23 sail cloths, Dutch Canvas, the wood above at the Head rope 26 ells, and is deep up and down 14¼ ells, it can be carried with or without Bonnet.

The Foresail is 20 cloths broad, above at the Head rope 22 ells, and deep 11¾ ells, the Bonnet underneath is deep 3¼ ells.

The Main Topsail below 21 cloths, and above broad 12 cloths, the wood 13½ ells, and is deep 18 ells.

The Fore Topsail below 18 cloths, and above 10¼ cloths, the wood 11½ ells, and is deep 14½ ells.

The Spritsail broad 18 cloths, French Canvas, the wood 17¼ ells, is deep 8¼ ells.

The Mizzen with its Bonnet is 16 cloths below, Dutch Canvas, aft deep 20 ells, the tack forward of the nock 2½ cloths, the wood to the lateen yard 24 ells.

The Spritsail Topsail below 16¾ ells, above 8 ells, deep 10¼ ells.

The Mizzen Topsail below 13 ells, above 7 ells, deep 9 ells.

The Main-Topgallant Sail below 13 ells, above 7 ells, deep 9 ells.

The Fore-Topgallant Sail below 9 ells, above 6 ells, deep 7¼ ells.

About the Bolt Ropes.

The Main Bolt Rope (that is the rope in which the Sails stand), thick 3½ inches, it is made of three strands, and 102 yarns, long without Bonnet 32 fathoms.

The Fore Bolt Rope thick 3¼ inches, is made of 84 yarns, long 42 fathoms.

The Main Topsail's Bolt Rope thick 3 inches, of 72 yarns, long 34 fathoms.

The Fore Topsail's Bolt Rope thick 2½ inches, is made of 52 yarns, long 28 fathoms.

The Spritsail Topsail's Bolt Rope, of 18 yarns, long 18 fathoms.

The Main-Topgallant Sail's and Mizzen Topsail's Bolt Rope has 15 yarns, together long 30 fathoms: the Fore-Topgallant Sail's Bolt Rope is made of 12 yarns, and long 12 fathoms.

The Mizzen and Spritsail's Bolt Rope thick 2 inches, the strand of 36 yarns, long 46 fathoms.

129. Sails

The pinas had ten sails: two to the bowsprit and the mizzenmast, three to the foremast and mainmast. Witsen does not give dimensions for any staysails, studding sails, or other light-weather kites.

The sails were made from canvas (flax) cloths with a width of 1⅛ ells (79 centimeters), or a finished width of 75 centimeters between seam centers after they were sewn together. The topsails were lighter, made of Irish linen woven to a width of 1 ell (69 centimeters). The bolt rope was stitched around the sails to reinforce the fabric. (For a detailed discussion of sail dimensions, see "Contracts for Sails" in chapter 3.)

A bonnet was a piece of sail that could be latched to the foot of a sail with a system of loops to enlarge the sail area. It could also be quickly removed or unlatched to reduce the sail area.

The total sail area of Witsen's pinas was about 900 square meters.

130. Parrels

(275 | 46) The parrel, being a ring of trucks and ribs, holds the yard against the mast: it rises and lowers with it along the mast. During battle chains are used with mousings below the parrel, for safeguarding against the breaking of the chains: so that the yard does not come falling down.

130. Parrels

The parrel was a system of ribs, pierced wooden balls (trucks), and rope, with which the yard was held against the mast. The larger yards often had a parrel of three rows of trucks; the smallest could suffice with one row.

The large parrels needed two pairs of single blocks and a fiddle block to be hoisted, plus a *bijvoet,* a tackle with similar blocks. The falls of these tackles were belayed on cleats on the mast.

The spritsail had a truss, a simple noose of rope without trucks.

131. Lanterns

(274 II 3) The lanterns, which stand aft, are fixed, or loose: the fixed ones are ornamental, and therefore gilded, and splendidly adorned.

(275 II 26) Sometimes an arm is stretched out behind the poop, on which a lantern is set at night.

(281 I 40) 3 Lanterns to stand aft.

131. Lanterns

Witsen includes the number of lanterns for his pinas in the inventory list (see section 137, "Loose Implements") but does not provide any descriptive details. From other sources we know that these lanterns were made of wood with a copper top; the sides were usually thin sheets of horn or mica (so-called Moscovia glass). They cannot have given much light.

132. Guns

(282 II 21) *Following is the war equipment, which these hundred men are to be given; with which they could defend themselves adequately on a voyage as proposed.*

		pounds.
2	Metal Guns of 12 lbs.,	1525
	taper-bored, weighing	1572
4	Metal Guns of 12 lbs.,	2470
	home-bored, weighing	2540
		2480
		2470
2	Metal cannon drakes of 6 pounds,	670
	weighing	664
8	these Metal Guns would weigh together	14391 lbs.

4	Iron guns of 12 pounds, weighing	13600 lbs.
8	Iron guns of 8 pounds, weighing	24000 lbs.
4	Iron guns of 4 pounds, weighing	6000 lbs.
16	Iron guns weighing	43600 lbs.

All these are to lie on adequate Gun carriages, which, when sailing in peacetime, are tied to the ship's side with ropes, going through holes, which are in the middle. The heaviest guns are stowed in the middle of the ship, because it is able to carry more there.
(283 I 16)

250	Cannonballs of 12 pounds weighing	3000 pounds
200	of 8 lbs. weighing	1600 lbs.
50	of 6 lbs. weighing	300 lbs.
100	of 4 lbs. weighing together	400 lbs.
600	Cannonballs, weighing	5300 pounds.

A number of large and small stones, of which use can be made in different circumstances.

40	Bar shot of 12 pounds weighing	480 pounds
40	of 8 lbs. weighing	320 lbs.
20	of 6 lbs. weighing	120 lbs.
20	of 4 lbs. weighing together	80 lbs.
120	Bar shot, weighing	1000 lbs.

In the past steel bogen, Rondaszen en Panoizen [?], were added, but these are no longer used.

132. Guns

The number and specifications of the guns are part of Witsen's inventory list for the pinas (see section 136, "Loose Implements").

As is to be expected for a merchantman, most of the guns are made of iron. Men-of-war generally had bronze guns (Witsen calls them "metal guns"). These were stronger and lighter, allowing for a greater quantity and larger calibers, but were also more expensive.

The pinas had eight bronze guns and sixteen iron guns. The bronze guns were preferred over the iron ones in the vicinity of the compass to avoid unwanted magnetic influence. They were light and long and were used for chasing. The taper-bored guns (called such because of their tapering chambers) had relatively short barrels and occupied the aftermost gunports in the side of the ship. Although their reduced length meant a shorter range of fire, they had the advantage of being easy to turn forty-five degrees to fire aft through the stern gunports. The home-bored guns had uniformly cylinder-shaped chambers.

The four iron twelve-pounders were positioned in the second and third gunports of the lower deck, the eight-pounders at the remaining gun positions, and the four four-pounders in the forecastle behind the drakes. According to the inventory list, the ship carried three tons of ammunition.

133. *Flags*

(65 II 5) *Names of the Flags.*

The *flag at the top:* stands on the mainmast. On the foremast and mizzenmast flags are set, called after the mast on which they stand.

The *Jack,* stands forward, above the spritsail, on the topmast.

A *Tzjouw* [no English equivalent] is a furled flag.

Pennants, are long narrow flags, forked at the end.

The white flag means peace; the black mourning; the red war; and fighting spirit. With the blue flag, the general council is signaled or called aboard.

Streamers are small flags, which are cut or forked, and which in general are seen at the masts.

. . .

(269 II 13) The flags aboard, which in this country are tricolored, white, orange and blue, or also yellow all over: on Dutch ships, they are used to tell the direction of the wind, and also as a sign of recognition of the land folk, sailing the ship; furthermore the seamen can signify special meanings with them.

A furled flag at the rear, means that someone, who has left the ship, is called for, or that there is much need of some cause or other: and this is called a *Tzjouw.*

The flag half-mast means a dead man aboard [. . .].

When the ship's crew is revolting, and has taken over the ship, unwilling to recognize the officers

Figure 2.233. Port view of the pinas model. (Courtesy Cees de Jonge, The Visual Art Box)

(which sometimes happens on ships making long voyages, such as to the East Indies), they usually lower all flags, only leaving the jack, or flag of the spritsail topmast.

. . .

(272 II 15) On their voyages ships are usually given two flags, two jacks, two pennants and 6 streamers. Besides the usual flags named above, the Admiral is given a flag of 12 cloths, and one of 9 clothes, one or two jacks, and also one or two pennants, a white flag, a blue pennant, a red pennant, a yellow pennant; and such to signal.

(272 II 36) Of the mizzenmast, in merchantmen, one often finds flags, on which the hometown of the skipper is drawn: and on the foremast, the arms of the hometown, of the owners of the ship.

Flags are for decoration, for splendor, in which the Dutchmen can be said to be the least outstanding [. . .]

The content of the flags is without law, and everyone draws on it, what he likes.

. . .

(269 II 52) On the aforesaid ship, shown in this book, a flag flies in the top: which is done for decoration only, without intending to designate a rank.

. . .

(272 I 45) The Men-of-War of this country, of the first and second Rate [*Certer*], have flags of 15 cloths, long 18 ells, Jacks of 6 cloths, long 7 ells: long pennants of 25 or 30 ells: short pennants of 5 or 4½ cloths, and 4 ells long.

The ships of the third Rate, have flags of 12 cloths, long 15 ells; jacks of 6 cloths, long 7 ells: Pennants as above: Streamers of 4 or 3½ cloths, long 3 ells.

The ships of the fourth and fifth Rate, are the same, as those of the third rate, in flags, pennants, jacks and streamers.

The ships of the sixth rate have flags of 9 cloths, long 10 ells: jacks of 4½ cloths, long 5 ells. The pennants are 25 ells. The streamers are 3½ or 3 cloths, and 2½ ells long.

The ships of the seventh Rate have flags of 7½ cloths, long 9 ells: jacks of 3 cloths, long 4 ells: pennants of 25 ells, and streamers of 3 or 2½ and 2 ells long.

(282 I 14)

1	Red Flag of 8 cloths.
1	White Flag of 8 cloths.
1	States Flag of nine cloths.
1	States Flag of 7½ cloths.
1	States Flag of 6 cloths.
1	Jack of 4½ cloths.
1	Top spindle.
6	New Streamers.

133. Flags

Flags not only served to indicate the direction of the wind; they were also meant for ornamentation and to signal or salute. Dutch ships were not elaborately ornamented with carvings and paintwork, but the dimensions of their flags were enormous.

The flags were made of flag cloth, woven at a width of three fourths of an ell (52 centimeters). A nine-cloth flag was thus 4.68 meters wide and 5.85 meters long.

Besides the ordinary flags there were pennants (certain types of which were allowed only on the admiral's ship), vanes attached to a vane stock, and forked flags.

A flag could indicate a number of things: the wish to parley, a dead man aboard, and war or peace. A flag incident was the immediate cause of the First Anglo-Dutch War.

134. Paintwork

(274 II 48) The ship is painted with snakes and monsters to one's own taste. Each will choose the paint, which pleases him. Yellow and red are much chosen, but also white; above, to the sides, clouds are made, or a painted Heaven, a sea and all sorts of implements of war. The tampions, closings of the guns are made red or white. The Cabin is sometimes painted all over on the inside with present-day history. The port lids are red on the inside, and also the washstrakes and the waistcloths, which are stretched around the ship; when in battle, often cloth or baize is used therefore. Ships are also gilded and silvered: but this excess is practiced by foreigners, rather than by Dutchmen.

. . .

(267 II 18) Ships are caulked for the sake of durability: because thus the water does not creep through the seams and cracks. This is done with old oakum, hair and moss, which, after having been driven in, must be

well tarred. All corners and scarfs of the planks, must join well and close off.

The Dressing, with which the ships are smeared, is made of bad resin, cod liver oil and sulfur: it is thought no worm penetrates through this. Sometimes it is smeared with bad tallow, mainly when the ships do not go west. This white paste protects ships for a long time from green growth of dirt. If they want to go far west, a good doubling is applied to the bottom, in which innumerable nails are hammered: it is made of cow hair: and also thin lead or copper is applied in between: and all this to avert the pests, which eat away the wood.

. . .

(269 I 13) When all the clefts are well filled, the ship is tarred above the water; because tar will not hold in water. It would be a good thing if it were possible to plaster ships, inside as well as outside, to prevent dirt and water, and also against fire.

134. Paintwork

In the merchant marine of the seventeenth century, there were certainly no rules about paint colors, although a few generalizations are possible. Gold leaf was seldom applied; gold or yellow paint was more often used. Carvings were tinted with colors as natural as possible. Cheap and durable paints were favored.

Based on his analysis of paint fragments from the works of the seventeenth-century Dutch and Flemish masters, A. Martin de Wild concluded that the following pigments and colorants were used in paints between 1600 and 1700:[16]

White:	lead white
	chalk (plaster)
Blue:	natural ultramarine
	azurite (copper blue)
	smalt (pulverized glass)
Yellow	ocher
	massicot (lead oxide)
Red:	vermilion
	madder
Green:	terre verte (earth color)
	verdigris (copper green)
Black:	lampblack
	ivory black
	bone black

Model builders working to achieve historical accuracy should be aware that the following pigments did not exist in the seventeenth century (the date indicates the year when the pigment was first introduced, according to de Wild):

Zinc white (1780)
Cobalt blue (1810)
Synthetic ultramarine (1810)
Prussian blue (1700)
Naples yellow (1750)
Chrome yellow (1800)
Cadmium yellow (1820)
Scheele's green (1780)
Emerald green (1810)
Zinc green (1780)

Below the waterline the ships were smeared with a mixture of low-quality resin, cod liver oil, and sulfur, a coating intended to prevent the accumulation of algae and barnacles. The combination of the ingredients gave the hull a grayish white appearance. Above the waterline the woodwork was tarred with thin brown tar to protect against the elements, darkening the wood.

135. The Boat and Sloop

(169 II 29) Next the *Boats* appear [. . .]. With the size of the parts in this vessel I give their names only, without pointing them out in the Boat, which I judge unnecessary.

. . .

(173 I 34) When Sloops and all kinds of other small vessels are built, the planks are forced with props, without the need of tongs.

When the Boats are secured in the Ships, holes in the floor are made to get rid of water, which can be closed with stopples.

. . .

(174 I 41) A boat may be 7, 8 or 9 inches wider than the third part of its length. The Bottom is often found to be wide three fifth parts of the length, for every 10 feet of length one foot for the depth, being afore the mast beam, being also 4½ inches wider and one inch deeper than aft.

(172 II 13) *Arrangement of a Boat, according to my Father's settlement.*

An able boat will be long between stem and sternpost 32 feet, wide 8 feet 9 inches, the bottom long 25 feet 5½ inches, the bottom wide 5 feet 3 inches, thick 2 inches, the floor timbers are thick 2 inches, wide 3 inches, are separated 1 foot 5½ inches.

The stem is high 6 feet 5 in. rakes 4 feet 9 inches, upward wide 13 inches, downward wide 10 inches, and inside thick 4 inches, on the fore side of the foot thick 3 inches.

The Sternpost high 5 feet 9 inches, rakes 1 foot 10 inches, downward wide 2 feet, upward wide 1 foot, inside thick 7 inches, outside thick 1½ inches.

The boat is deep at the underside of the wale 2 feet 3 inches, wide there 8 feet 9 inches: eleven feet from the fore perpendicular comes the leeboard bollard; where it is deep 3 feet ½ inch, wide 8 feet 2½ inches.

On the aft bench, wide 7 feet 8 inches, deep 3 feet 1 inch, and measuring ½ foot from the sternpost deep 6 feet 6 inches.

The bilge futtocks are from each other 16 inches, thick 3 inches and ½, wide 3 inches.

The bollards from each other 6 inches, are on the stem 2 feet.

The Back Bench is wide 8 inches, thick 3 inches and its front side is 8 feet from the sternpost.

The Stringer is wide 12½ inches, thick 1¾ inches.

The Windlass is 1½ feet from the center, a bit backward in the stringer.

The Mast thwart thick 3 inches, wide 1½ feet, on both ends wide 13 inches, and is roundish on the fore side.

The hole in which the Mast stands is wide 6½ inches, and deep 6 inches.

The mast step is wide 4 inches and ¾.

The Tabernacle thick 1 inch.

The Keelson is wide in the middle 18 in., on the aft side wide 12 inches, thick 2 inches.

There are 18 Frames, and one on the aft side of the bollard, 19 together.

The ring on the Stem is wide 7½ inches.

Through every frame a Bolt goes into the Wale.

The middle of the Wale wide 6 inches, thick 4 inches, aft wide 2½ in., thick 3 inches, afore a little bit more, rises 2 feet forward, 2½ feet backward.

The Gunwale wide 4 inches, thick 3 inches.

The sheer rail thick 1 inch, between the two wales wide aft 2¾ inches, in the middle 3 inches and ½.

There are 7 pairs of row cleats, 5 behind the mast, 2 fore: the holes wide 7 inches. On the front side of the leeboard Bollard, there will be a staple, which holds the rope.

The lowermost gudgeon on the stern, comes one foot from the heel: the uppermost 13 inches below the top.

(173 II 18) *Contracts of Boats and Sloops, built by the renowned Shipwright Dirck Raven.*

A Keelboat.

Long 40 or 41 feet, wide 10 feet, deep at its highest side 4 feet: the stem high 6½ feet, rakes 7 feet, is thick 6 inches: the sternpost high 6¾ inches, rakes 1½ feet: the Transom long 6½ feet, the Tuck stands at 3 feet, the bottom is wide 7½ feet: the keel deep 6 inches, wide 9 inches, the Futtocks deep 4½ inches, are from each other 9 inches.

. . .

(138 II 42) *Spritsail*

A Spritsail on a boat of 18 or 19 feet, should be wide upward 3½ cloths, downward 4 cloths, deep 7¾ ells, and afore 6¾ ells. The Mizzen's lower ends reach a little bit past the Yard, which is to be attached even easier, also because the narrowness there needs no wood to stand stiff, but stretched enough by the rope and can be held. The holes in the Spritsails are to get rid of the water if the Bowsprit bucks too much and the Sail scoops water at stormy weather.

. . .

(169 II 40) After this follows the *Sloop,* with a most perfect drawing [fig. 2.235] of the same. Also some Contracts of boats and Sloops built by the renowned Master Shipbuilder here in town.

(173 I 43) *The Sloop.*

A Sloop long between stem and sternpost 42 feet, is wide 9 feet: its bottom is wide 7 inches, deep below the wale 2 feet, the rest deep 3½ feet: the Stem is high 5¼ feet it rakes 6½ feet: the Sternpost is high 5¼ feet, it rakes 2 feet: the Wing Transom is long 6 feet, it rises fore 2½ feet: the Futtocks thick 3 inches, on the floor and in the turn of the bilge thick 2½ inches, planked with wainscot.

The Mainmast long 24 feet, the Yard long 12½ feet, the sail deep 21 feet: the Foremast long 15½ feet, the Foreyard long 11 feet, the sail deep 10½ feet.

The shape of this Sloop is depicted on two plates [figs. 2.235 and 2.236], in five different ways.

. . .

(173 II 32) *A Rowing Sloop.*

Long 42 feet, wide 9 feet, deep on its Wale 3 feet, a topside on top of that 8 inches: the bottom wide 6½ feet, rises 5 inches: the stem high 5¼ feet, rakes 6½ feet: the sternpost high 5 feet, the Tuck is on 2½ feet, the Wing Transom long 6 feet: the Futtocks 3 inches square, are 10 inches apart: the Wale thick 3 and ⅕ inches, wide 5 inches: the planking of Königsberg Barge planks, except 2 strakes below [probably above] the Wale, made of wainscot; with zoom-werck [*zoomwerk*: probably painted with decorative motifs].

135. The Boat and Sloop

In the seventeenth century the boat and sloop were integral parts of every seagoing vessel. They were carried on every ship that operated independently—on deck or towed and sometimes even sailed separately. Witsen does not give formulas for their sizes, but Van Yk states that the length of a ship's boat is equal to width of the ship or one quarter of its length (p 271) and that the length of the sloop is four fifths of the length of the boat (p. 274). The length-width ratio according to Van Yk was less than 4 to 1.The boat on Witsen's pinas should be a little over 33 feet long (134 ÷ 4), a size very consistent with the data for the vessel he inherited from his father.

The boat or barge was a flat-bottomed vessel with a curved, raking stem and a straight sternpost. Very early types also had a curved sternpost. It had a flat bottom, the sides were clinker-built, and it could be both rowed and sailed, for which it carried a sprit rig, or sometimes a bezan rig, like the boat of the Hohenzollern model of the Dutch two-decker. In the midship a heavy windlass was mounted, with which, for instance, an anchor could be brought out. In the stem was a sheave over which a rope ran, attached to the anchor, which was hung underneath the vessel.

The boat was mainly used to transport goods and water casks. Ships in those days were seldom moored to a quay wall; they usually lay at anchor in a harbor. The cargo always had to be brought aboard by means of lighters and the ship's own boat.

Sloops were basically for transporting people. The vessel had some decorative touches, often in the form of *prinsenwerk*—a band of traditional red, white, and blue triangles between the wales.

The sloop had a round carvel-built bottom but was clinker-built on top. It could be rowed or sailed. In contrast to the boat, the sloop was taken aboard and kept on deck. In the case of the pinas, the sloop was 27 feet long (⅘ × 33.5), which is considerably shorter than the sloops in Witsen's illustrations.

136. Costs

(157 I 38) As for the costs and the money, for which such a Ship could be had, it cannot be discussed with certainty, because I lack experience thereof, and it is also a matter inconstant, varying with the availability of the workmen, price of the wood and other materials, necessary in shipbuilding, and not appropriate for this treatise, but as not to be silent about it altogether, I have followed a short cost register of a recently built Ship in this town.

Register of Costs of a Ship long 165 feet, wide 43 feet, depth 16 feet, above which 8 and then again 7 feet.

	guilders
The Keel of 4 pieces will cost	2000
The Stem	300
The Sternpost 120, the Fashion Frames 200, the Wing Transom 60, together	380
2 Transoms 80, the Broeck-stuck 15, the Stern timbers 36, together	131
7 Planks in the bottom, 4½ planks long, together	2460
5 Bilge Strakes of 5 gank [planks] are 50 planks	2100

95 Floors up to 40 gldrs. 200 Bilge Futtocks and floors up to 20 gldrs. come to	7800
220 Futtocks up to 18 gldrs.	3960
The Keelson 200, three bilge ceiling planks of 5 inches 600	800
For the Ceiling planks on the Bottom and in the Bilges	2200
For two Deck Clamps superposed 7 inches	500
13 Riders up to 45, and 26 Futtocks up to 30 gldrs.	1365
For 7 rider floors, 4 Stern Knees aft in the Crutch, up to 40 gldrs., and 18 Futtocks up to 30	980
30 Main Deck beams 85, and 60 Knees up to 60 gldrs.	6050
For chocks and waterways	450
For Ledges and carlings	150
For Beams and bilge ceiling planks	560
For Ceiling planks in between	560
32 Upper Deck beams 50 gldrs., 64 knees up to 25	3200
Two breasthooks forward in the bow.	150
6 knees aft to the Stern gallery	400
For Binding Strakes	160

4 Strakes and Side Planking 5 planks long, being 40 planks	1400
2 Frame wales thick 9 inches	1000
90 Deck beams up to 9 gldrs.	810
The Washboard	250
230 Top Timbers up to 15 gldrs.	3500
Wood for the Filling Strake and Gunports	300
One Wale Timber, with 2 widths of fillings	280
A Sheer Rail and Washstrake	200
For Railings and Wainscot	70
For Channels, Chesstrees and cleats	120
For the Beakhead	300
For wood for all the carvings, inside and out	400
For all the inside paneling, Bulkheads and Counter Timbers	150
Wood for rooms, cots, galley, etc.	1600
For two-inch deals	900
For 2½-inch deals	700
For 1½-inch deals and battens	700
38 Third-deck beams up to 25 gldrs. and 76 knees up to 8 gldrs.	1558
For Deck Clamps and Bilge Ceiling planks	296

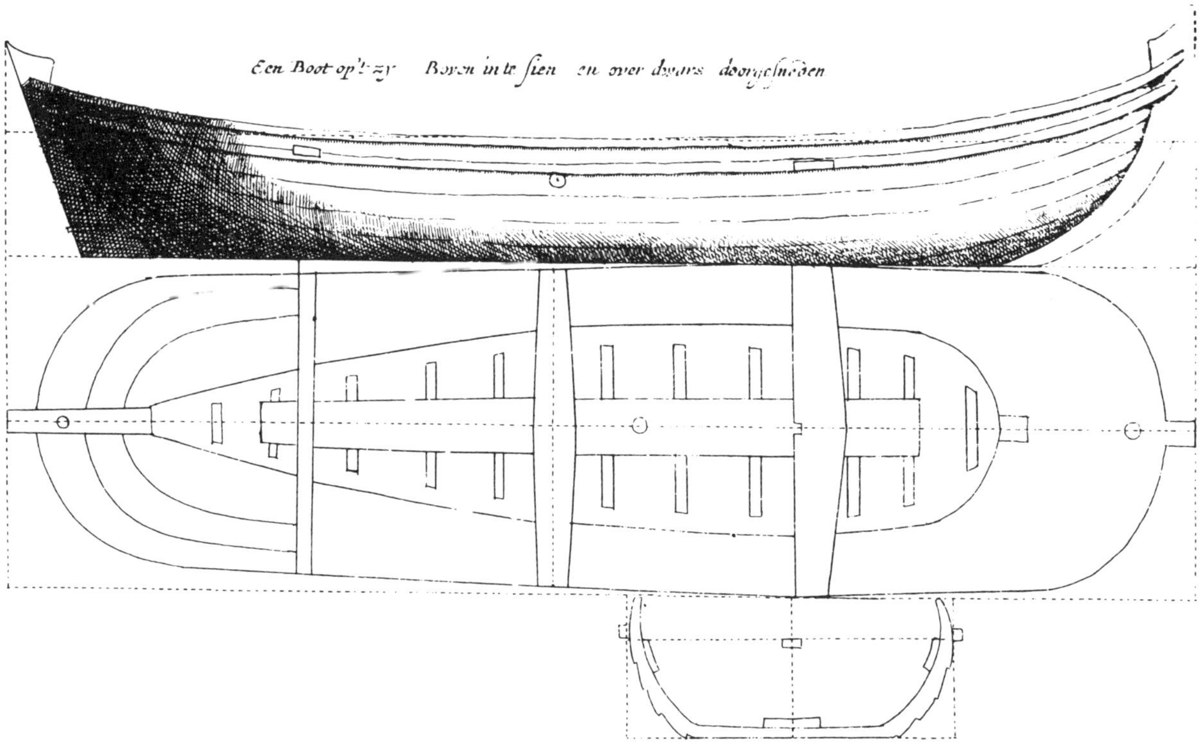

Figure 2.234. Plate LXVI. A boat to be seen from the side, from above, and in cross section.

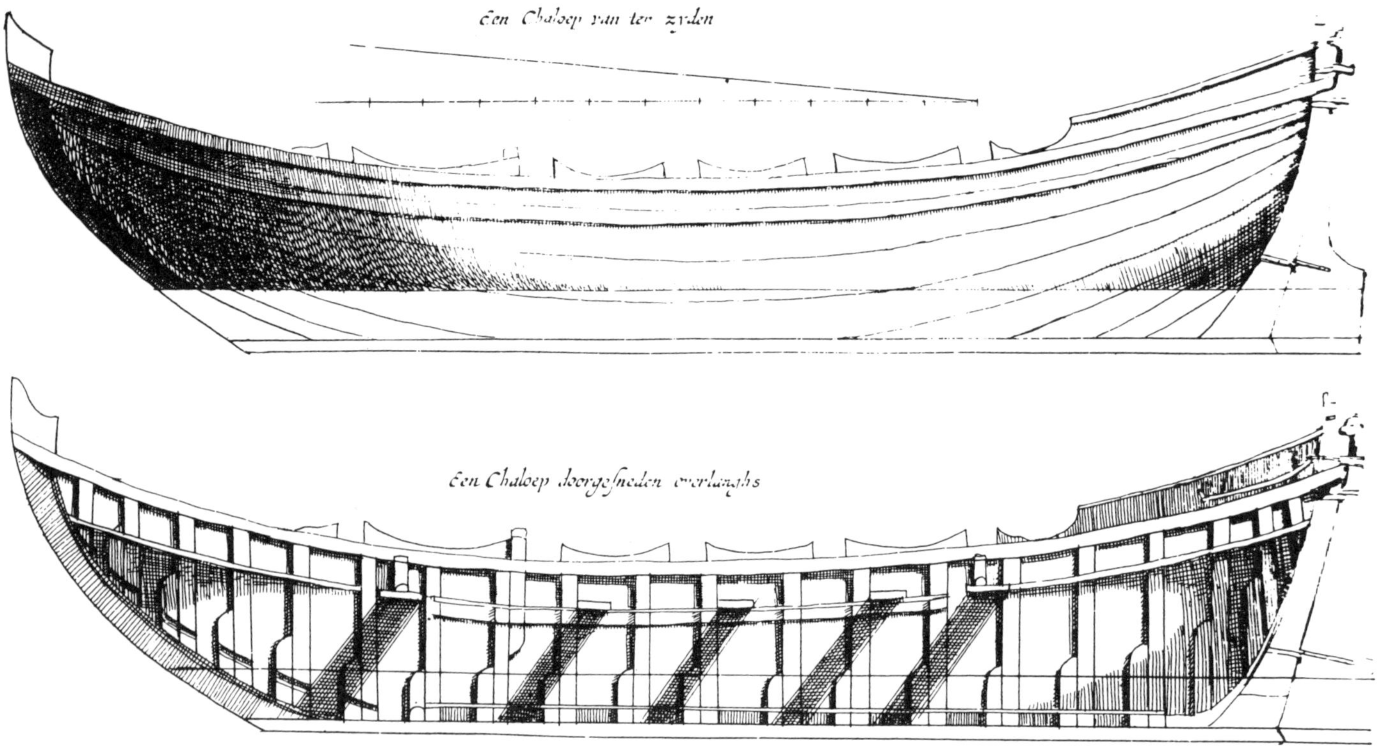

FIGURE 2.235. Plate LXVIII. A sloop seen from the side and a sloop in cross section.

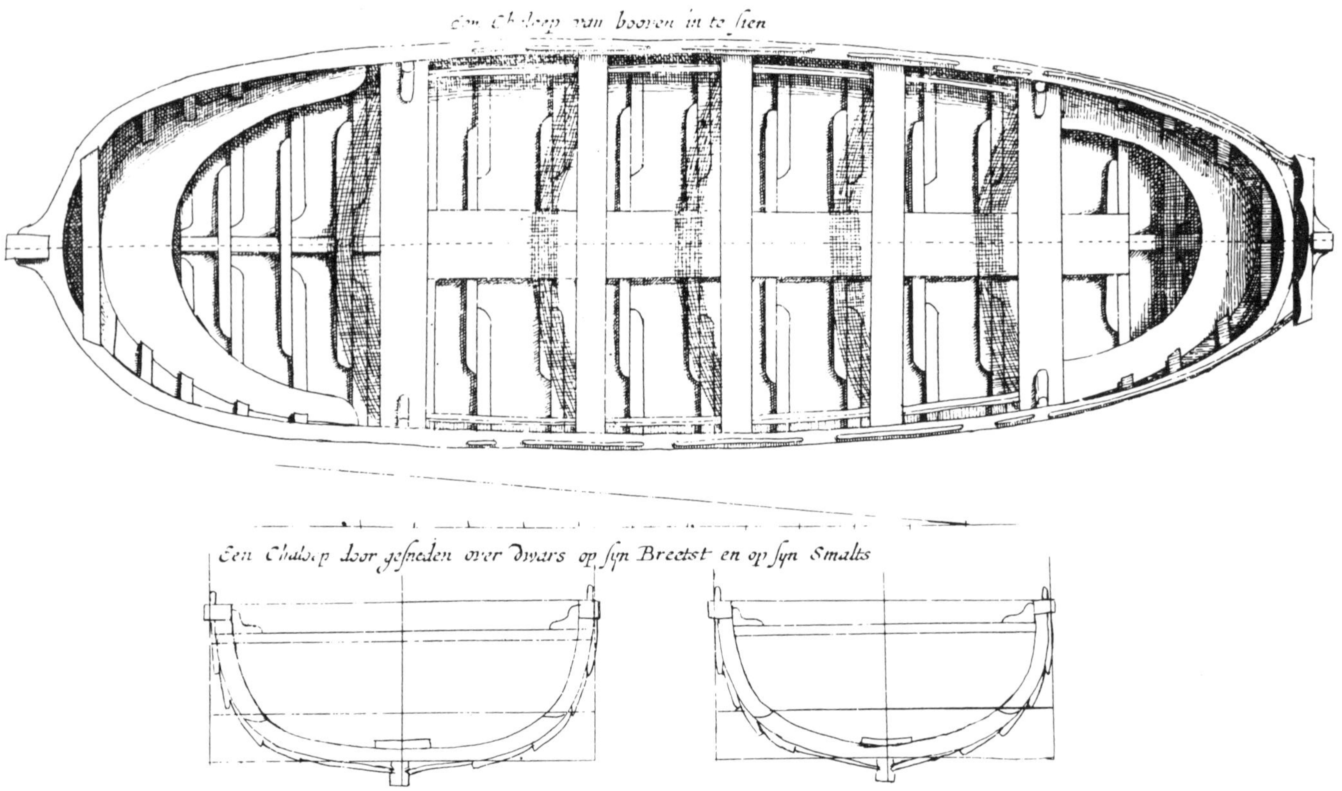

FIGURE 2.236. Plate LXVII. A sloop seen from above; section of a sloop at its widest and narrowest.

For binding strakes on the second deck, and Waterways 300
For the Gratings and Coamings. 150
For the Gratings above, and Coamings 190
For Waterboards and Bilge Ceilings 192
For Cabin Beams and Knees 120
For all knights large and small, 200
For all Bitts, bolsters, and bitt standards 200
For the shin-breaker and other Waterway. 60
For kevels, cleats, pinrails, and gunwales 400
2 Capstans and the Rudder 200
For pay of labor, 15000
For the Masts 4100
For a Frame-width of filling [?] 200
For Pitch, tar and oakum. 500
Treenails and scaffolding wood 600
Which costs together are 74152 glds.,

which were necessary for the woodwork of a Ship of the aforesaid dimensions, and so because is was made of the very best wood, because if one would take wood of lesser value and not the very best, a sum of 11070 guilders could be subtracted thereof. But apart from the prescribed wood, the ironwork which was applied into the Ship, comes to a cost of 7784.

The Cook's utensils, which will be required in a Ship of this size. 352
From the ropewalk one needs 35261 pounds of Rope, up to 45 gldrs. For 100 pounds together 5289
The Sails will at least cost 2827
The anchors weigh 6450 pounds together up to 3 farthings the pound, coming to 967
And for the trifles necessary on the ship a sum of 2264
So that the given Ship, apart from its war equipment, and victuals, before it goes to sea, will cost at least 93635 glds.

To which example, the costs of all others Ships can be guessed at.

136. Costs

Witsen, alas, does not supply us with the costs of his 134-foot pinas, but he does sum up the costs of a 165-foot ship: 93,635 guilders. Van Yk, who appropriated this list in his treatise, generally agreed with the costs for timber, although they could vary. For labor he arrives at an estimate 2,000 guilders higher than Witsen's, and for the ropework 6,000 guilders higher. For the anchors, blocks, and sails Van Yk also gives a higher price, driving the price of his ship up to 113,000 guilders: a difference of almost 20,000 guilders!

137. Loose Implements

(p. 280) *Loose Implements aboard the Ship.*

To this ship, which I have described, being 134 feet long between stem and stern, the following loose implements should be given along on a voyage, not being much shorter than that to the Far East, such as to *Curacao, Aleppo, Guinea* or elsewhere: which the men, being a hundred in number, which I trust will be enough for sailing and keeping the ship, can make use of at sea.

Cook's Utensils.

4 Kettles with their lids, weighing 208 pounds.
2 Copper cauldrons.
2 Flat pans.
2 Butter scoops.
1 Stewpan.
2 Serving spoons.
1 Butter spoon.
2 Skimmers, and oil and cotton thereto.
2 Gimbal Lamps.
4 Small Lamps.
6 Copper Taps.
4 Buttery Pumps.
4 Mauls.
1 Flint and steel.
1 Peppermill.
4 Wedges.
2 pairs of Dishes.
2 Pairs of Scales.
59 pounds of Weights.
2 Tongs.
2 Forks.
2 Two-pronged Forks.
1 Chimney Crook.
1 Bread Basket.
3 Augers
2 Guy ropes.
2 Fire trays.
1 Pair of Bellows.
2 Salt Funnels.
3 Vinegar Funnels.
2 Oilcans.
2 Wooden Lanterns.

2	Red Lanterns.
2	Small Lanterns.
1	Mustard stone.
1	Grill.
1	Trivet.
6	Pot Hangers.
10	Candle pricks [skewers or spits].
3	Large Scrapers.
8	Small Scrapers.
½	Piece of Mopping cloth.
2	Files.
25	Mengelen [1 *mengel* = 1.2 liters] of Linseed oil.
1	Barrel of Soap.
18	Casks of Lampblack.
10	Pounds of Red Lead.
8	Stone pitchers.
12	Small pots.
5	Rounds of Sulfur matches.
9	Brushes.
36	Hollow Troughs.
36	Flat Troughs.
48	Cups.
100	Small tablespoons
12	Dinner plates.
6	Wooden taps.
3	Ladles.
2	Mouden [possibly serving spoons?]
	Furthermore a parcel of floor thatch, moss, etc.
	Also some pounds of Spanish Soap, to wash the linen.
1	Spare Topmast.
1	Spare Boat Mast, and Sail Sprit.
1	Cheek.
5	Buoys.
30	Oars.
20	Handspikes and sticks.
4	Flagpoles.
6	Oars for the Barge.
4	Iron Sheaves.
	Capstan bars.
	Some pouring cans.

From the Ropewalk I would supply it well with:

		pounds
	Spun yarn	750
84	Bushels of Houseline.	84
84	Bushels of Marline.	42
3	Pairs of Sheets.	793
3	Pairs of tacks.	356
6	Iron straps.	1152
(p. 281)		
3	Anchor cables.	2615 lbs.
1	Hawser.	375
3	Lines.	375
1	Cargo gin tackle.	154
24	Bushel of White lines.	71
8	Lead lines.	55
2	Draglines.	40
12	Fishing lines, 12 gun lines.	50
2	Straps.	236
	Lashing for seizing.	
1	Boat Painter.	150
	Breechings as necessary, and also painters.	—
	Weighs together	35261 lbs.

Apart from these also ramshead blocks.
Parrel Lines.
Cathead Stoppers and Ratlines.
Halyards.
Runners
Buoy ropes.

From the Sailmaker.

20	Lines of twelve
6	Bushels of Houseline.
18	Pounds of Sail Twine.
30	Needles.
100	Ells old canvas.
2	Sail casks.
20	Thimbles.
	A party of loose clover and Haircloth.
	Lining cloths for the helm port.
	Tarred tarpaulin.
	Scupper hose.
	Hammocks.
	A Carpet to lay in the boat.
	A spare of every sail is taken along.

Of other trifles.

7	Barrels of Pitch.
24	Barrels of Tar.
200	Pounds of Tallow.
8	Pounds of Lard.
60	Fardels or cords Firewood.
1	Hide of Pump leather of about 26 lbs.
5	Hides of Covering leather.

26	Small Tarpaulins.
6	Large Tarpaulins.
2	Gin blocks, 9 Sheaves, 4 Bushes.
50	Pounds of Sulfur.
200	Balls of Oakum.
4000	pounds of Resin.
10	Splicing fids.
15	Marlin spikes.
	Scraping Irons.
	Tar and Pitch Cauldrons.
10	Buckets.
24	Pump plates.
10	Pump bolts.
6	Pump trunks.
4	Pump hooks.
1	Fish tackle pendant hook.
24	Different Hooks.
6	Large spindles.
15	Boat hooks.
2	Pendant hooks.
30	Different Thimbles.
1	Kruk met een Stel [no English equivalent].
2	Swivels for the Boat.
2	Grapnels weighing 94 pounds.
1	Hand grapnel, which weighs 236 lbs.
3	Pennant staves.
100	Yard cramps.
200	Yard hoops.
2	Bells, Squares, and Copper Wire.
6	Axes.
2	Locks.
4	Eel spears.
2	Harpoons.
2	Hei-hoeken [perhaps carpenter's squares?].
1	Crosscut saw.
1	Pit saw.
	Candlesticks.
	Carpenter's Grapnel.
	Different gun tackle rings and Yard hoops.
	Irons to suspend Lanterns.
	Fire hooks.
1	Jackscrew.
10	Chain plates.
7	Sounding leads weighing 150 pounds.
12	Servings.
2	Watering cans.
12	Tar brushes.
3	Lanterns to stand astern.
36	Fathoms old Rope.

12	Hearts.
12	Buckets.
300	Pounds of Sulfur.
18	Compasses.
1	Azimuth Compass.
1	Box with loose Compass Cards and Glasses.
18	Hourglasses.
1	Quarter Glass.
3	Pennant staves.
3	Gilded Trucks.
4	Scrapers.
	Coarse linen for cartridges
	Crowbars.
(p. 282)	
	Chains for Galley and yard.
	Plates.
	Grease horns
	Flints.
	Shears.
	Packing needles.
	Brooms, Starch, Glue, Chalk.
	Copper and Iron wire.
	One or two Sea clocks.
12	Bushels Rope Ends.
¼	Pound Moscovia Glass.
12	Shovels.
1	Whetstone.
1	Red Flag of 8 cloths.
1	White Flag of 8 cloths.
1	States Flag of nine cloths.
1	States Flag of 7½ cloths.
1	States Flag of 6 cloths.
1	Jack of 4½ cloths.
1	Top spindle.
6	New Streamers.
1	Rowle.
4	Pump hearts.
4	Pump boxes, 2 Commanders.
36	Belaying pins.
1	Double and one single block for the stay.
1	Sheave in an old block.
1	Cleat block.
10	Deadeyes for the boat.
2	Bushes, drilled.

Following are the spare Blocks, in case the others are irrecoverable or lost.

2	Double Topsail sheet blocks.
2	Pendants of 15 inches.

4	Tackle blocks of 15 inches.
2	Double Tackle blocks of 3 feet.
6	Blocks of 11 inches.
6	Blocks of 9 inches.
12	Blocks of 8 inches.
12	Blocks of 7 inches.
18	Blocks of 6 inches.
1	Snatch block of 2½ feet.
24	Braem-blox [perhaps "bram blocks," or topgallant blocks].
42	Belaying pins.
2	Serving mallets.
1	Main Parrel with 3 Rows of Trucks.
1	Topmast Parrel.
24	Bull's-eye fairleads
40	Spare Sheaves.
60	Spare block Pins.
12	Blocks of 10 inches.
6	Blocks of 12 inches.
2	Double-sheaved Tackle blocks.
5	Sheaves, 2 Pins for Clew blocks.
4	Sheaves, 2 pins.
5	Deadeyes, well mounted with iron hoops.
	Ram's heads and gin blocks.
	Ribs and Trucks for parrels.
	Trestletrees.
	Deadeyes.
	Buckets.
	Lantern casks.
	Bailers as necessary.
	Pennant staves
	Black lead brushes.
	Large oars, one to each gunport, for chasing or fleeing in a calm: these are stowed outboard between the channels.
	Some ship Ladders.
	Fishing nets.

Follows now the war equipment, which these hundred men are to be given; with which they could defend themselves adequately on a voyage as proposed.

		pounds
2	Metal Guns of 12 lbs., taper-bored,	1525
	weighing	1572
4	Metal Guns of 12 lbs., home-bored,	2470
	weighing	2540
		2480
		2470
2	Metal cannon drakes of 6 pounds,	670
	weighing	664
these 8	Metal Guns would weigh together	14391 lbs.

4	Iron guns of 12 pounds, weighing	13600 lbs.
8	Iron guns of 8 pounds, weighing	24000 lbs.
4	Iron guns of 4 pounds, weighing	6000 lbs.
16	Iron guns weighing	43600 lbs.

All these are to lie on adequate Gun carriages, which, when sailing in peacetime, are tied to the ship's side with ropes, going through holes, which are in the middle. The heaviest guns are stowed in the middle of the ship, because it is able to carry more there.

25	Handguns or Muskets, which
(p. 283)	on a Ship can be fairly heavier than on land.
4	Copper Muskets.
25	Swords.
25	Swords with Belts.
12	Boarding axes.
1	Snaphaunce gun.
10	Ship's pistols.
7	Long Pikes.
14	Trident Pikes.
21	Half Pikes.
3000	Pounds of Gunpowder.
400	Pounds of Fuses.
	Some wristlets, and one Halberd or two.

250	Cannonballs of 12 pounds weighing	3000 pounds
200	of 8 lbs. weighing	1600 lbs.
50	of 6 lbs. weighing	300 lbs.
100	of 4 lbs. weighing	400 lbs.
together 600	Cannonballs, weighing	5300 pounds.

A number of large and small stones, of which use can be made in different circumstances.

40	Bar shot of 12 pounds weighing	480 pounds
40	of 8 lbs. weighing	320 lbs.
20	of 6 lbs. weighing	120 lbs.
20	of 4 lbs. weighing	80 lbs.
together 120	Bar shot, weighing	1000 lbs.

In the past steel bogen, Rondaszen en Panoizen [?], were added, but these are no longer used.

12	Spits.
20	Spare measures.
20	Hand grenades.
8	Ship axes.
2	Mounted Shovels.
2	Spades.
10	Flints.
200	Pounds of scrap.
200	Pounds of Musket Balls.
25	Pounds of Bar lead.
25	Pounds of bar shot
1	Roll of sheet lead, weighing 150 lbs.
24	Iron Crowbars.
60	Rings and cramps for the guns.
10	Port lid hinges.
20	Gun-tackle hooks.
10	Spare Bolts.
90	Linchpins for the Gun carriages.
10	Pigtails.
1	Mallet.
24	Rings and Cramps for the gunports.
30	Washers.
20	Handspikes.
8	Hooks for the Gun carriages.
6	Plates for the same.
40	Iron Cringles.
2	Splicing fids.
3	Marlin spikes.
3	Flat Padlocks.
5	Round Nuremberg padlocks.
1	Bolt of 5 ft., with 8 foot-irons.
2	Bolts of 4 ft., with 6 foot-irons.
1	Bolt of 1 ft., with 2 foot-irons.

Hand and fish grapnels, of which the first have crooked barbed hooks to the flukes, and the second have a short shank.

Besides this, on board the ship there is also need for.

80	Fathoms second-laid [?] Shrouds.
30	Fathoms opstaende[?] Shrouds.
4	Second-laid [?] Straps.
½	Bundle of fine rope for Port ropes.
12	Bushels of Houseline.
10	Bushels of Marline.
1	Line of sixes.

2	Gunlines.
10	Leather Fire buckets.
3	Hides to cover up the powder.
2	Beurs-vaetjes [budge barrels?] of 50 pounds.
1	Beurs-vaetje of 25 pounds.
1	Bunch of Cork, to make wads.
1	Black hole Lantern.
2	Closed } Lanterns.
1	Open } Lanterns.
1	Magazine Lantern.
1	Powder measure for Bandoleers.
1	Powder measure for Cartridges.
1	Powder measure of half a pound.
1	Of a quarter pound.
24	Handspikes.
10	Rammers.
24	Stool beds.
24	Quoins.
10	Double Tackle blocks.
12	Single Tackle blocks.
2	Long Tackle blocks.
6	Chocks for Gunner's ladles.
10	Chocks for Ramrods.
10	Chocks for Rope Ramrods.
10	Chocks for Rope Sponges.
10	Large Carriage trucks.
10	Small Carriage trucks.
2	Commanders for the Master Gunner.
2	Wooden hammers.

(p. 284)

16	Wooden Sponges.
20	Rope Sponges.
1	Pound of red paint.
1	Pound of white Paint.
1	Pound of blue.
80	Lanterns for the guns.
20	Sheepskins for sponges.
8	Sponge rods.
16	Powder ladles.
10	Linstocks.
32	Cartridge boxes.
16	Gunner's Quadrants.
10	Draeplooden [?]

Master Gunner's tools.

1	Copper pair of Scales.
4	Pounds of Iron weights.
2	Pounds of Lead weights.
1	Powder sieve with Drum.

1	Hollow tray
2	Pounds of Cartridge twine.
3	Belts of Cartridge paper.
1	Pair of Tailor's Scissors.
500	Sponge Nails.
1	Small hammer with a shaft.
	Scrap iron to satisfaction.
	As well as crowbars.
4	Dozen Needles.
6	Tailor's rings.
1	Pair of Clipping Tongs.
1	Pair of Pliers.
1	Round File.
1	Half-round File.
1	Rough Flat File.
1	Wooden brace.
4	Bits [for the wooden brace?].
2	Master gunner's drills.
10	Copper Priming wires.
10	Powder horns.
	For each gun 3000 pounds of powder.
25	Copper Nails.
10	Pounds of Tallow.
5	Pounds of Soap.

Furthermore 10 pounds of sheet Lead.

6	Pounds of Lead white.
4	Pounds of black Paint.
2	Pounds of red Lead.
	Black Cork for outside wads, if there is enough.
10	Pounds of Linseed oil.
6	Pounds of Soot.
3	Jackscrews.
8	Double-headed shot.
2	Molds.
1	Beurs-vaetje.

A party of Tampions to put on the guns, made of good wood.

Corporal's tools.

1	Hand screw.
1	Cold chisel.
1	Large Lead spoon.
6	Small Lead spoons.
10	Small Bullet molds.
1	Scraper.
1	Iron Pounder.
1	Scratcher.
10	Wooden Pounders.
1	Can of oil to lubricate the gun.
1	Cask of Iron dross.

Ballast is brought aboard ship, as much as necessary, for the ship to acquire the right stiffness, the amount of which is known through experience. Ships sailing to hot countries, carry wooden laths, which can be raised, and sails stretched over them: to be protected from the Sun.

And these named tools I think will suffice, to bring the ship across the sea. But ships that are equipped for battle, carry more war implements, such as Grenades, Mast grapnels, Stench Pots, etc.

These hundred men, sailing the ship one year, are to be given livelihood on the voyage, in the manner I think is shown in the following list: which, when divided in days, and handed out to each person duly and in accordance with requirements, I estimate to be enough for everyone's fill.

List of Victuals for 100 heads, for one year.

16	Barrels of Meat.	
11	Barrels of Bacon.	
13	Aums of Spanish Wine.	
4	Aums of French Wine.	
6	Aums of Brandy.	
9	Hogsheads of Vinegar.	
4	Aums of Oil.	
1	Quarter of Cod Liver.	
(p. 285)		
2	Casks of fine	Salt each 1½ sacks.
5	Casks of rough	
80	Casks of small Beer.	
4	Casks of good Beer.	
150	Casks of water.	
8	Pair of Hams.	
6	Pair of smoked Meat.	
8	Pair of smoked Tongues.	
10816	Pounds of hard Bread.	
100	Sacks of Barley.	
50	Sacks of white	Peas.
25	Sacks of yellow	
25	Sacks of Beans, which are also served at times instead of Peas.	
4	Vats of Butter.	
2704	Pounds of Stockfish.	
330	Pounds of Cream Cheese.	
12	Pounds of Cow's cheese.	
½	Sack of Mustard seed.	

75½	Pounds of Wax, and	Candles.
50	Pounds of Tallow	

On the voyage every man needs a cask of Brandy, or some other strong liquor; which every man provides for at his own cost, to his own satisfaction.

On the Dutch men-of-war the sailors are provided for as follows.

General list, of the victuals on a ship, for one hundred heads a month.

2250	Pounds of hard bread, up 5 pounds each head per week.
40	Sacks of soft bread.
450	Pounds of cheese, is 1 pound for each head per week.
5	Tuns of Meat, is 1½ pounds each man a day: and 2 days in the week.
400	Pounds of Stockfish when it is herring, if not 700 pounds are needed.
4	Tuns of Herring a month.
1½	Vats of Butter a month.
5½	Pipes of barley a month.
4½	Narrow casks of white or green peas.
2½	Narrow casks of bean, or yellow peas a month.
½	Narrow cask of white salt a month.
½	Hogshead of Vinegar a month.
35	Barrels of Beer in winter, and
42	in Summer.
4	Fathoms of Firewood in winter, and in summer 3 fathoms. [A bundle of firewood one fathom long is a "cord."]

For the Cabin.

¼	Hogshead of French Wine a month.
4	Stopen of Spanish wine a month.
2	Stoop of Brandy a month.
1	Ham, and in summer a piece of smoked meat as well.
¼	Dutch butter for 4 months.
2	Pounds of Sugar a month.
2	Barrels of white biscuit for the journey.
	And sufficient water.
	Enough Cumin and fresh Cheese.
1	Furnished Spice box.
3	Stoop of Oil for the journey.
	Enough Mustard seed.

137. Loose Implements

Many interesting facts about life on board (and on land) can be derived from Witsen's inventory list. The crew's diet and eating habits could be examined as well as the ship's medical facilities, the spare parts needed, and so on. A good example of such research can be found in the *Hollandia Compendium,* which includes an inventory list based on the resolutions of the Dutch East India Company for equipping its ships.[17]

However, since the current study is focused on shipbuilding instead of equipment, we shall not go into the details of the in ventory such household utensils, side arms, and victuals.

Some interesting and important conclusions about shipbuilding can, however, be drawn from Witsen's list. The armament of the pinas, for instance, can be found only in this inventory. The same goes for the flags and the equipment of the boat (here called *boerenschuit*). What is especially interesting is that the pinas carried a large oar for every gunport, "for chasing or fleeing in a calm." These were stowed outboard below the channels.

The measures used by Witsen are of course obsolete; some of them are listed below with the contemporary equivalent:

1 *mengel* = 1.225 liters (according to Van Beylen, 1.2125 liters) = 18 kilograms
1 barrel = 4 hogsheads = 931.32 liters
1 aum = 128 *mengelen* = 155.22 liters
1 hogshead = 1½ aums = 232.83 liters
1 cask = 122.7 liters
1 sack = 83.5 liters
1 *stoop* = 2 *mengelen* = 2.45 liters
(So 1 barrel = 4 hogsheads = 6 aums = 384 *stoop* = 768 *mengelen*)

The weight of the total equipment, calculated as accurately as possible in modern units of measurement, would come to about 150 tons. On this ship of 600 tons, with its own weight of 300 tons, this means that its cargo capacity was about 150,000 kilograms. This seems a fairly small amount, relatively speaking.

138. Finally

(279 II 23) And with this the most important parts of the ship will have been explained: the less important left aside, such as there be seats, shelves, cabinets for glasses and pitchers, handrails, the shape of the berths, heads, sink, bottle-stands, fire-trays, tables, buckets, counters, peeping-windows, buckets, berth-doors, scoops, pins, on which lines are belayed, top-floors etc. because the shape and dimensions of all these, are shaped without rule as the master shipwright sees fit and approves. But it is true, that all the above-mentioned, is applied to the Pinas or Square-sterned Ship, is the same (changing, whatever is to be changed) in all ships, whatever shape they may be. But in all of them, the use to which the ships are ordered, is to be taken into account: whether they are built to carry wood, cereals, or piece-goods: the ones inclining to squareness, the others to roundness.

. . .

(158 II 28) A Ship built thus, can last long years, before it wastes from age, or needs to be broken up, as long as it is not damaged by severe and unusual disasters, up to 20, 30, 40, even 50 years; in which time it does need some looking after. I seem to remember having seen an English Ship which had reached the age of 70 years, but most Vessels perish sooner, in accidents of the sea, weather, wind and enemies.

. . .

(157 I 31) This Ship could well be finished within the time of four months, with twenty or twenty-two men. In this country a Ship long 180 or 150 feet can be completed by 50 men within five months, and built up.

. . .

(157 I 20) With this then, let enough have been said, about the building of my Pinas Ship, long 134 feet over stem and stern, which average size I have chosen with premeditation, so that bigger as well as smaller Ships can be fashioned proportionately to this example; with all due changes, as the use, for which the Ship is to be built, will require.

138. Finally

With this we have come to the end of the description of the building process of the 134-foot pinas, a ship that, according to Witsen, could be built by twenty to twenty-two men in the incredibly short time of four months and yet could last several decades.

In the next chapter we will delve into contracts and how they can still be understood if the reader is acquainted with the building method and terminology described in this chapter. It will also be clear that Witsen's explanations can be applied to contracts to see how the system worked.

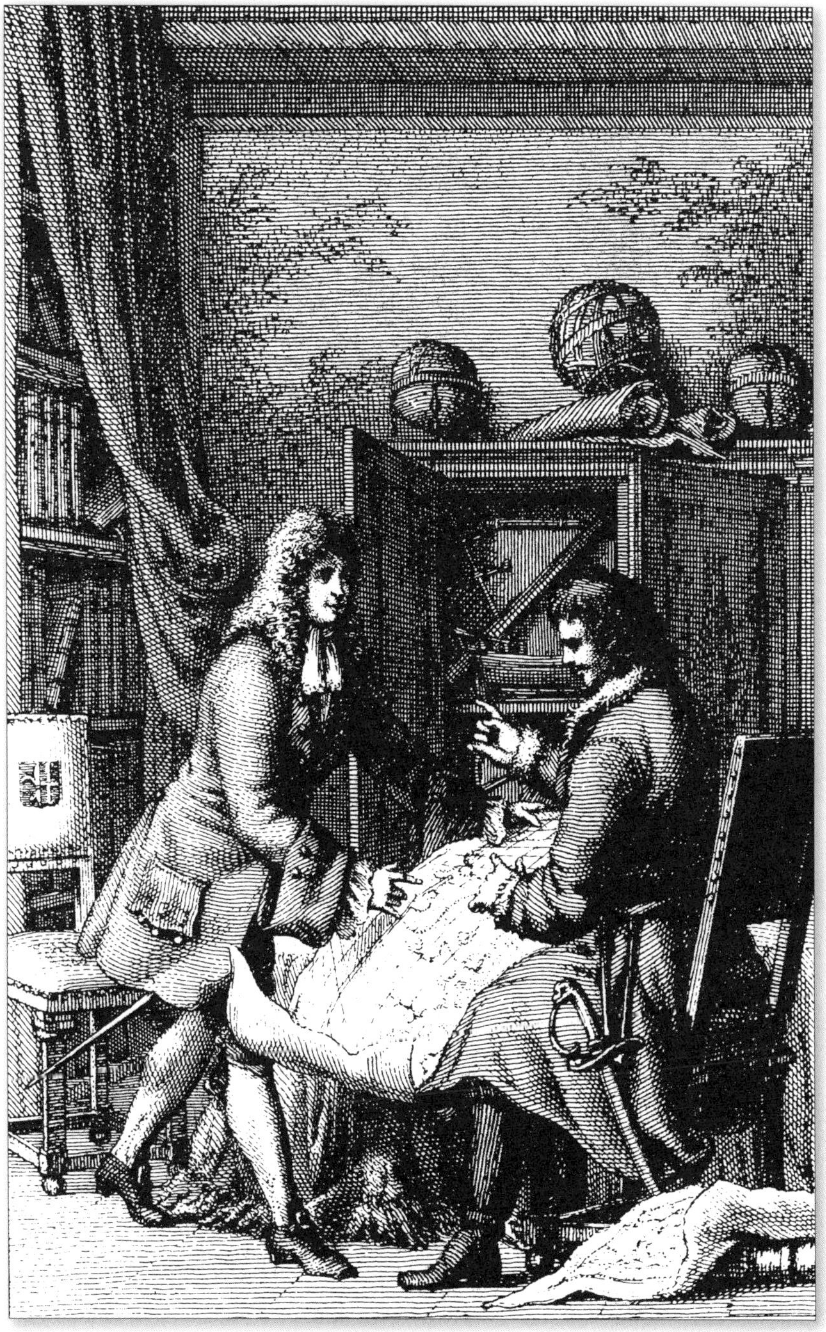

Peter de Groote met Burgemeester N. Witsen. . . . 1697, by Jacob Ernst Marcus, 1814. Courtesy Rijksmuseum Amsterdam.

THREE Contracts as Historical Sources

THE DATA ON THE PINAS, which we have seen in the previous chapter, could be called a contract. However, such elaborate contracts were rare. Strictly speaking, the contract was more than a collection of shipbuilding data. It was also a business contract in which costs and delivery date were stipulated.

The Contract

Witsen used the words *bestek* (contract) and *charter* (also *cherter* or *certer*, meaning "charter" but also "rate") indiscriminately. The difference between these terms is dealt with in chapter 1 (see the section "Contract Specifications"). Witsen describes a *certer* as a contract between client and contractor:

(98 I 49) When Ships, made in this country, are ordered, it is common that the client or buyer dictates rules, and conditions to the Master Shipwright and Seller, after which he wants the ship to be built: which rules and instructions are called, with a foreign word, *Certer.*

In his next chapter Witsen presents several contracts. However, a close examination reveals many differences. For instance, in the contract for the 125-foot ship in passage 98 II 5 (see "Contracts for Hulls" below), business agreements are included with the structural and technical details, whereas the short contracts "made by Mr. Dirk Raven" (see "Contracts for Hulls") are so concise that they appear to be no more than notes made by the shipwright for his own use. Yet this important difference is not conveyed by any significant difference in terminology in Witsen's text.

In general the main dimensions of the ship were the first issue to be mentioned in the contract: length, breadth, depth, and sometimes also the height between lower and upper deck. Occasionally the orlop deck is mentioned. Then the stem and sternposts are described: their height and rake, thickness and width, and sometimes (for the stem) also the curve (i.e., the distance between the inner face of the stem and the diagonal line drawn from the keel to the top of the stem).[1]

The contract for the *Mauritius* of 1639 in passage 108 I 33 (see "Contracts for Hulls") even states that the stem was to be more curved at the top and a little straighter below; in this case, a circle-arc is out of the question. In section 2 of chapter 2 Witsen describes the construction of a stem by the aid of a flexible batten to drawn curved lines. Sometimes the length of the stem, measured on the outer face, is given in the contract as well.

For the sternpost the contracts also mention the height of the garboard stern rabbet and parts of the construction of the tuck, such as the length, thickness, and curve of the wing transom, the height at which the fashion pieces touch the sternpost, and often, though not always, the height of the stern timbers and the distance between them. Sometimes the dimensions of the keel are given.

The thickness of the frames was stipulated in the contracts more often than the rise and breadth of the floor; this thickness was indicated on the keel, in the bilges, and at the height of the master ribband and the top of the top timbers. The strength of the construction was apparently considered more important than the shape of the hull. The dimensions of other structural elements, such as the ceilings, deck clamps, waterways, deck beams, and wales, also received more attention than their exact position.

In most cases these are all the dimensions that are given. Occasionally the size of the beakhead is mentioned as well as parts of the stern gallery.

Contracts, Formulas, Methods, Tools, and Materials

At first sight it would seem that a contract as brief as that described above could not possibly result in a reliable image of the intended ship. Too many details seem to be missing. Yet even the scarce information contained in short contracts like those provided by Raven must have been sufficient for the experienced shipbuilder, resulting in a ship: it would not have been written down otherwise.

Witsen himself confirms that, in a contract for a 160-foot ship, only the most important parts are indicated, the others easily following therefrom. And though it is larger than the proposed ship [the pinas], yet the proportions are the same, and will remain the same, even when the Ships are of much smaller size (97 I 21).

Even when contracts describe the rigging, general conclusions can be drawn: Following the example of these line measurements, the lines for all Ships can be found, of whatever length or width they may be, using a proportional calculation from which all lengths will follow (130 I 4). So we are obliged to acknowledge that more information can be extracted from the contracts than would seem possible at first sight.

But why were certain dimensions explicitly mentioned in the contract and others not? The obvious explanation is that the author of the contract recorded only such matters as those specific for the ship in question, simply assuming all the other dimensions were unnecessary, since they could be derived with the use of the shipbuilding formulas, as discussed at length in the previous chapter. But if we model builders are to form a more complete picture of a specific ship, the contracts and formulas together are not enough: in particular, information about the shape of the hull can seldom, if ever, be found in a contract. However, by combining the contracts and formulas with the original building method as described by Witsen, we can take a significant step toward the completion of the puzzle. The contract supplies the initial data, which are supplemented by the formulas. By adhering to the described building method and sequence, the builder will run into a number of practical problems; since he is working without drawings, he will need to solve these problems before continuing to the next step. Of course, there are no guarantees that the solutions we arrive at today are the same as or even similar to those of the seventeenth century.

But we have a fourth clue—to lay down a ship's bottom as was done in the seventeenth century. For instance, today's model builder will have to make use of tools similar to the ones used at the time, such as the planking tongs that kept the planks in position. Not only does this produce the attributes that are so typical of a hull fashioned by this method, but the builder will also inevitably encounter the same decision moments at which the seventeenth-century shipwright could influence the shape of the hull. These moments contribute decisively to the modern builder's understanding of the subject.

We do have the assurance that a ship built in this way will have the right proportions and characteristic features, that its parts will have the right dimensions, and that the midship section will have an authentic shape. The shape of the bow and stern of a hull thus reconstructed can of course be questioned, for these depended directly on the talent and experience of the individual shipwrights. But their secrets have accompanied them to their graves.

As for the variations in the shape of the bow and stern, these are much less important might be supposed. Wood is a material that, despite its wide-ranging possibilities, has clear restrictions. It can be bent, but only to a certain degree. It must be protected against all kinds of threats—decay, insects, deformation by heat and moisture, and so on. For this reason we can safely assume that if several shipwrights had built ships according to the same contract, the results would not have differed very much. Even though experience, talent, and taste all came into play, shipbuilding was essentially a profession exercised within strict technical parameters.

Thus, when contracts are relied on as information sources, they should be interpreted in conjunction with the formulas, building methods, tools, and materials used at the time. All these elements together are essential guiding the model builder to an enlightening end product, as they will undoubtedly still do if applied with some caution. The rest is a question of experience, common sense, and insight.[2]

Model Building as a Research Method

Shipbuilding is an expensive business. Even in early times a shipbuilder would be little inclined to experiment with his costly materials. Testing our present understanding of contracts at full scale would be horribly expensive.

However, model building presents a good alternative. This greatly underestimated method can lead, easily and cheaply, to a significant increase in knowledge and understanding. Of course, solutions that would be quite acceptable in model building do not necessarily translate to good practice at full scale, but documenting exactly what has been done on a small scale can solve this problem.

What follows is a description of two experiments in applying the shell-first method to specifications contained in two of Witsen's sample contracts. Accompanying photographs from the projects show the building process at various stages. The experiments revealed that convincing results can be obtained from contracts and can help us achieve a fairly good picture of the vessel described in the historical source.[3]

The Pleasure Yacht

The first experiment involved the construction of a model of a 42-foot pleasure yacht from the specifications in a contract provided by Witsen. This contract was chosen because it contains a number of coordinates for the master ribband, which is of paramount importance in defining the shape of the hull.[4]

(175 | 29) *Contract of a Pleasure yacht.*

Long over stem and stern 42 feet, is as follows: Wide inside the planking 9 feet 4 inches, deep on the Wale 3 feet 8¼ inches: the bottom wide 6 feet, rising 4 inches: the sternpost high in the perpendicular 6 feet 5½ in. It rakes 1 foot 5½ inches, inside thick 6 inches, aft thick 4 inches, at the top broad 8 inches, below broad 3 feet 5 inches: the Wing Transom long 6 feet, thick 5 inches, broad 6 inches: has a 7 inch curve. The transom deep 3 feet 5½ inches: the Fashion Pieces broad 7½ inches, thick 5 inches: the Sternpost high 6½ feet on the inside, rakes 10½ feet, at the top broad 2 feet, below 1 foot 2 inches, thick inside 6 in., forward 4 inches, it has a 15-inch curve: the keel long 30 feet, broad 6 inches, deep 5 inches aft, 4½ inches forward: the planking thick 2 inches: the Wale thick 4 inches, broad 5 inches: Planking above the Wale broad 12 inches, thick 1½ inches, the Floor timbers thick 4½ inches: on the wale thick 3½ inches.

The Wale high 5 feet forward on the inside of the stem:

5 Feet from forward, high 4 feet 3 inches, and wide 3 feet 6 inches, inside the planking.
9 high 3 feet 10 inches, wide 4½ ft.
14 high 3 feet 9½ inches, - - - 4 ft. 7½ in.
18 high 3 feet 9½ inches, - - 4 ft. 7½ in.
23 high 3 feet 10 inches, - - - 4 ft. 6¾ in.
28 high 4 feet 3 inches, - - 4½ ft.
33 high 4 feet 8½ inches, - - - 4½ ft.
38 high 5 feet 6 inches, - - 3 ft. 7½ in.

The mainmast tabernacle in the Binding Strake wide 9 inches, high 7 inches, but forward ½ foot: the poop long 16 feet. The Main Yard long 46 feet; the foreyard 34 feet: the leeboards 16 feet from forward, the ceiling broad 16 inches: the Wale broad 4 inches: the lower planking broad 12 inches, above 1 foot: the Cabin wide 2¼ feet: the camber of the bench without cover 6 feet: the Beer cabinet in the cabin long 4½ feet, the door wide 2 feet 2 inches: the hackebord is 19 inches above the planking, and wide 35 inches.

As noted, this contract had the advantage of specifying the measurements of the master ribband over the entire length of the hull. Furthermore, Witsen provides a drawing of the vessel (fig. 3.1). Laying down the bottom of the model proceeded without difficulty. As can be seen in the photographs, small brass planking tongs were used for this purpose.

Problems arose with the planking of the turn of the bilge. The shape in the midship section was no problem,

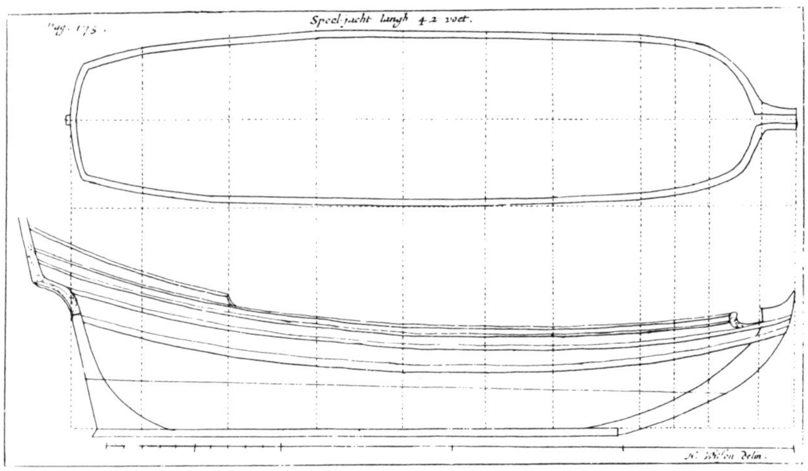

FIGURE 3.1. PLATE LXXI. A pleasure yacht 42 feet long. In Witsen's drawing it is striking that no frames are depicted; the shape of the hull was probably supposed to follow from the data in the contract.

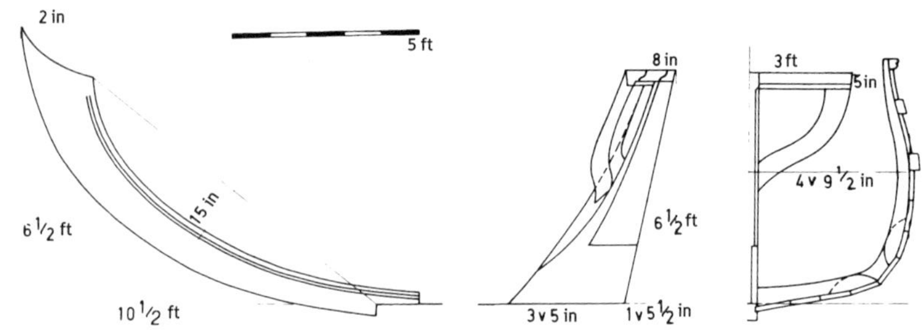

FIGURE 3.2. Preliminary drawings of the pleasure yacht. Prior to building the model, I made these drawings to elaborate on the contract data. The resulting model was 58 centimeters in length. (Drawing by A. J. Hoving)

but the torsion in the bilge planks in the bow and stern was extremely difficult. If they came out too level, they would form an ugly angle with the side planks at a later stage. If made too vertical, the result was even worse: a nasty hollow would remain visible. This stage had to be tried several times. It soon became clear that only experience could lead to a satisfactory result, and each attempt was a lot better than the previous one.

Measuring proved to be tedious. There are no straight lines in such a vessel. Witsen describes how a line was stretched from stem to stern, from which measurements were taken. I tried the same approach in this experiment, but it yielded but few hard results because of the impossibility of using a plumb line and a water level to go with it.

The problems ended once the bilges were planked and the master ribband was fitted at the right positions (because the master ribband in this case coincided with the wale, the wale itself could be used as a master ribband that would not have to be removed later on). The shape of the top timbers was practically the same all over, differing only in the bow and the stern; their shape had to be established experimentally. The ceiling on the inside then provided enough strength for the hull to be planked. Common sense and careful study of prints showing similar craft supplied the rest of the information necessary to complete the construction.

The most striking thing about the experiment was the speed with which a model was built with the shell-first method. It was of course a small, simple craft with relatively few frames, but even then it was surprising that only a couple of hours of work yielded a structure that revealed the eventual shape of the ship. Clearly Witsen's statement about the speed of Dutch shipbuilding (four months to build the 134-foot pinas with twenty-two workmen; five months to build a 175-foot man-of-war with fifty workmen) was no idle boast.

All other features of the shell-first method were present—the nearly flat bottom, the angle in the bilges, and the relatively arbitrary length of the frame timbers, which remained unconnected to one another and were fastened only to the ceiling and planking; there were also traces of where the floor planks and bilge planks had been temporarily joined with chocks spiked in place.

Subsequent checks revealed that the hull was not perfectly symmetrical, the result of faulty measuring. But in this respect the model followed the example of the seventeenth century, a time when almost all ships were asymmetrical in one way or another.

FIGURE 3.3. The first stage of the building process: keel, stem, and sternpost are joined to form the backbone of the ship. The first planks, the garboard strakes, have been fitted. (Courtesy A. J. Hoving)

FIGURE 3.4. The bottom planking is continued with the aid of planking tongs. Wooden chocks keep the planks in place after they have been pressed edge to edge with the hook and chain (see "Tools" in chapter 1). Small wooden poles (shores) support the construction. (Courtesy A. J. Hoving)

FIGURE 3.5. A floor timber and two bilge futtocks are raised at the location of the main frame. Because I was reluctant to believe Witsen at this stage, I even placed two floor timbers, although one would have sufficed. (Courtesy A. J. Hoving)

FIGURE 3.6. The bilges are now planked, guided by the bilge futtocks. At this stage, where the planks do not lie in one line next to each other, as in the previous stage, the planking tongs can only be used together with wedge-shaped filling chocks, filling the open space between planks and tongs. It is quite difficult to build a symmetrical hull shape with this technique. (Courtesy A. J. Hoving)

FIGURE 3.7. The shell formed by the bottom and the turn of the bilges is now filled with timbers. With a small vessel like this pleasure yacht, the frames do not lie close to each other as they do in larger vessels, but with wider spaces, as in launches and barges. (Courtesy A. J. Hoving)

FIGURE 3.8. Futtocks are raised at several locations. To check the overhang of the frames, measurements are taken from the line stretched between stem and stern and lines stretched between the port and starboard parts of the frames, combined with measurements taken with a plumb line. In model building this practice is almost impossible unless the model is mounted securely on a board. (Courtesy A. J. Hoving)

FIGURE 3.9. The master ribband is positioned from stem to wing transom along the futtocks. This temporary ribband was very important, as it indicated not only the maximum breadth of the hull but also the height of breadth and the sheer. (Courtesy A. J. Hoving)

FIGURE 3.10. The other futtocks and top timbers are added. Above the master ribbands, other ribbands were used—first, to find the right shape of the frame timbers and, second, to keep them in place as the building process continued. The frame timbers were not joined together in any other way. (Courtesy A. J. Hoving)

FIGURE 3.11. Inside the hull, the ceiling (inside planking) is applied; the keelson is placed above the keel, followed by the bilge ceiling in the turn of the bilges and finally the deck clamp on which the deck beams come to rest. As soon as the ceiling, as far as necessary, has been finished, the outside is planked. The ship is then ready to be launched and will be finished afloat. (Courtesy A. J. Hoving)

FIGURE 3.12. The finished model of the pleasure yacht, displaying the specific features of the Dutch shell-first building method: a nearly flat bottom, an angle where the turn of the bilges starts, frame timbers that are not joined to one another, and (only faintly visible) traces of the chocks with which the planks were held together in the first stage of construction. The inside was finished insofar as the scant contract data allowed. The rigging specified in the contract is a "lateen yard rig," the oldest type of rigging for this type of vessel; it seems to have been imported from Asia shortly after the first Dutch expeditions. The hoops and halyard that I fitted onto the model at the time are probably incorrect. (Courtesy A. J. Hoving)

FIGURE 3.13. Detail of a print by Claes Janszoon Visscher, c. 1635. On boats with the mastless "lateen yard rig" depicted here, the sails were probably never lowered but furled around the yard with the sheet. They were probably marled to the yards with a lacing line, and the yards would have to be lifted from the tabernacle to remove the sail altogether.

FIGURE 3.14. *Speeljachten,* by Willem van de Velde the Elder, 1660. Van de Velde depicts two lowering masts, rigged with an early version of the bezan rig with tiny gaffs. This type probably superseded the older "lateen yard rig" specified in Witsen's pleasure yacht contract. Indeed, this etching was made long after the probable date of the contract.

The Yacht de Brack

The second experiment had its inception in 1993 when I was commissioned by the Dutch community in Oakland, New Zealand, to build a model of Abel Tasman's *Heemskerck.* As no plans for the ship were known to exist, I used Witsen's contract for a yacht named *de Brack,* its specifications scaled up slightly, and applied his building method and formulas.[5] This experiment had some flair. The ship was larger than the pleasure yacht and the construction more complicated; yet the specifications were much less elaborate. But even from such a minimal contract it is possible to build a reconstruction if the rules described in chapter 2 are applied. Little more than the dimensions of the stem and sternpost, the tuck, the keel, and the main frame are provided in the contract:

(106 II 50) *A Yacht made, named de Brack, Anno 1631, November the 10th.*

Long 91 and ½ feet across stem and stern, wide 21 feet between the planking, deep 8 and ½ feet at deck level. One deck above that 5 and ½ feet high in the sides. The sternpost 13 and ½ feet high, thick 10 inches, at the back thick 5 and ½ inches, or 6 inches, raking 3 feet. The wing transom is 12 feet long, curving one foot, deep 10 inches. The tuck is raised 7 feet from the keel. The stem is 13 feet high in the right angle, raking 17 feet. The bottom is 14 feet wide, rising 10 inches, the turn of the bilge planked 4 feet deep, the top of the turn of the bilge there wide 19 feet. The floor timbers 6 inches thick. 5 inches at the bilge planking. At the master ribband 4 and ½ inches. The planking of Hamburg planks. The keel is 12 inches deep, and 14 inches wide at the midbody.

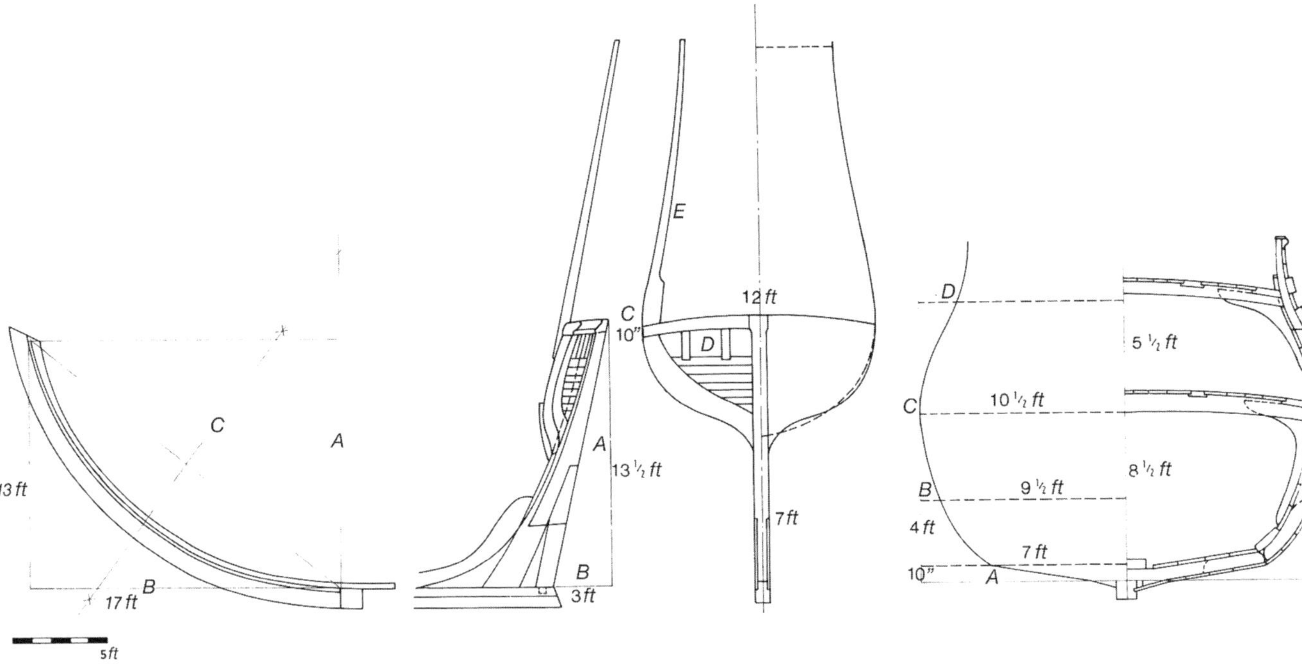

FIGURE 3.15. Preliminary drawings of the yacht *Brack*. *Left:* The stem derived from the contract data. *A* is the height of the stem, *B* the extent of its raking, and *C* the perpendicular on the diagonal going from the keel to the top of the stem. Where *C* crosses the vertical, one places the compasses and draws the circle. *Center.* The sternpost derived from the contract data. *A* is the height of the sternpost, *B* the extent of raking, *C* the wing transom, *D* the tuck, and *E* a stern timber. *Right:* Reconstruction of the main frame. *A* is the height and width of the bottom, *B* the bilge planking, *C* the height of the master ribband or height of breadth, and *D* the height of the main deck. (Drawing by A. J. Hoving)

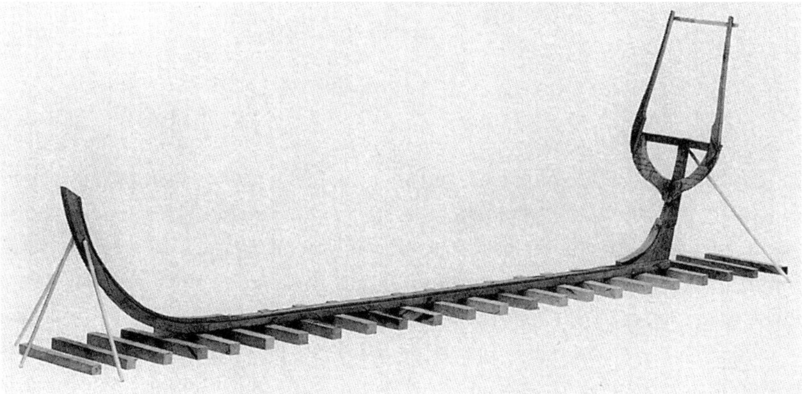

FIGURE 3.16. The keel has been laid and the stem and sternpost have been raised. The transoms still have to be applied in the tuck. (Courtesy Cees de Jonge, The Visual Art Box)

Figure 3.17. The bottom planking is finished. The structure is supported by small poles and held together with cleats (Courtesy Cees de Jonge, The Visual Art Box)

Figure 3.18. The floor timbers and two bilge futtocks have been placed at the position of the main frame. Their shape has been derived from the section of the main frame. (Courtesy Cees de Jonge, The Visual Art Box)

Figure 3.19. The turn of the bilge has been planked. The bilge futtocks of the main frame determine the shape. (Courtesy Cees de Jonge, The Visual Art Box)

This contract contains some peculiar features: the heights of the stem and sternpost, for instance, are much lower than in other contracts. This could only mean one thing: that the gunroom in the stern was to be made lower than the rest of the main deck. In the foreship there should also be a step-down in the deck layout, or the bowsprit would come too low in the hull. All this would have another consequence: the ship would have relative little rise in the sheer.

Before construction began, the dimensions of the frame parts of the main frame as well as parts of the stem and sternpost were taken from contract data and rendered as drawings (fig. 3.15). After the stem and sternpost were raised, the bottom planking was fitted (figs. 3.16 and

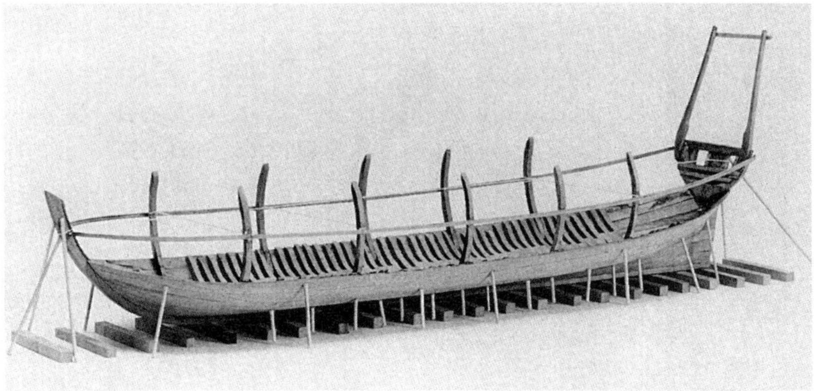

Figure 3.20. Most of the floor timbers and bilge futtocks have been filled in, and a couple of futtocks have been raised. The master ribband joins them. (Courtesy Cees de Jonge, The Visual Art Box)

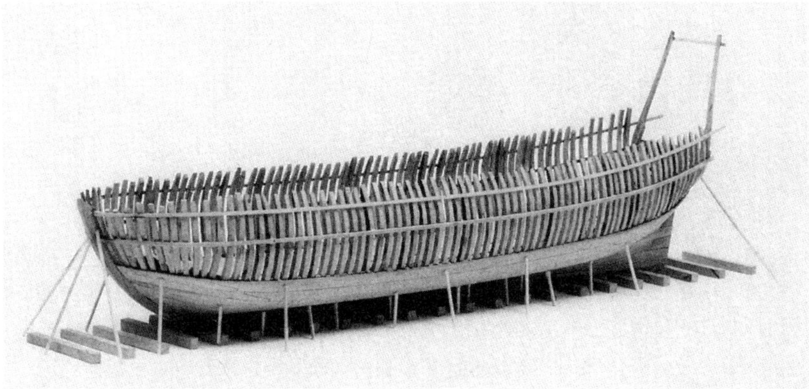

Figure 3.21. The remaining futtocks have been placed. An extra ribband has been applied above the master ribband to align the top ends of the futtocks. (Courtesy Cees de Jonge, The Visual Art Box)

3.17). This posed no difficulties after the experience I had gained from building the pinas and the pleasure yacht. Fitting the planking of the turn of the bilge likewise presented no difficulties owing to my experience with the previous projects, although some luck might have been involved (figs. 3.18 and 3.19).

The futtocks supporting the master ribband (fig. 3.20) were all identical in shape, and their dimensions and shape were again taken from the cross-section drawing.

Because the rise of the sheer was so little, it was hard to give the master ribband a satisfying look. The more rise we manage to give to the sheer, the better this works. In all honesty, I have to admit that the wales were eventually given more rise than the master ribband had at the beginning.

When all the futtocks had been raised (fig. 3.21), it also became clear why the frame lines in the drawing crossed at deck level (see the discussion in section 18 of chapter 2). When futtocks of the same or nearly the same form are placed on the shell, which already has some sheer, then the point of greatest width of these futtocks will also change the height forward and aft, making their widest points higher at these locations than at midships. The consequence is that the lines showing the shapes of the frames in the body plan will cross each other, which then is a consequence of the building method applied.[6] The shape of the futtocks, however, did require some adjustment in the foreship and aft—at those locations the frames curve slightly more. But this was so slight that the mold that the seventeenth-century shipbuilder undoubtedly used could have been adjusted with just a few strokes of a plane for each futtock.

In the meantime the ceiling could be fitted (fig. 3.22), and it was surprising how sturdy the construction became with that addition. Real strength, however, was added when the deck beams were fitted, notched into the deck clamp with dovetails (fig. 3.23). With the lower wales put in place, the model became so strong that it could support the weight of a grown man!

Since only the shape of the main frame was known from the contract data, placing the first top timbers proved difficult (fig. 3.24). However, after a ribband was applied from the stern timbers to the top timbers of the main frame, the intervening shape readily became apparent, and more top timbers could be placed. A few extra ribbands were a big help. Forward the same applies. At the location of the foremost top timber—the *verkeerde* (reversed) top timber—a temporary straight top timber was placed,

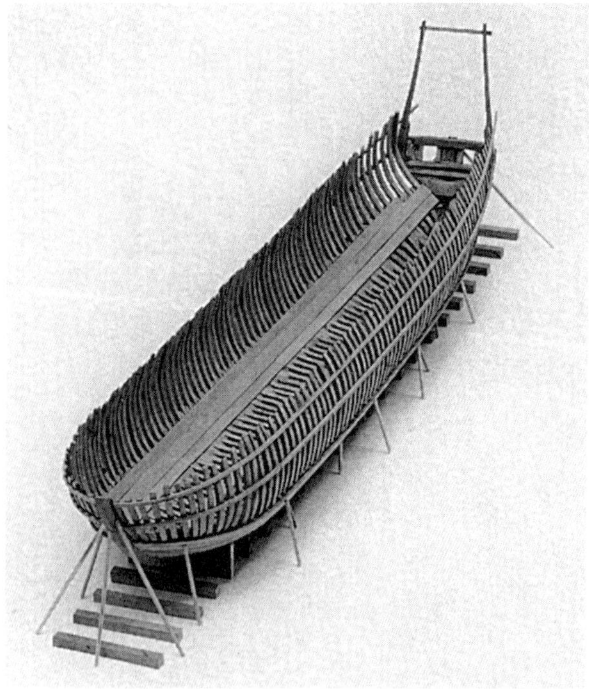

Clockwise from top left:

FIGURE 3.22. The ceiling planks are placed inside the hull. (Courtesy Cees de Jonge, The Visual Art Box)

FIGURE 3.23. Deck beams, waterways, and binding strakes have been fitted. The rear section of the deck (the gun room) is laid at a lower level to provide a flat platform for the guns. The stem and sternpost are extremely low, which has consequences for the rise of the sheer. Merchantmen had more sheer than men-of-war. (Courtesy Cees de Jonge, The Visual Art Box)

FIGURE 3.24. The first top timbers have been raised; the ribbands determine the shape of the upper structure. (Courtesy Cees de Jonge, The Visual Art Box)

FIGURE 3.25. The remaining top timbers have been filled in. The gunports could have been made at an earlier stage, when the futtocks were raised. (Courtesy Cees de Jonge, The Visual Art Box)

FIGURE 3.26. The upper deck has been laid and the planking applied. At this stage the vessel was launched. (Courtesy Cees de Jonge, The Visual Art Box)

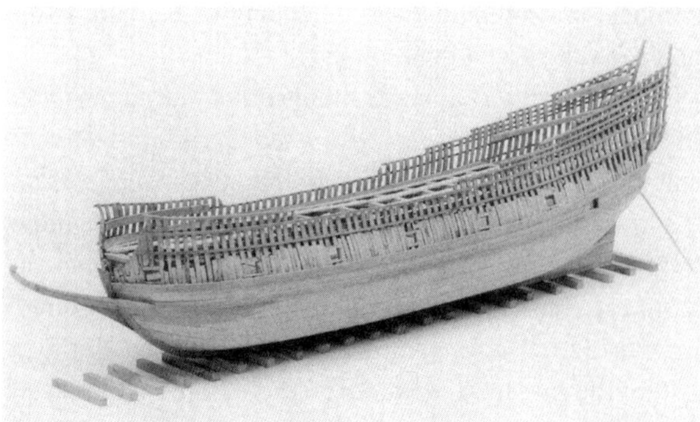

FIGURE 3.27. (*right*) The finished hull. (Courtesy Cees de Jonge, The Visual Art Box)

FIGURE 3.28. (*below*) The completed model. The contract was used for a reconstruction model of the war jacht *Heemskerck*, Abel Tasman's ship when he discovered New Zealand in 1642. The model is now part of the collection of the Voyager New Zealand Maritime Museum in Auckland, New Zealand (Courtesy Cees de Jonge, The Visual Art Box)

holding the ribband while the final shape was being established. More and more top timbers were placed, their shape derived as much as possible from the neighboring timbers (fig. 3.25). Applying a ribband for each hance is recommended, as was done on full-size ships. I discovered this rather late, which meant a lot of extra work (see the final position of the hances in fig. 3.26). At this stage the ship was launched and the shaping of the hull was completed.

Readers can judge the success of the experiment from the photographs of the final result (figs. 3.27 and 3.28). The model has some original features of the Dutch shell-first method: the straight rise of the floor; the angular bilges; the thoroughly disguised traces of the cleats and the arbitrarily positioned ends of the frame timbers; and the timbers joined together only by the planks and the ceilings, not directly to one another. The model was built almost entirely without plans and in less than 1,000 construction hours.

Sample Contracts

This section consists of various contracts for a number of ship types of Witsen's time. They all deal with vessels that were specified no further than "ships," which referred to three-masted, square-sterned vessels. I have grouped these sample contracts according to their main focus—hulls, masts and spars, rigging, blocks, and sails—and have added comments and simple drawings illustrating contract specifications if the contract offered sufficient data for such a drawing. The absence of a drawing for a particular contract does not mean that the document lacks sufficient data for constructing a model, however.

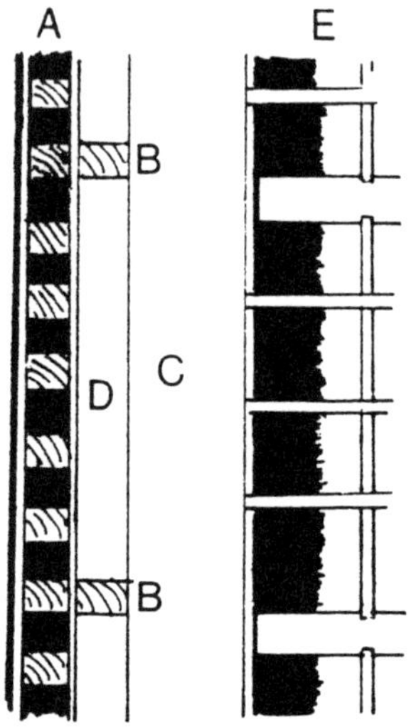

FIGURE 3.29. Waterway construction on ships with many futtock riders: *A*, the ship's wall; *B*, the futtock riders; *C*, the waterway; *D*, the chocks; *E*, deck beams, carlings, and ledges. (Drawing by A. J. Hoving)

Contracts for Hulls

The following contract has a legal character, although it contains many dimensions as well. The length of the ship is given as 125 feet, but that might be a slip of the pen, as most of the dimensions given would indicate a much larger ship—of at least 160 feet. It also has an orlop deck, which was normally a feature of big ships only. The heavy construction called for in the contract gives reason to believe that this ship was a man-of-war, so the orlop deck mentioned here would have had a much more decided function: providing space for the galley, sail and shot lockers, workshops for sailmakers and carpenters, and so on. In merchantmen the orlop deck changed depending on the load. The heavy construction is seen in the fact that each deck beam was to be provided with a futtock rider, complete with hold and bilge riders. In contrast, the pinas had only one futtock rider between each pair of gunports and hold and bilge riders only at the location of the mainmast.

With so many futtock riders it was easier to apply chocks on the deck beams between the riders and aligned with them, then just laying the waterways against them (see fig. 3.29), instead of having to cut out the waterways to fit them around the futtocks, thus weakening them. The construction described in this contract was applied on the *Wasa* and can still be observed.

This contract offers little information about the shape of the hull. Such details like the thickness of the frames are given; yet the dimensions of the bottom and the turn of the bilge are absent. On the other hand, the legal status of the contract is quite apparent.

(98 II 5) *Contract of a Ship, long across stem and stern 125 feet, wide inside the planking thirty feet, deep in the the hold at deck level twelve feet.*

Stem and stern will rake two and twenty feet altogether, or, forward eighteen and a half, and aft three and a half feet, forward no more than the wing transom, being 20 feet long, broad and deep in proportion, and as the work requires at least fifteen, and sixteen inches, with an upper deck of seven feet in the side. The main deck and the upper deck will run along straight, from the stem to the stern, with a long steering stand, on which four guns can easily be placed, with two gunports in the stern, in the cabin, and two aft, so that the guns on the steering stand may be used in the quarters and through the transom. Then the fore-scribed Ship will need a good keel, of two pieces, good, sound and square oak, broad, and thick twenty inches. The scarf of the keel will have to be long, nine feet, and the scarf of the stem seven feet with a good knee aft on the scarf: and the stem shall be two and twenty feet in right angle, thick sixteen inches inside the planking. The floor timbers will be eleven inches in the middle, some timbers forward and aft as necessary: the bilge futtocks nine inches: the futtocks eight inches, the top timbers at the bilge planking six inches, provided they turn back at the top, and fan out as the Client sees fit: the floor timbers, bilge futtocks and futtocks will also be well overlapping, at least five feet, even more, but not less. The planks from the lowermost bilge planking to the second wale shall be four inches thick, or three of one foot, the keelson in the hold shall be of two pieces, with a long scarf, covering at least three frames, in breadth being twenty inches, deep nine inches: the ceiling planks inside of good Königsberg planks, laid close together.

Furthermore there will be seven breasthooks in the bow, with a good fore step, and aft over the peak four stern knees with bilge riders, going past the knees, two knees to the ceiling, alongside the ship to the tuck. The tuck will come at eleven and a half feet: the wing transom will be secured with four knees, them being, two alongside the ship and two up and down: the stern timber will have a good knee to each transom, of the long sort, as well as good deck clamps, board two and a half feet, thick five and a half inches: the deck beams in the hold fourteen inches broad, and deep 15 inches, lying three and a half feet apart, provided that between each deck beam there are three oak ledges and a cross ledge, or carling. At each end of the deck beams there shall be a well-grown hanging knee, of the long kind, shooting down to the bilges, and turning back with the beams: to each beam a good futtock rider going up above the upper deck from the bilges, a hold rider and bilge rider to each futtock rider, a plank on the keelson, with bilge riders, around the main step.

The waterway will come against the futtock riders, and will be six inches thick, broad twenty inches, between each futtock rider a good chock will be driven, as well as inside and outside a good bilge stringer, thick five inches, at the waterway a ceiling of four and a half inches. The ceiling planks between the main deck and the upper deck will be planed smoothly, thick as required. The deck beams of the upper deck will be ten inches thick, and square, lying apart as underneath the main deck: the upper deck beams to be planed smoothly, the corners with a half circle, provided that between each beam there lie three ledges, and a carling, and a well-grown hanging knee at each end of the beams: beneath the black strake there will be two wales thirteen inches broad, thick seven inches, each end with long scarfs, covering four or five frames. The black strake to be made of three- and four-inch planks, gunports as many as the Client pleases, to have above the filling strake, two wales wide and broad as the construction requires, with a sheer rail and a planed half circle. The waterway of the upper deck will be four and a half inches thick, appropriately broad, and cut out one inch in the side, for a spirketing of three and a half inches to be placed on it; a plank above that, as high as required. Furthermore all well gunwaled, pinrailed, and the upperworks of sound thick dry wainscot, to apply as the work requires.

Railings and coatings of oak, with three broad channels to each side, with all the sheets, blocks, main and fore mizzen, topsail halyards, and topsail sheets, knights, kevels, pin racks, and pins, to the wishes of the Client: the main deck to be made of oak planks, thick 2½ inches, the orlop deck of sound deal, or Prussian planks, thick two and a quarter inches, the sapwood removed, planed smoothly underneath, forward, below and above binding strakes, in which the ledges lie, as the construction requires between the main deck and the orlop deck: two breast knees to the stem with a good breasthook, in which the bowsprit will be laid. The same to be laid before and abaft the foremast, the mainmast and the mizzenmast, a good partner, thick as the construction requires, with a good capstan, as well as a good bitt, well furnished, with a lining: and in each side two hawseholes and hawse pieces, a sound beakhead, catheads and cross pieces in the beakhead: a gallery to be made aft, with opening doors, a good cabin and quartermaster cabin, and all paneling in the cabin to be made; cupboards, berths, benches, as become, steering stands, cabins with opening doors, crooked timbers, wainscoting of the steering stand, and the cabins to be made with gun positions, also the deck beams in the captain's cabin, steering stand and cabins to be planed smoothly, with half a circle to the wainscoting of the cabin, two right-angle knees, captain's cabin, steering stand and cabins with oak ceiling, and to each beam a small knee above the main deck, as many gunports as the Client shall wish, with a bulkhead forward, with crooked timbers and gun positions with cleat rails and bulkheads. Furthermore the contractors shall supply a good oak rudder, as well as wood for carvings, and shall pay for the carving, as well as pitch, tar, oakum and moss, as well as have the Ship well caulked inside and out, doweled, and clenched with bolts, furthermore a galley, buttery, cable tiers and boatswain's room, pump wells, powder and bread room, berths, all to the wishes of the Client, furthermore whatever the work requires, because not everything can be thought of beforehand, to set to Sea. But the ironwork and nails the Client shall pay for himself, the chasing and doweling, the careening at the expense of the Client, all in Amsterdam feet and inches.

Whereas the above contract was of a legal nature, full of agreements concerning quality and soundness, the following contracts are clearly of a different nature—the financial responsibilities are not as important as the dimensions of the parts and the spaces. These contracts or *certers* appear as though they came straight from a shipwright's notebook.

(105 II 25) *But to get even more acquainted with the business of Shipbuilding: there will follow some Contracts of Ships, made by Mr. Dirk Raven, famous master Shipwright in Amsterdam.*

A Ship called Deventer,
Anno 1627 December the 11th.

Long 112 feet, wide 26 feet inside the planking, deep 11 and ½ feet at deck level. An upper deck above that high 6 and ½ feet in the sides. The stem 17¾ feet high in the right angle, is curved 5 feet, raking 21 feet, thick 12 inches, the front side thick 7 inches. The sternpost 19 feet high in the right angle: raking (but it is half a foot too many) 5 and ½ feet, thick 12 inches, the backside thick 7 inches, broad 6 feet below. In the right angle the tuck stands 10 feet from the upper side of the keel, the wing transom of the tuck is 15 feet 3 quarters long. The keel is thick 15 inches square. The stern timbers rise 13 feet above the wing transom, and there they stand 8 feet apart. The bottom is 17 feet wide, the turn of the bilge planked 5 feet, and the tops of the bilges there are 24 feet wide, then the sides hang out 14 inches [11 inches]. The wale rise 1½ feet forward, 9 feet aft.

(106 I 1) *A Ship named Campen,*
Anno 1628, March the 7th.

Long 112 feet, wide 26¼ feet, deep at deck level 11 feet ¼, the bottom is wide 17 feet, rising 15 inches, the bilges planked at 5 feet ¼, and there it is 24½ feet wide, the wale is rising 1¼ feet forward, aft 9 feet, the stem 18½ feet high, raking 20 feet, curving 4½ feet, thick 12 inches, the sternpost 20 feet high, raking 5 feet, thick 12 inches, the face aft 7 inches thick, the wing transom of the tuck 16 feet long, the stern timbers 21 feet long, where there are wide 8 feet 4 inches, the tuck is 10 feet ¾ high above the keel, the keel is 14 inches deep, broad 16 inches, the upper deck 6 feet and ½ high in the sides, the gunroom from the counter to the bulkhead is long, 20 feet, the beakhead at the upper side of the lower knee of the head is 20 feet long, the lion is 6 feet long, the trail board is deep 20 inches.

(106 I 24) *A Ship named Swol,*
Anno 1628 June the 20th.

Long 115 feet, wide 27 feet, deep at deck level 12 feet, the upper deck 6 feet ½ high in the sides, the stem 18 feet and ½ high in the right angle, the sternpost 21

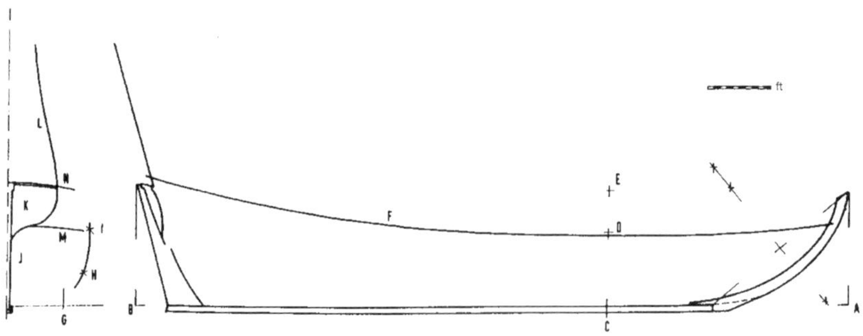

Figure 3.30. Basic dimensions of the *Deventer* (1627), 112 × 26 × 11½ feet. The ship was probably a man-of-war. *A-B,* the length between stem and stern; *C,* the main frame; *D,* the height of the main deck; *E,* the height of the upper deck; *F,* the upper face of the lowermost wale; *G,* the width of the bottom (the rise of the floor is not given in the contract); *H,* the bilge planking; *I,* the height of the master ribband; *J,* the height of the tuck; *K,* the wing transom; *L,* the stern timber; *M,* the main deck; *N,* the upper deck (Drawing by A. J. Hoving)

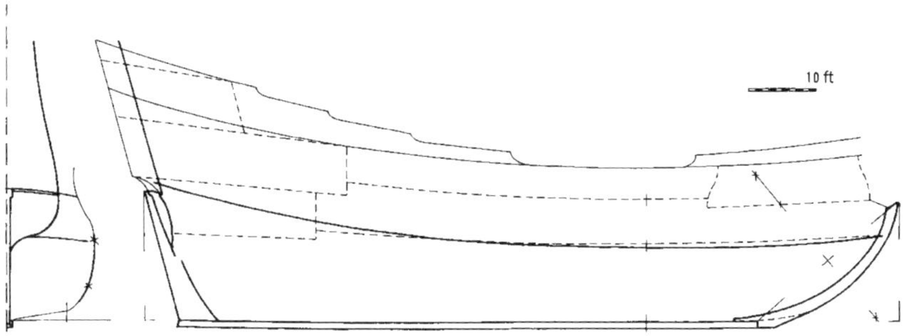

FIGURE 3.31. The *Deventer*'s hull completed with hypothetical lines. The wale gives us much to work from. If the height of the bulwark is estimated at 5 feet, then the line of the sheer rail can be traced: it has to be parallel, or almost so, to the wale. The position of the gun room, the height of the stem, and the height of the stern timbers give a reasonable indication as to the heights of the inner spaces (Drawing by A. J. Hoving)

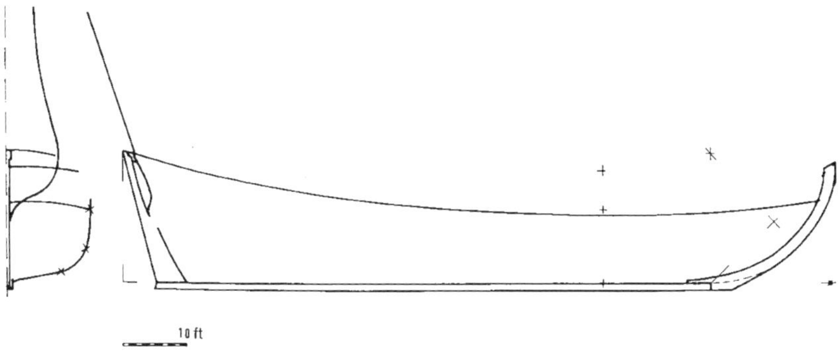

FIGURE 3.32. Hull shape of the *Campen* (1628), 112 × 26¼ × 11¼ feet. Like the *Deventer*, this ship was probably a man-of-war. The fairly steeply rising floor and strongly raking stem and sternpost, all providing speed, are good indications (Drawing by A. J. Hoving)

feet and ¼ high in the square, the tuck 11 feet high above the keel, the wing transom is long 16 feet.

(106 I 33) *A Ship named Over-yssel, Anno 1628 September the 2nd.*

Long 120 feet, wide 27 and ¼ inside the planking, or 28 feet, deep at deck level 12 and ¼ feet, the upper deck 6 and ½ feet high in the sides, the sternpost 21 and ¼ feet high in the square, raking 4 or 5 feet, the stem 18 and ½ feet high, or 21 feet, raking 17 feet, the tuck is 11 feet high, the wing transom is 16 feet long, or 16 feet and 8 inches, the bottom is 17 and ½ feet wide, or 18 feet, the top of the turn of the bilge planking 5 feet and ¼, the floor timbers thick 9 inches, at the turn of the bilge the planking 7 inches, at the first wale 5½ inches, the deck beams are 12 inches thick and square, the wale beneath the gunports 12 inches broad, the lowermost wale of the same breadth, both 6 inches thick, the beakhead is 20 feet long, to be measured on the upper face of the lower stud of the head, the lion is long 6½ feet.

(106 II 3) *A Ship made, named Prins Wilhelm, Anno 1630 January the 20th.*

Long 157 feet, wide 37 feet, deep below the lowermost continuous deck, which is the orlop deck, 13 feet, above that, being the gun deck 5 feet and ½, the upper deck above that 7 feet and ½ in the sides, the room 8 feet ¼ high, long 26 feet. The stem 28 feet in the square, raking 27 and ½ feet, curving from one end to the other 6 feet, 45 feet long measured along the curve, thick 17 inches, at the front 12 inches. The sternpost 28 feet high, raking 5½ feet, thick 17 inches, the back face 11 inches, broad as appropriate. The wing transom is 22 feet long. The tuck, from the keel, stands at 15 feet, the keel is 23 inches broad, deep 20 inches. The scarfs 9 feet long. The stern timbers stand 29 feet above the wing transom, 12 feet apart at the top. The bottom is 24 feet wide, rising 9 inches, the top of the turn of the bilge planking 5 feet and 4 inches, there it is 33 feet wide. The futtocks curve outward 2 feet. The floor timbers 13 inches deep, at the bilge planking 9 inches, at the top 5½ inches thick. The deck clamps 6 inches thick, the deck beams thick 15 inches square, lying 4 feet apart, thick 4½ inches, rising 3 feet to forward, aft 1 [missing digit; possibly "10" or "11"] foot ¾. The mainmast stands 3 and ½ feet, behind the middle of the lowermost deck. The upper deck beams deep 11 inches and ¼. The waterways on the lowermost deck thick 5 and ½ inches, on the gun deck 6 and ½ inches thick. 5 Stern knees in the peak aft, and riders. Forward in the bow, up to the deck, 9 breast knees. The keelson thick 12 inches, broad 3 feet. A futtock rider for each second deck beam, three pairs

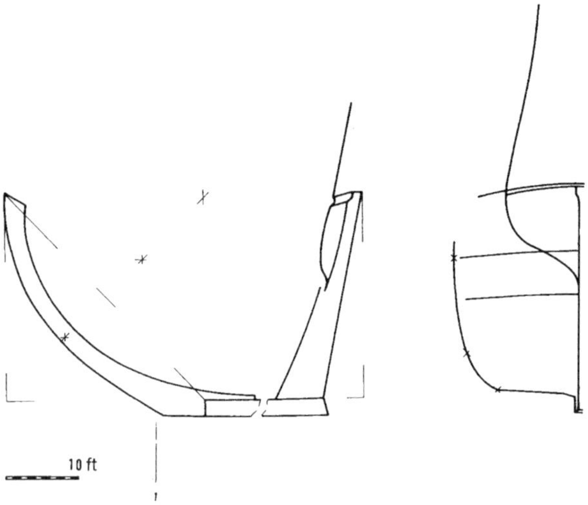

FIGURE 3.33. Stem, sternpost, and main frame of the *Prins Wilhelm* (1630), 157 × 37 × 18½. The orlop deck and the rising of the floor indicate that this ship was probably a merchantman. The details on the rising of the wales are obscure due to a printer's error: they could never have been 1¾ feet, as the contract stipulates; they must have been at least 10¾ or 11¾ feet. Thus only a portion of the hull has been drawn (Drawing by A. J. Hoving)

of wales beneath the gunports, thick 8 inches, broad 14 inches. The lower filling strakes broad 26 inches, the second 24 inches.

(107 II 45) *A Ship made, named Mauritius, Anno 1637 June the 20th.*

Long in the hold, 128 feet, wide 28 feet, deep at deck height 13 feet. Above that an upper deck, high 6 feet and ¼. Forward a forecastle long 30 feet, high 4 and /4 [¼? ¾?] feet. Aft a long steering stand. The stem high 25 feet, rakes 20 feet, thick 14½ inches, curves up to the keel joint 4 and ½ feet, is long, measured along the curve 38 feet, below broad 4 feet, above broad 2½ feet. The sternpost high 23 feet, rakes 4 feet and ¼, thick 14 and ½ feet. The wing transom is at 20 feet, and above that the stern protrudes 3 feet: the wing transom is long 16 feet, curved 1 foot, thick in the square [underneath the top] 13 inches. The tuck stands high from the keel 10 feet and ¼: the bottom is wide 20 feet, rises 3 and ¼ inches, height in the bilges 4¼ feet, is wide there 26 feet 2 inches, the sides rake 10 inches then. The bilge futtocks deep 9 inches, in the turn of the bilge 7½ inches: at the master ribband 6 inches: the keelson thick 9 inches, wide 20 inches, riders in the aft peak 3 inches, in the bow 8 inches. The two lower wales below the gunports are wide 12 and ½ inches, thick 6 inches. The filling strakes between them wide 27 inches, the filling strakes at the height of the gunports wide 3 and ¼ feet. The wale above that wide 11 and ½ inches, thick 5 inches, a filling strake above that one wide 1 foot. Above that a wale broad 9 inches, thick 4 and ½ inches. Above that a filling strake broad 12 inches. Above that a sheer rail broad 8 inches. Above upperworks as required.

(108 I 33) *A Ship made, named Mauritius, of the year 1639 November the 16th.*

Long between stem and stern 129 feet, deep 13 feet, wide 28¾ feet. The upper deck high in the sides 6¼ feet. A forecastle long 30 feet, lies 1 foot lower, high in the sides 6¼ feet. A long steering stand. The keel is deep 16 inches, wide 20 inches, long 107 feet, the scarf long 7½ feet. The stem high 24½ feet, rakes 19½ feet, is thick 12¼ inches, the front side thick 9 inches, below wide 3½ feet, above 2½ feet, is curved at 15 feet, 12½ inches, above a bit more bend, below a little straighter.

The sternpost high 23 feet, the transom is at 20 feet, then the portholes come above the wing transom at 20 feet. The tuck is high from the keel 10½ feet. The wing transom deep 16, thick 11 inches, long 17¼ feet. Curves 12 inches. The sternpost rakes 4½ feet, thick 13 inches, below wide 6 feet, above wide

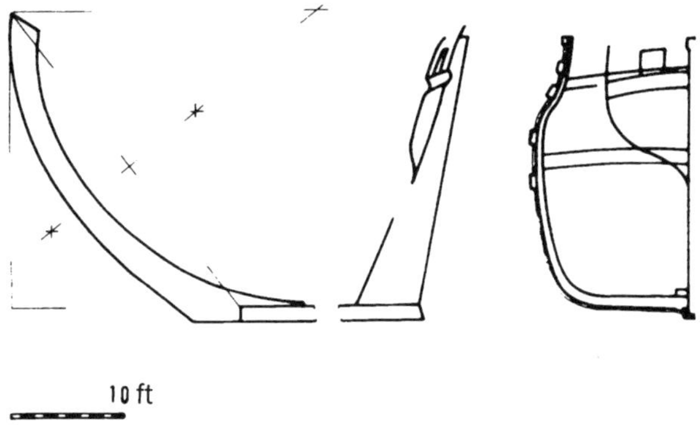

FIGURE 3.34. Stem, stern, and main frame of the *Mauritius* (1637). For this particular ship, measurements between stem and stern are not given in the contract, as in all the others. The contract simply says "long in the hold," the exact meaning of which is not clear. Stem, stern, and main frame sections are easy to construct. This is probably another merchantman. In this case the transom is not located at the top of the sternpost, as was standard. The gunports must have been made above the transom here (Drawing by A. J. Hoving)

18 inches, the aft side thick 7½ inches. The bottom is wide 19¼ feet, rises 4 inches, at the top of the bilges 4½ feet, wide there 26¾ feet, the sides rake 1 foot. The futtocks deep 9¼ inches, at the bilges 7 inches, at the master ribband 6 inches, above 4¼ inches. The mainmast is 5 feet abaft the middle of the ship. The deck clamps thick 4½ inches. The lower deck beams deep 12½ inches, wide from each other 3½ feet. Two pairs of wales below the gunports thick 6 inches wide 12½ inches, the filling strakes wide 28 inches, raised afore 3½ feet, aft 9½ feet. The pumps on the lower deck are measured from the stern 20½ feet. The beakhead is long 20½ feet, measured over the top of the lower knee of the head. The lion is long 7 feet. The trail board is deep 18 inches, measured at the ship and 8 inches afore.

(108 II 28) *A Ship made called the Burg, Anno 1640 July the 4th.*

Long 115½ feet, wide 27¾ feet, deep 11½ feet, above that 6¾ feet. The stem high 20 feet, the sternpost 20¾ feet, rakes 4 feet. The wing transom is long 17¼ feet, curves 12 inches. The bottom is wide 19 feet, rises 12 inches, at the top of the turn of the bilge 4 feet 7 inches high and wide 25½ feet, the sides rake 1 foot [and ⅛ foot]. The lower deck beams curve 7 inches at a length of 28 feet. The upper deck beams curve 13 inches at 22 feet length. The beakhead is long 20 feet, measured on top of the lower stud up to the fore end of the lion's paws. The lion is long 8 feet. The kam is deep 17 inches measured at the ship and 7½ inches afore.

(108 II 48) *A Ship made named Deventer, Anno 1640 December the 4th.*

Long 130 feet, wide 29¼ feet, deep at the lower deck in the main hatch 13½ feet, above that 6¼ feet, aft 7 feet, forward 6 feet, a forecastle long 30 feet, high from the deck, measured in the side 5½ feet, afore high 5¼ feet, aft a long steering stand long 16 feet. The deck of the captain's cabin is long from the transom 28 feet. The pumps are from the counter 21 feet, the beams of the cabin high in the side 6 feet 4 inches. The cabin is high afore 6 feet 2 inches, at the aft end high 6½ feet. The stem high 24¼ feet, rakes 23 feet, is curved up to the scarf 4½ feet thick 13 inches and is equally curved everywhere, afore thick 9½ inches, below wide 3¾ feet, above wide 2¾ feet and has a curve of 12 inches at 15 feet length. The sternpost high 23 feet, rakes 4¼ feet, thick 13 inches, aft face thick 9 inches, below wide 6 feet, above wide 18 inches. The wing transom is high at the stern 20 feet, is long 19 feet, deep 14 inches, thick 12 inches, curved on 19 feet length 13½ inches. The stern timbers are 25 feet above the transom, 10 feet apart from each other there. The keel is deep 16 inches, wide 19 inches. The scarf is long 8 feet. The bottom is wide 20¼ feet, rises 3 inches, at the top of the turn of the bilge 4¾ feet high and wide there 27 feet, the sides rake 1 foot [and

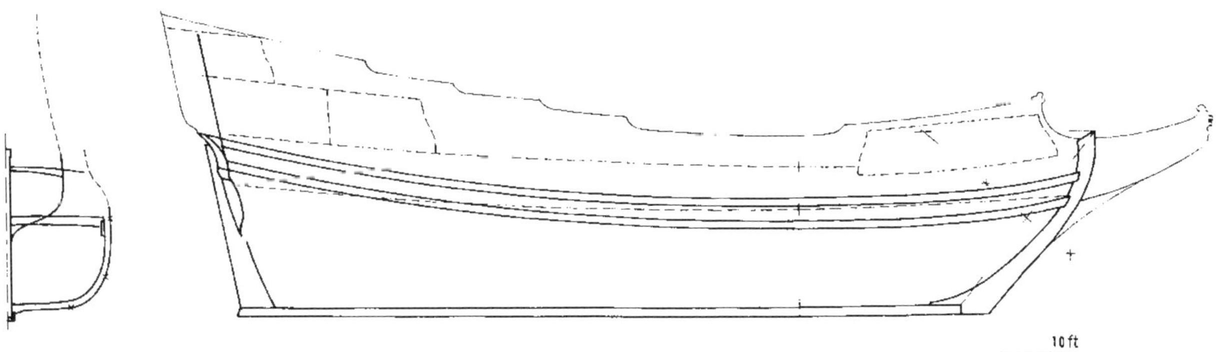

Figure 3.35. Stern and side views of the *Mauritius* (1639), 129 × 28¾ × 13 feet. The description of the stem is especially striking: "is curved at 15 feet, 12½ inches, above a bit more bend, below a little straighter." This is a clear example of a stern that does not match a quarter of a circle, struck from the intersection of the perpendicular on the diagonal and the height of the stem. Two wales are given, as well as the length of the beakhead. As in the preceding contract, the transom is not at the top of the sternpost. The hypothetical interior of the ship is indicated by dashed lines; thin lines indicate its supposed contours (Drawing by A. J. Hoving)

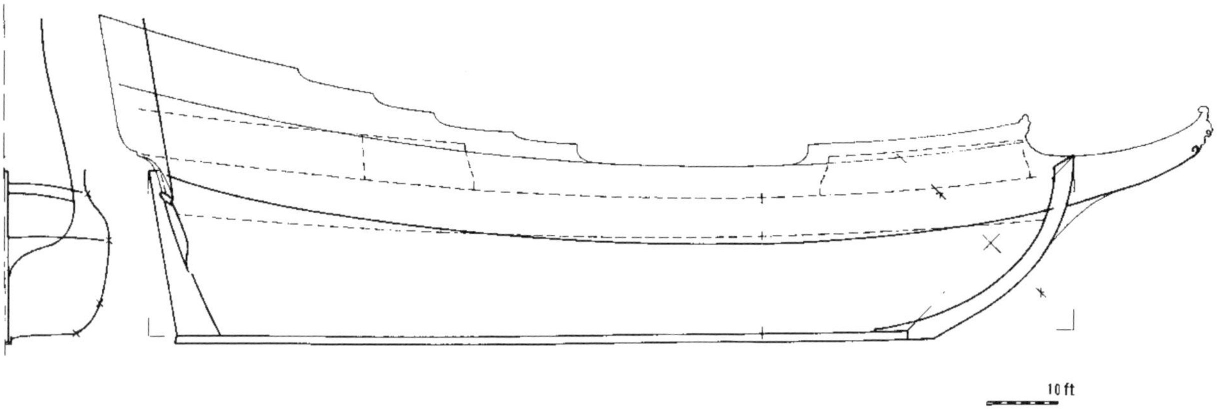

Figure 3.36. Stern and side views of the *Deventer* (1640), 130 × 29¼ × 13½ feet. The description of the shape of the stem is not completely clear; it probably means that at 15 feet from the top, the curve is 12 inches. The dashed lines and thin lines are hypothetical, indicating the ship's presumed interior and contours, respectively (Drawing by A. J. Hoving)

⅛ inch], raised afore 4 and ¼ feet, behind 9¼ feet. The foremast is from the stem 9 feet. The feet of the bitts are forward of the other 2¼ feet. The beakhead is long 22 feet. The hawseholes wide 10 inches. The main channel long 28 feet. The fore channel 22 feet. The galley wide 6½ feet from the side measured from the lower side of the gunports and along ship 4 feet. Below the foremast step 3 riders, between step and lower deck 2 and over the peak 3 riders. Two riders, and 4 futtock riders at the mast. Nine gunports in the side at the center of the lower deck. The gunports are high above the deck 25 inches, in square 2¼ feet. In the forecastle on one side 3 gunports. On the steering stand 2 gunports on one side, in the captain's cabin 1, in the upper cabin 1, together 32 gunports. The main deck beams are cambered 6 inches at 28 feet length. The upper deck beams 22 ft. 13 in.

(109 II 1) *A Ship made named the Haen, Anno 1641 March the 10th.*

Long 122½ feet, wide 27 or 28 feet, deep 12½ feet, above that 6¼ feet, behind a little bit higher 6¾ feet, afore with a beam long 30 feet, high from the deck in the side 5 feet, aft a steering stand. The stem high 23 feet, rakes 17½ feet, thick 12½ inches. The sternpost high 22 feet, rakes 4¼ feet. The wing transom is long 18 feet. The tuck high from the keel 11 feet. The lower deck beams cambered 7 inches at 28 feet length. The upper deck beams cambered 13 inches at 22 feet.

(109 II 30) *A Ship named the Elias, Anno 1641 October the 24th.*

Long between stem and stern 131½ feet, wide 29 feet, deep at deck height 13 feet, above that high 6 feet, aft 6½ feet high. Afore a forecastle long 32 feet, high from the lower deck 5 feet, above that just 5 feet. Aft a steering stand, with an adequate length so that the bars of the capstan fit, or from the wing transom 29 feet, high in front 6 feet, aft 6½ feet. The captain's cabin is high at the fore end 6 feet 2 inches, aft 6½ feet. The sternpost is high 23¼ feet, rakes 4 feet. The sternpost below wide 6 feet, above wide 17 inches. The tuck is at 12 feet. The fashion pieces 9 inches thick. The stem high 24½ feet, it rakes 24 and 3/ [¾] feet, is curved up to the scarf 5 feet, below wide 3¼ feet, above wide 2 feet 7 inches, inside thick 13 inches, forward face thick 10 inches. The keel is deep 16 inches, wide 19 inches. The stern timbers are above the wing transom 23 feet and apart from each other 11 feet. The bottom is wide 21 feet, rises 5 inches, at the top of the turn of the bilge 4¾ feet high and wide there 27 feet. So the sides rake at a depth of 13 feet, 1 foot, raised behind 9½ feet, afore 3 feet. The beakhead is long 22 feet. The lion is long 8¾ feet. The kam is deep 19 inches measured at the ship and 8 inches afore. The top of the capstan is thick 21 inches. In the beugel ["brace," probably in the partner] 15½ inches, below 14 inches, the head is long 5¼ feet. The lower deck beams have 7 inches curve on 28 feet length. The upper deck beams have 13 inches curve at 22 feet length.

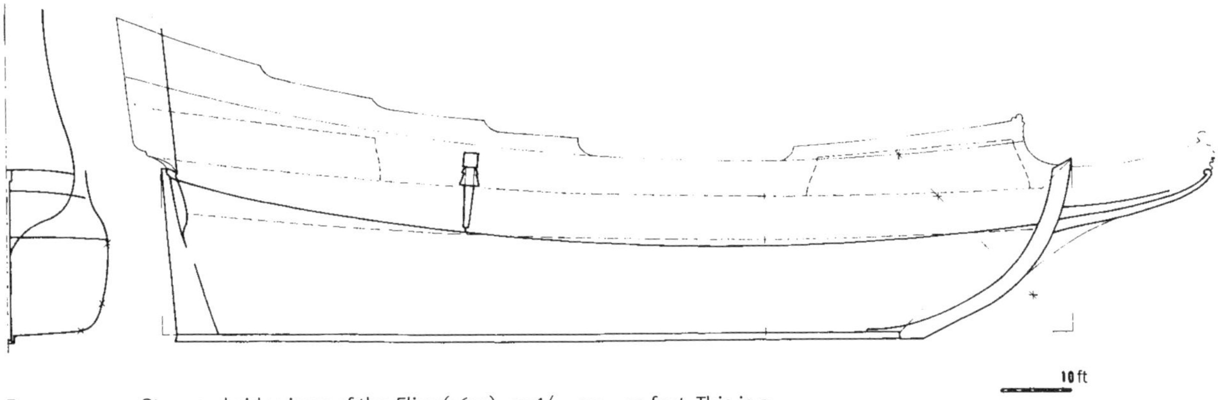

Figure 3.37. Stern and side views of the *Elias* (1641), 131½ × 29 × 13 feet. This is a remarkably complete contract. The thin lines indicate the supposed contours of the ship. Lengths and heights of the various compartments are all shown with dashed lines (Drawing by A. J. Hoving)

(110 II 13) *The Salmander.*

Long 135 feet, wide 33 feet, deep at deck height 14 feet. Two upper decks above that.

(110 II 17) *The Princes.*

Long 137 feet, wide 36 feet, deep at deck height 15 feet.

(110 II 20) *Gelderlandt.*

Long 133 feet, wide 32½ feet, in its first hold deep 12½ feet, above that 5½ feet, an upper deck on top of that 7 feet high. The futtocks thick 1 foot. The keelson 1 foot, wide 2¾ feet.

(110 II 27) *The Ship Leiden.*

Long 126½ feet, wide 28 feet: although too narrow, deep in the hold 12¾ above that 4 and ½ feet, above that 6 and ½ feet.

(110 II 32) *A Ship named the Eendracht, Anno 1660 February the 18th.*

Long 133 feet, wide 29 feet 4 inches, deep 12 and ½ feet. The upper deck high in the side 6 and ½ feet, afore a forecastle long 32 feet, high from the deck 5 feet, afore high 4 and ¼ feet. The captain's cabin afore high 7 feet, aft high 7½ feet. The cabin's floor long from the stern 16 feet, is ½ foot lower. The stem high 23½ feet, rakes 21 feet, curved to the scarf 4½ feet, thick 13 inches, forward face 10 inches, below wide 3½ feet, above 2½ feet. The sternpost high 23, rakes 4¼ feet. The wing transom is high at 20¼ feet. The sternpost below broad 6 feet, thick 13 inches, aft face 8 inches, the wing transom is long 18¼ feet, thick 12½ inches, deep 14 inches, curved 12½ inches. The keel is wide 20 inches, deep 17 inches, the tuck is at 11 feet, The bottom is 20 feet, rises 3 in., at the top of the turn of the bilge 4 feet 2 inches, wide there 26¾ feet, the sides rake 1 foot [plus 4 inches]. The lower deck beams camber 7 inches at 28 feet length. The upper deck beams 13 inches at 22 feet length.

(110 I 22) *A Ship made named Graef Enno, Anno 1642 September the 20th.*

Long 126 and ½ feet, wide 28 feet, deep in the hold at deck height 12¼ feet, above that 6¼ feet, the forecastle long 29 feet, high from the deck 5 feet. Aft a long steering stand so that the bars of the capstan can move free of the halyard knight. The stem high 22½ feet, rakes 20 feet, curves 4¼ feet, thick 12 inches, forward face 8½ inches, below wide 3½ feet. The sternpost high 21¼ feet, rakes 4¼ feet, thick 12 inches, aft face 7½ inches, below wide 6 feet. The tuck is high from the keel 10½ feet. The wing transom is long 18 feet. The stern timbers are above the wing transom 22 feet, above wide from each other 9½ feet. The bottom is wide 20 feet, rises 4 inches, at the top of the turn of the bilge high 4¼ feet, wide there 25 feet 7 inches, the sides rake 10 inches [12 inches], rises afore 3 feet, behind 9 feet. It is wide at the first wale 25 feet from the wing transom: 13½ inches less than in the middle, therefore at the dale as much narrower [a mysterious sentence and a most unusual remark in a contract; it seems to imply that the dale is at 25 feet from the wing transom]. The futtocks deep 9 and 1/2 inches, at the outside of the bottom 8 1/2 inches, at the top of the bilges 7 inches, at the master ribband 6 inches, above 4 inches. The top timbers which are turned outward stand from the stem 9 feet [these top timbers defined the forward end of the forecastle]. The hackebord stands off the sternpost 24 inches. The lower hackebord is wide at the inside 1 foot. The beams have 13 inches of camber at 22 feet: The ship has a sheer of 1 1/2 inch on a length of 23 feet. When it was launched from the slipway it was deep aft 7 feet and forward 5 feet.

Contracts for Masts, Spars, and Ropework

Shipyards gave work to many industries and trades—block and top makers, sailmakers, pump makers, ropewalks, smiths, and mast makers, all acting as subcontractors who executed portions of a shipbuilding contract.

In addition, woodcarvers and painters were often hired by the shipyards, as well as plumbers (for the pipe system that carried off the water from the bilge pumps and for the privies), braziers and blacksmiths (for covering the cheese and bread room and the rudder), and bricklayers (for laying the galley). In many cases the families of yard workers earned some extra money in their scarce free moments by cutting the wooden treenails that were used in shipbuilding. And there were the timber merchants, weavers, slaughterhouses, bakeries, fishmongers, brewers and brandy makers, grocers, chemists, and the other suppliers of such essentials as navigation instruments and all the other equipment and provisions that were necessary aboard a ship (for Witsen's inventory list, see chapter 2, section 137, "Loose Implements").[7]

So the shipyard contracted with other industries and trades to provide a number of things, which entailed separate contracts or *certers*. Witsen includes some contracts of this type, and a few of them are presented below as examples. As with his other contracts, it is not always clear whether they were purely legal documents or rather memoranda written by the craftsmen themselves for their own use.

Lengths of masts and spars were expressed in feet, their circumference, or thickness (*dick*), in *palms* (1 *palm* = 10 centimeters). Perhaps *palm* should be translated by "hand" (4 inches), but this English measurement was used in different circumstances, so Witsen's term has been retained. To obtain the diameter, the circumference must be divided by 3.14.

An "octagonal" refers to the upper part of a mast above the masthead. Close under the top the mast had a diameter that was three fourths of the size of the mast at deck level (see section 94, Step the Masts). An ell was 69 centimeters. We will return to the width of sailcloth in the section "Contracts for Sails" later in this chapter.

(115 I 1) As heretofore made clear the way in this country how Ships, to be made by shipwrights, were put out to tender, together with the laws and measurements they are held to, we will now show some Certers or regulations of masts and spars and blocks for the

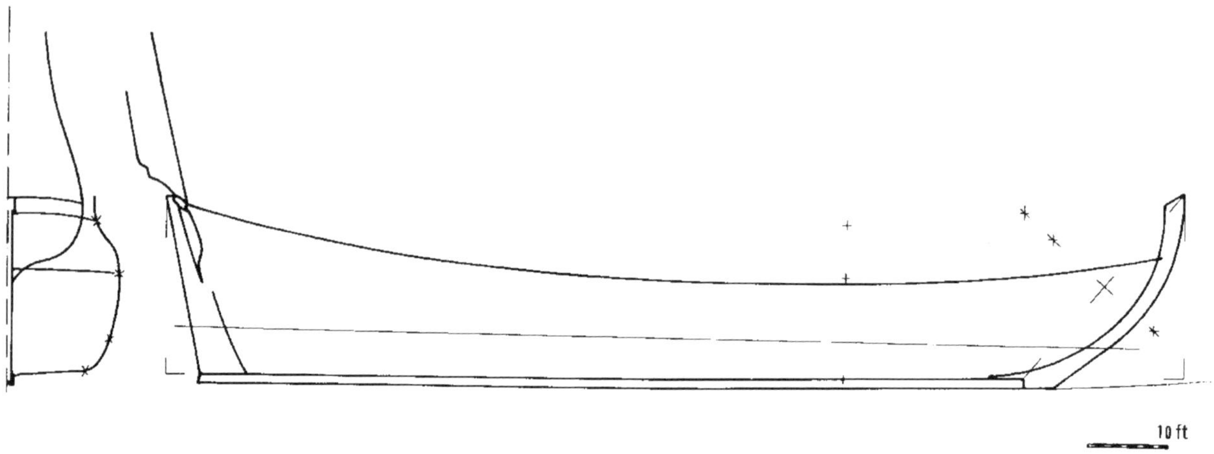

FIGURE 3.38. Stern and side views of the *Graef Enno* (1642), 126 1/2 × 28 × 12 1/4 feet. The vaguely formulated statement about the steering stand and the capstan bars means that the steering stand was so long that the bars of the capstan had sufficient room between the bulwark of the captain's cabin and the halyard knight. A similar arrangement was used in Witsen's pinas. Another interesting point is the measured depth fore and aft after launch. Apparently the ship was deeper in the stern because of its construction and shape, not because it was ballasted that way (Drawing by A. J. Hoving)

ships, of which the lengths are given as people in this country are used to do when they were contracted to deliver for a taken price.

The Masts and Spars for two Ships' rigs, long 136 feet, wide 34 feet.

The mainmast thick 21 palms, below its octagonal 16 palms, long 87 or 88 feet. The foremast 19 palms, under its octagonal ¼ less, long 77 or 78 feet. The bowsprit thick 21 palms, under the stay 16 palms, long 64 feet. The mizzenmast thick 14 palms, above, under the octagonal ¼ less, long 66 feet. The main topmast thick 13¼ palms, upward as required, long 55 feet. The fore topmast thick 12 palms, upward as required, long 45 feet. The main yard wide 27 cloths or 32 ells, thick 16 palms. The foreyard will be for 24 cloths, thick 14 palms. The spritsail yard wide 16 cloths, thick as required. The main topsail yard as required. The fore topsail yard as required. The spritsail topmast and yards. The topgallant masts and yards. The mizzen topmast and yards: the pennant staves: four studding sail booms: the mizzen yard: all long as required. Two lower studding sail booms. A flagpole at the poop for a flag of 12 cloths.

(115 I 41) *Following is a Cherter of the Masts and Spars of a Ship, long 130 feet, wide 32 feet.*

The mainmast thick 19¼ or 19½ palms, and below its octagonal ¼ less, long 84 feet. The foremast thick 17¼ or 17½ palms, and below its octagonal ¼ less. The mizzenmast thick 13 palms, and below its octagonal ¼ less, long 62 feet. The bowsprit thick 19 palms, below the stay ¼ less, long 62 feet. The main topmast thick 12½ palms, upward as required, long 54 feet. The fore topmast thick 10½ palms, long 44 feet, upward as required. The main yard wide 25 cloths which is 29 ells, thick 14 palms. The foreyard will be wide 22 cloths or 27 ells, thick 12¾ or 13 palms. The spritsail yard wide 15 cloths, thick as required. The main topsail yard wide 13 cloths, or 1 ell wider than half the main yard. The fore topsail yard wide 11 cloths. The mizzen yard long and thick as required. The spritsail topmast and yards: two topgallant masts and yards: two mizzen topmasts and yards: all as required. Four studding sail booms and two lower studding sail booms as required. Three pennant staves, a flagpole on the poop for a flag of 12 cloths.

Apart from these contracts, Witsen also claims to have taken measurements of masts and rigging on ships, and these results follow here. A fathom is 6 Amsterdam feet or 1.70 meters.

(128 II 13) *A Ship of 130 or 136 feet long, wide 31 or 32 feet.*

The mainstay thick 14 inches, with the collar long 26 fathoms. The forestay thick 11 inches, long 12 fathoms. The main shrouds thick 7 inches, every pair 20 fathoms. The fore shrouds thick 6 inches, every pair long 19 fathoms. The mizzen shrouds thick 4 inches every pair 16 fathoms. The mizzen stay thick 4 inches, long 11 fathoms. The main top shrouds thick 4 inches, every pair long 15 fathoms. The fore top shrouds thick 3½ inches, every pair long 14 fathoms. The main backstay thick 4 inches, long 34 fathoms. The fore backstays thick 3½ inches, long 32 fathoms. The main top rope thick 4½ inches, long 8 fathoms. The main topsail halyard long 36 fathoms. The foretop rope thick 3½ inches, long 7 fathoms. The main topsail lifts long 34 fathoms. The main top yard parrel line long 4½ fathoms. The main topsail clew lines long 36 fathoms. The main topsail braces long 28 fathoms. The main braces long 30 fathoms. The main halyard tie thick 8 inches, long 30 fathoms. The mainsail lifts long 28 fathoms. The main halyard rove through the ramshead long 46 fathoms. The main topsail sheets thick 5½ inches, long 25 fathoms. The main sheets thick 5 inches, long 28 fathoms. The main tacks thick 7½ inches, long 16 fathoms. The main top stay thick 4 inches, long 25 fathoms. The fore topsail braces long 27 fathoms. The fore topsail bowlines long 32 fathoms. The fore-topsail parrel line long 4½ fathoms. The fore halyard tie over the mast cap thick 7 inches, long 26 fathoms. The fore halyard long 36 fathoms. The pendant of the main tackle [hanging from the mainstay] long 12 fathoms. The lanyard long 38 fathoms. The foretackle long 11 fathoms. The fall long 36 fathoms. The tackle of the main parrel long 11 fathoms. The parrel line long 7 fathoms. The tackle of the fore parrel long 10 fathoms. The parrel rope long 6 fathoms. The fore lifts 26 fathoms. The foretopsail clew lines long 32 fathoms. The foretop stay long 12 fathoms. The fore-topsail sheets long 22 fathoms. Foresheets long 24 fathoms. Foretacks long 12 fathoms. Fore bowlines long 20 fathoms. Spritsail lifts long 20 fathoms. Spritsail braces long 22 fathoms. Spritsail lifts long 18 fathoms. Spritsail sheet pendant long 12 fathoms. Spritsail sheets long 15 fathoms. Fore braces long 22 fathoms. Mizzen sheet long 20 fathoms. Mizzen halyard long 36 fathoms.

The masts are in proportion, thick 20 palms, long 86 feet. The foremast long 77 feet, thick 18 palms. The bowsprit long 58 feet, thick 19 palms. The mizzenmast long 66 feet, thick 14 palms. The main topmast long 56 feet, thick 13 or 13½ palms. The foretopmast long 47 feet, thick 11 palms. The main yard should be 24 or 25 cloths, is long 78 feet. Furthermore with the topsail yard and spritsail yard as said heretofore.

Contracts for Blocks

Dutch contracts for blocks are always interesting, partly because of the colorful names of some of the blocks, whose double meanings are lost in translation.[8] Furthermore, as some parts of the rigging system were typically Dutch, they have no suitable English equivalent.

There is always a certain logic in these lists in that they usually begin with the biggest block on board: the ramshead. Typically Dutch, this huge block (almost three feet high) was connected to the knight (the top of which was nicely carved as a black moor) by the halyard, which was rove through both, and by the halyard tie to the main yard. The doubled tie ran over the half-rounded top of the mast cap; by hauling on the halyard, the yard was hoisted. The same system was employed for the foreyard with slightly smaller components.

Tackle pendants were put over the mast top before the shrouds were laid over. In Holland there were always two types on each side of the mast: one with a double-sheaved block stropped to the pendant (with sheaves of similar sizes next to each other) and one with a fiddle block (with a smaller sheave below a bigger one). The lower blocks of these tackles had iron hooks and were hooked into rings on the channels, just inboard of the deadeyes.

Next the contracts describe the pendants hung from the yardarms to hoist heavy loads, such as cargo or the ship's boat. Thereafter the sheaves in the catheads and topmasts are mentioned, followed by the blocks for the running rigging.

(95 II 32) *Sizes of Blocks, of a Ship long about 134 feet.*

The main halyard ramshead block, big 3 feet minus ¼ foot. The main knight at the mainmast, to hoist yard and sails, has 3 pockwood [lignum vitae] sheaves, just like the ramshead-blocks, or also metal sheaves. The fore knight has similar sheaves. 2 Double tackle blocks inboard the main shrouds, each 28 inches. 2 Single-Sheaved of 16 inches against double ones with the same mast. 2 Double-sheaved inboard the fore shrouds, of 25 inches each. 2 Single-Sheaved to go with them of 14 inches each. 8 Tackle blocks of 14 inches, fore and aft. 2 Pendants on the main yard, of 18 inches each. 2 Pendant blocks on the foreyard, 16 inches each. 2 Pendants of 18 inches under the top. 2 blocks of 18 inches to hang from the mast cap to hoist or lower the topmast in times of rough weather. 4 Topsail sheet blocks, one on each yardarm, the two big ones are 27 inches, the smallest 15 inches. A double Tackle Block, with a single (because a double one never goes without a single) to hoist goods, like beer, water, wine, etc., the double one of 26 inches, the single one 14 inches. The anchor tackle big 26 inches. Two bow rope blocks, big as they come. 3 Sheaves in each cathead, with which the anchor is raised, of uncertain size. 2 Main sheet blocks to the mainsail of 18 inches. 2 Fore ones of 16 inches. 1 Cat block, made with a swivel. Some have this block, others do not. 2 Sheaves of pockwood or metal in the topmasts. 2 Main sheet clamps of uncertain size. 2 Foresheet clamps of uncertain size. 2 Snatch Blocks, a sheave in each one. The main topsail halyard block has 18 inches in its width, namely the two-sheaved one, the single one 1 foot. The fore topmast halyard block 16 inches. The single one 10 inches. 4 lift blocks each 18 inches. 4 spritsail lift blocks, 2 of 16, 2 of 18 inches. 2 Bridle blocks on the bowsprit, of uncertain size. 1 Lizard, or deadeye with one hole, through which the spritsail sheet runs, which rope has the same name, of uncertain size. 1 Double block on the bowsprit, of 8 inches. 1 Spritsail halyard block of 20 inches, with 2 sheaves. The ribs of the main topsail parrel are 16 inches, combined with 10 Trucks of 8 inches. There are about 20 shroud bull's-eye fairleads, which are notched. For the topsail parrel, the ribs are 14 inches. For the mizzen parrel the ribs are 16 inches. A main parrel is 3 feet. A fore parrel is 2½ feet. 4 Sleepers for the two sails of guessed size. The snatch block, of uncertain size; against which comes a tackle block, double and single. A mizzen halyard block, with two sheaves of 18 inches. 8 sheet blocks, four main and four fore, of 10 inches. A metal or pockwood sheave in the mizzenmast. 80 clew-line blocks, each 10 inches. 50 Brace blocks, each of 8 inches. 70 Topgallant blocks, 5 inches each. 44 Topgallant deadeyes, 5 inches each, four in the main crosstree, four in the fore crosstree, four in the mizzen crosstree, four in the bowsprit. 4 Deadeyes for the boat, each 8 inches. A sheave fore in the boat's bow. 70 Belaying Pins, used to belay the ropes in the top of the sides. 4 Pins in the bitts. A pair of sheet blocks at the back of the ship. A small deadeye hooped with a hook on the mizzenmast, both of uncertain size. 2 cleats on each yard, one of 11 inches, a cross of which the sheave is big 20 inches, with a metal bush in it to unload corn. Furthermore to the given ship belong 30 double tackle blocks for the guns, of 18 inches. 30 porthole blocks, and furthermore stool beds, wedges, wheels, etc. The parrel fall blocks are of 16 or 17 inches, 12 in number, and are used for the parrels of the three masts.

To a Fluyt of 100 feet,

This all will be in the proportion of the ship, only differing in it, that some blocks mentioned above, will not be there.

The following contract contains information not only about the blocks, as announced, but also about many other things coming from other industries and trades: tops, caps, pumps, scuppers, sheaves in knights, halyard blocks and topmasts, bars for the capstan in the boat, and the wheels, stool beds, wedges, and crowbars for the gun carriages.

(115 II 17) *Contract of Blocks, of a Ship, long between posts 114 feet, wide 28 feet, deep 11½ feet, as it was put out to contract.*

This particular Ship will be fitted out with 4 tops, the main top wide 10¼ feet. The foretop wide 9¼ feet, to be made from planks 2½ inches thick, the two small tops 4½ feet, to be made from planks 1½ inches thick, the cap long 3½ feet, wide 2½ feet, high 16 inches: the fore cap long 3¼ feet, wide 2¼ feet, high as required. The main cap and furthermore all the other ones as required, everything made from the contractor's own wood and nails. 3 Pumps, 2 abaft, next to each other, and one with the mainmast, with all that belongs therewith. The pumps from good sound wood, for every pump 3 buckets and 6 haertjes. The scuppers should be well drilled, as will be the knights, supplied with pockwood sheaves. The sheaves in the ramshead block and the topmast top rope sheaves and the sheaves in the foot of the topmasts, should be delivered. The costs of the pockwood are for the commissioner's account. All the blocks to finish, as big as required and as many as required.

The Ship shall be fitted, at the mainmast with seven pairs, at the foremast with six pairs of shrouds. The main topmast and the mizzen with 4 shrouds.

The fore topmast with three shrouds: and for each a tackle for tightening. Fore and aft with Spanish burtons. The mizzen halyard with two double blocks and the topsail halyard with a double block above, as there will be for spare ones. A stay tackle and an anchor tackle, on the lower and upper yards studding sail halyard blocks, aft near the transom sheet blocks, and swifters, two winding tackle blocks, with a cat block to put in the guns, three snatch blocks with open sides, and backstays fore and aft.

The contractor will also be held to make pockwood sheaves in all topsail halyard blocks, topsail sheets and pendants below the yard, as well as all tackle blocks, and sheet blocks, mizzen halyard sheet blocks, and clew-line blocks, as well as the garnet and anchor tackle, all pockwood paid by the commissioner.

The contractor will also be held to let the spindles of the sheaves be of ironwood of the right size. Also to deliver 6 commanders and as many belaying pins as there are shroud fairleads, as will be necessary in fitting out the ship. Still there are port-lid blocks, and everything that might be forgotten, necessary in fitting out the ship. The burton lines fore and aft to tighten the shrouds, with a staysail halyard block, staysail deadeyes, eight boat bars for the boat. The backstays fore and aft above and below with double blocks. The main deadeyes wide 10½ inches, thick 4½ inches, the ones in the fore shrouds wide 9½ inches, thick 4 inches, wheels for 30 guns, tackle blocks, wedges, stool beds, crow bars of the mentioned 30 guns, of which 18 guns are on the lower deck, the wheels fore high 15, and aft 12 inches, thick 5 inches, on the upper deck there are 12 guns, of which the fore wheels are high 12 inches and aft high 10 inches, thick 4½ inches. All the tackle blocks, of the upper and lower deck, with double blocks.

Contracts for Sails

Sails consisted of canvas cloths sewn together. The lengths and widths of the sails were measured in ells of 69 centimeters, but sometimes, as we have seen, widths were expressed only in the number of cloths. The width of a cloth depended of the sort of material used and apparently was not standardized, which is why calculations of sail width are always uncertain. For instance, there was heavy Dutch canvas of 78 centimeters, lighter French canvas for topgallant sails of 52 centimeters, Flemish canvas, Melis canvas (from Meppel, in northeastern Holland), and "clover cloth" (shamrock?), also for upper sails. Each had its own width, which was of course related to the size of the loom on which it was woven. In a time when so many different measurements were used in the various trade centers, some variation in the widths of sailcloth is not surprising.

In chapter 2 the calculated widths of the cloths for the sails of Witsen's pinas vary from 76 to 78 centimeters for the mainsails and 67 centimeters for the lighter ones.

Van Yk says that the width of a cloth of Holland canvas, of which main, fore, and top sails were made, was 1⅛ ells, which is 78 centimeters, but after sewing, a finished width of only 72 centimeters was left. According to a contract for a 160-foot-long ship, each cloth is 3 feet wide, or 85 centimeters: The main yard is wide 30 cloths, each cloth taken to be 3 feet, so that 30 cloths is 90 feet (128 II 2). But this is too wide compared with the cloth widths of the sails for the pinas. In the earlier section of contracts for masts and spars, the contract for the 136-foot ship (115 I 13) mentions that 32 ells (69 centimeters each) equal 27 cloths, which would make the width of the cloth 81 centimeters (32 × 69 ÷ 27). The contract for the 130-foot ship (115 I 41) states that 29 ells equal 25 cloths, making one cloth 80 centimeters wide (29 × 69 ÷ 25); it further states that 27 ells equal 22 cloths, indicating a cloth width of 84 centimeters (27 × 69 ÷ 22).

The two sample contracts below are for sails for a boyer and a galliot, for which only sail sizes in ells and cloths are stipulated. Using Witsen's data, professional rigger Floris Hin has provided reconstructed sketches of the sails, figuring the widths of the cloths at 77 centimeters, with an overlap of 2 inches (5.2 centimeters).

Additional contracts are included for sails for different sizes of seagoing ships.[9]

(137 II 12) *For a Boyer of 60 Lasts.*

A Gaffsail is made, aft deep 29½ ells, forward deep 18 ells, head wide 13 cloths, below wide 15 cloths. A Topsail head wide 8½ cloths, below wide 15 cloths, deep 12½ ells; a Spritsail wide 11 cloths, deep 6 ells, a Staysail aft deep 19½ ells, forward deep 20½ ells, below wide 10 cloths; a Mizzen wide below 7 cloths, forward 10¾ ells, aft 12 ells.

(137 II 35) *A Galliot long 75 feet, wide 19, deep 10, loading 50 lasts.*

Has a Sail below wide 12½ cloths, upward 11½ cloths, aft deep 25½ ells, with its bonnet, and it has been cut on the front side, the largest cut is in the first gore, the other ones lessening, up to the two aftermost cloths, which should be straight cut. For under the Mast 4½ cloths should be gored and there are 5 straight-cut cloths, three cloths on the aft side are cut upward, here one or two bonnets were cut off as long as desired, the seams and gores should be drawn on the ells, measured with a piece of rope, until 8½ ells of gore are obtained. Its square Foresail is wide 11 cloths, deep 16 ells and has a Reef on top, which on one side is 6 ells, the other side to nothing; so that the Reef is oblique 6 ells, and in times of distress it can be used as a Gaffsail: below it has another Reef, deep 3 ells. Its Topsail is below wide 11 cloths, upward wide 7 cloths, deep 12 ells, and has a block on each sheet, and three cringles on each side. The Mizzen is wide 10 cloths, deep 12½ ells, and has a reef below deep 2 ells: the Boltrope runs half an ell above the Reef. The Jib is wide 9 cloths, deep aft 19 ells, and above one ell cut off to make it blunt, so that 18 ells remain, forward it is cut up 3 cloths, backward 2 cloths, so that 4 square cloths in the middle remain: it is seamed up from below and bolt-roped on the two most aftermost cloths: the first boltrope is thick 5 paer [no English equivalent], then forward a boltrope of 11 paer is taken. And upward two eyes are split, as is done with a Square Sail, it has a Reef with a bonnet deep 2½ ells. On the aft side of the Reef a tabling [reinforcing lining?], in which two Reef eyes, in which the cringle comes. The narrow Foresail wide 8½ cloths, deep aft 20 ells, forward 20½ ells: at the top another ell is taken off for the block so that 19 ells stay and forward 19½ ells: it has two reefs with linings, the lower Reef is deep 3 ells, the other one 2 ells. These Reefs are with holes, and at the back of the Reefs two tablings, and on each tabling 2 cringles, but forward no tablings are necessary; because there comes only one Cringle on two Reefs: these Cringles are struck through Reef holes. The spritsail is wide 4 cloths, deep 4 ells. The Mizzen [probably a storm sail] is wide 4½ cloths, deep aft 7 ells, fore at the Mast 6½ ells. The Foresail [probably a storm sail] wide 3 cloths, deep aft 5½ ells.

(135 II 23) *Sails for a Ship of 235 lasts.*

Its Mainsail without the bonnet was wide 22 cloths, deep 14 ells, total amount of cloth, tablings included: 328 ells. Its Foresail with a bonnet 19 cloths, deep without a bonnet 12 ells, the bonnet is deep 2½ ells, together 15 ells of cloth, with the tablings 309 ells. The Main Topsail above 12½ cloths, below 21 cloths, deep 18 ells of cloth, with the tablings 307 ells. The fore Topsail above wide 10¼ cloths, below 18½ cloths, deep 15 ells, with the tablings 244 ells. The Mizzen with its bonnet wide 16 cloths, deep aft 20½ ells, 170 ells of cloth. The spritsail wide 14 cloths, deep 18½ ells, 119 ells of cloth, of Dutch Canvas. The staysail was made of French Canvas, wide 16 cloths, deep aft 16 ells, 128 ells. The Main Top Staysail wide below 8 cloths, deep aft 13 ells, together 52 ells. Two studding sails for the Topsails, each wide above 1 cloth, below 5¼ cloths, deep 18 ells, together 114 ells, of Flemish canvas. Two Studding Sails for the main Yard, above 1¼ cloths, below 5½ cloths, deep 17 ells, together

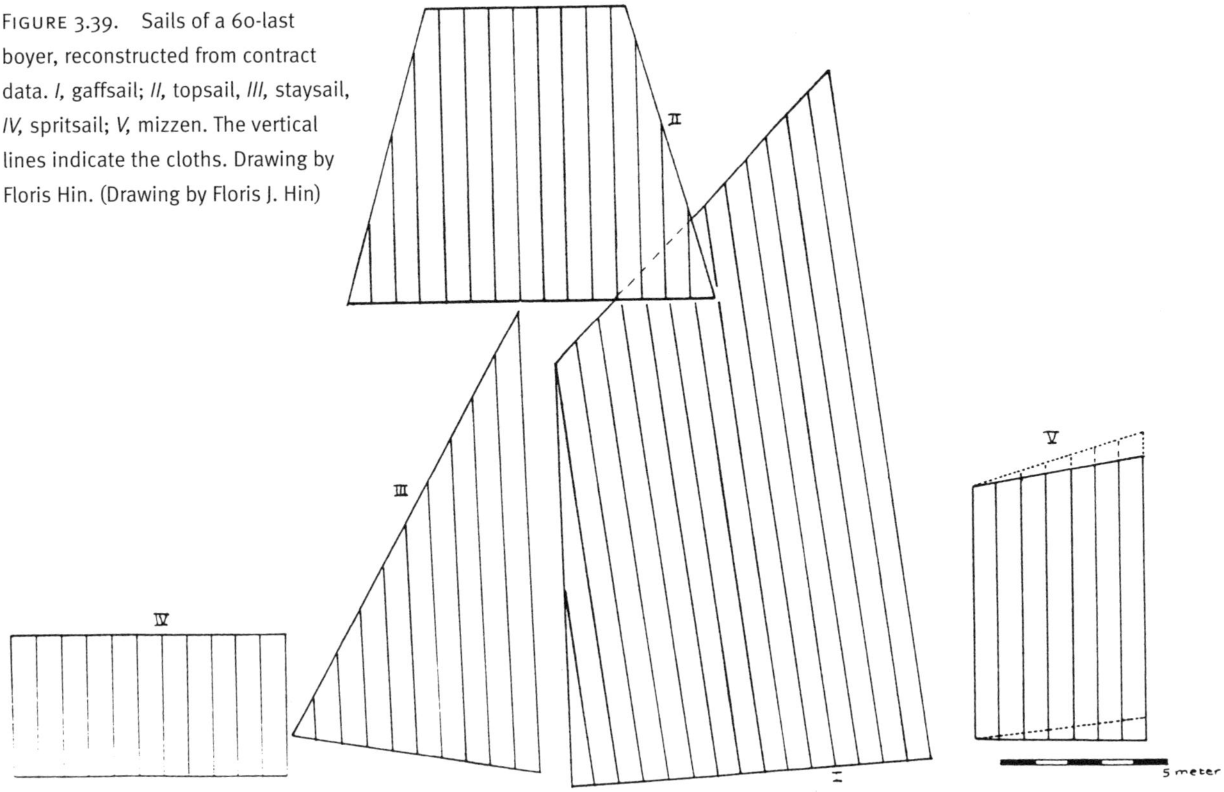

FIGURE 3.39. Sails of a 60-last boyer, reconstructed from contract data. *I*, gaffsail; *II*, topsail, *III*, staysail, *IV*, spritsail; *V*, mizzen. The vertical lines indicate the cloths. Drawing by Floris Hin. (Drawing by Floris J. Hin)

106 ells, of Melis canvas. Two poort-sails [no English equivalent], each wide 6 cloths, deep 6 ells, together 72 ells of cloth. The Spritsail topsail of Melis canvas, above wide 6 cloths, below 11½ cloths, deep 11½ ells, together 91 ells. A Main-topgallant sail above 4¾ cloths, below 8½ cloths, deep 8½ ells, together 58 ells. A Fore-topgallant sail above 3¼ cloths, below 7 cloths, deep 7½ cloths, together 42 ells. A mizzen topsail 4½ cloths, below 9 cloths, together 61 ells, of Flemish canvas. A Mizzen with the Foresail for the boat made of duck, requires 65 ells.

(136 I 15) *Sails of a Ship 126 feet long, wide 26½ feet, deep 13 feet, loading 225 lasts.*

A Mainsail of 21 cloths, deep 14 ells, requires cloth with the tablings 298 ells. A big bonnet of 21 cloths, deep 3¼ ells, together 69 ells of cloth. A foresail wide 18 cloths, deep 12 ells, with the tablings 219 ells of cloth. A Foresail bonnet wide 18 cloths, deep 3½ ells, 33 ells of cloth. A main Topsail, above wide 12 cloths, below 20 cloths, deep 18 ells: with the tablings 302½ ells of cloth. A fore Topsail, above wide 10 cloths, below a good 17 cloths, deep 15 ells, 206 ells of cloth. A Spritsail, wide 14 cloths, deep 8½ ells, 119 ells of cloth. A Mizzen wide 14½ cloths, deep aft 18 ells, 134 ells of cloth. A Mizzen [should be "bonnet"] wide 15 cloths square, deep 2½ ells. A Staysail wide 16 cloths, deep aft 17 ells, 147¼ ells of cloth. A Spritsail topsail, wide above 9 cloths, below 16 cloths, deep 11½ ells, 136 ells of cloth. A mizzen topsail wide above 7 cloths, and below a good 13 cloths, deep 9 ells, 92 ells of cloth. A Main-topgallant sail, wide above 6 cloths, below 11 cloths, deep 8½ ells, 73 ells of cloth. A Fore-topgallant sail above 5 cloths, below 9 cloths, deep 7½ ells, 55½ ells of cloth. Four Studding Sails above wide 2 cloths, below 9 cloths, 2 deep: 18 ells together: two wide above almost 2 cloths, below 7 cloths, deep 19 ells, together 394 ells. A Mizzen for the boat 6½ cloths, deep 11½ ells, above 2 cloths, the Foresail wide 4½ ells, deep 9 ells, 62 ells of cloth.

(136 II 1) *Ship sails of a Ship, long 164 feet, loading 135 Lasts.*

A sail wide 16 cloths, deep 14 ells, a Bonnet with it deep 3¼ ells, a Foresail wide 13 cloths, deep 11½ ells, with a Bonnet deep 3½ ells. A Mizzen wide 11 cloths, deep 16 ells, the bonnet wide downward 13 cloths, deep 2½ ells. A main Topsail below wide 15½ cloths, above 9¼ cloths, deep 15¼ ells. A fore Topsail wide below 12½ cloths, above 7½ cloths, deep 12 ells. A Spritsail wide 13 cloths, deep 8 ells. A Spritsail topsail, wide above 4½ cloths, below 7 cloths, deep 9¼ ells. A

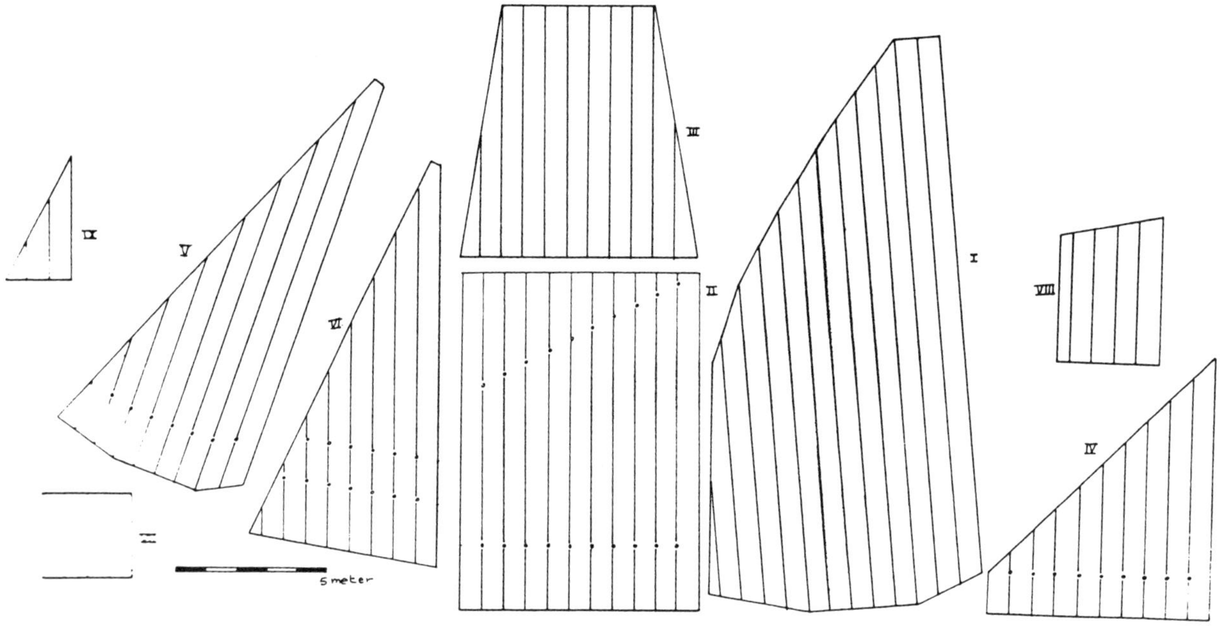

FIGURE 3.40. Sails of a 75-foot galliot, reconstructed from contract data. *I*, "sail"; *II*, foresail; *III*, topsail, *IV*, mizzen, *V*, jib, *VI*, "narrow foresail"; *VII*, spritsail; *VIII*, "mizzen" (storm sail?); *IX*, "foresail" (storm sail?). (Drawing by Floris J. Hin)

Main-topgallant sail, below 6 cloths, above 3¾ cloths, deep 7½ ells. A Fore-topgallant sail below wide 4¾ cloths, above 3 cloths. A Staysail wide 14 cloths, deep 13 ells.

For another one of similar shape, but shorter and wider.

The Mainsail wide 26 cloths, deep 14½ ells, 381 ells of cloth with the tablings, the Foresail wide 23 cloths, deep 12½ ells, the Bonnet deep 3½ ells, together deep 15¼ ells, 362½ ells of cloth, with the tablings. The main Topsail above wide 14 cloths, below wide 24½ cloths, deep 19 ells, with the tablings 378 ells. The fore Topsail above wide almost 12 cloths, below 22 cloths, deep 16¾ ells, 289 ells of cloth, with the tablings. The Mizzen wide 18 cloths, deep aft 22½ ells, with the bonnet 17 cloths, deep 9½ ells, 161½ ells of cloth.

(136 II 44) *Sails are also made with the following measurements, each fitted to her Ships and the wish of the Skipper.*

If a Mainsail is 17 cloths, deep 14 ells, the Bonnet may be deep 3 ells: together deep 17 ells, 302 ells of cloth, with the tablings, the Foresail 14 cloths, deep 11½ ells, the bonnet deep 3 ells, together deep 14½ ells, 210 ells of cloth with the tablings. The main Topsail deep 15½ ells, above wide 10 cloths, 205 ells of cloth with the tablings. The Mizzen and its bonnet wide 13 cloths, deep aft 17 ells. The fore Topsail above wide 10 cloths, below 16 cloths, deep 12½ ells. The Spritsail deep 8 ells, wide 14 cloths, 116 ells of cloth. The spritsail topsail wide 7 cloths, below 11 cloths, deep 10½ ells. The Topgallant sail above wide 6 cloths, below 9 cloths, deep 7½ ells.

And if the mainsail is wide 19 cloths, deep 14½ ells, then the bonnet is sometimes deep 3¼ ells; the Foresail wide 16 cloths, deep 12 ells, the bonnet deep 3½ ells; the main Topsail above wide 11 cloths, below 18½ cloths, deep 17 ells; the fore Topsail above wide 9 cloths, below 15½ cloths, deep 14 ells; the Mizzen with its bonnet wide 15 cloths, deep 18¼ ells; the spritsail wide 16 cloths, deep 9 ells; the spritsail topsail, above wide 6½ cloths, below 9¾ cloths, deep 11 ells; The Main-topgallant sail above almost 6 cloths, below 10 cloths, deep 8½ ells; the Fore-topgallant sail above almost 5 cloths, below 8 cloths, deep 7 ells.

And if a Ship is long 132 feet, wide 32 feet.

It also can have a Mainsail wide 24 cloths, deep without its bonnet 15 ells: a Foresail wide 21 cloths, deep with its bonnet 15½ ells: a main Topsail above wide almost 14 cloths, below wide more than 23 cloths, deep 19 ells: a fore Topsail above wide almost

12 cloths, below 20½ cloths, deep 15½ ells; a Mizzen below wide with its bonnet 18 cloths, deep aft 20½ ells; the spritsail wide 16 cloths, deep 8½ ells.

Otherwise.

If the Mainsail is wide 28 cloths, deep 14½ ells, then the bonnet may be 3 ells, together 17½ ells; the Foresail wide 24 cloths, deep 11¾ ells, the bonnet 3½ ells, together 15¼ ells; the main topsail wide above 13 cloths, below 23 cloths, deep 15 ells, the Mizzen wide 2 [one digit is missing] cloths, deep aft 20½ ells; the spritsail wide 18 cloths, deep 9 ells, the boat's sail deep at the Mast 9 ells, above wide 4½ cloths, below a good 5 cloths; the spritsail topsail, above wide 8½ cloths, below 15 cloths, deep 12 ells; the Fore-topgallant sail above wide 5 cloths, below 9 cloths, deep 7½ ells; the Main-topgallant sail above 6¾ cloths, below 10¾ cloths, deep 9 ells.

FOUR Conclusion

In the preceding chapters, the portion of Witsen's treatise devoted to seventeenth-century shipbuilding was rearranged and explained where necessary (and possible, of course). As noted in chapter 1, this translated material represents only a fraction of the wide-ranging work, which goes far beyond the subject of shipbuilding. Apart from lengthy discourses in which Witsen happily displays his knowledge of the ancient world or presents sections borrowed liberally from other works, there is much to be found about life on board as well—discipline, food, ranks and functions of the officers and crew, port facilities, and sea battles. All these topics, like naval architecture, were studied and recorded by Witsen firsthand and have prime historical importance. Witsen is clear and lucid in these sections, perfectly accessible for today's Dutch readers, so I refer them to Witsen himself. For foreign readers, these portions of Witsen's treatise should be tackled by someone who is especially knowledgeable about these subjects.

Some final remarks should be made here, the first of which concerns the traditional shell-first building method described in the previous chapters. At the beginning of the eighteenth century an influential group of naval officers, led by Admiral Cornelis Schrijver, sought to reform Dutch shipbuilding along British lines. By building from plans conceived at the drawing board, naval architecture would become more scientific, and experiences gained with former designs could be passed on to new designs, which would enhance progress in shipbuilding. Dutch shipbuilding could then catch up with other European countries.

It seems far from certain that the Dutch had suddenly slipped so far behind in those days, with foreign orders for ships to be built still coming in on a regular basis. More obvious is that the admirals wanted a tighter grip on the shipbuilding process, which had been entirely in the hands of shipbuilders in the seventeenth century. It is also dubious whether the introduction of mathematical equations was the key to progress, which was not easily acquired.

The translator of this book, Alan Lemmers, and I prepared a report about the transition from the shell-first method to the frame-first method of the admiralties using drafts.[1]

It describes the slow process by which drafting techniques were introduced in eighteenth-century Holland, aided by the ingenuity of (among others) a Rotterdam shipbuilding family, the Van Zwijndregts. With the aid of a recovered manuscript and the matching drafts found elsewhere, we were able to develop a picture of the drafting techniques that led to the first men-of-war in Holland built from an on-paper design process. The introduction of this process was slow indeed, for most shipwrights of those days were hardly convinced of the necessity to change their ways (and were not enthusiastic about sharing their knowledge with commissioners, either). It is therefore no surprise that when the drawing board was introduced at the yards of the Dutch Admiralties and (partly) of the East India Company, the old building method did not entirely disappear from the Netherlands. At smaller yards the building and design process remained integrated until the end of the wooden shipbuilding era, and shipbuilders continued working with the traditional shell-first method. As André Wegener Sleeswyk rightly remarks in the introduction, even today this method is still practiced on some yards in Friesland.

A second remark concerns Witsen's credibility. I hope this book has convincingly proved that Witsen's treatise is indeed a major historical source. As demonstrated in chapter 2, his pinas data unfailingly provide all the information required to build the entire ship, just as Witsen claimed. The model of the pinas can therefore be considered an archaeological find—albeit not excavated from the ground or raised from the sea, but recovered from a textual source. In combination with archaeological finds in the true sense and other textual information provided by Witsen, Van Yk, Van Dam, and other written sources, the pinas model represents a massive contribution to our understanding of seventeenth-century shipbuilding.

What began as an experiment in constructing a ship model evolved into an effort to "unveil" Witsen's book for other craftsmen like myself as well as for scholars. What use either of these will make of this book, only the future can tell.

Regarding my own approach, I will venture a final remark. The only way Witsen's book could be cracked was with my hands as well as my head. Only after isolating the pinas data from Witsen's more general observations did the rest of the treatise become comprehensible. Many obscure passages, for the pinas as well, became clearer after I tested them through practical application. For the stern and quarter galleries, for instance, there was no other way to comprehend their relationships except by doggedly executing in wood all the parts enumerated and described by Witsen. The construction emerged only after playing and puzzling with these parts. The fact that the quarter galleries for this pinas were open, cagelike structures came as a complete surprise to me—I did not even know such galleries had existed! It was much later that I observed in paintings and engravings that this kind of gallery was standard for all smaller and medium vessels. Only on large vessels was there enough space for these galleries to house the privy of the captain or cabinets and thus be closed.

The naval architecture described by Witsen was no science in the modern sense, as the reader will clearly have observed: there were no standard formulas with which ship properties could be calculated and predictions made about those properties. Dutch seventeenth-century shipbuilding was, in essence, an empirical trade. Witsen's book contains the sublimated experience of contemporary shipwrights, laid down in simple rules of thumb. To understand how a ship was built in those days, you must simply do it yourself. To understand the implications of the terms of a written contract, you must simply try them out. This line of research is not easily reported.

Some scholars have made disparaging remarks about the use and validity of archaeological experiments, particularly those pertaining to prehistoric settlements as well as Thor Heyerdahl's trials with his *Kon-Tiki* and *Ra* boats and contemporary reconstructions of ancient craft such as the Greek trireme. As far as my work can be numbered in this category, I think this criticism should not be automatic. There will always be situations in which the only avenue of research available is the experiment. In this case model building was a practical means.

Model building has played a far more important role in the past than is acknowledged today. Anyone who, like myself, spends most of his days surrounded by historical scale models such as the Dutch Navy Model Collection in the Rijksmuseum and is able to observe the craftsmanship and attention of detail that has gone into these models will realize that these creations were not made just for decoration or as a pastime, like so many are today. They provide a more detailed image of reality than could even be obtained from a drawing, text, or oral account; they instruct both layman and professional alike and convey ingenious ideas and propositions. In this book I hope to have demonstrated that model building can also clarify a hitherto incomprehensible and uncomprehended book.

With the publication of this book a period of my life has been completed. Witsen's book has changed my personal and professional life in considerable ways. My research into this subject culminated in the opportunity to leave my job as a teacher and exchange it for the post of chief restorer of ship models in the Rijksmuseum. My family moved with me from Groningen to Amsterdam. This book and its English translation is therefore a small monument celebrating almost thirty years of my life.

Appendix Variations on Witsen

Variants in the Two Editions (1671, 1690) of Witsen's Treatise on Shipbuilding

DIEDERICK WILDEMAN

NO MODERN STANDARD TEXTBOOK exists for the study of historical Dutch shipbuilding, and scholars with an interest in the subject have long felt this omission. This work by A. J. Hoving is the first serious attempt to explain and render accessible one of the foremost seventeenth-century sources, Nicolaes Witsen's *Aeloude en Hedendaegsche Scheeps-bouw en Bestier* (Ancient and Modern Shipbuilding and Management) of 1671. But no matter how much information we can glean from contemporary sources on seventeenth-century shipbuilding, our knowledge of the subject will never be complete. Shipbuilding of that time was a craft in which the master shipwrights did not entrust the details of their knowledge and methods to paper. The ships they built were no standard product, either. At present, only through underwater archaeology can we gain direct access to the practices of shipbuilding and compare, supplement, and, in some cases, correct the written and printed sources.

Nowadays we regard books as a standard product. If we have a certain edition of a text, we take for granted that every other copy is exactly the same. This is not the case for books from the preindustrial age. Like shipbuilding, the printing and publishing of books in the seventeenth century had many aspects of a craft. By looking at original copies we can learn much about the history of a text. The "archaeology" of books can reveal surprising information about their production and sometimes even their use.

Introduction

In the seventeenth century if a small error was discovered in the text during the printing process, it was corrected in the type. But paper was an expensive commodity, so the already printed, uncorrected pages were often still used to make the book. Sometimes the author or publisher wanted to add or correct entire passages or pages after the book was printed, or the authorities wanted to have certain passages removed. In that case, no new edition was printed, but pages were cut out and new ones inserted. These practices explain why even books with identical title pages can contain significant differences. These variations (variants, in bibliographical terminology) can explain much about the production of a book, too.[1] If certain copies contain the variant pages, they are usually referred to as a "state" of an edition.

The user of a modern facsimile of a seventeenth-century book should be aware of the possibility and particularities of variants. Every facsimile should therefore have an introduction describing the makeup of the original book and a statement explaining why a certain copy and a particular state was used as a basis. Preferably, other variant pages should appear in an appendix to the facsimile. Without this information the status of the book as a source is not clear and therefore unreliable. Practically everyone studying Witsen nowadays will be using a facsimile edition, as original copies of the *Aeloude en Hedendaegsche Scheeps-bouw en Bestier* have become rare and very expensive. The original 1671 edition is available only in the rare book department of large or specialized libraries. An original copy of the second, 1690 edition is an extremely rare book, bearing the new title *Architectura navalis et regimen nauticum ofte Aaloude en Hedendaagsche Scheeps-bouw en Bestier . . .* (Shipbuilding and Management, or Ancient and Modern Shipbuilding and Management), which adds Latin to the original Dutch title. Only five copies are known today, and just three are accessible in libraries. A facsimile of the 1671 edition was published in 1979. Some years before, in 1970, a facsimile of the 1690 edition had already seen the light.[2] Neither facsimile edition of these books contains an introduction.

This article aims to provide the user of these facsimiles with some information about the nature of each book and its makeup. It also identifies the variant states and tries to explain why they were made. But before turning our attention to the study of the books themselves, it is useful to elaborate a little on its author.

Nicolaes Witsen as an Author on Naval Architecture

Nicolaes Witsen (1641–1717) was thirty years old when *Aeloude en Hedendaegsche Scheeps-bouw en Bestier* was published. Witsen was certainly not a shipwright himself but rather the son of a wealthy merchant with a leading position within the most important city in the Netherlands, Amsterdam.[3] Nicolaes' father, Cornelis Witsen, was a man with substantial financial interests in commerce and shipping. He was a director of the West India Company, had a seat in the Admiralty of Amsterdam, and was a member of the Pilotage Commission. Other members of the family were also employed in commerce. Amsterdam in the seventeenth century was by far the richest and most powerful city of the Netherlands and the commercial capital of Europe. The international businessmen formed the ruling class; nobility and religious leaders played no role in the city's affairs. The Amsterdam officials had many tentacles in the provincial and the national government. Their voices could reach far beyond the city walls, and they had a very significant impact on Dutch foreign policy. The young Nicolaes Witsen had all the prospects to become an important man in his native city and, as a consequence, in the Dutch republic. He made excellent use of these opportunities. Between 1682 and 1706 Witsen was one of the mayors of Amsterdam, special ambassador to England in 1689, and beginning in 1693 a director of the mighty Dutch East India Company.

In his youth Witsen must have learned about commerce, including a great deal on shipping affairs. But shipbuilding is quite a different trade from shipping. Witsen himself hints at the circumstances that drew him to this subject. In the first edition, of 1671, he specifically mentions his father, who had died only shortly before in 1669: "Never would I have dared take up the part of this work on the modern shipbuilding, had not come into my hands some basic principles and drawings made by my late father Cornelis Witsen in the past." Apparently Cornelis' interest in ships had gone beyond the value of their cargo. In the second edition Witsen writes: "and that which I have added, I have derived from the mouth and design of fine Masters: because I do not claim myself to be a shipwright or sailor in practice."[4]

Nicolaes Witsen was indeed not a shipwright or sailor, but he was first and foremost an inquisitive man. Throughout his life he studied all kinds of subjects with great tenacity. As a law student in Leiden, he displayed more interest in subjects like Oriental languages and ethnology and in mathematics and physics rather than law. Nevertheless, he successfully completed his law studies and graduated in 1664. As became a patrician's son, he made a long journey after his studies. While most young men of his age went on a grand tour to the south of Europe, Witsen joined an official legation of the Dutch States General to Moscow as a so-called *edelman van staat*, a position given to privileged young men to gain some experience in international affairs. This position afforded him the opportunity to look around and meet all kinds of people. He returned in 1665 and compiled an extensive travel account of his voyage in manuscript.[5] Shortly after his return from Russia, Witsen departed again, this time for France and Italy, and in 1668 and 1671 he visited England. In France and England he studied the local shipbuilding practices.[6]

The rest of his life Witsen's interest in Russia never abated, and he collected an enormous amount of information on the country and its people and remained in touch through correspondence with many leading figures in the country. This enabled him to edit the best map of his day of eastern Russia, including the largely unknown region of present-day Siberia and Central Asia. This map is, in fact, a large wall map in six sheets, printed in 1687 at Witsen's own expense.[7] Not long after, in 1692, he finished a 600-page treatise on the same area and had it printed, also at his own expense. Written as a companion volume to the wall map, *Noord en Oost Tartarije . . .* (North and East Tartary) is now also extremely rare: only five copies are known to exist.[8] Although the book and the map were clearly meant for publication, the printed sheets were never bound and put on the market. Perhaps this was because Witsen wanted to add more information to the book. He claimed he had trouble with the engravers and wanted additional maps drawn up, which were not yet complete. He also complained that he never seemed to have the time to really finish this magnum opus.[9]

Witsen's influence in the Netherlands and his interest in Russia made him an ideal contact for the Russian czar Peter the Great, who wanted to modernize his country and gain ideas by traveling abroad. When the czar toured western Europe, Witsen organized his lengthy stay in the Netherlands from August 1697 until June 1698.[10] This provides another clue to the rarity of the book: the large wall map and the first edition were dedicated to the czar, and Witsen clearly wanted to present it to him in person during the visit. Peter also urged him to continue to expand his study. Witsen received so much new information from the delegates of the Russian legation that he probably held back the publication of the 1692 edition himself.[11] In 1705 a much enlarged edition of 986 folio pages was printed. But again Witsen waited, wanting to add more material and complaining about the lack of time. In the end, the stock of printed sheets of the book remained in his house until Witsen died in 1717. These sheets survived and were finally published, with a new title page and introduction, by M. Schalekamp in Amsterdam in 1785.[12]

Some of Witsen's character traits have clearly left traces in his writing. He was a precise and careful man, sometimes even too careful. He had an inclination to substantiate points of view with an abundance of arguments, sometimes undermining the effect of his reasoning process. Witsen strove for completeness and did not want to omit anything that might be relevant to the subject. However, he often lacked the ability to analyze his material thoroughly and to provide the reader with a well-structured overview. In many cases he wanted to let his sources speak for themselves and provides lengthy quotations. As a result, *Aeloude en Hedendaegsche Scheeps-bouw en Bestier* and his other writings are difficult to interpret. Many commentators have praised the contents of the books but criticized their structure.[13]

After completing *Aeloude en Hedendaegsche Scheeps-bouw en Bestier,* Witsen embarked on his political career and confined his scholarly interests to his spare time. He continued to display as much interest in shipping and shipbuilding as he did for Russia. From a young age, he was also an ardent collector of curiosities. As a director of the Dutch East India Company and through his many other contacts, he was able to acquire valuable information, drawings, and objects from all over the globe. Witsen corresponded quite extensively with scholars in Europe about his collection and findings. During a visit to England in 1689 he made contact with British scholars, showed them his map, and discussed his information on Russia and East Asia. His knowledge was appreciated, and he was elected a fellow of the Royal Society in the same year.[14]

Witsen's scholarly interests were those of a rich private collector and learned amateur. He always remained very much a part of the ruling class in his native town and far above the ordinary crowds. He was deeply hurt when, in a discussion on maritime trade with Prince William III of Orange, he was cut short with the remark "Sailors have no notion of politics." The prince was sneering not only at Witsen as a member of a group of his political rivals, the Amsterdam merchant class, but also at Witsen's rather suspect interest in the lowly business of seafaring and shipbuilding.[15]

The First Edition, 1671

Aeloude en Hedendaegsche Scheeps-bouw en Bestier aims to provide a comprehensive overview of building, outfitting, and operating ships from antiquity up to Witsen's own time. The result of Witsen's ambition for completeness was a large and expensive book, covering 574 pages and containing 114 engravings. The first section, concerned with antiquity, has little value for us today, for Witsen simply reproduced the theories of others and provided no original insights. This section even contains illustrations of fantastic ships carrying an entire basilica or other large buildings. Copied from drawings made by Pirro Ligorio (1510–83) and obtained by Witsen from an Italian correspondent, these illustrations (marked "P. L." by Witsen in his book) derived from the vivid imagination of Ligorio rather than antiquity.[16] Modern knowledge of Greek and Roman shipping renders Witsen's comments obsolete.[17]

Witsen skips the Middle Ages entirely. In the seventeenth century this period was regarded as a time of darkness and barbarity and thus unworthy of much study. With a few remarks Witsen leaps from antiquity to about AD 1500 to pursue to his own era. This last period provides the most rewarding information for us today, mainly due to the sources Witsen used. Where he obtained his information is largely unknown, but it is unique material and in many parts first-hand. As noted earlier, Witsen's father was interested in shipbuilding and had amassed a collection of drawings and other material. Obviously the son made use of this material, but he also ventured out on his own: he claims to have had many conversations with shipwrights. With the information he obtained he tried to capture the trade on paper, which makes the book a prime source for shipbuilding in the seventeenth century. The book also had a glossary (which in fact was the first Dutch maritime lexicon), an appendix with supplementary information, and finally an index.

FIGURE A.1. Engraved portrait of Witsen, 1688. From Jan Wagenaar, *Vaderlandsche Historie,* 2nd ed. (Amsterdam, 1770).

The allegorical engraving on the frontispiece (fig. 1.1) points to the encyclopedic structure of the book: naval architecture is symbolized by the woman in the center, flanked by the gods of war and trade, Mars and Mercury, symbolizing naval warfare and merchant shipping. In the background we can see a number of men carrying ship models from all parts of the world and from various historical periods. The engraving is an early work from a very prolific engraver, Romeyn de Hooghe (1657–1709). The other engravings in the book were produced in large part by Witsen himself; as a young man he had learned to draw and knew how to produce acceptable engravings and etchings.[18]

Publishing *Aeloude en Hedendaegsche Scheeps-bouw en Bestier* was a costly business. Witsen managed to interest two publishers in his book, the Appelaer brothers and Casparus Commelin, who produced it as a joint publication after applying in July 1671 for the publishing rights. This protected them against pirated copies for a period of fifteen years, or until 1686.[19] Although both publishers had their own presses, they chose to leave the actual printing to someone else. The book was produced on the presses of Christoffel Conradus, one of the most important printers in Amsterdam at the time. Witsen would have received quite a number of copies for himself to give away to friends and other interested parties. Authors were rarely paid for their efforts in those days, but a man like Witsen would certainly have deemed it below his station in society to receive payment for such an activity.[20] The book's size, many illustrations, and luxurious quality suggest a high price. We have a few indications of its cost. In 1671 it came on the market for twelve guilders, and in 1683 it was sold at auction for nearly seventeen guilders. In contemporary accounts there are complaints about the difficulty of finding a copy, and in 1698 the French minister of the navy asked his ambassador in the Hague to buy one, which he managed only by paying the hefty sum of forty-three guilders.[21] An ordinary shipwright earned around twenty-five guilders per month, a captain on an East Indiaman sixty to eighty guilders, and a mate thirty to fifty guilders.[22] Very few sailors and shipwrights would be inclined to spend such an amount of money on the book, but it was not meant for them. Witsen had attempted to lay down his findings in an encyclopedic and scholarly manner, mainly for the use of other scholars and authors.

We have no figures for the print run, but even without contemporary reports we have other evidence pointing to a reasonably large number. The initial rights expired in 1686, but in December 1685 Casparus Commelin obtained a renewal for another fifteen years. The Appelaer brothers were not mentioned this time. In 1675 they had run into serious financial difficulties and were forced to sell their business. Commelin's business had also changed a lot since 1671. He had acquired the rights to publish Amsterdam's official newspaper, which provided such steady income that he ceased his other publishing activities. Thus, Commelin was no longer publishing books at the time he obtained the extension of the rights for Witsen's book. Commelin probably applied for the renewal not because he intended to bring out a new edition of the work but because he still had copies of the 1671 edition in stock and wanted to continue to protect his assets.[23] It is difficult to give an estimate of the number of copies printed. For a book of average size and quality, the print run was about 1,000 to 1,500 copies in seventeenth-century Holland.[24] But Witsen's book was more for the high end of the market, so the print run could have been lower than these figures. Another indication is the number of surviving copies. Nearly all important maritime libraries with historical collections and the major general libraries worldwide seem to own a copy. Even today the work, although rare, is occasionally available on the

FIGURE A.2. An illustration from Witsen's section on antiquity. Witsen made use of images found on coins and reliefs, which explains the squat form of many of the illustrations.

antiquarian book market. You don't have to wait years to find an original copy for sale somewhere.

The Second Edition, 1690

After 1671 Witsen did not rest. He had continued collecting material to enlarge his work, and in 1690 a second edition was printed. Taken as a whole, the structure of the second edition is somewhat better than the first. Many subjects were expanded and improved, and the new section on Russian shipbuilding with its beautiful illustrations is especially remarkable.[25]

This second edition, with the new title *Architectura navalis et regimen nauticum ofte Aaloude en Hedendaagsche Scheeps-bouw en Bestier . . . ,* came from the presses of the famous publishing house of Blaeu. For a long time the existence of the second edition was completely unknown, as no authors on shipbuilding or bibliographers mentioned it. The first copy came to light in 1913, when it was discovered by the antiquarian book dealer Wouter Nijhoff.[26] Since then, the existence of four other copies has been confirmed. The production of the 1690 edition differs in one important aspect from that of 1671: whereas the first edition was the legal property of publishers who had put it on the market, the second edition was printed at the author's expense, just like the two editions of *Noord en Oost Tartarije* of 1692 and 1705. From his correspondence with Cuper and others it is clear that Witsen by now had complete control over the production and distribution of his publications. As with his other books, Witsen kept on collecting material, waiting for letters containing new information from his extensive network of correspondents and always intending to add this information to his work. In the end Witsen grew old and lacked the energy to finish the book. After his death in 1717 the description of his estate includes mention of "Affgedrukte Exemplaren van 't boeck van de Scheepsbouw" (printed copies of the book on shipbuilding), with an estimated value of 2,500 guilders. Peters calculates this as between 500 and 1,500 copies.[27] As Witsen owned the stock

FIGURE A.3. Russians vessels as depicted in the 1690 edition of Witsen's treatise.

of the 1690 edition, it is highly unlikely this refers to the earlier 1671 edition. Further study has revealed that the 1690 edition appears only in two eighteenth-century auction catalogues. Witsen's close friend Cuper had three copies, and Witsen's heir, his nephew Nicolaas Lambersz. Witsen, owned four.[28] Witsen died childless, and his eighteenth- and early nineteenth-century heirs were not careful with his collections and papers, including the undoubtedly extensive source material for his books. It is very likely the unbound copies were at some point thrown away or sold for scrap.[29]

Influence

Witsen's work on shipbuilding had a significant influence and was soon used by others. In 1678 J. Robijn compiled and published a small booklet on rigging called *Hollandsche Scheepsbouw,* which was largely based on Witsen. The *Dictionnaire de Marine,* written by Nicolas Aubin in 1701, also draws heavily on Witsen. Aubin, a French protestant minister who sought refuge in the Netherland, notes that the definitions "sont tirées du livre incomparable de M. Witsen ou l'on trouve tout ce qui regarde la Marine des Anciens et celle d'aujoud'hui" (the definitions are taken from the incomparable book of Mr. Witsen where one finds everything regarding the navigation of the ancients and of today).[30]

The second author of a major work in the Netherlands on shipbuilding was Cornelis van Yk, a master shipwright, who published his *De Nederlandsche Scheeps-bouw-konst Open Gestelt* in 1697. He also made extensive use of the work of his predecessor. He even copied a number of tables from the first edition verbatim. When the second edition of Witsen surfaced in the early twentieth century, it became clear that Van Yk was probably the only author to make use of the rare second edition. I shall pursue this later. The illustrations from Witsen were very influential, too, and were frequently copied—for example, in Van Yk's treatise (1697), *Nieuwe Hollandsche Scheepsbouw* (1695 and later editions), *L'art de batir les vaisseaux* (1719), and the dictionary of Aubin. More than a century after its publication the German author J. H. Röding, in his well-known multilingual dictionary *Allgemeines Wörterbuch der Marine,* still called Witsen's book "das vollständigste Werk, wass man von dieser Art hat" (the most comprehensive work of its kind).[31]

The book was acquired not only by readers and libraries in the Netherlands. People like Samuel Pepys (1633–1703), secretary of the British Navy, owned a copy, and, as we shall see, the Swedish admiral Count Carl Gustaf Wrangel (1613–76) ordered the book when it came out.[32] The book was in a number of royal collections as well: Charles II of England, Louis XIV of France, Christina of Sweden, and Peter the Great of Russia all obtained a copy.[33]

In the seventeenth and eighteenth centuries the Dutch held a dominant position in trade and shipping, especially along the coasts of northern Germany, Scandinavia, and the Baltic region. On Dutch merchant vessels there were many sailors in all ranks from these countries. The Dutch language was widely understood, read, and written by people involved in trade and shipping activities in these countries. As the Dutch were also dominant in publishing, most books on maritime affairs used in the region came from Dutch presses and were printed in Dutch. In fact, the first textbook on the art of navigation in German was not published until the end of the eighteenth century.[34] Before this time German sailors had to learn navigation from Dutch books. In the wake of this flourishing sea trade, the knowledge of the Dutch language spread to other parts of society. In the seventeenth century theater companies from the Netherlands toured the Baltic countries and even performed at the Danish and Swedish courts.[35] Witsen's book could certainly be read by a much wider audience than his countrymen alone, and it was named as a source by almost every author writing on continental shipbuilding before 1800.

Variants in the 1671 Edition

Studying the variants within copies of the same edition is often the key to interesting information about the book's birth. With both editions of Witsen's book this is the case. The first variant in the 1671 edition is not spectacular, but it reveals the identity of the printer. On the title page of most copies of the first edition, only the names of the publishers appear; but there is a unique copy in which the names of the publishers are not given—only the name of the printer. Through this earlier version of the title page we know that the book was printed by Christoffel Conradus in Amsterdam.[36]

The other variants are of a completely different nature. A considerable number are found in the texts on pages 345–46 and 463–79, which deal with recent maritime events and conflicts. Quite a number of changes concern events of the Second Anglo-Dutch War (1665–67), which had only recently ended. The differences mainly concern the offensive tone directed at the English adversaries: some strongly anti-English remarks in the oldest version have been softened in the next version. Many sentences and paragraphs have even been removed or altered. For example, on page 345 the description of the Dutch commemorative medal celebrating the raid on the River Medway in 1667 is simply deleted on the reprinted page. A clear example of these changes is found on page 475, which contains the following passage: "But even in defeat did the cruelty of the English enflame. Because the captain of a fire ship, after having been captured, had his head and manhood cut off, even after having been pardoned. Dutchmen do not act in such a way, because their word is their seal, because they are content at seeing their enemy conquered." In a later version this part has been replaced by a phrase excusing the episode: "The crew of some fire ships fell into the hands of the English, and, in the habit of war, were robbed of their lives in the fury." In short, the pages with anti-English passages were cut out of the book and replaced with an expurgated version. Yet even this was not enough. In the next variant entire pages were removed. In the majority of the surviving copies, pages 475–76, again dealing with the successful Dutch raid on England's southeastern coast in 1667, were left out. During this raid the Dutch had sailed up the Medway from the Thames estuary, attacked English ships at Chatham's naval docks, and towed one of the flagships of the English fleet away to the Netherlands. This first successful attack on British soil, immensely humiliating to the English, was now omitted, with of course adjustments to the text to ensure continuity.[37]

In two letters to the Dutch scholar Vossius, Witsen explains some of the reasons for these changes: "On my return from England I agreed to omit the spiteful history of the raid on the Medway entirely, which could easily be done because only a few copies had been sold." Witsen had already presented Vossius with a copy of the first version, and this letter is in fact a note he sent with the reprinted pages. Less then a month later he writes that the changes are the result of his visits to England: "I hate all this sharpness, all the more because this nation had treated me so kindly."[38] Perhaps Witsen received advice to mitigate the tone, but the letters clearly demonstrate that he changed his mind about the text.[39] External pressure certainly played a role, too. The earlier variant contains a description of the Battle of the Sound (the Øresund, the sea strait between Sweden and Denmark), which was fought between the Dutch and Swedes in 1658. On pages 463–65 Witsen writes in rather unfavorable terms on the role of the Swedish admiral, Count Wrangel, after he received a broadside and was about to be boarded: "Whereupon Wrangel turned his arse and anchored under Elsinore Castle because he received various shots aft and was himself wounded by a splinter in the face. . . . although his ship was called Victoria he abandoned the fight." Unknown to Witsen, Count Wrangel had received the book soon after its publication through his Dutch business agent, De Geer. The old admiral must have been absolutely furious. In a letter to De Geer he demanded a full apology from Witsen for his impertinence and insisted that the pages be removed from all remaining copies; Witsen complied. With these changes in the text, both the Anglo-Dutch war and the Battle of the Sound are described in neutral terms.[40]

Some authors have claimed that Witsen's work was banned because of these vexing passages—mainly because of a remark in a work by the French author Jerôme Lalande in 1793: "Le Traité de N. Witsen eut beaucoup de réputation en Hollande, mais les États-Généraux en ordonèrent la suppression" (The treatise by N. Witsen had a considerable reputation in Holland, but the States General suppressed it).[41] However, there is no evidence that the work was ever officially banned, although it clearly stirred some emotions. Voorbeijtel Cannenburg notes that no record of such a ban has been found in the archives of the States General or the City of Amsterdam. Moreover, as noted earlier, the book remained on the market for decades after its initial publication.[42]

In summary, we can distinguish four variants in the 1671 edition:

1. Title page with the name of Christoffel Conradus; unexpurgated text (one extant copy, University Library, Amsterdam).
2. Title page with the names of Casparus Commelijn and Broer and Jan Appelaer; unexpurgated text.

3. Pages 345–46 and 463–78 reprinted with changes in the text.
4. Pages 475–76 removed; pages 477–78 reprinted.

Most copies also have a portrait of Witsen, made in 1677, when he was thirty-six years old. Some of these portraits are incorrectly dated 1671.[43]

Variants in the 1690 Edition

Only five copies of the second edition, the *Architectura navalis,* are known to have survived beyond Witsen's time. Four are now kept in Amsterdam: one copy is in the library of the University of Amsterdam, two are in the library of the Scheepvaartmuseum (Maritime Museum) in Amsterdam, and one is in private hands. The private owner obtained this copy in the mid-1990s from a Norwegian dealer who had acquired it from a Norwegian collector. The fifth copy, now lost, used to be in the library of the Seefahrtschule (School of Navigation) in Bremen, Germany; a nineteenth-century handwritten inventory of the library clearly lists the book by author and title, along with the year of publication and folio size.[44] However, when the old and rare books of the school were incorporated into the Staats- und Universitätsbibliothek Bremen (State and University Library of Bremen) in the 1980s, the 1690 edition of Witsen was not among the books transferred, and it is not known when the volume was lost. S. P. l'Honoré Naber mentions it in his 1914 article, as does W. Voorbeijtel Cannenburg in a 1956 article. Cannenburg, however, probably copied the information from the earlier article, not having seen the book himself.[45] During the Second World War, under the threat of Allied bombing, the school, with its library, was moved and incorporated in the navigation school in Wustrow on the Baltic Sea.[46] After the war the school reopened in Bremen, but the town of Wustrow became part of the newly formed German Democratic Republic and the books remained there. Only in the 1990s, after the German reunification, was the remaining collection in Wustrow returned to Bremen to become part of the holdings at the Staats- und Universitätsbibliothek. But, as noted earlier, the *Architectura navalis* was not among the books transferred.[47]

The copies in the University of Amsterdam library, Scheepvaartmuseum in Amsterdam (inventory number S.0147), and the private collection are the same—with one exception: the one in the university library has an engraved portrait, by the engraver Petrus Schenk, of a sixty-year-old Witsen wearing a sumptuous wig. The second copy in the Scheepvaartmuseum's collection (inventory number A.2992) contains significant differences compared with the other three. This copy has a new introduction, completely reset and with many differences in text. For instance, more sources are mentioned, and pages have been removed and replaced by others, as in the first edition. Pages 29–38 were reprinted with new text, but this text is two pages longer, so between pages 36 and 37 are two pages numbered [35] and [36]. After the first variant of the 1690 edition was printed, Witsen apparently received some additional information, a good deal of which he took—as mentioned in the book itself—from a letter he received in 1693 from his good friend Gijsbert Cuper. All these variants occur in the section on antiquity.

But the most important change in this copy (A.2992) is of a different sort altogether: it contains an appendix of no less than thirty-one pages, providing additions of every kind to all parts of the book. This appendix can be dated. An important collection of Witsen's letters has survived, an interesting part of which concerns his correspondence with the aforementioned Cuper. In one letter Witsen writes, "I am becoming of a mind to use the figure of Charon and add him to the appendix of my work referring to that which Your Lordship has said in his Consecratio Homeri on page 235." The quotation can be found in the appendix. As this letter from Witsen to Cuper is dated August 3, 1693, the appendix could not have been printed before the fall of that year.[48]

The text of the appendix also reveals that it was originally meant to contain five illustrations. These are absent from the Scheepvaartmuseum copy but can be identified. Long after Witsen's death, in the late eighteenth century, the publisher M. Schalekamp came into possession of the copper plates of Witsen's work. With these plates he compiled a book with the title *XXX Platen van Vreemde Vaartuigen* (XXX Plates of Foreign Ships). Among these engravings there are five with the text "in the appendix p. . . ." Comparison with the text has established that indeed they belong to the appendix of the 1690 edition. All the plates mentioned in the text of the appendix are present in Schalekamp's book.

Van Yk, in writing his *De Nederlandsche Scheeps-bouw-konst Open Gestelt*, clearly had the first edition of Witsen's book at his disposition, for he borrowed entire tables from it. And he must have had access to the 1690 edition as well, or at least parts of it—which places Van Yk among the very few who had this opportunity. A list of books in Van Yk's introduction is largely identical to the one in Witsen's introduction to the second variant of the 1690 edition.[49]

Two known variants of the second edition can be summarized as follows:

1. Without appendix (University of Amsterdam library; Scheepvaartmuseum, Amsterdam, inventory number S.0147; private collection, Amsterdam).
2. With changes in the introduction on pages 28–38; with appendix of 31 pages (Scheepvaartmuseum, Amsterdam, inventory number: A.2992).

The Facsimile Editions

By comparing the list of variants with the two facsimile editions of Witsen's book, we can draw some conclusions.[50]

The 1671 facsimile was printed from an unknown copy of variant 2, with the names of Casparus Commelin and the brothers Broer and Jan Appelaer on the title page. It contains the unexpurgated version of the text dealing with the Battle of the Sound and the Raid on the Medway. The Witsen portrait bears the incorrect date of 1671 instead of 1677. It is a facsimile of a good and complete copy. It is therefore a useful substitute for the original in the study of seventeenth-century Dutch shipbuilding; for other types of research it would have been necessary to reproduce the variant pages as well.

The facsimile of the 1690 edition of Witsen's book is based on variant 1, a copy of which is in the library of the University of Amsterdam.[51] A far better choice would have been variant 2, as found in the Scheepvaartmuseum. The changes concerning antiquity might have been of interest to some scholars, and the revisions in the introduction and especially the addition of the appendix are changes of such importance that they should not have been overlooked. With the illustrations in *XXX Platen van Vreemde Vaartuigen* a complete copy could have been produced. Neither of these facsimiles reproduces the variant pages from the other known copies.

Conclusion

Witsen's interests and personal wealth made it possible for him to pursue a number of subjects and publish his findings in voluminous books. We are fortunate that one of these interests included contemporary shipbuilding. He has left us with a very interesting but abstruse source. Even with Dutch readers with an interest in the subject, Witsen is notorious for the unstructured presentation of his material. But to users who have the stamina to delve deeper, it is a rich lode of information. Scholars and others who want to use this source turn to one of the facsimile editions and only rarely use the original. The publishers of facsimile editions of seventeenth-century works are not always aware of the fact that they need to research their material carefully. Any facsimile used for scholarly purposes should present the readers with the result of such a study. In the case of Witsen's books, by pure chance, the original used for the publication of the 1671 facsimile turned out to be a good choice, at least for those interested in shipbuilding. However, this is not the case with the 1690 facsimile, which is clearly defective. Here other choices should have been made. The users of Witsen's works in modern facsimile should also have been made aware of the fact that the 1671 edition was a published work available to many in its day, whereas the 1690 edition was never published and almost the entire stock was probably destroyed—therefore this edition could not have reached interested readers. Book archaeology, or bibliographical research, is vital in preventing modern-day publishers from making costly mistakes and in providing readers and scholars with a trustworthy source for further use and study.

Tables

TABLE 1. Comparison of Witsen's and Van Yk's shipbuilding formulas.

A key dimension in Witsen's formulas is the width (thickness) of the inboard face of the stem, which was calculated as one inch for every ten feet of the ship's total length. Almost half of Witsen's formulas (about 46 percent) are based on the width of the inside stem. Van Yk's formulas also use this reference point but to a much lesser degree (about 10 percent). Compared with Witsen, Van Yk uses a wider range of criteria to obtain the dimensions of components, with reference points like the keel, wing transom, and the main dimensions of the ship. While many of the formulas in both Witsen's and Van Yk's treatises remained constant throughout the seventeenth century, some changed in response to the fashion of the day. With this second group of formulas (indicated with italics in the table below), we should therefore pay considerable attention to the period in which they were applied.

Key

L = Length of ship between outsides of stem and stern
W = Width (beam) of ship, measured over frames (without planking)
D = Depth in hold, measured between top of keel and top of lower deck beams without camber
i.s. = width of inside stem = $\frac{1}{10}$ inch per 1 foot L (or 1 cm per 112 cm L)

Witsen's Formulas	Van Yk's Formulas
Ship's dimensions: L : W : D = 20 : 5 : 2	Main dimensions of ships: W = $\frac{1}{4}$ L D = $\frac{3}{4}$ W – (height of bulwark + height "'tween decks") Or: D = 40%–50% W
Keel: Width = $1\frac{1}{2}$ i.s. Height = $1\frac{1}{4}$ i.s.	Keel: Length = $\frac{9}{10}$ L Width = 1 inch per 7 feet W Height = a little bit less than its width Length of joints = 5 inches per 1 inch width of keel (two bolts per 2 feet length of joint)
Stem: Width outside = $\frac{3}{5}$ to $\frac{3}{4}$ i.s. Breadth = in the middle, 3 × i.s. Height = D + deck rising + "'tween decks" + forecastle deck Or: $\frac{2}{11}$ L Or: $\frac{11}{60}$ L *Rake = $\frac{28}{29}$ height of stem*	Stem: Height = D + deck rising (1 inch per 10 feet) + height "'tween decks" + anything above that Width = $\frac{1}{10}$ inch per 1 foot L Width outside stem = width i.s. – 1 or 2 inches Breadth = in the middle, 2 × width; above and below, a bit more Rake = not specified

Witsen's Formulas	Van Yk's Formulas
Sternpost:	Sternpost:
Length (height) = W + deck rising + "'tween decks" (or height of upper deck)	Length = trim ($^1/_{50}$ foot per 1 foot L) + D + deck rising (2 inches per 10 feet) + height "'tween decks" + height of wing transom
Or: = height i.s.	
Width = stem	Width = i.s.
Breadth top = $1^1/_5$ i.s.	Breadth top = $1^1/_4$ i.s.
Breadth foot = 5 × i.s.	Breadth middle: $1^1/_3$ breadth top
Top square = wing transom	Breadth foot = $^1/_4$ foot per 1 foot length
Rake = 1 foot per 6 feet length (height)	Raking = $^1/_7$ foot (fluyts $^1/_{14}$ foot) per 1 foot length
Wing transom:	Wing transom:
Length = $^2/_3$ W	Length = $^3/_4$ W
Or: = width of bottom	Height = height of keel
Width, height, and camber = $^1/_{11}$ length (for every foot, 1 inch) [doubtful]	Width = $^4/_5$ height
Or: = i.s.	Camber = 1 inch per 2 feet length
Fashion pieces:	Fashion pieces:
Thickness = $^5/_6$ width of wing transom	Thickness = $^2/_3$ wing transom
Or: $^1/_2$ i.s.	Width = a minimum of 2 × thickness
Width = 2 × width of wing transom	
Transoms:	Transoms:
$^2/_3$ wing transom	Thickness = $^3/_4$ wing transom
Or: $^4/_5$ i.s.	
Stern timbers:	Stern timbers:
Length above transom = length of stern timber	Length above transom: = height of sternpost
Thickness = $^2/_3$ i.s.	Thickness = $^1/_2$ i.s.
Tumble home: $^3/_5$ transom	Width = a bit more than the thickness
	Tumble home = 2 inches inward per 1 foot of length
Bottom:	Bottom:
Width = $^2/_3$ W	Width = length of wing transom ($^3/_4$ W)

TABLE 1 (*continued*)

Witsen's Formulas	Van Yk's Formulas
Planking: Thickness = 1/4 i.s.	Planking: Thickness = according to table: Length of shipThickness of planks 40–60 feet 2 inches 60–80 feet 2[1/2] inches* 80–100 feet 3 inches 100–120 feet 3 1/2 inches 120–140 feet 4 inches 140–160 feet 4 1/2 inches 160–170 feet 4 1/2 inches (*There are some discrepancies in Van Yk's table. The dimension "2 1/2 inches"—for ships of 60–80 feet—is difficult to read in the text and appears to be "2 1/8 inches," although that does not fit the pattern of the list. Furthermore, it seems strange that the same thickness of planking is given for the last two categories: 140–160-foot ships and 160–170-foot ships. If these dimensions are what Van Yk intended, it is unclear why he did he not combine the two lines in a single category for 140–170-foot ships.)
Floors: Height = 3/4 i.s.	Floors: Height = 1 inch per 3 feet W Width = a little less
Bilge futtocks and futtocks: 1/2 i.s. *Tumble home = 1/3 "'tween decks"*	Bilge futtocks and futtocks: Thickness at bottom = 9/10 of floors Thickness at top = 3/4 of floors
Top timbers: 2/5 i.s.	Top timbers: Thickness = 3/4 floor futtocks *Tumble home: 4 inches per 1 foot of height*
Turn of the bilge: *Width = W – 1/10 inch per 1 foot L* Height of the turn of the bilge = 1/3 D	
Keelson: Thickness = 2/3 i.s. Width = wider than the keel	Keelson: Height = 2/3 i.s. Width = as much as the wood provides
Lower deck beams: Thickness and width = 1 1/8 × i.s. Camber = 1/10 inch per 1 foot L	Lower deck beams: Height = 2 inches per 5 feet W Width = a bit more Camber = 6 or 7 inches
Ceiling planks: Thickness = planking (1/4 i.s.)	Ceiling planks = outside planking

Witsen's Formulas	Van Yk's Formulas
Bilge stringers: Thickness = a little more than the ceiling planks	Bilge stringers: Thickness = $1/3$ i.s.
Lower deck clamp: Thickness = $1/2$ i.s.	Lower deck clamp: Thickness = $2/5$ i.s.
Lower deck beam knees: Thickness = $2/3$ i.s. Riders, etc. = $2/3$ i.s.	Lower deck beam knees: Thickness in the middle = about the thickness of the deck beam
Waterways, lower deck: Thickness = $1/3$ i.s. Width = $1\,2/3$ i.s.	Waterways, lower deck: Thickness = $1/3$ i.s.
	Riders: Width = $7/8$ i.s. Thickness = $2/3$ i.s.
Binding strakes: Thickness = $1/4$ i.s. Width = $1\,1/4$ i.s.	Binding strakes: Thickness = $1/3$ i.s.
Ledges: Thickness = $1/4$ i.s.	Ledges: Width and thickness = thickness of planking
Carlings: Thickness = $1/3$ i.s.	Carlings: Width and thickness = thickness of planking
Lower wales: Thickness = $1/2$ i.s. Width = i.s.	Lower wales: Width = 12 inches per 100 feet L; – 1 inch per 10 feet shorter, + 1 inch per 10 feet longer Thickness = $1/2$ width Length of joints = $1/2$ foot per 1 inch width of the wale, over at least 5 frames
Gunports: Height = $1/3$ "'tween decks" Width = a little more	
Upper deck beams: Height = $2/3$ i.s. Width = $2/3$ i.s Camber = $5/6$ i.s.	Upper deck beams: Height = $2/3$ lower deck beams Width = $2/3$ lower deck beams Camber = a little more than the camber of the lower deck beam
Upper deck clamp: Thickness = $1/5$ i.s.	Upper deck clamps: Thickness = $1/3$ i.s.
Waterways, upper deck: Thickness = $1/4$ i.s.	
Lower deck planking: Thickness = $1/6$ i.s.	

TABLE 1 *(continued)*

Witsen's Formulas	Van Yk's Formulas
Coamings of hatches on upper deck:	
Thickness = 1/3 i.s.	
Inner frame thickness and width = 2/5 of the coaming	
Groove width = 1/8	
Ledges = 1/4 i.s.	
Hatch beam = 1/3 i.s., the others less	
Forecastle and quarterdeck clamps:	
Thickness = 1/5 i.s.	
Forecastle and quarterdeck beams:	Forecastle and quarterdeck beams:
Thickness = 1/2 i.s.	Height = 2/3 upper deck beams
	Width = 2/3 upper deck beams
	Camber = a little more than the camber of the upper deck beams
Forecastle and quarterdeck knees:	
Thickness = 2/5 i.s.	
Or: 1/2 i.s.	
Forecastle and quarterdeck waterways:	
Thickness = 1/4 i.s.	
Poop deck beams:	Poop deck beams:
Thickness = 1/3 i.s.	Height = 2/3 quarterdeck beams
	Width = 2/3 quarterdeck beams
	Camber = a little more than the camber of the quarterdeck
Poop deck clamps:	
Thickness = 1/6 i.s.	
Poop deck waterways:	
Thickness = 1/6 i.s.	
	Sheer rail:
	Width = 9/10 upper wales
	Thickness = 1/2 width of sheer rail
Lower spur of the beakhead:	Lower spur of the beakhead:
Length = 1/5 L	*Length = 1/8 L*
Width afore = 1/4 i.s.	Width = outside stem
Width aft = 3/5 i.s.	Depth = i.s.
Height afore = 1/2 i.s.	
Top rail of the beakhead:	Top rail of the beakhead:
Width aft = 1/2 i.s.	Width: = width of sheer rail
Width afore = 1/4 i.s.	Thickness = 1/2 width
Thickness = 1/2 width	
Standard anchor bitt:	Standard anchor bitt:
Thickness = 1 1/3 i.s.	Thickness = height of keel
	Width = 9/10 height of keel
Bolster:	Bolster:
Thickness = 1 1/3 i.s.	Thickness = width of standing part
Thickness of deal cheek = 1 1/2 i.s.	Width = 9/10 width of standing part

Witsen's Formulas	Van Yk's Formulas
Main knight: Width = 1/8 inch per 1 foot L.	Main knight: Thickness = 9/10 height of keel Fore knight: Thickness = 9/10 thickness of main knight
Counter timbers: Thickness = 1/2 i.s. Lower *hackebord:* Width in the middle = i.s. Thickness = 1/8 i.s. Cover on top of lower *hackebord:* Width = i.s.	Counter timbers: Thickness = 5/4 thickness of planking Lower *hackebord:* Thickness = 5/4 thickness of planking Width = 3/4 width of wing transom Lower cover on top of lower *hackebord:* Thickness = 3/4 thickness of *hackebord* Width = 1 1/2 thickness of *hackebord*
Capstan: Thickness = 2 1/2 inches per 10 feet L Holes for beams = 1/6 thickness	
Rudder: Width = 4 inches per 12 feet L Thickness = 3/5 i.s. Nails: Thickness = 1 inch per 100 feet L	Rudder: Width = 3 inches per 10 feet L Width = 2 × thickness
	Top railings: Thickness = 3/4 thickness of planking
Location of masts: *Divide the ship into 5 1/2 parts; beginning aft, locate the foremast on the first part, the mainmast on the third part, and the mizzenmast on the fifth part.*	Location of masts: *Divide L into 6 equal parts; mizzenmast on the first part from aft; mainmast in the middle; foremast on a half part from fore.*
Main channel: Length = 1/8 L Thickness = 1/3 i.s. Width = i.s.	Main channel: Length = 1/5 L Thickness outside = thickness of planking Thickness inside = 5/4 thickness of planking
Fore channel: A little less than the main channel	Fore channel: Length = 5/6 length of main channel Thickness outside = planking Thickness inside = 5/4 planking
Mizzen channel: Length = 1/3 main channel Width = 1/3 main channel Thickness = like fore channel	Mizzen channel: Length = 1/2 main channel Thickness = 3/4 main channel
Location of the chesstree: 2/5 L from fore	Location of the chesstree: 3/4 W before the mainmast

TABLE 1 (*continued*)

Witsen's Formulas	Van Yk's Formulas
Rake of the main mast: $1\frac{1}{8}$ inches per 10 feet D Rake of the foremast: 2 inches per 10 feet D Rake of the mizzenmast: 1 inch per 10 feet D	
Mainmast: *Length* = $2(W + D)$ Thickness = 1 foot per 6 feet D	Mainmast: *Length* = $3 \times W$ *(for bigger ships,* 3×25 + *as many feet as W is more than 25)* Thickness = $\frac{3}{4}$ W in inches
Foremast: *Dimensions* = $\frac{7}{8}$ *main mast*	Foremast: *Length* = $\frac{9}{10}$ *main mast* Thickness = $\frac{9}{10}$ main mast
Mizzenmast: *Dimensions* = $\frac{3}{4}$ *main mast*	Mizzenmast: *Length* = $\frac{3}{4}$ *main mast* Thickness = $\frac{3}{4}$ main mast
Bowsprit: *Dimensions* = $\frac{3}{4}$ *main mast*	Bowsprit: *Length* = $\frac{3}{8}$ *L* Thickness = between main and foremast
Main-topmast: *Length* = $\frac{5}{8}$ *mainmast* Thickness = $\frac{5}{6}$ mast top	Main-topmast: *Length* = $\frac{5}{3}$ *W* Thickness = $\frac{4}{5}$ mainmast
	Fore topmast: *Length* = $\frac{9}{10}$ *main-topmast* Thickness = $\frac{4}{5}$ foremast
Mizzen topmast: *Length* = $\frac{1}{3}$ *main-topmast* Thickness = not specified	Mizzen topmast: *Length* = $\frac{11}{10}$ *main-topgallant mast* Thickness = $\frac{4}{5}$ mizzenmast
Spritsail topmast: *Length* = $\frac{1}{3}$ *bowsprit + 2 feet* Thickness = not specified	Spritsail topmast: *Length* = $\frac{5}{6}$ *main-topgallant mast* Thickness = main-topgallant mast
Main-topgallant mast: *Length* = $\frac{1}{3}$ *main-topmast* Thickness = not specified	Main-topgallant mast: *Length* = $\frac{3}{4}$ *W* Thickness = $\frac{7}{8}$ topmast
Fore-topgallant mast: *Length* = $\frac{8}{9}$ *fore topmast* Thickness = not specified	Fore-topgallant mast: *Length* = $\frac{5}{6}$ *main-topgallant mast* Thickness = like main-topgallant mast
	Mainmast flagpole: *Length* = $\frac{8}{7}$ *main-topgallant mast* Thickness = a little thicker
	Foremast flagpole: *Length* = $\frac{8}{7}$ *fore-topgallant mast* Thickness = a little thicker
	Mizzenmast flagpole: *Length* = $\frac{5}{6}$ *mizzen topmast* Thickness = half of mizzen topmast

Witsen's Formulas	Van Yk's Formulas
	Bowsprit flagpole: *Length* = $\frac{3}{4}$ *spritsail topmast* Thickness = $\frac{1}{2}$ spritsail topmast
	Poop deck flagpole: *Length* and thickness = mainmast flagpole
Main yard: *Length* = $2 \times W + 1 \times D$ Thickness = $\frac{3}{4}$ mainmast	Main yard: *Length* = $\frac{6}{16} \times (L + W)$ Thickness = 1 inch per 4 feet length of the yard
Foreyard: *Length* = $\frac{6}{7}$ *main yard* Thickness = not specified	Foreyard: *Length* = $\frac{6}{7}$ *main yard* Thickness = 1 inch per 4 feet length of the yard
Mizzen yard: *Length = length of mizzenmast + 1 or 2 feet* Thickness = not specified	Mizzen yard: *Length = between mainyard and foreyard* Thickness = 1 inch per 4 feet length of the yard
Spritsail yard: *Length* = $\frac{3}{4}$ *bowsprit* Thickness = not specified	Spritsail yard: *Length* = $\frac{5}{8}$ *main yard* Thickness = 1 inch per 4 feet length of the yard
Main topsail yard: *Length* = $\frac{1}{2}$ *main yard plus* Thickness = not specified	Main topsail yard: *Length* = $\frac{4}{7}$ *main yard* Thickness = 1 inch per 4 feet length of the yard
Fore topsail yard: *Length* = $\frac{1}{2}$ *foreyard minus* Thickness = not specified	Fore topsail yard: *Length* = $\frac{4}{7}$ *foreyard* Thick = 1 inch per 4 feet length of the yard
	Crossjack yard: *Length = main topsail yard* Thickness = 1 inch per 6 feet length of the yard
	Mizzen topsail yard: *Length* = $\frac{4}{7}$ *crossjack yard* Thickness = 1 inch per 4 feet length of the yard
Spritsail topsail yard: *Length* = $\frac{1}{3}$ *bowsprit* Thickness = not specified	Spritsail topsail yard: *Length* = $\frac{4}{7}$ *spritsail yard* Thickness = 1 inch per 4 feet length of the yard
	Main-topgallant yard: *Length* = $\frac{1}{2}$ *main topsail yard* Thickness = 1 inch per 4 feet length of the yard

TABLE 1. *(continued)*	Witsen's Formulas	Van Yk's Formulas
		Fore-topgallant yard: *Length = ½ fore topsail yard* Thickness = 1 inch per 4 feet length of the yard
	Main top: Diameter = 9 inches per 10 feet L Upper wales: Width = 9/10 lower wales Thickness = ½ width of upper wales	

TABLE 2. Main dimensions, number of lasts, and last factors of early ships.

Year / *Name of Ship*	Length	Beam	Depth	Lasts	Last Factor
1590 / *Vergulde Leeuw*	90	23 or 24	10	75	276 or 288
1592 / *Tonijn*	94	26	13	100 or 120	264 or 318
1592 / *Dolphijn*	100	25.5	13.5 or 11.5	100 or 140	218 or 358
1593 / *Neptunus*	102	33	13.5	250	181
1593 / *Valk*	90	29	12	80	350
1598 / *Tijger*	90	25	12	80	338
1601 / *Zilveren Ster*	104	25	10	130	200
1604 / *Gelderland*	90	27	11	150 or 180	150 or 187

Note: Dimensions are in Amsterdam feet
Source: B. E. van Bruggen, "Beschouwing over het aangeven van de hoofdafmetingen van de Nederlandse zeeschepen en de daarbij gebruikte maateenheden (1600–1800)," *Mededelingen van de Nederlandse Vereniging voor Zeegeschiedenis / Tijdschrift voor Zeegeschiedenis,* no. 20 (March 1970): 34. Original data from the National Archive, The Hague, 1335–36.

Year / *Name of Ship*	Length	Beam	Depth	Lasts	Last Factor
1603 / VOC ship	130	35	16	300	242
1614 / VOC ship	150	33	13	300	215
1614 / VOC ship	138	36	14	300	232

Source: Van Bruggen, "Beschouwing," 34. Original data from Pieter van Dam, *Beschryvinge van de Oostindische Compagnie, 1639–1701.*

Year / *Name of Ship*	Length	Beam	Depth	Lasts	Last Factor
1625 / *Goes*	125	29	11.5	200	208
1632 / *De Zon*	128	31.5	12	250	194
1632 / *Aemilia*	144	34	14.3	300	233

Source: Van Bruggen, "Beschouwing," 34. Original data from J. E. Elias, *De vlootbouw in Nederlund in de eerste helft der 17de eeuw, 1596–1655.*

Year / *Name of Ship*	Length	Beam	Depth	Lasts	Last Factor
	125	24.5	14	146	294
	122	24.5	13.5	138	292
	120	24	13.5	130	299
	118	24	13	123	299
	116	24	13	118	306
	114	23.5	13	112	311
	124	25	13	195	224
	126	26	12	176	223
	85	21	11.5	60	342
	90	21.5	12	68	341

Source: Van Bruggen, "Beschouwing," 34. Original data from Jacobus Robijn, *Hollandsche Scheepsbouw.*

Table 2. (*continued*)

Year / *Name of Ship*	Length	Beam	Depth	Lasts	Last Factor
	95	21.5	12	73	335
	106	24	9	90	254
	90	22	8.5	80	210
	135	28	10.5	150	265
	100	20	7.5	50	300
	150	33	13	300	215
	138	36	14	300	232
	138	36	14	300	232

Source: Pieter van Dam, *Beschrijvinghe van de Oostindische Compagnie,* chap. 17, "De schepen van de Compagnie," 450–554.

Year / *Name of Ship*	Length	Beam	Depth	Lasts	Last Factor
	120	24.25	13.5	140	281
	132.2	24.5	11.75	130	293
	116	23.25	13.25	120	301
	125	23	11	140	226
	118	25	11	150	216
	120	22	13.5	130	274
	135.5	29	12.25	270	178
	118	23	13	120	294
	112	22	15	120/140	64/308
	112	22.5	13	125	262

Source: Loading contracts in "Statistieken van de Nederlandse nijverheid uit de eerste helft der negentiende eeuw," *Amsterdamse bevrachtingscontracten, wisselprotesten en bodemer en van de notarissen Jan Franssen Bruyningh, Jacob Meerhout e.a.:1617–1625,* ed. P. H. Winkelman, Rijks Geschiedkundige Publicatiën, Grote Serie 168 (The Hague: Nijhoff, 1983), 40.

Year / *Name of Ship*	Length	Beam	Depth	Lasts	Last Factor
	100	22	11	100	242
	115	23.5	11	150	198
	125	24	12	200	180
1647	125	25	14	155	276
	123	24.5	14	146	289
	122	24.5	13.5	138	292
	120	24	13.5	130	299
	118	24	13	123	299
	114	23.5	13	112	310
	95	21.5	12	73	335
	90	21.5	12	68	341
	85	21	11.5	60	342
	124	25	13	162	249
	122	24.5	13	153	254

Year / *Name of Ship*	Length	Beam	Depth	Lasts	Last Factor
	120	24.5	13	145	264
	118	24	12.5	134	264
	130	28	12	195	224
	126	26	12	176	223
	124	25	11.5	166	215
Average factor					274

Source: Nicolaes Witsen, *Aeloude en Hedendaegse Scheeps-bouw en Bestier,* 160; appendix, 33, 34.

TABLE 3. Witsen's measurements for eight locations along the hull of the pinas.

Total hull length = 37.92 meters

Frame locations:

Frame 1	main frame, 12.62 m from outside of stem
Frame 2	12 feet (3.40 m) forward: 9.22 m
Frame 3	11 feet (3.11) forward: 6.11 m
Frame 4	12 feet 5 inches (3.53 m) behind frame 1: 16.15 m
Frame 5	11 feet 8 inches (3.32 m) behind frame 4: 19.47 m
Frame 6	11 feet 10 inches (3.37 m) behind frame 5: 22.84 m
Frame 7	18 feet 5 inches (5.09 m) behind frame 6: 27.93 m
Frame 8	14 feet 6½ inches (4.13 m) behind frame 7: 32.06 m

Note: Witsen's pinas is not divided in equal parts. The first frame is at the location of the main frame (at one third of the total length of the ship), and for the other frames the locations are stipulated, first forward, then aft. The distances are measured between the last and the next frame in the list, except for frame 4, which is measured from frame 1.These frame locations are shown in fig. 2.48. Shape and measurements of the stern can be found in chapter 2, sections 21, 22, 24, 38, and 39.

Measurements of the bottom:

	Width		Rise
	feet	inches	inches
Frame 1	21	—	5
Frame 2	20	9½	6 (?)
Frame 3	20	7½	4
Frame 4	20	9	6
Frame 5	20	7	—
Frame 6	20	3	—
Frame 7	19	5	—
Frame 8	—	—	—

Measurements of the top of the turn of the bilge:

	Width		Height	
	feet	inches	feet	inches
Frame 1	13	6	4	5½
Frame 2	13	5	4	8
Frame 3	12	7	4	4
Frame 4	13	6	4	5½
Frame 5	13	5½	—	—
Frame 6	13	5½	5	1
Frame 7	13	1½	7	½
Frame 8	11	6½	9	9

Measurements of the height of the master ribband, to be added to those of the top of the turn of the bilge:

	Width		Height	
	feet	inches	feet	inches
Frame 1	14	7	12	9
Frame 2	14	6	12	8
Frame 3	14	4	12	11
Frame 4	14	6½	13	½
Frame 5	14	6½	9 +	?
Frame 6	—	—	14	5
Frame 7	14	2½	16	6
Frame 8	13	4½	19	7

The curve of the futtock between the top of the turn of the bilge and the master ribband:

	Curve in inches
Frame 1	3
Frame 2	3½
Frame 3	4½
Frame 4	3
Frame 5	4
Frame 6	4
Frame 7	6
Frame 8	7

Amount of overhang of the master ribband at the top of the turn of the bilge:

	Overhang	
	feet	inches
Frame 1	1	
Frame 2	1	1
Frame 3	1	8
Frame 4	1	⅔
Frame 5	1	⅔
Frame 6	1	⅔
Frame 7	1	¼
Frame 8	1	9

TABLE 3. (*continued*)

The tumble home of the top timbers, measured from the master ribband:

	Height	Tumble home	
	feet	feet	inches
Frame 1	6½	2	
	8½	3	
Frame 2	—	—	
Frame 3	6	1	10
	9	3	
	12	3	6
Frame 4	6½	2	4
	8½	3	
Frame 5	6½	2	6
	8½	3	
Frame 6	—	—	
Frame 7	unclear2	8½	
Frame 8	7	2½	
	10	4	
	14	5	

Table 4. Key to parts shown on the plans of the pinas (see plan drawings 1–5 following table).

Ship compartments:

A hold
B forepeak
C protective bulwark for the pumps
D powder room
E cheese and bread room
F peak
G gunroom (also called the tiller flat)
H lower deck or main deck
I cable tier
J forecastle
K forecastle deck or forecastle head
L upper deck
M steering stand
N captain's cabin
O quarterdeck
P forward compartment of the cabin
Q aft compartment of the cabin
R poop deck

Note: All dimensions in the Thick and Wide columns are in centimeters. The listed parts are shown on drawing 2 unless a different drawing is indicated in the Drawing column.

Number	Part	Thick	Wide	Drawing
General				
1	keel (L = 29.43 m)	42	57	2
2	stem	35	70	2
3	sternpost	35	110	2
4	sternpost knee	16	35	2
5	keelson	23	67	2
6	wing transom	35	38	5
7	fashion piece	21	62	5
8	rabbet	8	9	2
9	*broekstuk*	18	100	2
10	transom	24	35	5
11	bottom planks	10	50	2
12	bilge planks	10	50	2
13	side planking	8	50	2
14	*tingel*	2	16	2
15	wale	17	34	3
16	wale	16	34	3
17	wale	12	25	3
18	wale	10	25	3
19	sheer rail	10	22	3
20	filling strakes	6	3	2
21	upper-works plank	4	50	3
22	washstrake	4	50	3
23	rail	4	13	2
24	hance	4	13	2
25	fore channel	8	42	3
26	main channel	9	44	3
27	mizzen channel	7	23	3
28	chesstree	20	21	3
A: Hold				
29	floor timber	23	24	2
30	bilge futtock	23	24	2
31	futtock	16	23	2

TABLE 4. (*continued*)

Number	Part	Thick	Wide	Drawing
32	top timber	16	23	2
33	bilge stringer	12	49	2
34	lower deck clamp	14	65	2
35	ceiling plank	10	50	2
36	breasthook, rider floor	20	25	2
37	bilge rider	20	25	2
38	knee on the *broekstuk*	20	20	2
39	foremast step	26	71	2
40	stemson	26	—	2
41	lower deck beam (camber 21 cm)	31	34	2
42	binding strake	12	44	2
43	waterway	16	49	2
44	ledge	9	9	2
45	carling	10	10	2
46	deck plank	7	50	2
47	hanging knee	20	25	2
48	rider	26	30	2
B–F: Storeroom and Peak				
49	post	10	10	2
50	plank	3	50	2
G: Gunroom				
51	chock	10	10	2
52	rudder	26	75	2
53	tiller	5	15	2
54	whipstaff	10	10	2
55	sweep	12	12	2
56	mizzenmast step	21	36	2
57	wing transom knee	28	28	2
58	transom knee	20	20	2
59	spirketing	9	39	2
60	beam clamp	8	47	2
61	ceiling plank	8	42	2
62	hanging knee	18	21	2
63	deck beam (camber 14 cm)	23	26	2
64	ledge	7	9	2
65	carling	9	9	2
66	waterway	8	42	2
67	binding strake	8	36	2
68	deck plank	5	50	2
69	spirketing	7	26	2
H: Lower Deck				
70	upper deck clamp	8	47	2
71	ceiling plank	7	50	2
72	hanging knee	8	21	2
73	upper deck beam (camber 23 cm)	21	23	2

Number	Part	Thick	Wide	Drawing
74	ledge	7	9	2
75	carling	9	9	2
76	waterway	10	49	2
77	binding strake	12	44	2
78	plank	5	50	2
79	capstan step	21	36	2
80	main knight	43	46	2
81	bitt	26	28	2
82	galley plank	4	50	2
83	galley post	7	8	2
84	butlery plank	4	50	2
85	butlery post	7	8	2
86	mast partner	16	131	2
87	hatch coaming	10	10	2
88	hatch	3	—	2
89	bulwark post	10	10	2
90	bulwark plank	4	50	2
I: Cable Tier				
91	spirketing	7	26	2
92	beams clamp	9	47	2
93	deck beam (camber 18 cm)	23	26	2
94	hanging knee	21	21	2
95	ledge	7	9	2
96	carling	9	9	2
97	waterway	12	57	2
98	binding strake	8	36	2
99	plank	5	50	2
J: Forecastle				
100	spirketing	8	47	2
101	beam clamp	4	25	2
102	ceiling	4	50	2
103	deck beam (camber 31 cm)	13	21	2
104	ledge	5	5	2
105	carling	5	5	2
106	waterboard	10	26	2
107	binding strake	5	20	2
108	plank	4	50	2
109	hanging knee	13	16	2
110	foremast partner	12	100	2
111	bitt	39	42	2
112	bolster	30	40	2
113	bitt knee	26	50	2
114	hatch coaming	6	20	2
115	bulkhead stanchion	10	10	2
116	bulkhead plank	4	50	2

TABLE 4. (*continued*)

Number	Part	Thick	Wide	Drawing
K: Forecastle Head				
117	fore knight	39	41	2
118	bitt standard	26	26	2
119	bolster	14	14	2
120	hatch coaming	10	21	3
121	ledge	4	5	3
122	batten	2	10	3
123	forecastle head rail	21	24	3
124	stanchion	5	8	2
125	rail	4	13	2
126	pinrail	11	21	3
L: Upper Deck				
127	ceiling or inwale	7	—	2
128	gunwale or capping rail	4	8	2
129	hatch coaming	14	26	2
130	ledge	5	6	2
131	carling	2	10	2
132	kevel	8	28	2
133	pin rail	8	14	2
134	bitt	14	18	2
135	capstan partner	10	64	2
136	capstan	—	52	2
M–N: Steering Stand and Captain's Cabin				
137	spirketing	7	26	2
138	deck clamp	4	25	2
139	ceiling	4	50	2
140	deck beam (camber 31 cm)	13	21	2
141	hanging knee	13	16	2
142	ledge	5	5	2
143	carling	6	6	2
144	waterboard	8	42	3
145	binding strake	5	18	3
146	deck plank	5	50	2
147	mizzenmast partner	11	70	2
148	stanchion	10	10	2
149	plank	3	50	2
O: Quarterdeck				
150	rail	4	13	2
151	stanchion	5	18	2
152	mizzen knight	21	26	2
153	kevel	5	14	2
154	pinrail	10	21	2

Number	Part	Thick	Wide	Drawing
P: Forward Compartment of the Cabin				
155	ceiling	5	50	2
156	deck beam (camber 34 cm)	13	21	2
157	ledge	5	5	2
158	waterboard	5	24	3
159	plank	3	50	2
160	hanging knee	10	13	2
Stern Gallery				
161	counter timber	12	14	3
162	first *hackebord*	10	35	3
163	first cover	5	34	3
164	second *hackebord*	5	38	3
165	second cover	5	35	3
166	third cover	5	26	3
167	third *hackebord*	10	50	3
168	cross of the flat of stern	6	14	3
169	post	9	12	3
170	central stanchion	16	18	3
171	stern timber	15	28	3
172	plank	3	50	2
173	cornice	7	21	3
174	herm figures	25	25	3
175	quarter console	40	40	3
176	statues in the side	21	26	3
177	pediment of the quarter gallery	5	17	2
178	cover	5	25	3
179	quarter gallery architrave	5	26	3
180	lodging knee	10	14	3
Beakhead				
181	lower beakhead spur	36	44	2
182	upper beakhead spur	26	36	2
183	beakhead knee	14	33	2
184	beakhead top knee	11	25	2
185	cheek	15	28	2
186	kam	8	25	2
187	*es*	26	38	2
188	lion	53	—	2
189	first headrail	11	37	2
190	second headrail	11	31	2
191	third headrail	3	31	2
192	cathead	21	26	3
193	cathead bracket	26	28	2
194	crosspiece	14	15	2
195	beam	6	7	2

TABLE 4. (*continued*)

Number	Part	Thick	Wide	Drawing
196	batten	3	4	2
197	rib	14	19	2
Masts and Spars: Mainmast				
198	top	9	283	5
199	trestletree	12	20	5
200	rim	6	13	5
201	border	1	9	5
202	lubber's hole	147	—	5
203	knee	7	10	5
204	masthead	39	198 (L)	4
205	mainmast	57	2264 (L)	4
206	main-topmast	37	1443(L)	4
207	crosstree and trestletree	6	10	5
208	topmast masthead	10	85 (L)	4
209	topgallant mast	14	623	4
210	crosstree	3	5	4
211	topgallant masthead	35	184 (L)	4
Foremast				
212	top	8	255	5
213	crosstree and trestletree	11	18	5
214	rim	6	13	5
215	border	1	9	5
216	lubber's hole	119	—	5
217	knee	6	9	5
218	fore masthead	35	184 (L)	4
219	foremast	51	1981 (L)	4
220	fore topmast	30	1160 (L)	4
221	crosstree and trestletree	5	9	4
222	fore top masthead	20	99 (L)	4
223	fore-topgallant mast	13	509 (L)	4
224	crosstree and trestletree	3	5	4
225	fore-topgallant masthead	9	35 (L)	4
Bowsprit				
226	bowsprit	57	1698 (L)	4
227	top	6	127	5
228	crosstree and trestletree	5	9	5
229	knee for the spritsail topmast	13	—	4
230	rim	6	12	5
231	border	1	9	5
232	knee	4	5	5
233	spritsail topmast	13	566 (L)	4
234	crosstree and trestletree	3	5	4
235	spritsail topmasthead	10	57 (L)	4

Number	Part	Thick	Wide	Drawing
Mizzenmast				
236	mizzenmast	57	1698	4
237	top	6	127	5
238	crosstree and trestletree	5	9	5
239	rim	6	12	5
240	border	1	9	5
241	lubber's hole	—	63	5
242	knee	4	5	5
243	mizzen masthead	38	127 (L)	4
244	mizzen topmast	14	736 (L)	4
245	crosstree and trestletree	3	5	4
246	mizzen top masthead	10	85 (L)	4
Flagpoles				
247	flagpole	12	736	4
248	flagpole	12	712	4
249	flagpole	11	582	4
250	flagpole	7	614	4
251	flagpole	7	426	4
Yards				
252	main yard	41	1953	4
253	main topsail yard	23	1019	4
254	main-topgallant yard	14	623	4
255	foreyard	37	1714	4
256	fore topsail yard	17	906	4
257	fore-topgallant yard	13	453	4
258	mizzen yard	24	1755	4
259	crossjack yard	13	1019	4
260	mizzen topsail yard	13	538	4
261	spritsail yard	23	1274	4
262	spritsail topsail yard	14	566	4
Standing Rigging (Drawing 4)				
	anchor cable	11		2
1	mainstay	11		2
2	forestay	8		2
3	mizzen stay	4		2
4	main-topmast stay	4		2
5	fore topmast stay	3		2
6	mizzen topmast stay	1		2
7	main-topgallant stay	1		2
8	fore-topgallant stay	1		2
9	sprit topmast backstay	2		2
10	main shroud	5		2
11	main-topmast shroud	3		2
12	main-topgallant mast shroud	2		2

TABLE 4. (*continued*)

Number	Part	Thick	Wide	Drawing
13	fore shroud	4		2
14	fore topmast shroud	2		2
15	fore-topgallant mast shroud	1		2
16	mizzen shroud	3		2
17	mizzen topmast shroud	2		2
18	manrope	2		2
19	bowsprit topmast shroud	2		2
20	gammoning	3		2
21	main-topmast backstay	3		2
22	fore topmast backstay	3		2
23	mizzen topmast backstay	3		2
24	main pendant tackle	4		2
25	fore pendant tackle	4		2
26	mizzen pendant tackle	2		2
Running Rigging (Drawing 4)				
27	main halyard tie	5		2
28	parrel truss pendant	3		2
29	main lift	2		2
30	main brace	2		2
31	main sheet	4		2
32	main clew garnet	2		2
33	main buntline	2		2
34	main bowline	2		2
35	main topsail halyard	3		2
36	main topsail brace	2		2
37	main topsail sheet	3		2
38	main topsail clew line	2		2
39	main topsail buntline	2		2
40	main topsail bowline	2		2
41	main-topgallant halyard	2		2
42	main-topgallant lift	1		2
43	main-topgallant brace	1		2
44	main-topgallant sheet	2		2
45	main-topgallant clew line	1		2
46	main-topgallant sail bowline	1		2
47	fore halyard tie	5		2
48	fore parrel truss pendant	3		2
49	fore lift	2		2
50	fore brace	2		2
51	fore sheet	4		2
52	fore clew garnet	2		2
53	fore buntlines	2		2
54	fore bowline	2		2
55	fore topsail halyard	3		2
56	fore topsail brace	2		2
57	fore topsail sheet	3		2

Number	Part	Thick	Wide	Drawing
58	fore topsail clew line	2		2
59	fore topsail buntlines	2		2
60	fore topsail bowline	2		2
61	fore-topgallant halyard	2		2
62	fore-topgallant lift	1		2
63	fore-topgallant brace	1		2
64	fore-topgallant sheet	2		2
65	fore-topgallant clew line	1		2
66	fore-topgallant sail bowline	1		2
67	spritsail sling	3		2
68	spritsail outhaul	2		2
69	standing lift	2		2
70	spritsail brace	2		2
71	spritsail sheet	3		2
72	spritsail clew line	2		2
73	spritsail topsail halyard	1		2
74	spritsail topsail lift	1		2
75	spritsail topsail brace	1		2
76	spritsail topsail sheet	2		2
77	spritsail topsail clew line	1		2
78	mizzen halyard tie	3		2
79	mizzen parrel truss pendant	2		2
80	mizzen lift	2		2
81	mizzen brails	1		2
82	mizzen sheet	2		2
83	mizzen tack	2		2
84	mizzen heel tackles	2		2
85	mizzen topsail halyard	2		2
86	mizzen topsail lift	1		2
87	mizzen topsail brace	1		2
88	mizzen topsail sheet	2		2
89	mizzen topsail clew line	1		2
90	mizzen topsail bowline	1		2
91	crossjack yard jeer	2		2
92	crossjack yard lift	2		2
93	crossjack brace	1		2
94	main tack	3		2
95	foretack	3		2
Sloop				
263	keel	4	9	5
264	stem	—	11	5
265	sternpost	—	11	5
266	frame	4	4	5
267	wale	6	9	5
268	wale	8	8	5
269	stringer	3	15	5

TABLE 4. (*continued*)

Number	Part	Thick	Wide	Drawing
270	mast thwart	9	28	5
271	aft thwart	9	21	5
272	keelson	4	3	5
Boat				
273	bottom	6	141	5
274	floor timber	9	10	5
275	stem	—	13	5
276	sternpost	—	13	5
277	planking	—	—	5
278	futtock	8	10	5
279	bollard	18	18	5
280	mast thwart	10	35	5
281	aft thwart	10	26	5
282	stringer	3	18	5
283	wale	9	17	5
284	keelson	5	30	5
285	windlass	—	25	5
286	filling strake	—	—	5

Pinas Drawings

follow on pp. 276–285

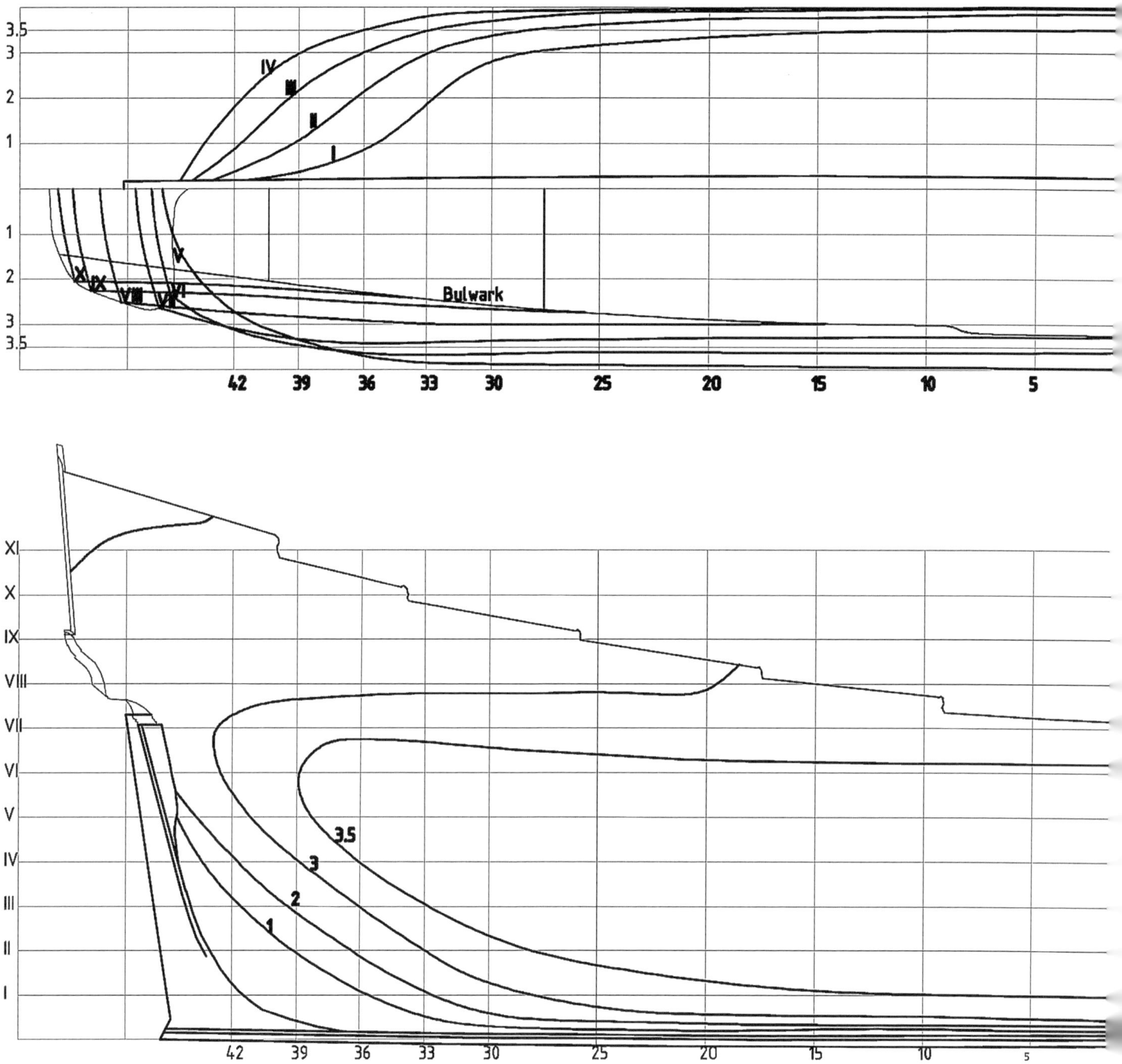

DRAWING 1 Lines plan. (Drawing by A. J. Hoving; redrawn in AutoCAD by Cor Emke)

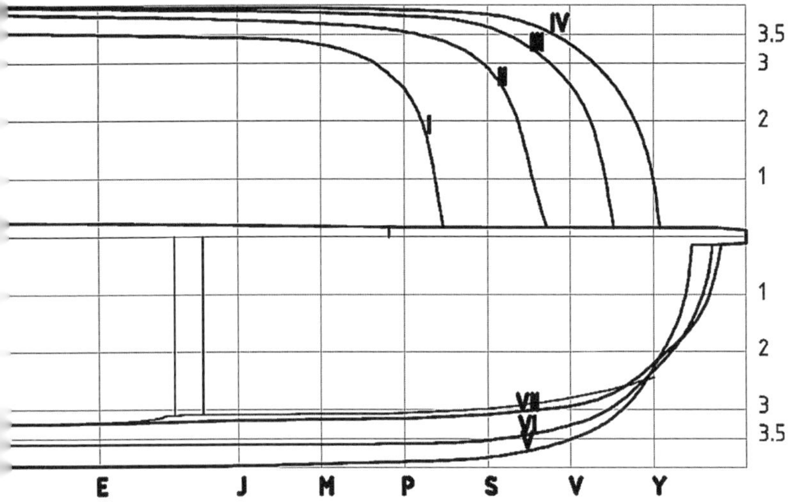
IV
III
II
I
3.5
3
2
1
1
2
3
3.5
VII
VI
V
E J M P S V Y

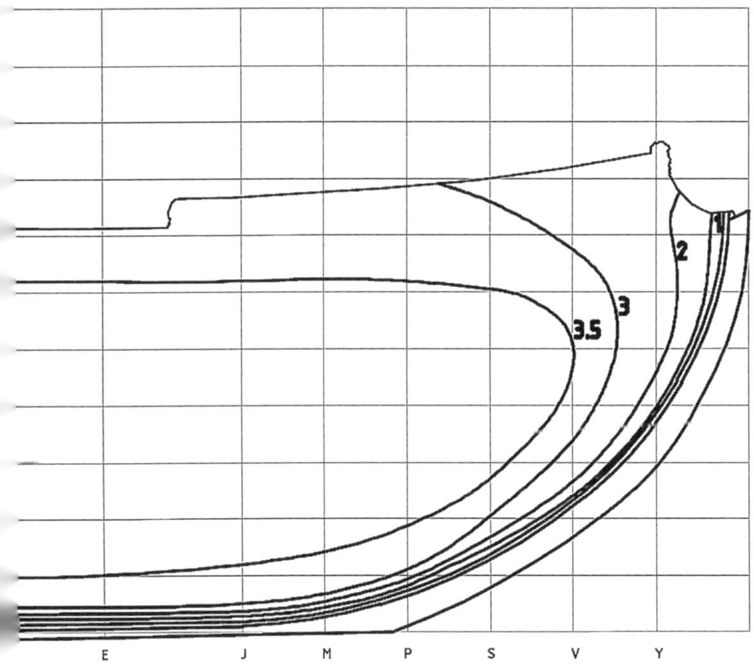
1
2
3
3.5
E J M P S V Y

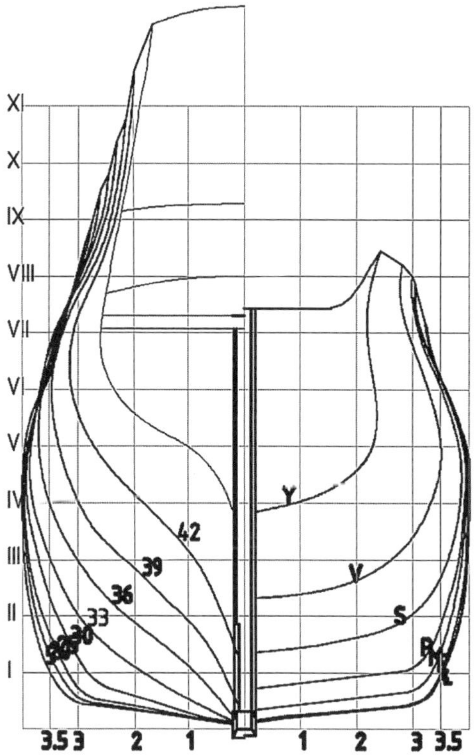
XI
X
IX
VIII
VII
VI
V
IV
III
II
I
42
39
36
33
30
Y
V
S
3.5 3 2 1 1 2 3 3.5

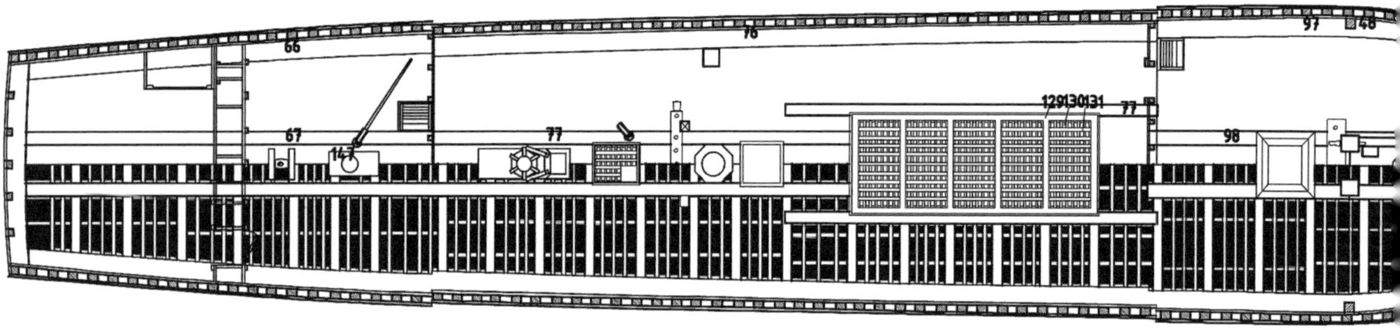

Upper deck

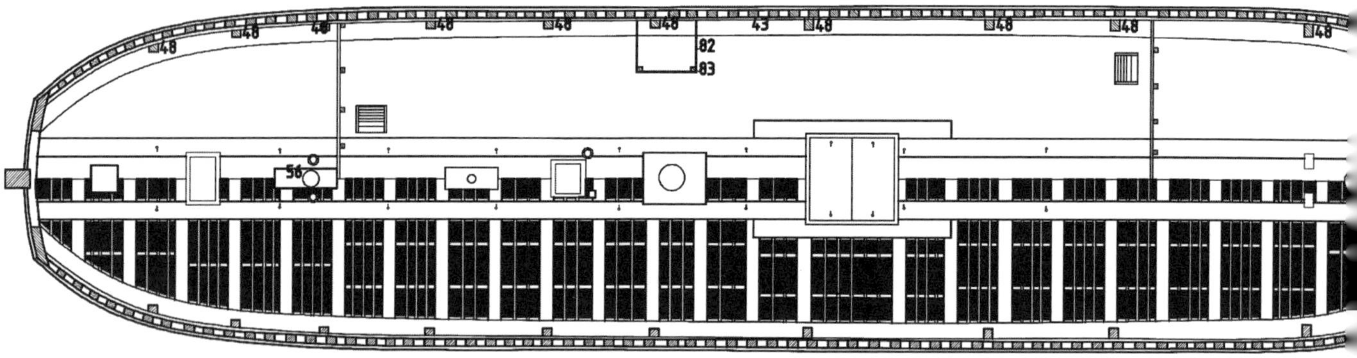

Lower deck

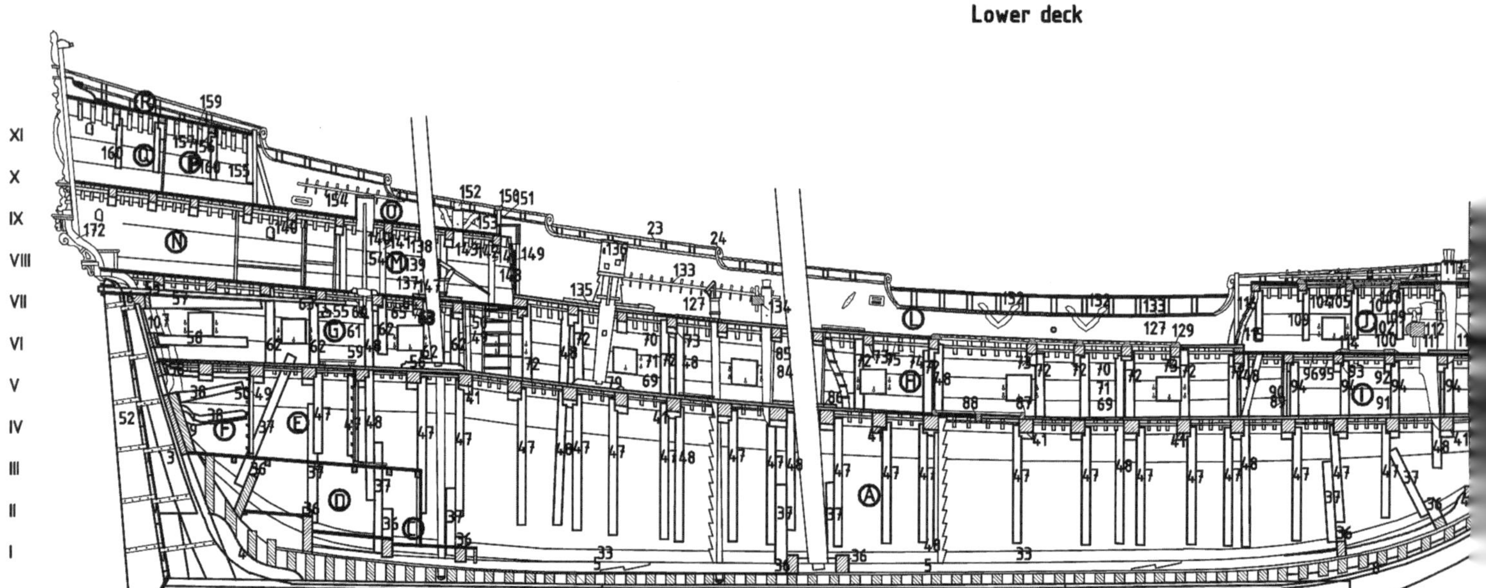

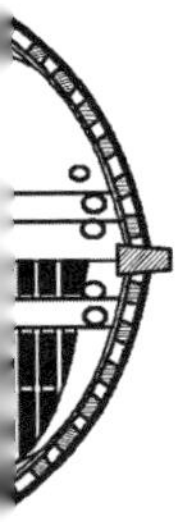

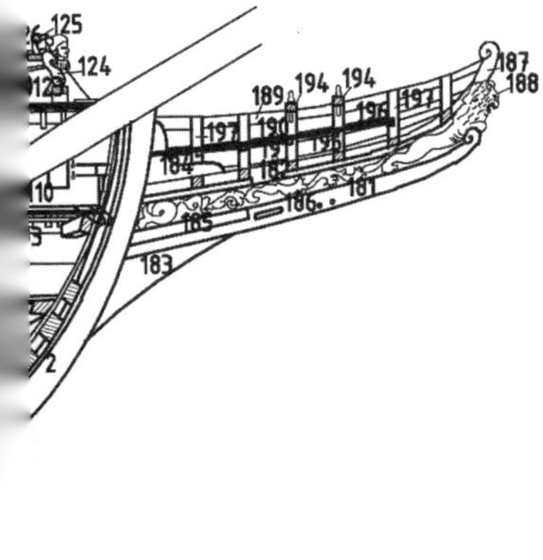

Drawing 2 Sections and decks. (Drawing by A. J. Hoving; redrawn in AutoCAD by Cor Emke)

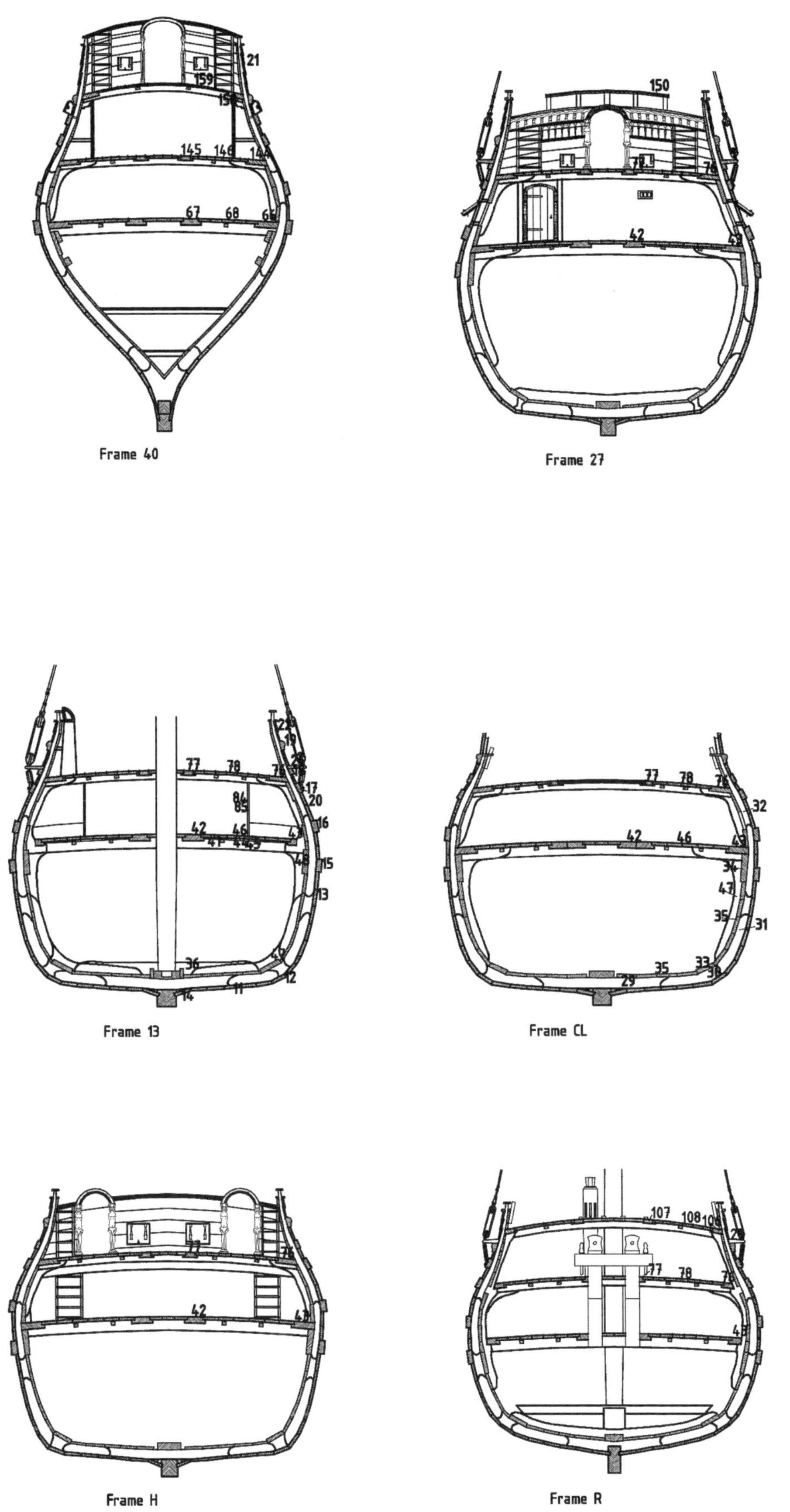

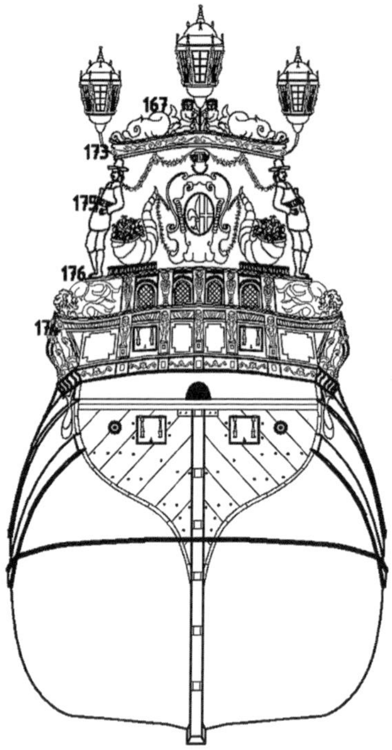
167
173
175
176
174

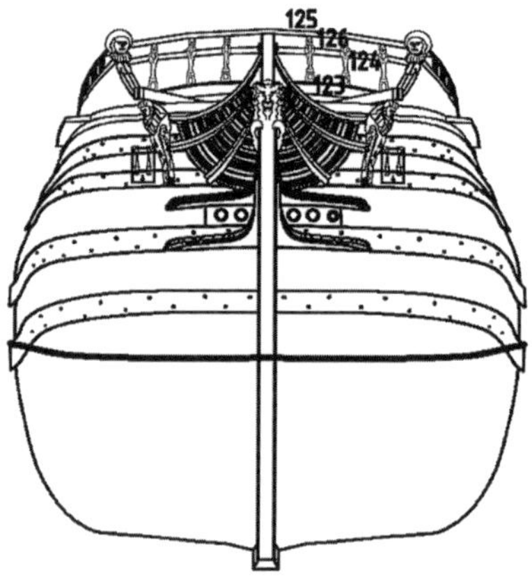
125
126
124
123

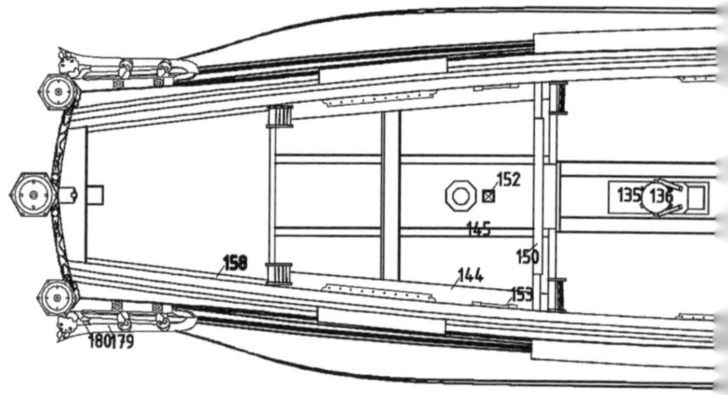
152
135
136
145
150
158
144
153
180
179

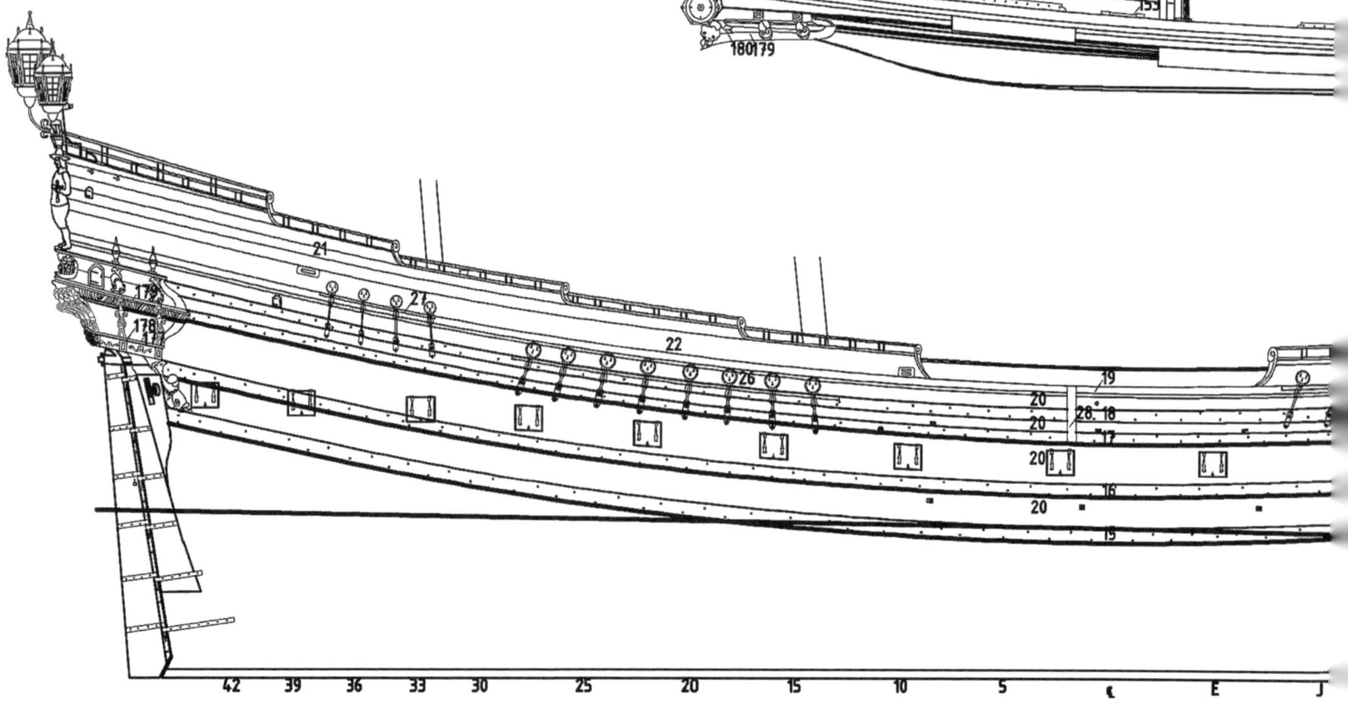
21
179
178
21
22
26
19
20
28
18
20
17
20
16
20
15
42
39
36
33
30
25
20
15
10
5
E
J

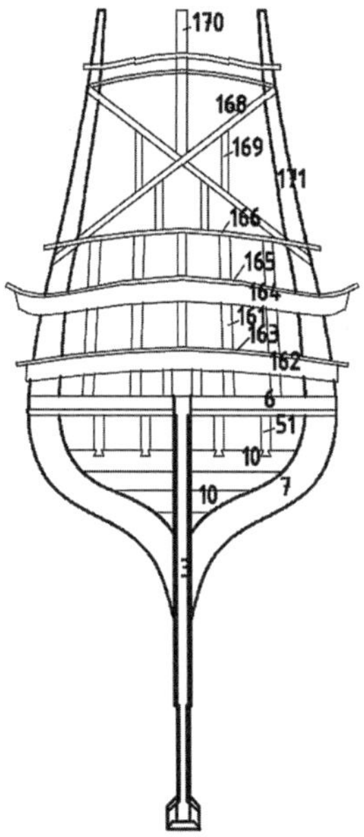

Drawing 3 Side view. (Drawing by A. J. Hoving; redrawn in AutoCAD by Cor Emke)

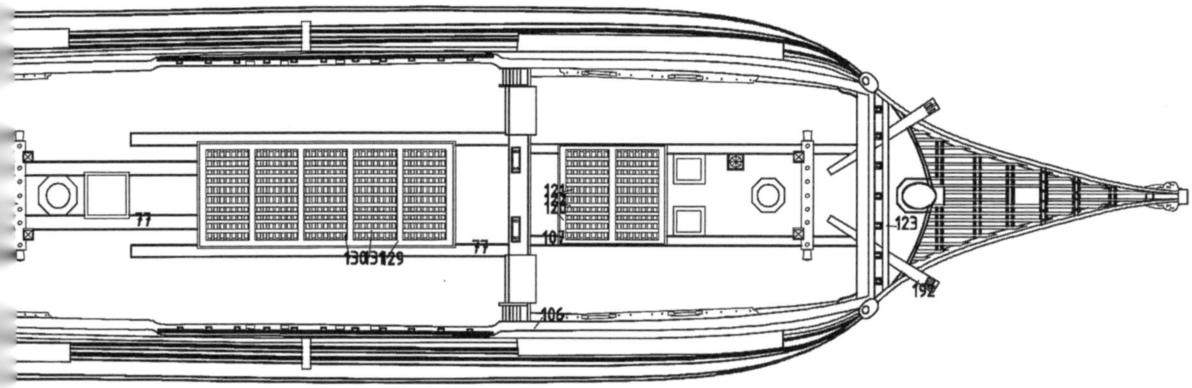

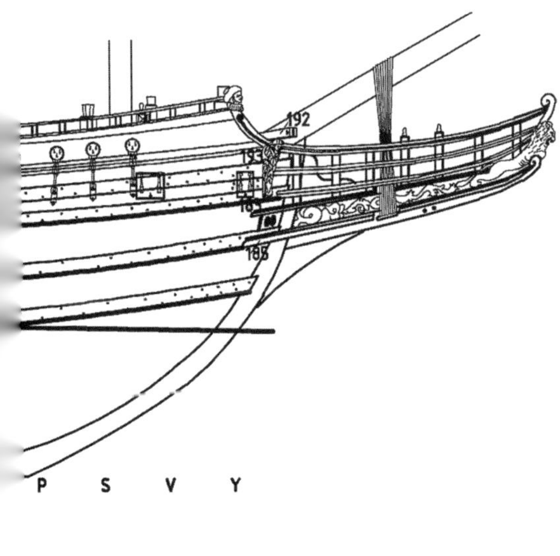

42 39 36 33 30 25 20 15 10 5 ℄ E J M P S V

Drawing 4 Rigging plan. (Drawing by A. J. Hoving; redrawn in AutoCAD by Cor Emke)

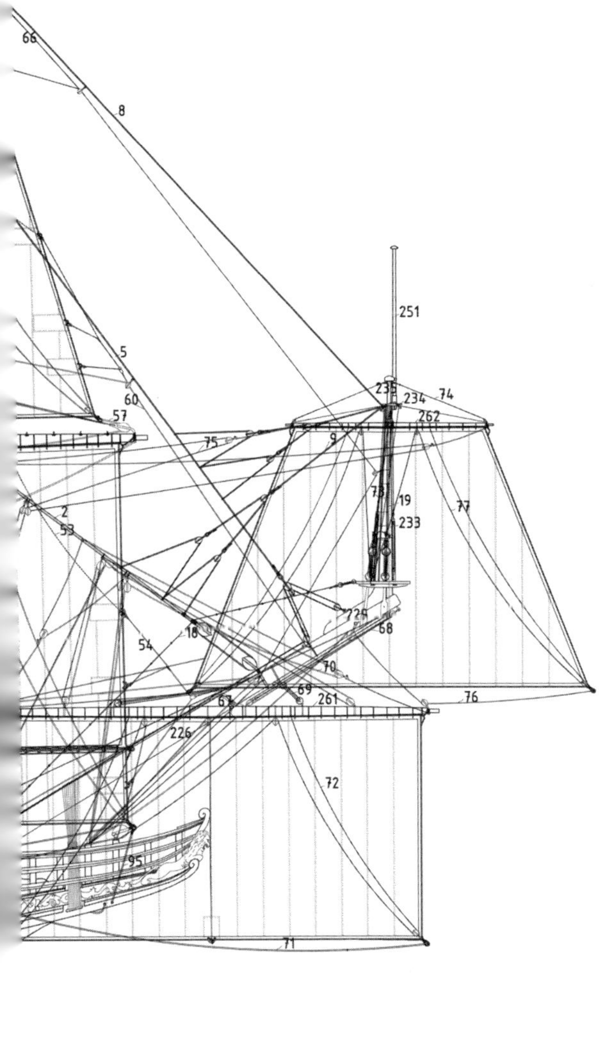

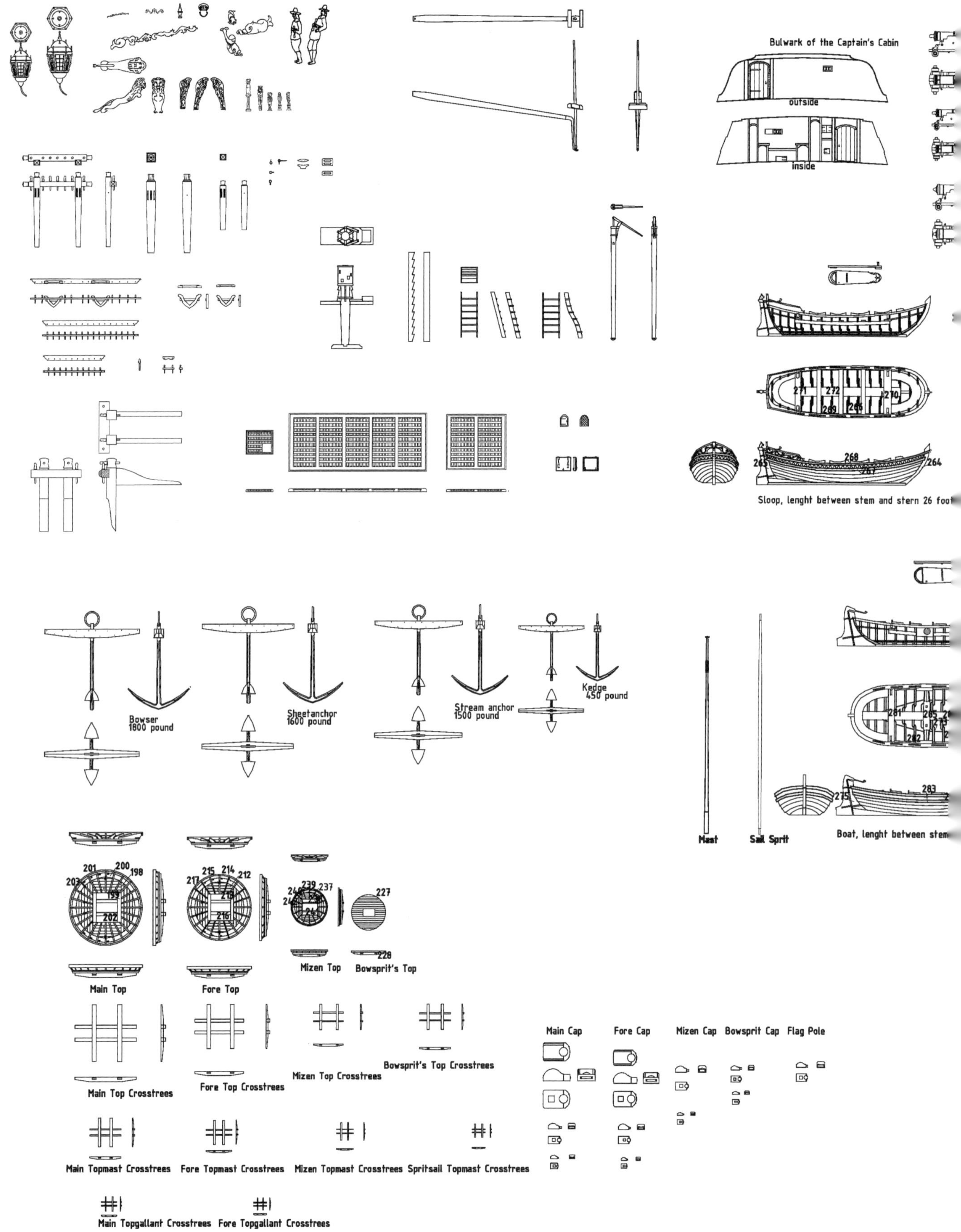
Bulwark of the Captain's Cabin
outside
inside
271
272
270
269
265
268
265
267
264
Sloop, lenght between stem and stern 26 foot
Bowser
1800 pound
Sheetanchor
1600 pound
Stream anchor
1500 pound
Kedge
450 pound
281
282
283
275
Mast
Sail Sprit
Boat, lenght between stem
201
200
198
203
199
202
215
214
212
217
213
216
239
237
227
228
Main Top
Fore Top
Mizen Top
Bowsprit's Top
Main Top Crosstrees
Fore Top Crosstrees
Mizen Top Crosstrees
Bowsprit's Top Crosstrees
Main Cap
Fore Cap
Mizen Cap
Bowsprit Cap
Flag Pole
Main Topmast Crosstrees
Fore Topmast Crosstrees
Mizen Topmast Crosstrees
Spritsail Topmast Crosstrees
Main Topgallant Crosstrees
Fore Topgallant Crosstrees

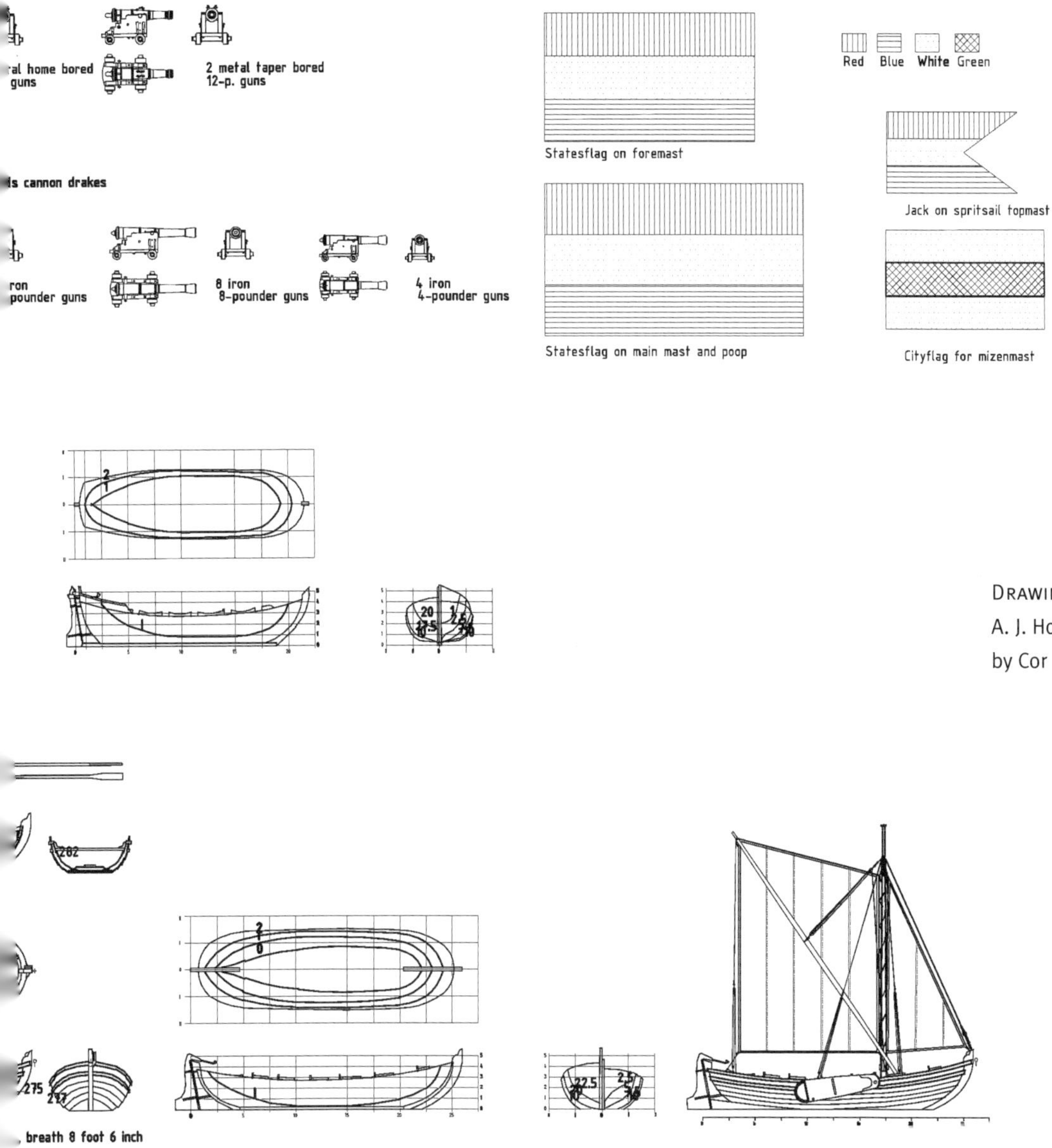

Drawing 5 Details. (Drawing by A. J. Hoving; redrawn in AutoCAD by Cor Emke)

Notes

Foreword

1. James Hornell, *Water Transport: Origins and Early Evolution* (Cambridge: Cambridge University Press, 1946), 194.

2. Reinder Reinders and Kees Paul, eds., *Carvel Construction Technique: Fifth International Symposium on Boat and Ship Archaeology, Amsterdam 1988,* Oxbow Monograph 12 (Oxford: Oxbow, 1991).

3. A. H. J. Prins, *A Handbook of Sewn Boats: The Ethnography and Archaeology of Archaic Plank-Built Craft* (Greenwich: National Maritime Museum Trust, 1986), 33.

4. Conrad Engelhardt, *Nydam Mosefund, 1858–63* (Copenhagen, 1865).

5. Lionel Casson, *Ships and Seafaring in Ancient Times* (London: British Museum Press, 1994), 145.

6. Homer *Odyssey* 5.247–53.

7. André Wegener Sleeswyk, "Phoenician Joints, *coagmenta punicana,*" *International Journal of Nautical Archaeology and Underwater Exploration* 9, no. 4 (1980): 243.

8. S. Müller, *Verslag van de Provinciaal Utrechtsch Genootschap voor Kunsten en Wetenschappen* (Utrecht, 1895), 1, 131.

9. Casson, *Ships and Seafaring,* 106.

10. Michelangelo Muraro, *Paolo da Venezia* (Philadelphia: Pennsylvania State University Press, 1970).

11. J. van Beylen, "Scheepstypen," in *Maritieme Geschiedenis der Nederlanden,* ed. G. Asaert et. al. (Bussum, Unieboek, 1976–78), 1:108–51.

12. André Wegener Sleeswyk, "The Engraver Willem A. Cruce (WA) and the Development of the Chain-Wale," *Mariner's Mirror* 76, no. 4 (November 1990): 345–61.

13. Heinrich Winter, *Das Hanseschiff im ausgehenden 15. Jahrhundert* (Rostock: Hinstorff, 1961), 21.

14. Lionel Casson, *Ships and Seamanship in the Ancient World* (Princeton, N.J.: Princeton University Press, 1971), fig. 162.

15. Winter, *Das Hanseschiff,* 20.

16. Athenaeus *Deipn.* 207a.

17. Ibid. 207b.

18. Ibid. 207b.

Chapter 1

1. The year 1672 was disastrous for the Dutch Republic. England, France, and the German cities Munster and Cologne simultaneously attacked the country. Secretary of State Johan de Witt and his brother Cornelius were killed by a mob in The Hague. If not for the defense of the seaside by Admiral Michiel de Ruyter, the country would surely have been conquered.

2. See J. E. Elias, *De vroedschap van Amsterdam, 1578–1799* (1903–1905; repr., Amsterdam: Nico Israel, 1963), 1:544–45.

3. See Marion Peters, "Nicolaes Witsen and Gijsbert Cuper: Two Seventeenth-Century Dutch Burgomasters and Their Gordian Knot," *Lias* 16, no. 1 (1989): 111–51; J. F. Gebhard Jr., *Het leven van Mr. Nicolaes Cornelisz. Witsen, 1641–1717,* 2 vols. (Utrecht: Leeflang, 1881–82).

4. Jan van der Waals, "Wankelend wereldbeeld: Onderzoek naar taal, geloof en tijd in rariteitenkabinetten," in *De Wereld binnen Handbereik, Nederlandse Kunsten Rariteiten verzamelingen 1585–1735,* ed. E. Bergvelt and R. Kistemaker, 135–52 (Zwolle: Waanders; Amsterdam: Amsterdams Historisch Museum, 1992).

5. Fernando Oliveira, *Livro da fábrica das Naos* (facsimile ed., Lisbon: Academia de Marinha, 1991); Joseph Furttenbach, *Architectura navalis: Das ist: von dem Schiff-Gebäw, auff dem Meer und Seekusten zugebrauchen . . .* (Nuremberg: J. Saum, 1629); Robert Dudley, *Dell'Arcano del Mare* (Florence: Francesco Onofri, 1646); Bartolomeo Crescentio, *Nautica Mediterranea* (Rome: Bartolomeo Bonfadino, 1607); Edward Hayward, *The Sizes and Lengths of Riggings and the Hayward-Kendal Pamphlet Controversy,* ed. R. C. Anderson (1655; repr., London: Francis Edwards, 1967).

6. The manuscript is in the collection of the library of the Nederlands Scheepvaartmuseum Amsterdam (Amsterdam Maritime Museum) and has been transcribed by Herman Ketting.

7. For example, Fabrica di Galere, or Libro di marineria, a mid-sixteenth-century manuscript from the Archivo di Stato di Venezia, is known to be a copy of the manuscript of Michele da Rodi (Libro di Michele da Rodi), dated to 1436.

8. See Lars Bruzelius, *A Bibliography of Books on Naval Architecture, Rigging, and Seamanship Printed 1600–1919* (Upsala: privately printed, 1990).

9. For a modern edition, see Anthony Deane, *Deane's Doctrine of Naval Architecture, 1670,* ed. Brian Lavery (London: Conway Maritime Press, 1981).

10. A second edition of the book was published in 1690, of which only a few have been preserved. In this later edition, the information on the pinas has hardly been touched; a few building specifications have been removed, a "shipbuilding recipe" has been added, and a few figures have been depicted differently. For a discussion of the variants in the two editions, see Diederick Wildeman's "Variations on Witsen" in the appendix.

11. Reinder Reinders, *Cog Finds from the IJsselmeerpolders,* Flevobericht 248 (Lelystad: R ksdienst voor de sselmeerpolders, 1985); Peter Throckmorton, *History from the Sea: Shipwrecks and Archaeology* (London: Mitchell Beazley, 1987).

12. Thijs J. Maarleveld, "Archaeology and Early Modern Merchant Ships: Building Sequence and Consequences. An Introductory Review," in *Rotterdam Papers,* vol. 7, *A Contribution to Medieval Archaeology: Teksten van lezingen, gehouden tijdens het symposium "Handel, handelsplaatsen en handelswaar vanaf de Vroege Middeleeuwen in de Lage Landen" te Rotterdam van 2 t/m 3 novem-*

ber 1990, ed. A. Carmiggelt (Rotterdam: Bureau Oudheidkundig Onderzoek van Gemeentewerken Rotterdam, 1992).

13. Cornelis van Yk, *De Nederlandse Scheeps-bouw-konst Open Gestelt* (1697; repr., Rotterdam: Langerveld, 1980), 55. Subsequent references to this work are given in the text.

14. See also A. J. Hoving, "Away from the Drawing Board" (pts. 1 and 2), *Model Shipwright* 69 (1989): 43–51; 71 (1990): 47–50; A. J. Hoving and Robert Parthesius, "Hollandse scheepsbouwmethoden in de zeventiende eeuw," in *Batavia Cahier,* vol. 3, *Herbouw van een Oostindiëvaarder,* ed. Robert Parthesius, 5–11 (Lelystad: Stichting Nederland bouwt V.O.C.-retourschip, 1991).

15. See Reinder Reinders and Kees Paul, eds., *Carvel Construction Technique: Fifth International Symposium on Boat and Ship Archaeology, Amsterdam 1988,* Oxbow Monograph 12 (Oxford: Oxbow, 1991).

16. A. Vreugdenhil, A. *Swhips of the United Netherlands 1648–1702,* part 4 of *Lists of Men-of-War, 1650–1700,* Society for Nautical Research, Occasional Publications 5 (Cambridge: Cambridge University Press, 1936).

17. The Nederlands Scheepvaartmuseum in Amsterdam has a beautiful, although somewhat primitive model of this type of ship (inventory no. A.0115[02]).

18. Pieter van Dam, *Beschryvinge van de Oostindische Compagnie, 1639–1701,* vol. 1, ed. F. W. Stapel (The Hague: Martinus Nijhoff, 1927), 467, 470, 493–504.

19. J. E. Elias, *De vlootbouw in Nederland in de eerste helft der 17de eeuw, 1596–1655,* Werken uitgegeven door de Commissie voor Zeegeschiedenis 1 (Amsterdam: Noord-Hollandsche uitgeversmaatschappij, 1933), 58.

20. For the sake of completeness—which, alas, does not make this terminological jungle any simpler—I should mention that the same word, *pinas,* was also used for the second-largest launch of a man-of-war, a rowboat that could also be sailed.

21. In his famous diary Samuel Pepys declares, in his entry for May 19, 1666, his deep respect for master shipwright Anthony Deane for being "the first that hath come to any certainty before-hand, of foretelling the draught of water of a ship before she is launched" (cited by Lavery in Deane, *Deane's Doctrine,* 25). If this is true—and we may assume as much, as experts state that in the sixteenth century Matthew Baker was already capable of such calculations—then England presents an entirely different picture of technical know-how in shipbuilding than in Holland. It was Pepys, in fact, who invited Deane to write his *Doctrine of Naval Architecture,* the first page of which reads: "written in the year 1670 at the instance of Samuel Pepys Esq." (ibid., 21).

22. G. Timmerman, "Das Eindringen der Naturwissenschaften in das Schiffsbauhandwerk," *Deutsches Museum Abhandlungen und Berichte* 30, no. 3 (1963): 5–54.

23. Jan Dirkzwager, *Dr. B. J. Tideman, 1834–1883: Grondlegger van de moderne scheepsbouw in Nederland* (Leiden: Brill, 1970).

24. A. J. Hoving and Alan A. Lemmers, *In tekening gebracht: De achttiende-eeuwse scheepsbouwers en hun ontwerpmethoden* (Amsterdam: Bataafsche Leeuw, 2001).

25. Probably Simon Stevin's *De beghinselen des waterwichts* (Leiden: Françoys van Raphelinghen, 1586).

26. The last as a unit of measure varied considerably, depending not only on what was being measured (ships, grain, wood, etc.) but also on the locality; thus it could represent different weights. Originally a measure of volume, the last was generally equal to 4,000 Amsterdam pounds, or 1,975 kilograms. See

B. E. van Bruggen, "Beschouwing over het aangeven van de hoofdafmetingen van de Nederlandse zeeschepen en de daarbij gebruikte maateenheden (1600–1800)," *Mededelingen van de Nederlandse Vereniging voor Zeegeschiedenis / Tijdschrift voor Zeegeschiedenis,* no. 20 (March 1970): 25–34.

27. Van Dam, *Beschryvinge,* 460.

28. A. J. Hoving, *Het oorlogsjacht* De Heemskerk: *Het schip waarmee Abel Tasman Nieuw-Zeeland ontdekte* (Amsterdam: Rijksmuseum, 1992).

29. We cannot know this for certain, as the first statutory regulations for the minimum of freeboard date from c. 1875.

30. Oral communication from Robert Parthesius, March 1991.

31. For a discussion of specialization on the yards, see Jerzy Gawronski, "*De Hollandia* en *De Amsterdam:* Twee Schepen en Een Bedrijf: Materiële Cultuur en Organisatie VOC Amsterdam 1740–1750" (PhD diss., Leiden University, 1994).

Chapter 2

1. B. E. van Bruggen, "Beschouwing over het aangeven van de hoofdafmetingen van de Nederlandse zeeschepen en de daarbij gebruikte maateenheden (1600–1800)," *Mededelingen van de Nederlandse Vereniging voor Zeegeschiedenis / Tijdschrift voor Zeegeschiedenis,* no. 20 (March 1970): 25–34.

2. Communication from Robert Grenier, Parks Canada, April 6, 1992.

3. Leendert van Zwijndregt and Cornelis de Ruiter, *Verhandeling van den Hollandschen Scheepsbouw . . .* (The Hague: De Hondt and H. Scheurleer, 1757).

4. This measurement bears no relation to the method for calculating the hold of a ship for determining toll rates. In that case, the distance between the keelson and the planks of the lower deck was measured; the width was then taken between the ceiling, either on the lower or upper deck.

5. Witsen states that this thickness is found from the length of the ship: example, 10 feet length, 1 inch thick (66 I 14). See 2. The Stem later in this chapter.

6. Cornelis van Yk, *De Nederlandse Scheeps-bouw-konst Open Gestelt* (1697; repr., Rotterdam: Langerveld, 1980), 56. Subsequent page references are given in the text.

7. Nick Burningham, personal communications regarding the translation of technical terms, April 2009.

8. Another passage, recommending that the keel be laid with an upward bend, is in complete contradiction with this practice: The keel, which sinks deeper than the ship itself, keeps the ship from drifting: it is raised in the middle, which is because, the load being heaviest in the middle the ships will sag there, and the keel will straighten again: if it is made straight from the onset, it will bend outward, which will produce much trouble. Others are of the opinion that one should make the keel straight from the beginning: because it will then serve the ship best: reasoning that the handicap which a stooped ship has does not weigh against the advantage of a straight keel, although it may sag afterward. A bent keel will keep the course of a ship as well as a straight one, but will drift sooner (263 II 40). The answer to this paradox probably lies in the sort of ships Witsen is writing about in this passage: presumably long, narrow ships with straight stems, like some types of inland craft.

9. Anthony Deane, *Deane's Doctrine of Naval Architecture, 1670,* ed. Brian Lavery (London: Conway Maritime Press, 1981), 57.

10. Herman Ketting, Prins Willem: *Een zeventiende-eeuwse Oostindievaarder* (Bussum: Unieboek, 1979).

11. Heinrich Winter, *Der holländische Zweidecker von 1660/1670* (Bielefeld: Delius, Klasing & Co., 1978).

12. Van Zwijndregt and Ruiter, *Verhandeling.*

13. Winter, *Der holländische Zweidecker.*

14. Jules van Beylen, *Zeilvaart Lexicon: Maritiem Woordenboek* (Weesp: De Boer Maritiem, 1985), 173.

15. Thanks to Nick Burningham and Menno Leenstra for supplying this information.

16. A. Martin de Wild, *The Scientific Examination of Pictures: An Investigation of the Pigments Used by the Dutch and Flemish Masters from the Brothers Van Eyck to the Middle of the 19th Century,* trans. Leonard Cecil Jackson (London: G. Bell & Sons, 1929).

17. Jerzy Gawronski, B. Kist, and O. Stokvis van Boetzelaer, eds., *Hollandia Compendium: A Contribution to the History, Archaeology, Classification and Lexicography of a 150 ft. Dutch East Indiaman (1740–1750)* (Amsterdam: Rijksmuseum, 1992).

Chapter 3

1. This again indicates that the shipwright did not always use a pair of compasses, as I mentioned in chapter 2 (see section 2, "The Stem"). It might be argued that instead of compasses a piece of string and a chalk might also do the job, but this does not explain why the curve is dictated in inches or feet and measured from the straight connection line between the two extremities, not with the desired length of the string.

2. In a personal communication of April 2009, Nick Burningham observes: "I think the builder of a full-size ship, or a model, has to know what the result should look like. The several English texts on 17th C ship building that lack illustrations are of little use in reconstructing the ships' designs. Building plank-first is essentially a form of sculpture. If the sculptor has the design clearly conceptualised/visualised, and he is a talented sculptor, the result will be good."

3. It would be a fascinating experiment if a number of well-trained model shipwrights were to build a model on their own using the same contract and Witsen's building method. I am convinced that the resulting variations among the individual models would not be much greater than those that were no doubt manifested in real ships built by different shipwrights in Witsen's time.

4. For details about this experiment, see A. J. Hoving, "Away from the Drawing Board" (pts. 1 and 2), *Model Shipwright* 69 (1989): 43–51; 71 (1990): 47–50.

5. A. J. Hoving, *Het oorlogsjacht* De Heemskerck: *Het schip waarmee Abel Tasman Nieuw-Zeeland ontdekte* (Amsterdam: Rijksmuseum, 1992).

6. Nick Burningham: "I think the same thing can happen with the English system. If the height of breadth rises to the stern, there is little reduction of beam through a long midbody, and there is tumblehome, the lines will cross. It might be seen as a consequence of the change to a long midbody and a

relatively wide stern which took place during the 17th century" (personal communication, April 2009).

7. See also Jerzy Gawronski, "*De Hollandia* en *De Amsterdam:* Twee Schepen en Een Bedrijf: Materiële Cultuur en Organisatie VOC Amsterdam 1740–1750" (PhD diss., Leiden University, 1994).

8. For example, a *kontwachter* was a block through which the sheet of the spritsail passed; the literal translation of the term is "ass-watch." A very protruding sail carried the name *lul,* a slang word for the male organ.

9. A comment about the size of these ships is in order. In the contract for the 225-last ship (136 I 15), the factor is 192 (L × W × D = 43,407 ÷ 225 = 192). As we saw in the section "Ship Measurement" in chapter 1, this is probably the factor for a fluyt. In the contract for the 164-foot ship (136 II 1) the number of lasts (135) is not proportionate to the length: the factor would total 821 here. Thus, one of the measurements is not correct—either the length is too great or the number of lasts is too small, or the ship was a man-of-war and had some additional loading capacity.

Chapter 4

1. A. J. Hoving and Alan A. Lemmers, *In tekening gebracht: De achttiende-eeuwse scheepsbouwers en hun ontwerpmethoden* (Amsterdam: Bataafsche Leeuw, 2001).

Appendix

1. John Carter, *ABC for Book Collectors,* 5th ed. (London: Hart-Davis, MacGibbon, 1974), 204–25.

2. Nicolaes Witsen, *Aeloude en Hedendaegsche Scheeps-bouw en Bestier . . .* (1671; facsimile ed., Alphen aan de Rijn: Canaletto, 1979; facsimile repr., Franeker: Canaletto and Van Wijnen, 1994); Nicolaas Witsen, *Architectura navalis et regimen nauticum ofte Aaloude en Hedendaagsche Scheeps-bouw en Bestier . . .* (1690; facsimile ed., Amsterdam: Graphic, 1970).

3. These biographical details have been taken from Marion Peters, "Mercator Sapiens (De Wijze Koopman). Het wereldwijde onderzoek van Nicolaas Witsen (1641–1717), burgemeester en VOC bewindhebber van Amsterdam" (PhD diss., University of Groningen, 2008; a revised, published version is forthcoming); J. F. Gebhard Jr., *Het leven van Mr. Nicolaes Cornelisz. Witsen, 1641–1717,* 2 vols. (Utrecht: Leeflang, 1881–82); and J. T. Bodel Nijenhuis, "Verspreide bijzonderheden over Mr. Nicolaes Cornz. Witsen, burgemeester van Amsterdam," *Bijdragen voor Vaderlandsche Geschiedenis en Oudheidkunde* 10 (1856): 1–31.

4. Nicolaes Witsen, "Aen de Lezer" (To the Reader), in *Aeloude en Hedendaegsche Scheeps-bouw en Bestier . . .* (Amsterdam: Casparus Commelijn, Broer and Jan Appelaer, 1671), [6]; and Nicolaas Witsen, "Aan de Lezer," in *Architectura navalis et regimen nauticum . . .* (Amsterdam: Pieter and Joan Blaeu, 1690), [6].

5. Only a copy of his report has survived; it has been published as Nicolaas Witsen, *Moscovische Reyse, 1664–1665,* 3 vols., ed. J. G. Locher and P. de Buck, Werken uitgegeven door de Linschoten-Vereeniging, 66–68 (The Hague: Martinus Nijhoff, 1966–67).

6. Peters, "Mercator Sapiens," 34 n. 135.

7. Thirteen copies (mounted or in separate sheets) are now known; see ibid., 127 n. 547.

8. Peters (ibid., 141) mentions two copies in the Netherlands (at university libraries in Amsterdam and Utrecht) and two in St. Petersburg, Russia (at the Academy of Science library and the Russian National Library). There is, however, a fifth copy whose present location is unknown. It was on loan from the antiquarian bookseller A. W. M. Mensing (1866–1936) to the Nederlands Scheepvaartmuseum in Amsterdam in the 1920s; then Mensing withdrew a part of his collection from the museum, including this book. After his death it was auctioned by Sotheby's in London. See *Verzameling-Mensing in het Scheepvaartmuseum te Amsterdam: Bibliotheek* (Amsterdam: Frederik Muller & Co., 1923), 2:365, no. 2762; and *The Mensing Library . . . formed by the late Mr. Ant. W. M. Mensing of Amsterdam,* 2 vols. (London: Sotheby's, 1936–37), 1:169, lot no. 611.

9. Peters, "Mercator Sapiens," 139.

10. The stay in the Netherlands was interrupted by a three-month visit to England in early 1698.

11. Peters, "Mercator Sapiens," 138–39.

12. Nicolaas Witsen, *Noord en Oost Tartaryen. Behelzende eene beschryving van verscheidene Tartersche en nabuurige gewesten . . .* (Amsterdam: M. Schalekamp, 1785).

13. Gebhard, *Het Leven,* 2:159, 161, 197.

14. Peters, "Mercator Sapiens," 60–61.

15. Gebhard, *Het Leven,* 1:350.

16. Witsen, *Aeloude en Hedendaegsche Scheeps-bouw,* 2, and appendix 9, col. 2.

17. Louis Th. Lehmann, *Witsen en de Oudheid* (Amsterdam: De Gouden Reaal, 1995), and Lehmann *(On?)kritische wetenschap in de 17e eeuw* (Amsterdam: De Gouden Reaal, 1995).

18. A complete overview of Witsen's prints, maps, and book illustrations are in Christiaan Schuckman, Ilja M. Veldman, and Jeroen de Scheemaker, *Hollstein's Dutch and Flemish Etchings, Engravings, and Woodcuts, ca. 1450–1700,* vol. 53, *Frederick de Wit to Lieven de Witte,* ed. Dieuwke de Hoop Scheffer (Rotterdam: Sound & Vision, 1999).

19. Peters, "Mercator Sapiens," 114.

20. Marieke van Delft and Clemens de Wolf, *Bibliopolis: Geschiedenis van het gedrukt boek in Nederland* (Zwolle: Waanders, 2003), 72–73. Available in English at www.bibliopolis.nl/handboek; see "1585–1725: Hey-day as centre of world trade," sec. 2.2.3, "Relation between publisher and author" (accessed August 27, 2008).

21. Peters, "Mercator Sapiens," 116.

22. G. Asaert et. al., eds., *Maritieme Geschiedenis der Nederlanden,* 4 vols. (Bussum: Unieboek, 1976–78), 2:141, 144.

23. Isabella H. van Eeghen, *De Amsterdamse boekhandel, 1680–1725,* 5 vols. (Amsterdam: H. J. Duyvisfonds, 1960–87), 3:78; and M. M. Kleerkoper and W. P. van Stockum, *De Boekhandel te Amsterdam, voornamelijk in de 17de eeuw,* 2 vols. (The Hague: Van Stockum, 1914–16), 2:1234.

24. Van Delft and de Wolf, *Bibliopolis,* 82. Available in English at www.bibliopolis.nl/handboek; see "1585–1725: Hey-day as centre of world trade," sec. 2.2.10, "Financing, print-runs and prices" (accessed November 12, 2010).

25. The best commentaries on the second edition are in S. P. l'Honoré Naber, "Een tweede uitgave van Nicolaes Witsen's 'Scheepsbouw en Bestier' ontdekt!" *Het Boek* 2 (1914): 81–95; and W. Voorbeytel Cannenburg, "Witsen's *Scheepsbouw*," in *Jaarverslag Nederlandsch Historisch Scheepvaart Museum 1954–1955* (Amsterdam, Nederlandsch Historisch Scheepvaart Museum, 1956), 28–34.

26. L'Honoré Naber, "Een tweede uitgave," 81.

27. Peters, "Mercator Sapiens," 119.

28. Marion Peters, "Nicolaes Witsen and Gijsbert Cuper: Two Seventeenth-Century Dutch Burgomasters and Their Gordian Knot," *Lias* 16, no. 1 (1989): 148 n. 70; and Peters, "Mercator Sapiens," 118.

29. Peters, "Mercator Sapiens," 119–20. Peters also remarks here that the 1705 edition of *Noord en Oost Tartarije* was probably stored elsewhere or passed on to others, which would explain how it survived and, as mentioned, was finally published in 1785.

30. Cited in Eric Rieth, "Quelques remarques à propos du Dictionnaire de Marine (1702) de Nicolas Aubin," *Neptunia: Revue de l'association des Amis du Musée de la Marine,* no. 190 (June 1993): 28.

31. Johann Hirich Röding, *Allgemeines Wörterbuch der Marine in allen Europæischen Seesprachen nebst vollstændigen Erklærungen* (Hamburg, 1793–96), 1:58.

32. N. A. Smith et. al., *Catalogue of the Pepys Library at Magdalene College, Cambridge,* vol. 1, *Printed Books* (Cambridge: D. S. Brewer, 1978), 193; R. Prud'homme van Reine, *Jan Hendrik van Kinsbergen 1735–1809: Admiraal en Filantroop* (Amsterdam: Bataafsche Leeuw, 1990), 39; and Arne Losman, "Drei schwedische Büchersammler des 17. Jahrhunderts: Per Brahe d. J., Carl Gustaf Wrangel und Magnus Gabriel De la Gardie," in *Arte et Marte: Studien zur Adelskultur des Barockzeitalters in Schweden, Dänemark und Schleswig-Holstein,* ed. Dieter Lohmeier (Neumünster: Karl Wachholtz, 1978), 171–72.

33. Peters, "Mercator Sapiens," 115–16.

34. Ernst Crone, *Cornelis Douwes 1712–1773: Zijn leven en zijn werk. Met inleidende hoofdstukken over navigatie en zeevaart-onderwijs in de 17de en 18de eeuw* (Haarlem: Tjeenk Willink, 1941), 86–88.

35. C. Bordewijk, "Strolling Players along the North Sea Coasts," in *The North Sea and Culture (1550–1800): Proceedings of the International Conference held at Leiden, 21–22 April 1995,* ed. Juliette Roding and Lex Heerma van Voss (Hilversum: Verloren, 1996), 436–49.

36. This copy is now in the library of the University of Amsterdam (shelf no. OF 93–5). It was formerly in the possession of the well-known American book collector Harrison D. Horblit (1912–88) of Ridgefield, Connecticut; see: H. P. Kraus, *Catalog 168. The History of Science, including Navigation: A First Selection of Books from the Library of Harrison D. Horblit* (New York: H. P. Kraus, c. 1976) 108, no. 202; and H. P. Kraus, *Catalog 178. Voyages, Travels, and Ancillary Sciences: Including Geography, Cartography, Navigation, etc.* (New York: H. P. Kraus, 1988), 93, no.192. The book remained in the trade until it was eventually sold to the university library in 1992 by the Amsterdam rare book dealer Nico Israel.

37. Wouter Nijhoff, "De Anglofobie van Nic. Witsen en de verschillende redactiën van zijn *Scheepsbouw,*" *Het Boek* 13 (1925): 88–96.

38. Witsen to Isaac Vossius in London, October 11 [1672] and November 6 [1672], Leiden University Library, Burman Manuscript Collection, F 11, vol. 2, Let-

ters to Vossius, nos. 805 and 806. I wish to thank Dr. Marion Peters for pointing out these letters and providing transcriptions.

39. Peters, "Mercator Sapiens," 119.

40. Losman, "Drei schwedische Büchersammler," 171–72.

41. Jerôme Lalande, *Abrégé de navigation* . . . (Paris: privately printed, 1793), 9.

42. Cannenburg, "Witsen's *Scheepsbouw,*" 32.

43. Frederik Muller, *Beschrijvende catalogus van 7000 portretten van Nederlanders* (Amsterdam: Frederik Muller, 1853), 297, no. 6201e.

44. Katalog der Bibliothek der Seefahrtschule Bremen, vol. 2, "19th Century," sachgruppe F, manuscript msa 0353, Staats- und Universitätsbibliothek Bremen. I wish to thank Dr. Armin Hetzer of the Staats- und Universitätsbibliothek Bremen for this information on the Bremen copy of 1690 edition. Letter to the author, Nederlands Scheepvaartmuseum Archive; letters received 1995; no. 2043 (includes a photocopy of the manuscript catalog).

45. L'Honoré Naber, "Een tweede uitgave," 83; and Cannenburg, "Witsen's *Scheepsbouw,*" 28–34.

46. Christof Marcus and Dieter W. F. Schoppmeyer, eds., *200 Jahre Seefahrtschule Bremen* (Bremen: H. M. Hauschild, 1999), 21.

47. Letter to the author from Dr. Armin Hetzer of the Staats- und Universitätsbibliothek Bremen, Nederlands Scheepvaartmuseum Archive; letters received 1995; no. 286.

48. Witsen's letter to Cuper is printed in Gebhard, *Het Leven,* 2:288.

49. Wouter Nijhoff, "Een variant in Van IJk's *Scheeps-bouw-konst* in verband met Witsen's *Scheepsbouw,*" *Het Boek* 9 (1921): 139–44.

50. The facsimile of the 1671 edition, published in 1979 by Canaletto in Alphen aan den Rijn, was jointly reissued in 1994 by Canaletto and Van Wijnen in Franeker.

51. Published by Graphic in Amsterdam in 1970 in the series Monumenta nautica historica selecta; no. 2.

Glossary

Arch 1. **(*Boog*):** Ornamental arch over the entrance of the steering stand. 2. **(*Boogje*):** Curved part of the quarter gallery connecting the ship's side to the architrave of the quarter gallery.

Backstay (*Backstag*) 1. Part of the rigging that supports a mast backward. 2. Line to support the spritsail topmast backward.

Battens (*Latten*) Open planking in the beakhead and on the hatches.

Beakhead (*Galjoen*) A partly ornamental, partly functional construction on the fore part of the ship. It served as a platform for working the rigging of the foresail and spritsails and as a privy for the crew.

Beam 1. **(*Balck*):** Athwartship support for the decks. 2. **(*Grootspant*):** Width of the ship, measured over the midship frame under the planking.

Belaying pin (*Karvielnagel* or *korvijnagel*) Wooden pin on which ropes are belayed.

(*Bengel*) Iron hoop around the top of the rudder or the foot of the capstan.

Bezan rig (*Bezaantuig*) Type of rigging of one-masted, inshore vessels carrying a mainsail with a forked gaff on top and a boom below.

Bilge, bottom (*Vlak*) Bottom of the ship; in the beginning consisting of planks only, without frame timbers.

Bilge ceiling (*Kimweger*) Part of the inside planking at the turn of the bilge.

Bilge futtock (*Zitter* or *sitter*) Frame timber fitted in the turn of the bilge.

Bilge planking (*Boeisel*) Bilge strakes that plank the turn of the bilge.

Bilge strake (*Kimgang*) Row of planks in the bilge.

Bill (*Neb*) The short arm of a knee; the horizontal arm of a hanging knee.

Binding strake (*Schaarstok*) Thick strake that formed part of the typical Dutch construction of the deck planking, running fore and aft on both sides of the hatches, mast partners, bitts, etc.

Bitt (*Beting*) A strong, upright frame in the fore part of the ship on which the anchor cable is belayed.

Bitt knee (*Beting knie*) A large knee supporting the bitts on their forward side.

Bitt standard (*Beting speen*) The vertical part of the bitts.

Bolster (*Betingbalk*) A heavy, horizontal timber connecting the bitts.

Bolt (*Bout*) An iron rod that fastens together heavy parts of the ship.

Bordered hatch cover (*Stulpluik*) A hatch cover for small hatches, constructed as an open, inverted box with sides that fit around the hatch coamings to make the hatch watertight.

Bow (*Boeg*) The entire front part of the ship, including the planking on both sides of the stem.

Bowline (*Boelijn*) Part of the rigging that helps to pull the weather leech of the sail forward to prevent the luff from collapsing.

Bowsprit (*Boegspriet*) A spar protruding forward at an angle over the stem, carrying the spritsail yard and the stays of the foremast.

Brace (*Bras*) A rope attached to both ends (yardarms) of the yard of a square sail; used to "brace up" or pull the yard so that it lies with one yardarm angled forward, and to "brace in" or pull the yard so that it lies square across the ship at right angles to the fore-and-aft line.

Bread room (*Brootkamer*) Compartment below the gun room for storing food that was not sealed in casks. To keep out vermin, the interior of the room was entirely covered with tinplate.

Breasthook 1. **(*Band*)**: A piece of compass wood placed in the forward section of the ship to strengthen the construction. 2. **(*Twil*)**: Reinforcement timber fitted inside the ceiling in the bow and tying the two sides of the ship together. The *twil* is distinct from the *band* in that it also carried the ends of deck planks.

(*Broekstuk*) A large piece of wood that connected the fashion pieces and the sternpost.

Bullseye, fairlead (*Kloot*) A ball-shaped wooden part with a hole for guiding a rope, usually seized to a shroud.

Bush (*Bos*) A piece of squared wood, drilled lengthwise. They were placed in the ship's side as scupper pipes to drain water from the ship's decks.

Butlery (*Bottelarij*) Storeroom on the lower deck for storing sufficient food for daily use.

Cabin (*Hut*) Sleeping quarters for the officers, located above the captain's cabin.

Cable tier (*Kot*) Compartment afore on the main deck where the anchor cable is stored.

Calve (*Kalf*) Vertical timbers in the tuck between the upper transom and the wing transom, forming the vertical sills of the portholes in the tuck.

Camber (*Bocht*) A measurement indicating the curvature of the top side of a beam, especially a deck beam.

Capstan (*Kaapstander*) A device to hoist heavy loads.

Captain's cabin (*Kajuit*) Compartment for the captain, located aft on the upper deck.

Cathead (*Kraanbalk*) A beam with sheaves that protrudes from the beakhead or the forecastle head and allows the anchor to be hoisted, or "catted."

(*Cardinaels-hoet*) (lit., "cardinal's hat") A circular wooden rim surrounding the holes in the tuck next to the gunports; it served as a guide for the rope with which the boat was towed.

Carling (*Karvielhout* or *klamaai*) A short supporting timber between the beams and underneath the ledges.

Cathead knee (*Drukker*) A carved figure supporting the cathead.

Caulk (v) (*Kalfaten*) To fill the seams between planks with moss or old rope, covered with pitch, to make the ship watertight.

Ceiling plank, stringer (*Weger*) Planking inside the frames.

Chain plate (*Putting*) A metal strap that attached the deadeye on the channel to the ship's hull.

Channel (*Rust*) A heavy plank, fitted horizontally along the side of the ship, on which the deadeyes were fixed to hold the shrouds and lanyards clear of the ship's sides.

Chesstree (*Halsklamp*) A carved piece of wood to lead the main tack through the ship's side.

Chine (*Hoekige kim*) The angle between the planks of the bottom and the first planks of the turn of the bilge.

Chock (above a gunport) 1. **(*Kalf*):** The upper sill of a gunport. 2. **(*Klos*):** General term for a short piece of wood to close an opening.

Cymatium (*Cimatium*) A piece of decorative molding with an S-shaped profile, often carved for the first cover; also called an *ovolo,* with an egg-and-dart pattern.

Cleat 1. **(*Klamp*):** General term for a piece of wood holding two parts tog ether. Early in the building process cleats held the bottom planking together temporarily; horizontal cleats on doors hold the vertical planks together. 2. **(*Oorhout*):** A small chock used in a pair of planking tongues if the tongue held planks together at an angle.

Clew line, clew garnet (*Geitouw*) Rope on the aft side of a sail to pull the clew up to the yard.

Collar eye (*Kraag*) Part of the mainstay, consisting of a strop around stem and foremast, with a block stropped in to set up taut the mainstay.

Cornice (*Vaes*) Rail molding at the top of the taffrail (and elsewhere).

Counter (*Wulf*) The curved area between the wing transom and the flat of the stern or upper transom.

Counter timber (*Wulfstut*) Vertical timbers that supported the counter and upper transom, with their lower ends resting on the wing transom.

Cover (*Deck*) Profiled plank on top of a *hackebord,* sealing off the ends of the deck planks behind it.

Covering board (*Schanddek*) A plank that covers the openings between the top timbers on the forecastle and the quarter deck.

Cross (*Kruis*) Supporting construction to facilitate planking (such as diagonal bracing in the flat or the stern or upper transom). . Also used in bulwarks.

Crossjack yard (*Begijnera*) The lower yard on the mizzenmast; it spread the foot of the mizzen topsail and did not carry a sail.

Crosstree (*Dwarszaling*) A piece of wood let in to the trestletree to form the support for the top and to hold the heel of the topmast.

Crutch (*Zogstuk*) A Y-shaped piece of wood forming the floors of the aft frames.

Cyma (*Scima*) A molding planed with a curved, S-shaped profile, a decorative feature of the third cover in the stern gallery.

Dale (*Dael*) A sluice or channel that conveyed water from a pump and discharged it through the side of the ship.

Deadeye (*Juffer*) A round block with three holes, used for tightening the lanyard and the shroud.

Deal (*Deel*) Thin pine planking.

Deck clamp (*Balkweger*) The thicker part of the ceiling into which the deck beams were dovetailed.

Depth in the hold (*Holte*) The distance between the top of the keel and the height of the lower deck, measured at the sides and thus without the camber of the beam. Also the height of the scupper or watershed.

Dottel (*Dolfje*) A small wooden or iron nail hammered into treenails to make them watertight.

Dowel (v) (*Nagelen*) To fasten the underwater strakes to the frames with wooden dowels (treenails or trunnels).

(*Es(se)*) The top of the upper beam of the beakhead, ending in a carved curl above the lion's head, to which the forward ends of the railings of the beakhead are attached.

Fair, dub off (v) (*Slechten*) To make the surface of frames fair by dressing it smooth, to create an even surface for faying planks.

Fashion piece (*Rantsoenhout*) Piece of compass wood that formed the curved outline of a flat tuck. The fashion piece ran from the ends of the wing transom to the sternpost.

Filling strake (*Breegang, schut-vulling, vulling*) Planks between the gunports, also between two wales.

Floor, floor timber (*Buikstuk*) Lowest part of a frame, lying across the keel.

Forecastle (*Bak*) The fore part of the upper deck where the crew performed some of their daily duties; the forecastle was covered by the forecastle head.

Forecastle head (*Bak, bakdek*) The deck covering the forecastle.

Foremast (*Fokkemast*) The mast closest to the bow on ships having two or more masts.

Foremast step (*Fock-spoor*) A half-moon-shaped piece of wood in which the foremast is stepped.

Frame (*Spant*) A construction unit consisting of one floor timber, two bilge futtocks, two futtocks, and two top timbers.

Futtock (*Oplang*) Lower vertical part of a frame.

Futtock rider (*Steunder*) A heavy vertical timber running from bilge to upper deck on top of the ceiling, reinforcing the hull between gunports.

Galley (*Kombuis*) The compartment of a vessel where food is prepared.

Garboard stern rabbet (*Streek*) The part at the foot of the sternpost that was chopped off to receive the lowermost planks. This feature was characteristic of Dutch shipbuilding.

Garboard strake (*Kielgang, zandstrook*) The first strake, fitted in the rabbet on both sides of the keel.

Gin block, shear tackle block (*Jein-blok*) A block with three or more sheaves, used when hoisting the stem and stern.

Grapnel (*Dreg*) An anchor with four flukes, used in sloops and boats.

Grating (*Tralieluik*) A type of hatch cover consisting of short pieces of wood athwartship on which battens are nailed in a way that the spaces between them admit air and light.

Groove (*Borst*) The faying face or table of a scarf.

Gun carriage (*Rolpaard*) Wheeled mounting for a gun.

Gunport (*Geschutspoort, poort*) Opening in the ship's hull through which a gun can be fired.

Gun room (*Constapelskamer*) Aft part of the lower deck where the master gunner had his quarters.

(*Hackebord*) A carved plank in the stern gallery that supported a cover.

(*Haertje*) A valve inside a pump; it kept water from flowing back into the ship.

Halyard (*Val*) A rope for hoisting or lowering a yard.

Hance (*Gilling*) A step created where the railing of the ship dropped to a lower level.

Hance piece (*Gilling, gillinghout*) A carved piece, often with curved lines, that filled the hance and produced a more finished look.

Hatch (*Luik*) An opening in the deck.

Hatch beam (*Merker*) The removable part of a deck beam, placed in the main hatch to support the two grids covering the hatch. The topside was grooved to drain water to the sides.

Hatch cover (*Luik*) A cover for an opening in a deck.

Hawsehole (*Kluis*) An opening in the bow for the anchor cable.

Hawser (*Paardelijn*) Heavy line used for various purposes.

Head of the mast (*Top*) Square part at the highest part of the mast on which the mast cap is placed.

Heel of the keel (*Hieling*) Aftermost part of the keel.

Height of breadth ribband. *See* Master ribband.

Herm piece (*Terme, hoekman*) A carved support underneath the second deck between the quarter consoles, often in the shape of a stylized figure.

Hook and chain (*Hel*) Tool for pressing planks of the bottom sideways to each other during construction.

Hounds (*Achtkant, hommer, krans*) Octagonal framing at the foot of the masthead where the trestletrees are attached.

(*Huiddicht*) Planking between the top of the bilges and the lower wale.

Jackscrew (*Dommekracht*) A tool for lifting heavy objects.

Joint (*Las*) Connection between two pieces of wood. There are many variants: hook and butt scarf, boxing, sidescarf, lapscarf, etc.

(*Kam*) An ornamental connection between the lower and the upper beam of the beakhead.

(*Katteblok*) Block provided with a hook and one or more sheaves that served as a component of the tackle.

Keel (*Kiel*) The backbone of the ship, carrying the stem, stern, and frames.

Keelson (*Kolsem*) A major component of the ship's structure, fitted parallel to the keel, over the floor timbers, and fastened to the keel through the floor timbers, creating a girder with the keel.

Kevel (*Kruishout*) A type of large cleat fitted to the waist bulwark on the forecastle and near the poop to belay thick ropes such as tacks and sheets.

Kevel rail (*Halsmast*) The plank inside the bulwark of the waist onto which both midship kevels were mounted.

Knee (*Knie*) A piece of compass wood used to reinforce angled joints.

Knight (*Knecht*) A vertical post with sheaves that projected through the deck and was used to hoist yards and other heavy loads.

Lanyard (*Reep*) A piece of rope rove through the deadeyes, stay blocks, and other devices to set up taut the standing rigging.

Ledge (*Rib*) Support for the deck planks between the waterway and the binding strakes, laid parallel to the beams.

Leech (*Gording*) One of the vertical edges of a square sail.

Leech line 1. **(*Dempgording*):** Line for pulling the leeches up to the yard. 2. **(*Nokgording*):** Line used to haul the leech up to the yard when furling the sail.

Length (*Lengte*) The distance from the outside of the stem to the outside of the sternpost. Use of these two reference points was the only available method for defining a ship's length before it was constructed.

Lift (*Toppenant*) A rope to lift or hold up the yardarm, particularly when the yard is lowered.

Limber hole (*Lokgat*) A space between the upper face of garboard strakes and the bottom face of floor timbers beside the keel that allowed water to drain toward the pumps.

Lining, lining cloth (*Bout*) An extra piece of cloth stitched to a sail wherever extra wear and tear is expected.

Lintel (*Drempel*) The upper sill of a gunport.

Lion (*Leeuw*) A figurehead shaped like a lion.

Lizard (*Kontwachter*) A dumb sheave and pendant, hung from the fore channel to hold the spritsail's sheet pendant up out of the water.

Lower deck, main deck (*Overloop*) First deck running from fore to aft. Related to the English term *orlop deck* but different in meaning.

Main frame (*Hals*) Widest part of the ship, usually measuring one-third of the ship's length.
Widest part of the ship, usually measured at a point one third of the ship's length from the outside of the stem.

Mainmast (*Grote mast*) The middle mast of a three-masted ship.

Main yard (*Grote ra*) The lower yard on the mainmast.

Master ribband (*Scheergang, scheerstrook*) A temporary plank used during construction to indicate the sheer and the height of the ship at its widest point. The locations of the deck beams, hatchways, masts, gunports, and so on could be marked on the ribband.

Mast cap (*Ezelshooft*) A semicircular block on top of a mast for supporting the topmast and carrying the halyard tie that holds the yard.

(Mast) partner (*Visser*) A heavy wooden plate on the deck above the mast step through which the mast fits. Placing the partner forward or aft determined the rake of the mast.

Mast step (*Spoor*) A heavy piece of wood for securing the foot of a mast.

(Mast) top (*Mars*) A platform on top of a mast, supported by crosstrees and trestletrees, that served as a working and fighting platform. Its rim had openings for the deadeyes of the topmast shrouds.

Mizzenmast (*Bezaansmast*) A ship's aftermost mast.

Mizzen topmast (*Kruissteng*) An extension of the mizzenmast.

Mizzen topsail yard (*Kruiszeilsra*) Yard on the mizzen topmast, carrying the mizzen topsail.

Mizzen yard (*Bezaansroede*) Lateen yard on the mizzenmast, carrying the triangular mizzen sail.

Mizzen yard (topping) lift (*Bekajer*) A rope used to peak up the lateen mizzen yard, usually attached to the yard by means of a running bridle.

Molding (*Biesje*) A decorative strip of wood that covered the edges of planking.

Open strake (*Rokgang*) An open space in the top of the topsides where the uppermost plank was left out, to show the tops of the top timbers.

Orlop deck (*Koebrug*) Extra deck in the hold of a ship, often temporary and with limited headroom; used in traders to keep goods dry. The beams of this structure also add considerable strength to the ship's hull.

Parceling (*Servings*) The act of wrapping a wormed rope with strips of tarred sailcloth to prevent chafing.

Parrel (*Rak*) A device consisting of rope, wooden trucks, and ribs for holding the yard to the mast. The rotating trucks (timber beads) allowed the parrel to run up and down the mast when the yard was raised and lowered.

Pawl (*Pal*) Ratchet cleats on both sides of the capstan.

Peak (*Piek*) A storeroom for the master gunner, located below the gun room.

Pendant (*Schenkel*) A rope with a block or tackle that hung from the mast top or yardarm.

Pendant tackle, burton (*Takel*) Part of the rigging to hoist loads and stay masts before shrouds are set up.

Pinrail (*Bosbank*) A piece of wood with holes for belaying pins; located near the mainmast on the inside of the bulwark. Other types of rails (*beting-balkjes*), such as cleat rails and fife rails, were located near the other masts.

Pintle (*Vingerling*) Hinge between sternpost and rudder.

Pivot (*Geck*) Vertical fork on top of a pump to allow the handle to pivot.

Plank (*Boeiplank*) A length of sawn timber to cover the frames.

Plank (v) (*Boeien*) To cover the skeleton of a ship with planks.

Planking tongs (*Boeitang*) A tool for holding the bottom planks of the floor in alignment during construction of a ship with the shell-first method.

Putlog (*Juck*) Horizontal part of the scaffolding on which planks were laid.

Quarterdeck (*Halfdek*) A deck abaft the main mast.

Quarter figure (*Hoekman*) A carved figure serving as a support between the first and second *hackebord*.

Quarter gallery (*Galerij*) Ornamental balcony on the sides of the stern.

Quarter gallery architrave (*Scheergangetje*) Part of the quarter gallery, connected to the second cover.

Rabbet (*Sponning*) A groove in the keel, stem, and sternpost into which the planks were fitted.

Railing, (capping) rail (*Regeling*) A narrow plank covering the tops of the top timbers.

Rake (*Vallen*) The inclination of a vessel's mast from its upright angle with the keel. The rake may be either forward or aft.

Ramshead block (*Kardeel blok*) A heavy halyard block for hoisting the lower sails.

Ratline (*Scheerlijn*) One of the lines attached horizontally between the shrouds to serve as rungs of a ladder.

Ribband (*Zent, cent, sent*) Batten to temporarily define the shape of the ship above the master ribband during construction.

Rider (in the hold) (*Kattespoor*) An assembly of several timbers, constructed like a frame but fitted inside the ceiling, which doubled the number of frames and provided reinforcement.

Rise of the sheer (*Opzetten*) The amount by which the wales or decks were higher at their ends than at the main frame (i.e., the curved line of a ship from fore to aft).

Rising wood (*Slemphout*) A piece of wood connecting stem and keel. (*Slemphout* is also used for the cheek of the beakhead and for the support for the foremast step.)

Rise, dead rise (*Rijzen*) The amount by which the ship's bottom and floor timbers curve upward.

Roband (*Raband*) Piece of rope connecting a sail to a yard.

Rocker (*Stapelen*) The deliberate upward curvature at the ends of a longitudinal timber, such as the keel. If a keel is laid on the stocks, the middle part is lower than the ends—it is given rocker. In the water buoyancy compensates for this difference, as the upward force of the water is greater in the mid-body than at the ends.

Rowle (*Bril*) A rotating piece of wood with a hole through which the whipstaff was let. It allowed the whipstaff to be moved not only from left to right but also in and out to increase the working angle of the rudder.

Rudder (*Roer*) The ship's steering mechanism, hung from the sternpost with pintles.

Scarf (*Las*) The joint between two pieces of wood.

Scarf table (*Lip*) The faying surface of a scarf.

(*Schild(je), schilt*) (lit. "shield") 1. A coat of arms on the taffrail or on the bulwarks of the forecastle and quarter deck. 2. The carved part in the deck that holds the rowle, through which the whipstaff pivots. 3. The forward bulkhead of the forecastle.

Scupper hose, scupper valve (*Mammiering*) A leather or linen hose nailed to the scupper outboard to prevent water from running into the ship through the scupper when the ship is heeling.

Shear mast (*Stut*) A wooden pole for lifting stem and stern during construction of the ship.

Sheer rail (*Rahout*) The uppermost wale, planed with a molding.

Sheet (*Schoot*) A rope to control the lower corner of a sail.

Shore (*Schoor*) A wooden pole for temporarily supporting the stem and stern while the ship is being built.

Shroud (*Hoofdtouw*) A heavy rope that supports the mast laterally. Ratlines were rigged across the shrouds to facilitate climbing aloft.

Sill (*Drumpel*) The lower edge of a gunport.

Skeg (*Scheg*) An extension of the sternpost, to straighten the aft side. Also part of the stem to act as a gripe or cutwater.

Snatch block (*Kinnebaksblok*) A single block with an opening in one cheek to receive the bight of a rope; it was attached to the foot of the foremast to tighten parts of the running rigging, if necessary.

Spall (*Zwieping*) A batten to temporarily connect parts of the scaffolding.

Spirketing (*Wegering*) Plank to the inside of the bulwark.

Spline (*Rey, rij*) A flexible batten for drawing curved lines and taking measurements while planking.

Spritsail topmast (*Blinde steng*) A small vertical topmast on the bowsprit that carried the spritsail topsail; a typical feature of seventeen-century ships.

Spritsail yard (*Blinde ra*) Yard on the bowsprit, carrying the spritsail.

Spritsail topsail yard (*Bovenblinde ra*) Yard on the spritsail topmast, carrying the spritsail topsail.

Spur of the beakhead (*Uitlegger*) The basic assembly of the Dutch beakhead, consisting of two curved beams (upper and lower) separated by a carved trail board (*Kam*) and ending in the figurehead.

Stay (*Stag*) Rigging component that serves as a forward support of a mast.

(*Steeker*) A frame timber aft; specifically, a futtock with an S-shaped curve.

Steering stand (*Stuurplecht*) The covered location where the helmsman stands, usually forward of the captain's cabin.

Stem (*Voorsteven*) The main structural element extending upward from the forward end of the keel.

Boxing scarf, boxing of the stem (*Kinnebak,* lit., "jawbone") The joint between keel and stem.

Stern, tuck, transom (*Spiegel*) The flat aft end of the hull of a pinas.

Stern gallery (*Galdery*) A gallery built on both parts of the transom, sometimes open, sometimes closed, for the skipper's toilet.

Stern timbers (*Hekstutten*) Vertical parts attached to the stern in an early stage of construction to indicate the shape of the upper aft part of the ship.

Sternpost (*Achtersteven*) A more or less vertical timber placed aft on the keel.

Stock, keel block (*Stapelblok*) One of the wooden blocks on which the keel rests while the ship is being built.

Stopwater (*Scheinagel*) A hole drilled through the faying surfaces of a scarf, filled with moss, and plugged at both ends with dowels,a technique for making the scarf watertight.

Strake (*Gang*) A row of planks running from fore to aft.

Strut (*Stander*) Support underneath the putlog; part of the scaffolding.

Studding sail (*Lijzeil*) Auxiliary sail to be hoisted on the end of a yard.

Sweep (*Luiwagen*) Support for the tiller.

Tack (*Hals*) Part of the rigging that pulls the lower corners of the lower sails forward.

Tarpaulin (*Presenning*) Tarred sailcloth used for sealing hatches and other openings and for making mast collars. *Pap-doeken* (lit., "drenched cloths") was possibly a synonym for "tarpaulins."

(*Tingel*) A batten, triangular in cross section, that closed the outboard part of the space between the garboard strake and the floors.

Tiller (*Roerpen*) A beam for controlling the rudder.

Top. *See* (Mast) top.

Topgallant mast (*Bramsteng*) An extension of the topmast, carrying the topgallant yard.

Topgallant yard (*Bramra*) Yard on a topgallant mast, carrying a topgallant sail.

Topmast (*Steng*) An extension of the mast.

Topmast yard (*Marsera*) Yard on the topmast, carrying the topsail.

Top timber (*Stut*) The top part of a frame, shaping the topside of the ship.

Torus (*Thorus*) A molding planed into a convex shape to form the second cover in the stern.

Transom (*Worp*) Horizontal reinforcement of the stern situated between the wing transom and the *broekstuk,* connecting the fashion pieces and the stern.

Treenail (*Nagel*) Wooden dowel used to fasten planking to the frames.

Trestletree (*Langszaling*) One of a pair of wooden pieces on top of the sides of the hounds to support the top.

Truck (*Kloot*) A ball-shaped piece of wood forming part of the parrel.

Turn of the bilge (*Kim*) Area of the ship's hull where the horizontal bottom curves toward the vertical.

Tumble home (*Invalling*) The inward curve of the ship's sides as viewed in cross section.

Twisted cornice (*Slingerlijst*) A term for the second cover, in reference to its complex shape: high in the middle, sloping downward toward the sides, and rising again at the end, where the plank was twisted to align with the rail of the quarter gallery.

Upper deck (*Verdek*) The second deck of a ship, running from bow to stern.

Upper works (*Vertuining*) The uppermost, clinker-planked part of the hull above the sheer rail.

Wainscot (*Wagenschot*) Radially sawn or cleft planks.

Wale (*Berghout*) One of the thick strakes above and below the gunports, following the sheer of the ship. Very important for longitudinal strength.

Washboard (*Zetplank*) A removable plank in the waist of the ship above the sheer rail.

Washstrake (*Zetgang*) The first plank of the upper works.

Waterboard (*Waterbord*) The waterway of the uppermost decks of the ship.

Watershed (*Uitwatering*) Literally, the place where the lower deck drains. In the shipbuilding procedure described by Witsen, the height of the watershed is the same as that of the master ribband.

Waterway (*Lijfhout, legwaring*) The heavy part of the deck planking, the part fitted against the frames in the sides of the ship.

Whelp (*Scheen*) Tapering piece of wood radially mounted to the drum of a capstan.

Whipstaff (*Kolderstok*) A timber bar or rod attached with a bearing to the forward end of the tiller; used as a lever to control the tiller at small helm angles and also as a rod to push or pull the tiller at larger helm angles.

Windlass (*Spil*) An apparatus for lifting heavy loads, usually a type of winch with a horizontal or vertical shaft and a rotating drum around which a rope can be wound.

Wing transom (*Hek, hekbalk*) A horizontal beam on top of the sternpost that determined the width of the stern. All the planking of the flat tuck ended there; thus, it defined the breadth of the tuck.

Yard (*Ra*) A spar to which a sail is attached.

Bibliography

Anonymous. *De volmaakte bootsman. . . .* Amsterdam: J. van Keulen, 1680.

———. Evenredige Toerusting van Schepen ten Oorlog Bijder See. 1660–65. Nederlands Scheepvaartmuseum Amsterdam. Manuscript Collection, shelf no. Hs-0551. Typescript transcription by Herman Ketting, 1974.

———. *Hollandsche Scheepsbouw.* Amsterdam: Jacobus Robyn, 1678.

———. *L' art de bâtir les vaisseaux et d'en perfectionner la construction.* Amsterdam: David Mortier, 1719.

———. *Leven en Bedrijf van den vermaarden Zeeheld Cornelis Tromp.* Amsterdam: printed for the author, 1692.

———. *Nieuwe Hollandsche Scheepsbouw.* 3rd ed. Amsterdam: Carel Allard, 1695.

———. *Taackelasie boeckje van groote en kleine schepen.* Amsterdam: Jacobus Robyn, c. 1690.z

Asaert, G., et al., eds. *Maritieme Geschiedenis der Nederlanden.* 4 vols. Bussum: Unieboek, 1976–78.

Baay. F. *Bouwbeschrijving van een 17de-eeuws Statenjacht.* Rotterdam: Bosman, n.d.

Beylen, Jules van. *Schepen van de Nederlanden, van de late Middeleeuwen tot het einde van de 17de eeuw.* Amsterdam: P. N. van Kampen, 1970.

———. *Zeilvaart Lexicon: Maritiem Woordenboek.* Weesp: De Boer Maritiem, 1985.

Bosscher, Philip M. *Een nuchter volk en de zee.* Bussum: De Boer Maritiem, 1979.

Bruijn, J. R., F. S. Gaastra, and I. Schöffer. *Dutch-Asiatic Shipping in the 17th and 18th Centuries.* 3 vols. Rijks Geschiedkundige Publicatiën, Grote Serie 165, 166, 167. The Hague: Martinus Nijhoff, 1979–87.

Bruzelius, Lars. *A Bibliography of Books on Naval Architecture, Rigging, and Seamanship Printed 1600–1919.* Upsala: privately printed, 1990.

Burningham, Nick. "Learning to Sail the *Duyfken* Replica." *International Journal of Nautical Archaeology* 30, no. 1 (2001): 74–85.

Chapman, Fredrik Henrik af. *Architectura Navalis Mercatoria: A Facsimile of the Classic Eighteenth-Century Treatise on Shipbuilding*. 1777. Reprint, London: Adlard Coles, 1968.

Crescentio, Bartolomeo. *Nautica Mediterranea.* Rome: Bartolomeo Bonfadino, 1607.

Crone, G. C. E. *De jachten der Oranjes: Historisch en scheepstechnisch overzicht van de jachten der stadhouders en vorsten uit het Huis van Oranje-Nassau van af het einde van de 16de eeuw.* Amsterdam: Swets & Zeitlinger, 1937.

——. *Nederlandsche jachten, binnenschepen, visschersvaartuigen en daarmee verwante kleine zeeschepen 1650–1900.* 1926. Reprint, Schiedam: Schiepers, 1978.

Deane, Anthony. *Deane's Doctrine of Naval Architecture, 1670.* Edited by Brian Lavery. London: Conway Maritime Press, 1981.

Decquer, Hendrick. *Middelen om uit te vinden de ware ladinge der schepen na hare groote.* N.p.: printed for the Dutch East India Company, c. 1690.

Dik, G. Cees. *De 7 Provintiën: Een poging tot reconstructie.* Franeker: Van Wijnen, 1993.

Dirkzwager, Jan. *Dr. B. J. Tideman, 1834–1883: Grondlegger van de moderne scheepsbouw in Nederland.* Leiden: Brill, 1970.

Dudley, Robert. *Dell'Arcano del Mare.* Florence: Francesco Onofri, 1646.

Elias, J. E. *De vlootbouw in Nederland in de eerste helft der 17de eeuw, 1596–1655.* Werken uitgegeven door de Commissie voor Zeegeschiedenis 1. Amsterdam: Noord-Hollandsche uitgeversmaatschappij, 1933.

——. *De vroedschap van Amsterdam, 1578–1799.* 2 vols. Haarlem: Loosjes, 1903–1905. Reprint, Amsterdam: Nico Israel, 1963.

Fournier, Georges. *Hydrographie contenant la théorie et la pratique de toutes les parties de la navigation.* Paris, 1646.

Furttenbach, Joseph. *Architectura navalis: Das ist: von dem Schiff-Gebäw, auff dem Meer und Seekusten zugebrauchen. . . .* Nuremberg: J. Saum, 1629.

García de Palacio, Diego. *Instrucción Náutica.* Mexico City: Pedro Ocharte, 1587.

Gawronski, Jerzy. *De Equipage van de* Hollandia *en de* Amsterdam*: VOC bedrijvigheid in de 18de-eeuws Amsterdam.* Amsterdam: Bataafse Leeuw, 1996.

——. "*De Hollandia* en *De Amsterdam:* Twee Schepen en Een Bedrijf: Materiële Cultuur en Organisatie VOC Amsterdam 1740–1750." PhD diss., Leiden University, 1994.

——, B. Kist, and O. Stokvis van Boetzelaer, eds. *Hollandia Compendium: A Contribution to the History, Archaeology, Classification and Lexicography of a 150 ft. Dutch East Indiaman (1740–1750).* Amsterdam: Rijksmuseum, 1992.

Gebhard, J. F., Jr. *Het leven van Mr. Nicolaes Cornelisz. Witsen, 1641–1717.* 2 vols. Utrecht: Leeflang, 1881–82.

Green, Jeremy. *The Loss of the Verenigde Oostindische Compagnie Retourschip* Batavia, *Western Australia 1629: An Excavation Report and Catalogue of Artefacts.* B.A.R. International Series 489. Oxford: B.A.R., 1989.

Groot, Irene de, and Robert Vorstman. *Zeilschepen: Prenten van de Nederlandse meesters van de 16de tot de 19de eeuw.* Maarssen: Gary Schwartz, 1980.

Hayward, Edward. *The Sizes and Lengths of Riggings and the Hayward-Kendal Pamphlet Controversy.* 1655. Edited by R. C. Anderson. Reprint, London: Francis Edwards, 1967. First edition published as *The Sizes and Lengths of Riggings for All His Majesties Ships and Frigats.*

Hoving, A. J. "Away from the Drawing Board." Pts. 1 and 2. *Model Shipwright* 69 (1989): 43–51; 71 (1990): 47–50.

——. *Het oorlogsjacht* De Heemskerck: *Het schip waarmee Abel Tasman Nieuw-Zeeland ontdekte.* Amsterdam: Rijksmuseum, 1992.

——. "A 17th-Century Dutch 134-Foot Pinas: A Reconstruction after *Aeloude en Hedendaegse Scheepsbouw en Bestier* by Nicolaes Witsen 1671." Pts. 1 and 2. *International Journal for Nautical Archaeology* 17, no. 3 (1988): 211–22; no. 4 (1988): 331–38.

——, and Alan A. Lemmers. *In tekening gebracht: De achttiende-eeuwse scheepsbouwers en hun ontwerpmethoden.* Amsterdam: Bataafsche Leeuw, 2001.

——, and Robert Parthesius. "Hollandse scheepsbouwmethoden in de zeventiende eeuw." In *Batavia Cahier,* vol. 3, *Herbouw van een Oostindiëvaarder,* edited by Robert Parthesius, 5–11. Lelystad: Stichting Nederland bouwt V.O.C.-retourschip, 1991.

Huitema, T. *Ronde en Platbodemjachten.* 5th ed. Amsterdam: Van Kampen, 1977.

Ketting, Herman. Prins Willem: *Een zeventiende-eeuwse Oostindievaarder.* Bussum: Unieboek, 1979.

Landström, Björn. *Regalskeppet* Vasan *fran borjan till slutet: En bok om byggandet och utsmyckandet, riggningen och bestyckningen av* Vasan *som förliste i Stockholms hamn. . . .* Stockholm: Interpublishing, 1980.

Lehmann, Louis Th. *De galeien: Een bijdrage aan de kennis der zeegeschiedenis.* Amsterdam: De Bezige Bij, 1987.

L'Honoré Naber, S. P., ed. *Reizen van Willem Barents, Jacob van Heemskerck, Jan Cornelisz. de Rijp en anderen naar het noorden, 1594–1597.* 2 vols. Werken uitgegeven door de Linschoten-Vereeniging 14–15. The Hague: Martinus Nijhoff, 1917.

L'Hour, Michel, Luc Long, and Eric Rieth. Le Mauritius: *La Memoire engloutie.* Grenoble: Casterman, 1989.

Maarleveld, Thijs J. "Archaeology and Early Modern Merchant Ships: Building Sequence and Consequences. An Introductory Review." In *Rotterdam Papers,* vol. 7, *A Contribution to Medieval Archaeology: Teksten van lezingen, gehouden tijdens het symposium "Handel, handelsplaatsen en handelswaar vanaf de Vroege*

Middeleeuwen in de Lage Landen" te Rotterdam van 2 t/m 3 november 1990, edited by A. Carmiggelt, 155–73. Rotterdam: Bureau Oudheidkundig Onderzoek van Gemeentewerken Rotterdam, 1992.

Murphy, Henry C., ed. *Henry Hudson in Holland: An Inquiry into the Origin and Objects of the Voyage Which Led to the Discovery of the Hudson River.* The Hague: Martinus Nijhoff, 1909.

Oliveira, Fernando. *Livro da fábrica das Naos.* Manuscript, c. 1580. Facsimile ed., Lisbon: Academia de Marinha, 1991.

Pepys, Samuel. *Dagboek 1660–1669: Een selectie uit het volledige dagboek.* Translated by A. Alberts. Bussum: Unieboek, 1981.

Peters, Marion. "Nicolaes Witsen and Gijsbert Cuper: Two Seventeenth-Century Dutch Burgomasters and Their Gordian Knot." *Lias* 16, no. 1 (1989): 111–51.

Petrejus, E. W. *Oude zeilschepen en hun modellen: Binnenschepen, jachten, en vissersschepen.* De Boer maritieme handboeken. Bussum: Unieboek, 1971.

Ploeg, J. "Buizen en Hoekers, Schepen voor de Hollandse Noordzeevissers : Historisch onderzoek naar schepen en tuigages die zes eeuwen lang typerend waren voor de visserij op haring en rondvis." Manuscript, 1991–92. Author's files.

Posthumus Meyjes, R., ed. *De reizen van Abel Tasman en Franchois Visscher ter nadere ontdekking van het Zuidland in 1642/43 en 1644.* Werken uitgegeven door de Linschoten-Vereeniging 17. The Hague: Martinus Nijhoff, 1919.

Rålamb, Åke C. *Skeps byggerij eller Adelig öfnings tionde tom.* 1691. Faksimileditioner, vol. 1. Stockholm: Sjöhistoriska museet, 1943.

Reinders, Reinder. *Cog Finds from the IJsselmeerpolders.* Flevobericht 248. Lelystad: Rijksdienst voor de Ijsselmeerpolders, 1985.

———, and Kees Paul, eds. *Carvel Construction Technique: Fifth International Symposium on Boat and Ship Archaeology, Amsterdam 1988.* Oxbow Monograph 12. Oxford: Oxbow, 1991.

Robijn, Jacobus. *Hollandsche Scheepsbouw.* Amsterdam, 1695.

Rouffaer, G. P., and J. W. IJzerman, eds. *De eerste schipvaart der Nederlanders naar Oost-Indië onder Cornelis de Houtman, 1595–1597.* Werken uitgegeven door de Linschoten-Vereeniging 7. The Hague: Martinus Nijhoff, 1915.

Timmermann, G. "Das Eindringen der Naturwissenschaften in das Schiffsbauhandwerk." *Deutsches Museum Abhandlungen und Berichte* 30, no. 3 (1963): 5–54.

Throckmorton, Peter. *History from the Sea: Shipwrecks and Archaeology.* London: Mitchell Beazley, 1987.

van Bruggen, B. E. "Beschouwing over het aangeven van de hoofdafmetingen van de Nederlandse zeeschepen en de daarbij gebruikte maateenheden (1600–1800)." *Mededelingen van de Nederlandse Vereniging voor Zeegeschiedenis / Tijdschrift voor Zeegeschiedenis,* no. 20 (March 1970): 25–34.

van Dam, Pieter. *Beschryvinge van de Oostindische Compagnie, 1639–1701.* Vol. 1. Edited by F. W. Stapel. Rijks Geschiedkundige Publicatiën, Grote Serie 63. The Hague: Martinus Nijhoff, 1927.

van der Waals, Jan. "Wankelend wereldbeeld: Onderzoek naar taal, geloof en tijd in rariteitenkabinetten." In *De Wereld binnen Handbereik, Nederlandse Kunsten Rariteiten verzamelingen, 1585–1735,* edited by E. Bergvelt and R. Kistemaker, 135–52. Zwolle: Waanders; Amsterdam: Amsterdams Historisch Museum, 1992.

van Yk, Cornelis. *De Nederlandse Scheeps-bouw-konst Open Gestelt.* Amsterdam: Jan ten Hoorn, 1697. Reprint, Rotterdam: Langerveld, 1980.

van Zwijndregt, Leendert, and Cornelis de Ruiter. *Verhandeling van den Hollandschen Scheepsbouw. . . .* The Hague: De Hondt and H. Scheurleer, 1757.

Versteeg, W. K. *Scheepsmodellen, 1700–1900: Bevattende modellen van visschersschepen, vrachtschepen, schepen voor groote vaart, yachten.* Antwerp: De Nederlandsche Boekhandel, 1947. Reprint, Amsterdam: Van Kampen, 1972.

Vreugdenhil. A. *Ships of the United Netherlands, 1648–1702.* Part 4 of *Lists of Men-of-War, 1650–1700.* Society for Nautical Research, Occasional Publications 5. Cambridge: Cambridge University Press, 1936.

Winter, Heinrich. *Das Hanseschiff im ausgehenden 15. Jahrhundert.* Rostock: Hinstorff, 1961.

———. *Der holländische Zweidecker von 1660/1670.* Bielefeld: Delius, Klasing & Co., 1978.

Witsen, Nicolaes. *Aeloude en Hedendaegse Scheepsbouw en Bestier. . . .* Amsterdam: Casparus Commelijn, Broer and Jan Appelaer, 1671.

IJzerman, J. W., ed. *De reis om de wereld door Olivier van Noort, 1598–1601.* 2 vols. Werken uitgegeven door de Linschoten-Vereeniging 27–28. The Hague: Martinus Nijhoff, 1926.

Index

(Numbers in italic indicate tables or illustrations)

Ed Rachal Foundation Nautical Archaeology Series

Yassi Ada: Volume I, A Seventh-Century Byzantine Shipwreck
George F. Bass and Frederick H. van Doorninck Jr.

The Porticello Shipwreck: A Mediterranean Merchant Vessel of 415–385 B.C.
Cynthia J. Eiseman and Brunilde S. Ridgway

Athlit Ram
Lionel Casson, and J. Richard Steffy

The Western River Steamboat
Adam I. Kane

The Philosophy of Shipbuilding: Conceptual Approaches to the Study of Wooden Ships
Frederick M. Hocker and Cheryl A. Ward

Serce Limani: An 11th Century Shipwreck Vol. I, The Ship and Its Anchorage, Crew, and Passengers
George F. Bass, Frederick H. van Doorninck Jr., Sheila Matthews, and J. Richard Steffy

Homeric Seafaring
Samuel Mark

The Pepper Wreck: A Portuguese Indiaman at the Mouth of the Tagus River
Filipe Vieira de Castro

Ships' Fastenings: From Sewn Boat to Steamship
Michael McCarthy

The Life and Times of the Steamboat Red Cloud *or How Merchants, Mounties, and the Missouri Transformed the West*
Annaliese Corbin

Seagoing Ships and Seamanship in the Bronze Age Levant
Shelley Wachsmann

Sea of Galilee Boat
Shelley Wachsmann

Serçe Limani, Vol 2: The Glass of an Eleventh-Century Shipwreck
George F. Bass, Berta Lledo, Sheila Matthews, and Robert H. Brill